Social-ecological Systems of Latin America: Complexities and Challenges

Luisa E. Delgado • Víctor H. Marín
Editors

Social-ecological Systems of Latin America: Complexities and Challenges

Springer

Editors
Luisa E. Delgado
Escuela de Ciencias
Facultad de Ciencias
Universidad de Chile
Santiago, Chile

Fundación Centro Transdisciplinario
de Estudios FES-Sistémicos
Santiago, Chile

Víctor H. Marín
Laboratorio de Modelación Ecológica
Departamento de Ciencias Ecológicas
Facultad de Ciencias
Universidad de Chile
Santiago, Chile

ISBN 978-3-030-28451-0 ISBN 978-3-030-28452-7 (eBook)
https://doi.org/10.1007/978-3-030-28452-7

This Springer imprint is published by the registered company Springer Nature Switzerland AG
The registered company address is: Gewerbestrasse 11, 6330 Cham, Switzerland

Preface

No doubt it was necessary to civilize man in relation to man. That work is already advanced and is making progress every day. But man must be civilized also in relation to nature.[1]

The book *Social-ecological systems of Latin America*[2]*: complexities and challenges* resulted from the multi- and transdisciplinary efforts of several social actors, scientists, resource managers, and members of international organizations. Our main goal was to describe our contemporaneous society–nature relationships. Throughout the writing process, the desire to contribute with knowledge and experiences, both individuals and collectives, always prevailed. This process was marked by the solidarity among authors, a fundamental value that should be the basis for the relationships between Latin American peoples and the entire world in the twenty-first century[3]. This book is dedicated to the future generations, a future that we want and wish for them.

The concept of social-ecological systems (SES) was the theoretical umbrella for this book since it allows for the interconnected study of ecosystems and human beings. We, humans, are an indissoluble part of earth, cohabitants or "ecosystem's people," sharing the biosphere's ecosystems with other biotic and abiotic components. Human beings, complex and reflexive components of nature, are a powerful force that alters historical ecosystems. This ongoing process requires new, already existing, interdisciplinary approaches to study the different relationships between ecosystems and human societies (e.g., bioethics, environmental economy, among others). A key issue of these studies is to accept and embrace the contextual (both in time and space) inherent system's complexity. One consequence of this approach is that environmental management should be thought in such a way that it may contribute to coadaptation and human well-being.

---

[1] Victor Hugo (1898) The Alps and Pyrennes. Bliss, Sand and co., London.

[2] Latin America is an ethnic-geographic concept, identifying a region of the American continent where languages derived from the Latin are spoken https://es.wikipedia.org/wiki/América_Latina

[3] In the year 2016, the United Nations proclaimed December 20th as the "International Human Solidarity Day." https://www.un.org/en/events/humansolidarityday/

Fig. 1 Map of Latin America showing the different ancestral cultural areas on top of the countries of the region. (Source: Museo Chileno de Arte Precolombino, Santiago, Chile)

The sociocultural and ecosystemic complexity, diversity, and heterogeneity of Latin America motivated us all to write our experiences contextually. The origins of our region include multiple cultural, ancestral, and current roots. The syncretism between nature and human beings has always existed in Latin America and currently persists in ecological and traditional cultures (Fig. 1). For example, it is still possible to see in the high Andes sacred Inca sites and paths and in low lands, such as the Amazon forests, historical society–nature relationships (e.g., Kayapo culture) that include tropical forest's conservation. In this way, we Latin-Americans have generated our environmental history in the last 16,000 years, starting from hunters–collectors in the pre-Inca period to our days, adapting to the provision of services and resources provided by the natural ecosystems.

The main objective of this book is to increase the knowledge about science for the people from a Latin American perspective. That is, with a postnormal scientific focus oriented to action. This approach accepts that each individual may have multiple valuations and perspectives about reality, influencing in the study, understanding, and management of the social-ecological systems. This book can be considered one of the first in its style and shape, not only due to the participation of experts from several Latin American countries (Argentina, Brazil, Chile, Colombia, Ecuador, Mexico, Paraguay, Peru) and invited experts from other regions, but also due to, as stated by Jay Inslee, Governor of Washington State: "We are the first generation to feel the impact of climate change, and the last generation that can actually do something about it."[4] Consequently, we the authors not only want to share our knowledge but also to express our deepest concerns about the future social-environmental challenges for mankind.

This book is divided into four sections. In the first section, we discuss concepts and methods related to the study of social-ecological systems in four chapters. Chapter "Postnormal Science and Social-ecological Systems" analyzes the SES concept from a postnormal perspective; Chapter "Simplifying the Complexity of Social-ecological Systems with Conceptual Models" describes how to simplify the study of SES using conceptual models based on the narratives of social actors; Chapter "Social Actors and Participation in Environmental Issues in Latin America" reflects on the differences between stakeholders and social actors, with a gender perspective and its effects on environmental issues; Chapter "Social-ecological Systems and Human Well-Being" discusses the complex issue of human well-being and its multiple approaches, proposing that the concept is different in Latin America than in other areas of the world.

The second section, divided into six chapters, discusses the challenges that SES represent for Latin America. Although theoretical concepts are considered as "universal" (i.e., applicable in the same way to any system on earth), several have evolved contextually in different regions. In this section, we have focused on these developments. Chapter "Studying Social-ecological Systems from the Perspective of Social Sciences in Latin America" analyzes the evolution of the SES concept from the perspective of social sciences in Latin America, Chapter "Environmentalism of the Poor: Environmental Conflicts and Environmental Justice" discusses the poor people ecologism; Chapter "A New Environmental Governance" proposes new environmental governance for Latin America; Chapter "A Hierarchical Approach for the Evaluation of Multiple Ecosystem Services" develops the idea of a hierarchical approach to value ecosystem services; Chapter "Social-ecological Complexities and Novel Ecosystems" discusses the complexities of novel ecosystems and Chapter "Social-ecological Systems and the Economics of Nature: A Latin American Perspective" analyses the nature's economic management and its biophysical effects and threats.

The third section, case studies, is an invitation for the readers to travel to eight countries of Latin America with the help of 45 authors who will show them the social-ecological complexities of our region. The main idea, behind the 12 chapters, is to fascinate readers with Latin America, its ecosystems and peoples.

---

[4] See Chapter "Social-ecological Systems and the Economics of Nature: A Latin American Perspective" for details.

In the last chapter, the editors reflect on the future challenges for Latin America and its complex social-ecological systems, based on the information presented and discussed by the authors in the previous 22 chapters.

The authors of this book would like to thank our institutions and our families for understanding our tireless desire to share science and our life experiences. We give special thanks to the Chilean Museum of Pre-Columbian Arts (Museo Chileno de Arte Precolombino) for granting us permission to use the Cultural Map of the American Continent (Fig. 1).

Santiago, Chile
June 2019

Luisa E. Delgado
Víctor H. Marín

The original version of this book was revised. The correction to this book is available at https://doi.org/10.1007/978-3-030-28452-7_24

Acknowledgments

This publication was generated within the framework of Proyecto FONDECYT Regular 1170532 (CONICYT-Chile), awarded to Luisa E. Delgado, entitled: *Conceptual models of the society-nature complex system: opening roads towards new forms of construction of rural development in Chile*. We thank FONDECYT to support the preparation of this work and to give us the chance to invite colleagues from Latin American and Australia that share the theoretical developments built within the project.

Contents

Part I
Concepts and Methods to Study Social-ecological Systems

Postnormal Science and Social-ecological Systems

Víctor H. Marín, Ignacio A. Marín, and Luisa E. Delgado

Abstract Postnormal science is an epistemological approach to understand and hopefully help in the solution of societal problems accepting the perspectives of the many social actors that participate in them. Its fundamental characteristics are the acceptance of the uncertainty of facts, disputed values, and the urgency of the decision-making processes. The social-ecological system concept, on the other hand, was proposed as a better way to recognize, study, and manage the relationships between human beings and ecosystems. In this chapter, we discuss both concepts, based on the idea that social-ecological systems should be studied using a postnormal perspective. We provide an example of the study of a Latin American social-ecological system: the Rio Cruces wetland.

Keywords Social-ecological systems · Latin America · Complexity · Postnormal science · Río Cruces · Transdiscipline

V. H. Marín (✉)
Laboratorio de Modelación Ecológica, Departamento de Ciencias Ecológicas, Facultad de Ciencias, Universidad de Chile, Santiago, Chile
e-mail: vmarin@uchile.cl

I. A. Marín
Ingeniería Comercial, Universidad Adolfo Ibañez, Santiago, Chile
e-mail: imarin@alumnos.uai.cl

L. E. Delgado
Escuela de Ciencias, Facultad de Ciencias, Universidad de Chile, Santiago, Chile

Fundación Centro Transdisciplinario de Estudios FES-Sistémicos, Santiago, Chile
e-mail: ldelgado@uchile.cl

L. E. Delgado, V. H. Marín (eds.), *Social-ecological Systems of Latin America: Complexities and Challenges*, https://doi.org/10.1007/978-3-030-28452-7_1

1 Science, Society, and the Environment

In the twenty-first century, human societies are constantly enquiring science for solutions to our self-generated problems (e.g., global warming, extinction of emblematic species, water pollution, overfishing). Science, not long ago, used to respond to these challenges and questions in different ways, depending upon the discipline of the scientist answering them. For example, if the question was: do we have a problem with the conservation of some key species? (other than human, of course), ecologists would have replied by describing the population size of the species, its birth and death rates, and its probability of extinction. Sociologists would have, most likely, discussed issues related to whether or not the species was of importance for social actors, especially in terms of well-being. Finally, political scientists would have pointed toward the governmental structures that should be in charge of answering the question and whether or not the question was even a valid one or not, from a political perspective. However, human societies started needing more integrated answers from science, perhaps due to our increased awareness of the problematic condition of ecosystems worldwide and human well-being (Carpenter et al. 2005), requiring the mixture of several disciplines. We can take the problem of pain suffering in fibromyalgia patients as a first example, where current treatments require experts from different fields (e.g., medicine, psychology, physical therapy) and terms such as interdisciplinary and multidisciplinary are frequently used (Giusti et al. 2017). However, somebody could claim that in the previous example, we are still within the general field of medicine! What if we now turn to a different type of problem? One that may push the issue of scientific disciplines to a limit. Let's discuss current environmental problems!

Declaring that humanity is facing severe environmental problems is perhaps the understatement of the century. Indeed, our current situation has been clearly identified by United Nations Environment Programme, UNEP, (https://www.unenvironment.org/environment-you): "*We are depleting the Earth's natural resources, polluting its air and water, destabilizing the climate, and driving many of its species to extinction.*" One of the characteristics of all the issues identified by UNEP is that working on them not only require scientists from different disciplines (natural, social, political, medical, etc.) working together, which would correspond to a multidisciplinary team, but to generate synthesis and harmony between them in order to reach a coherent view (i.e., interdisciplinary synthesis). However, not even that seems good enough and society is demanding the integration of natural, social, political, and health sciences in the context of human needs transcending disciplinary boundaries; that is, transdisciplinary science (Delgado et al. 2009; Pohl 2008).

Transdisciplinary science is frequently associated with complex problems, requiring knowledge integration (Morandín Ahuermal et al. 2018; Urquiza et al. 2018; Schoon and van der Leeuw 2015). Still, for certain issues (especially those within science–policy interfaces) not even transdiscipline is a solution by itself (Marín and Delgado 2013). We shall return to that point later in this chapter. For now, let us suggest that interdisciplinary and transdisciplinary science are more and more required when addressing complex environmental problems (e.g., Balvanera et al. 2012). Disciplinary and multidisciplinary science, on the other hand, are key concepts when developing scientific projects oriented mostly to advance knowledge for knowledge's sake. Indeed, the history of science shows that many advances, that have later generated practical applications to human societies, have been based on disciplinary studies (Geymonat 2006). The point here, however, is that time is running against us and waiting for this mechanism from disciplinary studies to practical applications is no longer a viable alternative.

The contemporary need for a transdisciplinary science (centered on problems rather than conceptual developments within science) that may help to solve our environmental complications is perhaps better understood if we compare the human population of Latin America prior to the arrival of Spaniards and now. Although there is much controversy about it, an average, conservative, estimate for pre-Columbian population size is in the order of 57 million people (Denevan 1992). CEPAL (United Nations Economic Commission for Latin America and the Caribbean) estimated that by the year 2016 the population size of Latin America was 625 million people (CEPAL 2016), which represents an increase in one order of magnitude. Consequently, Latin American ecosystems now have to provide services for a much larger human population. Several authors and international programs have highlighted the severe environmental problems facing Latin American countries. For example, the synthesis from the Millennium Ecosystem Assessment (MEA 2005) shows that half of the Latin American population suffers from "inadequate water and sanitation" (p. 62). Balvanera et al. (2012) further point out to the severe inequities among the diverse members of Latin American societies in terms of income, health, and negative environmental impacts from the unsustainable use of natural resources. One result of this turmoil is that we can no longer analyze human societies and nature and its ecosystems separately. We need a transdisciplinary concept that integrates them both. We discuss such a concept in the next section.

2 Neither Society Nor Nature But Both

Most Latin Americans (80% according to CEPAL 1999) live in urban cities, where contact with nature and its ecosystems are sporadic. Thus, it should not come as a surprise that viewing society and nature as separated entities is easily accepted. For example, Chilean national parks cover a surface close to 186×10^3 km^2, representing 21.1% of the Chilean continental territory (CONAF 2019). Yet, aside from their use for scientific research on native species (disciplinary research), Chileans use them for summer vacation purposes only. In fact, during the year 2017 national parks received over three million visits (CONAF 2019). The remaining 20% of the Latin American population (rural people) lives in close contact with nature, heavily depending on their ecosystems for their well-being (Delgado et al. 2013). Thus, we need a concept that may serve to understand both situations: ecological systems being used exclusively as touristic attractions versus living environments. Twenty-one years ago, Berkes and Folke (1998: 4) stated that separating social and ecological systems was "artificial and arbitrary," generating the social-ecological concept. The issue was also discussed by Latour (1999) from a philosophical perspective. Since then, 6018 articles have been written using the concept as the basis of the analysis on the interrelations between human societies and nature (source: Web of Science; http://apps.webofknowledge.com). However, only 74 articles are from Latin America, all written in the last 10 years and where only 58% have been led by a Latin American scientist. Thus, we still have a long road ahead regarding humans–ecosystems interactions and how it is affecting our well-being and future.

A social-ecological system (SES) can be defined as structured by two interacting subsystems (Delgado and Marín 2016): a social subsystem (humans, social actors, and their organizations) and an ecological subsystem (or ecosystem). The multiplicity of interactions, both within each subsystem as between them, makes social-ecological systems complex. One of the classical strategies used in science is to provide simple explanations for most phenomena. However, as discussed by Rogers et al. (2013) such a method may "disfigure" reality when facing complex systems, which means that complexity has to be embraced or our ideas, hypothesis, and explanations for social-ecological systems may be entirely wrong. In this book, we have included chapters that discuss methods to analyze the complexity in social-ecological systems.

Complex systems share many general characteristics. One that is particularly relevant to SES is contextuality (Chu et al. 2003; Preiser et al. 2018). That is, the property of different systems to be organized in alternative ways, based on similar or sometimes the same components. The same idea has been proposed for ecosystems (Schmitz 2010), but under the name of "contingency." Thus, if we want to learn about Latin American SES, we have to study them, since knowledge acquired in other places on earth may help with general ideas and concepts on SES dynamics and interactions among their components, but specific results (e.g., importance of different interactions within and between subsystems) could also be entirely different. That is why case studies are so important. Part III of this book contains 12 SES case studies from several Latin American countries (Argentina, Brazil, Chile, Colombia, Ecuador, México, Paraguay, and Perú).

Therefore, how do you study a system that is complex and contextual? Several authors have proposed that we may want to do it using theoretical, conceptual frameworks (McGinnis and Ostrom 2014; Holland 2014; Binder et al. 2013; Gunderson and Holling 2002; Holling 2001). Frameworks can be used to describe complex systems because they offer a common language that "offers a substantial advantage in making comparisons" (Holland 2014: 75). For example, Quiñones et al. (2017) use the DPSIR framework from European Environmental Agency (EEA 1999) to analyze the ecosystem services of a watershed dominated by commercial forestry. Lewison et al. (2016) review the use of the DPSIR (Drivers-Pressures-State-Impact-Responses) framework to study coastal social-ecological systems, pointing out that the framework is still under-utilized. Campuzano et al. (2013) use the same framework to analyze three Latin American coastal zones.

Delgado-Serrano et al. (2015) use Ostrom's sustainability framework (2009) to compare local perceptions of local actors in three Latin American SES (México, Argentina, and Colombia). We further discuss the use of conceptual models for SES studies in chapter "Social Actors and Participation in Environmental Issues in Latin America" of this book.

In summary, social-ecological systems are complex and contextual requiring a transdisciplinary team of ecologists, sociologists, political scientists, and at times experts in human health and lawyers. The members of such a team should all make their best efforts to go beyond their disciplines and to build a conceptual SES model that may serve as a guide not only for future research and decision-making but to communicate with social actors and request information from them as an invitation for participation (see chapter "Social Actors and Participation in Environmental Issues in Latin America"). Now, let's suppose that two different teams, with similar composition in terms of disciplines, analyze the same social-ecological system without ever exchanging information. Should we expect them to arrive at similar results? Does science have mechanisms to decide which one is right or wrong? Is the last question valid? How do social actors enter in this knowledge system? We discuss some of those issues in the next section.

3 Social-ecological Systems: A Postnormal Challenge

Is science in crisis? Some of us could argue that of course, it is in crisis! It has to be, otherwise, how our knowledge would evolve in the first place. Nevertheless, that is a Kuhnian perspective (paradigms and their examination) of how science changes through time (Kuhn 1962). However, when Saltelli and Funtowicz (2017) answered this question, their perspective was different. According to them, contemporary science lives a crisis centered on public trust. Oreskes (2018), on the other hand, discussing the use of science by the Environmental Protection Agency of the United States, proposes that there is indeed a crisis, but it has to do with "the attempt to discredit scientific findings that threaten powerful corporate interests." Thus, as long as scientific knowledge is discussed within its inner circles (e.g., scientific journals, scientific congresses and meetings), having disputed hypothesis and crisis (e.g., multiple hypotheses coexisting at a given time) is not only normal but highly welcomed. The problem is when science is supposed to inform societies to help decision-making processes that will then impinge on future scenarios affecting human well-being (e.g., global climate change, critical species extinctions, the sustainability of coastal fisheries, what to do with the water scarcity problem). The problem with social-ecological systems (SES) is that they always include, by definition, human societies and their relationships with nature.

If we use the term "normal science" to define the historical or classical, problem-solving, way to develop knowledge about our world and the mechanisms to test them (Marín and Delgado 2013): is this the correct epistemological approach to deal with social-ecological systems? We propose that the answer is no and that people embarking in the analysis of these systems should work from a postnormal perspective of science.

What do we do in normal science that may not be applicable to SES's analysis and governing proposals? Let us start from a classical definition of science, extracted from the eighth edition of Oxford Advanced Learner's Dictionary[1]: "*knowledge about the structure and behavior of the natural and physical world, based on facts that you can prove, for example with experiments.*" The first comment about this definition of science is that SES analyses are fully affected by the value that different social actors may give to social/natural components. For example, Ainscough et al. (2018) describe how multiple values play a role when analyzing ecosystem services and that in complex systems "*the boundary between values and facts is fuzzy*" (p. 98). Second, it should be clear to anyone that SES experimentation is both impossible and unethical and, as a result, our understanding of these systems comes from observations, both quantitative and qualitative according to Berkes et al. (2003). The consequence of such a way to analyze them is that uncertainty of the data, the techniques utilized to gather it, and the epistemic approach of scientists, may indeed be very high (Ainscough et al. 2018; Marín et al. 2018). Third, if values play a role in relation to how scientists define and give priority to different social/natural components, then multiple legitimate perspectives may arise. Funtowicz and Ravetz (1992) then proposed a new way of understanding science (postnormal science) when: "facts are uncertain, values in dispute, stakes high and decisions urgent." This way of understanding science accepts that there may be several legitimate perspectives for a given problem and that as a result, the dialogue is crucial, not only between scientists but among all social actors. This, from our perspective, is the main point when using a postnormal approach to SES studies, the participation of social actors. We shall return to this issue in chapter "Social Actors and Participation in Environmental Issues in Latin America", but as a final section to this chapter, we succinctly describe a social-ecological conflict that was triggered in southern Chile 15 years ago, the Rio Cruces conflict, from a postnormal point of view.

4 The Rio Cruces Conflict: A Social-ecological, Postnormal, Perspective

A conflict is a situation in which people disagree over a given issue. This type of situation is common in science. A review of the term "conflicting hypotheses" using the WoS,[2] produced 273 articles with an average of 11 articles per year in the last 20 years. However, it is even more common for social issues. Indeed, if we used the same procedure but this time using the term "social conflict," the result is more than

[1] https://www.oxfordlearnersdictionaries.com/us/definition/english/science?q=science

[2] http://webofknowledge.com

Fig. 1 Black-necked swans (*Cygnus melancoryphus*) in the Chilean coast (Chiloé, January 2019; Credits: Víctor H. Marín)

2000 articles for the period 1975–2018 (over 46 articles per year). One of the most passionate Chilean social-ecological conflicts was the so-called Rio Cruces conflict, involving the Valdivia society (an urban zone located in the southern Chilean coast), a cellulose plant, several scientists, and an emblematic species: the black-necked swan (Jaramillo et al. 2018; Marín et al. 2018). The conflict was triggered by the sudden decrease in the local population of black-necked swans (Fig. 1), from 8000 individuals in April 2004 to 249 in August 2005.

Rio Cruces is a wetland, located near the city of Valdivia. The wetland was generated in May 1960 as the result of one of the strongest earthquakes recorded in recent human history (9.6 on the Richter scale), transforming cattle prairies into a 2–3 m water ecosystem. Thus, it is a new ecosystem if we considered that most historical ecosystems on earth are several centuries old (e.g., McClenachan et al. 2015). Soon after its formation, the wetland was invaded by a Brazilian macrophyte (*Egeria densa*) that became the preferred food for swans. The fast growth of the invader generated also a fast increase in the local swan population, soon becoming the dominant bird species (Delgado and Marín 2013). The swan became a local social symbol, with people renting boats for summer bird watching to the wetland and even giving its name to a local beer. Although the wetland is part of a watershed dominated by forestry and, as such, it could be modified in its water quality given the frequent changes in forest coverage (Delgado et al. 2014), it was accepted by the Ramsar Convention as a wetland of international importance (Ramsar 1998). Still, within the text of the information sheet for the Convention, authors called the attention to the fact that watershed deforestation may affect water provision and sediments entering the wetland. Furthermore, even though the wetland was accepted by the Ramsar Convention in 1998, with the status of "Sanctuary of Nature" under the Chilean law (CONAF 2019), there was no ecosystemic monitoring aside from birds (Delgado and Marín 2013). Thus, when swans started dying in the wetland, scientists did not have the benefit of a data set that could be used to understand its causes. This generated the first postnormal condition to the problem: facts uncertainty. The main result was the generation of several hypotheses to explain the sudden change within the ecosystem but without the benefit of before–after data (Marín et al. 2018). The next postnormal condition of this conflict had to do with the urgency of the decisions. The Rio Cruces conflict found the Chilean government applying for membership on the Organization for Economic Co-operation and Development (OECD 2019), and the unsettled condition of the conflict was interpreted by some social actors as a negative sign (Marín and Delgado et al. 2013). In fact, some authors have argued that even swans became political actors (Sepulveda-Luque 2018). One result of this urgency was a governmental lawsuit, filed on April 2005, against the pulp mill company for environmental damage, which ended on July 2013 with the Chilean judicial system ruling against the company; a verdict accepted by all parties that stopped social concerns for a period of time (Marín et al. 2018). Were stakes high? Yes! On one side, there was an ecosystem that either due to human-generated causes or not, suddenly changed its ecological structure (Jaramillo et al. 2018; Marín et al. 2009). On the other, an environmental conflict that threatened with breaking the Chilean environmental system (Sepulveda-Luque 2018). Finally, wetland's valuation had a diversity of perspectives. For some, it was the intrinsic value of any species within the system (i.e., ecocentrism; Kortenkamp and Moore 2001) represented by environmentalists (Marín and Delgado 2013). For others, mostly people related to micro- and macro-industries, the wetland generated an industrial image that at the time of the conflict was negative and for tourism operators, it represented a change in their income since people were less interested in visiting the wetland (Delgado et al. 2009).

What is the current ecological condition of the wetland? Jaramillo et al. (2018) propose that the ecosystem has returned to conditions prior to 2004, CONAF (2018) show that the swan population, up until March 2018, has grown beyond its historical values to reach a maximum, never recorded before, of 14,130 individuals in February 2018 and swan-based tourism has returned.[3] If we use classic Kuhnian science, we could indeed say that "industry guilty" hypotheses were correct, that the social-ecological system has recovered after changes in the residual water of the pulp mill plant. However, we also know that information was rather uncertain, that ecosystems may respond catastrophically to small changes in controlling factors (i.e., regime shifts; Marín et al. 2009), that the dominant invader plant, *Egeria densa*, is sensitive to changes in water availability (Yarrow et al. 2009), and that given the size and flying capabilities of black-necked swans, emigrating to different locations within the Chilean coast (e.g., Chiloé island; see Fig. 1) should not and has not represented a problem for the maintenance of its population[3]. A postnormal approach to issues related to the Rio Cruces wetland and its social actors should include opening science to social actor's participation,[4] evaluate scientific uncer-

[3] https://deskgram.net/explore/tags/cisnecuellonegro

[4] http://www.comunidadhumedal.cl/

tainty and ignorance (Saltelli and Funtowicz 2017) and share the evaluation with all social actors, promoting open discussions on "what if" scenarios. One interesting result of the conflict was the recently (end of 2016) created "*Centro de Humedales Rio Cruces*".[5] The center incorporates citizen's science and it visualizes the wetland as a social-ecological system. In this respect, science will have to play a different role; discussing different alternatives when analyzing data in terms of ecosystem functioning and generating all the necessary "narratives" (sensu Kay 2000), so social actors may be properly informed of our uncertainties and levels of ignorance.

Acknowledgements The work contained in this chapter was financed by CONICYT-Chile FONDECYT Grant N° 1170532 awarded to L. Delgado.

References

Ainscough J, Wilson M, Kenter JO (2018) Ecosystem services as a post-normal field of science. Ecosyst Serv 31:93–101

Balvanera P, Uriarte M, Almeida-Leñero L et al (2012) Ecosystem services in Latin America: the state of the art. Ecosyst Serv 2:56–70

Berkes F, Folke C (1998) Linking social and ecological systems for resilience and sustainability. In: Berkes F, Folke C (eds) Linking social and ecological systems. Management practices and social mechanisms for building resilience. Cambridge University Press, Cambridge, pp 1–25

Berkes F, Colding J, Folke C (2003) Introduction. In: Berkes F, Colding J, Folke C (eds) Navigating social-ecological systems. Building resilience for complexity and change. Cambridge University Press, Cambridge, pp 1–29

Binder CR, Hinkel J, Bots PWG et al (2013) Comparison of frameworks for analyzing social-ecological systems. Ecol Soc 18(4):26. https://doi.org/10.5751/ES-05551-180426.

Campuzano FJ, Mateus MD, Leitão PC et al (2013) Integrated coastal zone management in South America: a look at three contrasting systems. Ocean Coast Manag 72:22–35

Carpenter SR, Pingali PL, Bennett EM et al (2005) Ecosystems and human well-being: scenarios, vol 2. Island Press, Washington, DC

Chu D, Strand R, Fjelland R (2003) Theories of complexity: common denominators of complex systems. Complexity 8:19–30

Comisión Económica para América Latina y el Caribe (CEPAL) (1999) Latin America: total, urban and rural population and urban percentage, by quinquennium and countries. Demographic Bulletin N° 63. https://www.cepal.org/celade/publica/bol63/Bd6311.html. Accessed 8 Jun 2019

Comisión Económica para América Latina y el Caribe (CEPAL) (2016) La población de América Latina ancanzará 625 millones de personas en 2016, según estimaciones de la CEPAL. https://www.cepal.org/es/noticias/la-poblacion-america-latina-alcanzara-625-millones-personas-2016-segun-estimaciones-la. Accessed 8 Jun 2019

Corporación Nacional Forestal (CONAF) (2018) Informe monitoreo avifauna Marzo 2018. Santuario de la Naturaleza Carlos Anwandter. http://www.conaf.cl/cms/editorweb/rio_cruces/informes/Marzo-2018.pdf. Accessed 8 Jun 2019

Corporación Nacional Forestal (CONAF) (2019) Parques de Chile. http://wwwconafcl/parques-nacionales/parques-de-chile/. Accessed 8 Jun 2019

[5] http://www.cehum.org/english/

Delgado LE, Marín VH (2013) Interannual changes in the habitat area of the black-necked swan, *Cygnus melancoryphus*, in the Carlos Anwandter sanctuary, southern Chile: a remote sensing approach. Wetlands 33:91–99

Delgado LE, Marín VH (2016) Well-being and the use of ecosystem services by rural households of the Río Cruces watershed, southern Chile. Ecosyst Serv 21:81–91

Delgado LE, Tironi A, Vila I et al (2014) El humedal del Río Cruces, Valdivia, Chile: una síntesis ecosistémica. Lat. Am. J. Aquat. Res. 42:937–949

Delgado LE, Marín VH, Bachmann PL et al (2009) Conceptual models for ecosystem management through the participation of local social actors: the Río cruces wetland conflict. Ecol Soc 14(1):50. http://www.ecologyandsociety.org/vol14/iss1/art50. Accessed 8 Jun 2019

Delgado LE, Sepúlveda MB, Marín VH (2013) Provision of ecosystem services by the Aysén watershed, Chilean Patagonia, to rural households. Ecosyst Serv 5:102–109

Delgado-Serrano M, Oteros-Rozas E, Vanwildemeersch P et al (2015) Local perceptions on social-ecological dynamics in Latin America in three community-based natural resource management systems. Ecol Soc 20(4):24. https://doi.org/10.5751/ES-07965-200424. Accessed 8 Jun 2019

Denevan WM (1992) The native population of the Americas in 1492. University of Wisconsin Press, Madison

European Environment Agency UE (EEA) (1999) Environmental indicators: typology and overview. Technical report n. 25, Copenhagen

Funtowicz SO, Ravetz JR (1992) Three types of risk assessment and the emergence of post-normal science. In: Krimsky S, Golding D (eds) Social theories of risk. Praeger, Westport, pp 251–273

Geymonat L (2006) Historia de la filosofía y de la ciencia. Crítica, Barcelona

Giusti EM, Castelnuovo G, Molinari E (2017) Differences in multidisciplinary and interdisciplinary treatment programs for fibromyalgia: a mapping review. Pain Res Manag 2017:1. https://doi.org/10.1155/2017/7261468. Accessed 8 June 2019

Gunderson LH, Holling CS (2002) Panarchy. Understanding transformations in human and natural systems. Island Press, Washington, DC

Holland JH (2014) Complexity. A very short introduction. Oxford University Press, Oxford

Holling CS (2001) Understanding the complexity of economic, ecological and social systems. Ecosystems 4:390–405

Jaramillo E, Lagos NA, Labra FA et al (2018) Recovery of black-necked swans, macrophytes and water quality in a Ramsar wetland of southern Chile: assessing resilience following sudden anthropogenic disturbances. Sci Total Environ 628-629:291–301

Kay J (2000) Ecosystems as self-organizing Holarchic open systems: narratives and the second law of thermodynamics. In: Jorgensen SE, Muller F (eds) Handbook of ecosystems theories and management. CRC, Boca Ratón, pp 135–160

Kortenkamp K, Moore CF (2001) Ecocentrism and anthropocentrism: moral reasoning about ecological commons dilemmas. J Environ Psychol 21:261–272

Kuhn T (1962) The structure of scientific revolutions. University of Chicago Press, Chicago

Latour B (1999) Pandora's hope. Harvard University Press, Cambridge

Lewison RL, Rudd MA, Al-Hayek W et al (2016) How the DPSIR framework can be used for structuring problems and facilitating empirical research in coastal systems. Environ Sci Policy 56:110–119

Marín VH, Tironi A, Delgado LE et al (2009) On the sudden disappearance of Egeria densa from a Ramsar wetland site of Southern Chile: A climatic event trigger model. Ecological Modelling 220 (15):1752–1763

Marín VH, Delgado LE (2013) From ecology to society and back: the (in)convenient hypothesis syndrome. Int J Sustain Dev 16:46–65

Marín VH, Delgado LE, Tironi-Silva A et al (2018) Exploring social-ecological complexities of wetlands of international importance (Ramsar sites): the Carlos Anwandter Sanctuary (Valdivia, Chile) as a case study. Wetlands 38:1171–1182

McClenachan L, Lovell S, Keaveney C (2015) Social benefits of restoring historical ecosystems and fisheries: alewives in Maine. Ecol Soc 20(2):31. https://doi.org/10.5751/ES-07585-200231. Accessed 8 Jun 2019

McGinnis MD, Ostrom E (2014) Social-ecological system framework: initial changes and continuing challenges. Ecol Soc 19(2):30. https://doi.org/10.5751/ES-06387-190230. Accessed 8 Jun 2019
Millennium Ecosystem Assessment (MEA) (2005) Ecosystems and human well-being: synthesis. Island Press, Washington, DC
Morandín Ahuermal I, Contreras-Hernádez A, Ayala Ortiz DA et al (2018) Complexity and transdiscipline: epistemologies for sustainability. Madera y Bosques. https://doi.org/10.21829/myb.2018.2431673. Accessed 8 Jun 2019
OECD (2019) Organization for Economic Co-operation and Development. http://www.oecd.org
Oreskes N (2018) Transparency rule is a Trojan horse. Nature 557:469
Ostrom E (2009) A general framework for analyzing sustainability of social-ecological systems. Science 325:419–422
Pohl C (2008) From science to policy through transdisciplinary research. Environ Sci Policy 11:46–53
Preiser R, Biggs R, De Vos A et al (2018) Social-ecological systems as complex adaptive systems: organizing principles for advancing research methods and approaches. Ecol Soc 23(4):46. https://doi.org/10.5751/ES-10558-230446. Accessed 8 Jun 2019
Quiñones D, Caro J, Marín VH et al (2017) Manejo resiliente de cuencas forestales de Chile: La cuenca de Cayucupil (Cañete) como caso de estudio. Boletín Nahuelbuta Natural 1:7–30
Ramsar (1998) Ficha Informativa de los Humedales de Ramsar. Santuario de la Naturaleza e Investigación Científica "Carlos Anwandter" del río Cruces, Valdivia. https://rsisramsarorg/ris/222. Accessed 8 Jun 2019
Rogers KH, Luton R, Biggs H et al (2013) Fostering complexity thinking in action research for change in social–ecological systems. Ecol Soc 18(2):31. https://doi.org/10.5751/ES-05330-180231. Accessed 8 Jun 2019
Saltelli A, Funtowicz S (2017) What is science's crisis really about? Futures 91:5–11
Schmitz O (2010) Resolving ecosystem complexity. Princeton University Press, Princeton
Schoon M, van der Leeuw S (2015) The shift toward social-ecological systems perspectives: insights into human-nature relationship. Nat Sci Soc 23:166–174
Sepulveda-Luque C (2018) Bringing animals within political communities: the citizens/swans association that fractured Chile's environmental framework. Soc Mov Stud 17(3):333–352
Urquiza A, Amigo C, Billi M et al (2018) Metálogo como herramienta de colaboración transdisciplinaria. Cinta Moebio 62:182–198
Yarrow M, Marín VH, Finlayson M et al (2009) The ecology of Egeria densa Planchón (Liliopsida: Alismatales): A wetland ecosystem engineer? Revista chilena de historia natural 82:299–313

Simplifying the Complexity of Social-ecological Systems with Conceptual Models

Luisa E. Delgado, Daniela C. Pérez-Orellana, and Víctor H. Marín

Abstract If you need to understand a social-ecological problem and an expert tells you "it is too complex," please go with someone who knows how to generate conceptual models. Conceptual models serve to simplify a problem, to help in research activities, and to develop management strategies. Furthermore, they are transdisciplinary tools that help to communicate different social actors and experts from different disciplines. The main objective of this chapter is to discuss conceptual modeling from a social-ecological perspective.

Keywords Social-ecological systems · Latin America · Complexity · Conceptual models · Scales · Scientific and social knowledge

1 Introduction

The Millennium Ecosystem Assessment (MEA 2005) stated that the speed of ecosystem changes has increased in the last 50 years affecting human well-being. What took centuries now develops in decades or even less. In 20 more years, children born today will have finished their education and will go out to work in a radically different world than the one we currently know. Are we preparing them to face that unknown future? We suggest that it is time to teach human societies how to deal with the unknown by accepting that we may not have all the required knowledge, but we could help ourselves by exchanging ideas and perspectives using transdisciplinary strategies (Delgado et al. 2019).

L. E. Delgado (✉)
Escuela de Ciencias, Facultad de Ciencias, Universidad de Chile, Santiago, Chile

Fundación Centro Transdisciplinario de Estudios FES-Sistémicos, Santiago, Chile
e-mail: ldelgado@uchile.cl

D. C. Pérez-Orellana · V. H. Marín
Laboratorio de Modelación Ecológica, Departamento de Ciencias Ecológicas, Facultad de Ciencias, Universidad de Chile, Santiago, Chile
e-mail: vmarin@uchile.cl

L. E. Delgado, V. H. Marín (eds.), *Social-ecological Systems of Latin America: Complexities and Challenges*, https://doi.org/10.1007/978-3-030-28452-7_2

These strategies, according to Marín et al. (2015), include the development of conceptual models as one of the best options to develop transdisciplinary science and to improve social communication on complex issues. Conceptual modeling has also been used to show that reality can be seen from diverse points of view, depending on the perspective used to sense it, generated and reproduced through diverse operations within the social systems (Arnold-Cathalifaud 2003; Marín and Delgado 2008). In some cases, visions are the result of strong links between humans and the natural environment; so, experts should respect each point of view, accepting from the beginning that when generating conceptual social-ecological models, there may be several outcomes and that there is not, a priori, a correct model but different visions about a given system.

There is currently a renowned interest in the generation of participative spaces to achieve social empowerment. Rincón-Ruiz et al. (2019) suggest that current challenges include the political will and the academic support to generate a transdisciplinary research culture. One of the key issues in this context is the ability to understand each other using tools that may help us in the process. The generation of conceptual models, the subject of this chapter, is one of them.

2 Human Beings and Their Ways to Know

The Science Field

The scientist, from the perspective of postnormal science (PNS), is just one more narrator of the phenomena occurring in social and natural systems (Marín et al. 2018). One of the most important PNS challenges is to blend scientific knowledge with that coming from society (e.g., ancestral and traditional) as a way to understand reality. Furthermore, reality from a postnormal constructivist perspective does not correspond to an ontologically objective phenomenon but as the organization of different worlds generated in our experience (see chapter "Postnormal Science and Social-ecological Systems"). This epistemological perspective corresponds to a new way to understand science and its relations with human society in the twenty-first century. There is no longer a single scientific truth when explaining social-ecological problems, and available truths are transitory in nature. Indeed, the transitory condition of scientific truths was recognized even by Popper (1962), a widely used epistemological reference by many scientists in the past and even today. Indeed, scientific truth from a Popperian perspective will only last during the time it resists attempts to falsify it. Rozzi (2019) states that all types of scientific knowledge (e.g., normal science, postnormal science) coexist affecting the meaning of rejecting or accepting hypotheses. He further adds that it also influences the way in which they should be transmitted to society and social actors.

Biocultural ethics emphasizes that human history is nonlinear since multiple ethics, values, and cosmologic cultural visions occur simultaneously in different places on earth (Rozzi 2019). This society-nature proposal can also be found in Holling's (2001) conceptual model. In this case, the author represents it as a four-phase adaptive/recursive framework that can be tracked through social-ecological history. Thus, reality perception will be different for different social actors depending on their degrees of awareness about nature, education, and social responsibility. It will further depend on whether its relationship with nature is close and continuous (directly affecting the quality of life and well-being) or not (e.g., scenic beauty, nature sports). Consequently, conceptual scientific models used to represent human–nature relationships are necessarily contextual.

These models generated by science influence ecosystem management and conservation. They are also useful when analyzing perception and valuation of ecosystem services and components by different social actors (see chapters "Social-ecological Systems and Human Well-Being" and "Ecosystem Services from a Multi-Stakeholder Perspective: A Case Study of a Biosphere Reserve in Central Chile") and their impact on the sustainability of social-ecological systems (Marín and Delgado 2013; Marín et al. 2018). Finally, including social narratives in scientific conceptual modeling can be an interesting way to reconstruct old human visions (Guerrero-Gatica et al. 2019).

The Social Field

The social reconstruction of reality defines an observation mode centered on the relationships between subjectivity, daily experiences, and concrete practices by individuals. In this sense, it projects a subjective quality as related to the construction of social reality but understood as an objective condition through the practices and concrete experiences of individuals and social groups (Berger and Luhmann 1968). Peluffo and Catalán Contreras (2002) define knowledge production as the capacity to relate in a highly structured way, data, information, and knowledge of a given object allowing to effectively act upon it based on specific values and contexts, which Berger and Luhmann (1968) relate to social groups. Thus, using the ideas discussed by these authors, a subjective meaning becomes a factual reality for a given social group (Fig. 1).

The main idea is that a human being will build objects and its reality subjectively as he/she goes through life. The main consequence of this way of understanding reality is that although many human beings may experience a single social-ecological system, there shall be several narrations about it. This, in turn, can be considered as one of the origins of the complexity of these systems which emphasizes the need to share these narrations through conceptual models, with the goal of producing a common understanding of the system.

Folchi (2001) has proposed that this subjectively experienced reality will become "normal" if there are not profound changes in day-to-day routines. For example, the life of social groups that have lived in proximity with a given ecosystem will necessarily include a mental image of what the "normal ecosystem" looks like (Marín and

Fig. 1 Plato myth of the cave as explanation for social truth. Men in chains within the cave consider as "the truth" the shadows of the objects that other men bring before the entrance of the cave reflected in the cave's wall due to the sunlight

Delgado 2013). These perceptions, configured as social representations, become common sense through three phases (Piñero 2008): (a) naturalization: its meaning appears as given without a specific origin but as a collective expression; (b) iconization: social perceptions become material objects even with a graphic expression and (c) anchorage: perceptions get inserted in pre-existing semantic networks.

Thus, the process of the social construction of reality, when referred to human beings and the natural environment, may go from symbolic (e.g., quality of life) to utilitarian components (e.g., ecosystem services) where changes will depend upon social, economic, and political factors (Piccolo 2017). In this regard, one special type of experience is ancestral-ecological and traditional knowledge, where human beings transform ecosystems and/or adapt to their changes to maintain their quality of life and well-being (see chapter "Environmentalism of the Poor: Environmental Conflicts and Environmental Justice").

3 Social-ecological Conceptual Models

If the reader goes through this book chapters, especially those related to case studies, he/she may see the need for a conceptual infrastructure to help to explain social-ecological problems. The same principle applies when communicating the results on inter- and transdisciplinary work to local social actors so they may understand their interactions with nature. Indeed, they are important to be considered when generating local management plans for conservation, sustainability, and environmental governance (Delgado et al. 2009). What happens when there is no information to generate conceptual models? This is today a rather common situation in

Latin America. A recent example, discussed in chapter "Postnormal Science and Social-ecological Systems" of this book, is what was known as "the Rio Cruces wetland social-environmental crisis." In this case, the Chilean State Defense Council requested the ecological restoration of the ecosystem through a lawsuit filed in 2005. However, at that time there was not a synthesis about the ecosystem's structure, let alone a social-ecological one, that could have served as guidance for the system's recovery or restoration. Thus, when the judge included in the verdict a sentence stating that "nature has done its part" (Bio-Bio 2013: 160), consequently rejecting restoration as requested by the Chilean State, nobody objected.

A postnormal, constructivist generation of models for social-ecological systems requires the participation of experts (e.g., scientists) and social actors (Delgado et al. 2009; Marín et al. 2008). Accordingly, conceptual models are observer dependent, spatially explicit, system's descriptions where components (biotic, abiotic, and social) and spatial limits will depend upon the questions asked, the people that formulated them and the social context in which they were asked (Marín et al. 2018).

Marín et al. (2015) propose that conceptual models built by experts and social actors share five characteristics:

1. *A model only reflects the state of knowledge at the time of its generation.* Therefore, it should be continuously checked if new information becomes available. That is, it should be conceived as an adaptive structure.
2. *A model is the image of those who built it.* Thus, it should be analyzed by a wide variety of social actors, especially if it will be used for ecosystem management.
3. *A model only serves for the objectives for which it was created.* So, they should be clear from the beginning, being agreed among social actors.
4. *A conceptual model is a communication tool for social actors.* It does not replace the necessary dialogue among them, it is instrumental to it.
5. *A model cannot solve uncertainties, it only incorporates them.* In the absence of information about the social-ecological system, processes and relationships can only be hypothetical to be considered for future research.

Furthermore, as stated by Lawrence and Després (2004), conceptual models are transdisciplinary. They should be built accepting complexity and the need to go beyond the classic disciplinary academic structure. One of the characteristics of transdisciplinary knowledge is that it is problem-solving oriented. Wickson et al. (2006: 1048) eloquently express this idea when they state that transdisciplinary research starts with a problem that "exists in the world and it is real." That is why conceptual modeling is an appropriate methodology when analyzing social-ecological problems.

4 Social-ecological Complexity: Levels and Spatial-Temporal Scales

The social-ecological concept is very appropriate when dealing with transdisciplinary issues such as changes in environmental management paradigms, social actors' participation, and institutions and ecosystems (Delgado et al. 2019; Ávila and Perevochtchikova 2018). Partial views of complex social-ecological systems (e.g., those separating nature and human beings) generate erroneous interpretations of contemporary environmental problems. However, if looked from a systemic, transdisciplinary, and postnormal perspective (Marín et al. 2018), it is possible to develop a conceptual infrastructure (e.g., Delgado et al. 2014; for the Río Cruces wetland) that can then be used to manage and monitor social-ecological changes after an ecosystem perturbation (e.g., Jaramillo et al. 2018). Guerrero-Gatica et al. (2019) propose that it is necessary to supplement current scientific management gaps through social actors' participation, as a way to download large scale (e.g., Convention on Biological Diversity) and national scale (public policies) proposals, into local scale actions.

Social-ecological systems behave as complex adaptive systems (CAS) since their interactions are adjusted and self-organized continuously in time (Maass 2018). One of the characteristics of CAS is that their components, interactions, and responses change when viewed from different temporal and spatial scales (Delgado et al. 2019; Prieto-Barbosa 2013). Consequently, they should be studied at multiple scales (Ostrom 2009). One example of this multi-scale behavior is that national public policies behave as indirect forcing functions in relation to local economic development (e.g., land use changes and ecosystem's modifications). They, in turn, are forced by the world's economic trends and agreements which affect the export of natural products. Thus, a local conceptual model for a social-ecological system necessarily has to incorporate large-scale processes. Ávila and Perevochtchikova (2018) suggest that social-ecological analyses should be done at all scales (micro, meso, macro, and global environment) necessary to understand people–environment relationships. Gunderson and Holling (2002) have proposed a conceptual framework (panarchy) as a way to understand and rationalize changes and persistence in social-ecological systems at several scales. Another alternative is to generate different conceptual models for different scales (Delgado et al. 2019). We next explore some modeling alternatives.

5 Conceptual Models Generated by Experts

Holling (2001) generated a cycling-adaptive framework that Delgado et al. (2019) modified to incorporate ecosystem services as key variables for social-ecological systems. Holling's main idea is that ecological, economic and social systems behave adaptively through four phases that repeat sequentially in time: (1) r phase of fast growth, (2) K phase of conservation, where changes are slow, (3) Ω phase of collapse, where either exogenous or endogenous perturbations generate crises inside the system, and (4) α phase of reorganization where, after a perturbation, innovations will generate a new adaptive cycle.

The framework, when applied to social-ecological systems, implies that perturbations applied to one subsystem (e.g., ecological) will generate responses in the other subsystem (e.g., social). The relationships between the subsystems are then analyzed through their connectivity (high or low) and capitals; union and bridge for the social subsystem and natural for the ecological subsystem (Fig. 2).

The ecological subsystem is built from natural capitals that sustain planet earth and generate the services used by humans (Costanza et al. 1997; Berkes and Folke 2000). In this context, the utility of the ecosystem service concept lies in its dependency on biophysical conditions and the changes produced by humans, both locally (e.g., land use changes) and globally such as climate change (Burkhard et al. 2012).

The social subsystem is formed from four interconnected capitals: economic, social, human, and cultural. Their interconnection allows the auto-organization of the subsystem (Kay et al. 1999). The economic capital allows immediate access to goods and services, being institutionalized as property rights (Bourdieu 2001). It is generated through economic activities, human ingenuity, technological means, and other production modes, all included as a sum in the concept of financial capital (Cinner et al. 2009). The social capital represents several characteristics of the social organization such as trust relationships and networks (Coleman 1990) and institutional capitals (Ostrom 1990). The social capital is further divided into two subtypes: (a) union social capital, corresponding to communication networks and

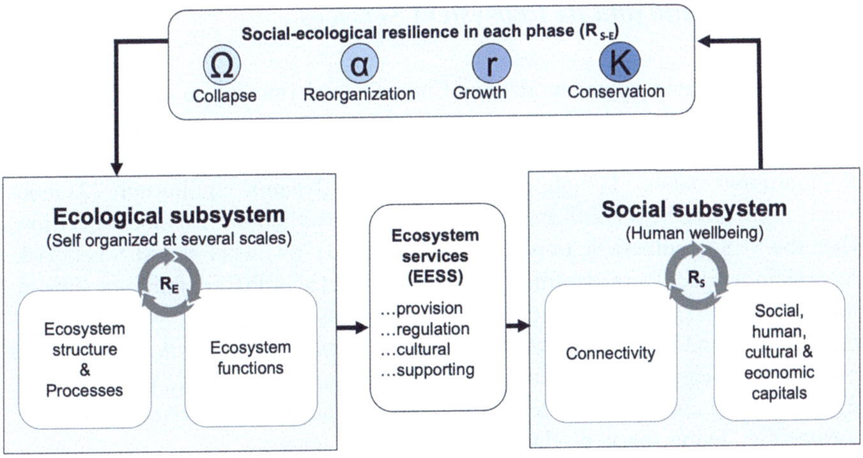

Fig. 2 Conceptual (SES-ES-adaptive) framework linking the ecological subsystem and its support and provision of ecosystem services to the social subsystem. The temporal dynamic will change as they move through the four adaptive phases (modified and integrated from the models of Müller and Burkhard (2012) and Holling (2001). $R_{S\text{-}E}$ social-ecological resilience; R_E ecological resilience; R_S social resilience

trusted friends (e.g., friends, neighbors, social networks), and (b) bridge social capital, corresponding to institutional socially guaranteed networks (e.g., municipalities). The human capital is understood as the productive capacities that an individual acquires accumulating general and/or specific types of knowledge (Pomeroy et al. 2004). Finally, cultural capital corresponds to all the details provided by human societies that have generated adaptations to deal with the natural environment and its active modifications, which in some cases may even be done unconsciously (Bourdieu 1986).

Therefore, the conceptual framework is based on the resilience concept but understood from three perspectives (Fig. 2): ecological, social, and social-ecological. Ecological resilience corresponds to the capacity of an ecosystem to absorb perturbations, maintaining its structure, functions, and feedback mechanisms (Holling 1973). Social resilience is the capacity of human beings to deal with stress (Adger 2000). High social resilience implies strengthening social networks, social abilities, social leadership, and knowledge among others which will allow people to face system's perturbations, at times even changing the state of the social-ecological system (Berkes and Ross 2013). Finally, social-ecological resilience, also known as system's adaptive capacity, corresponds to the changes of human actions related to changes in ecosystem so its generation to provide services remains unchanged (Anderies et al. 2004; López-Angarita et al. 2009). In the next subsection, we show an example of the application of this framework to a Chilean social-ecological system: The Chiloé Island.

A Social-ecological, Experts-Based, Conceptual Model for Chiloé Island and its Ecosystem Services

We generated a conceptual model for Chiloé Island (located in southern Chile) social-ecological system (SES) based on the previously described framework (Fig. 2). For the purpose of this model, we have divided social-ecological resilience ($R_{S\text{-}E}$) into four stages: (1) high, when the SES is in dynamic equilibrium, (2) moderately high, if the SES is still stable when facing perturbations, (3) moderately low, when the SES is vulnerable to perturbations, and (4) low, when the SES is at risk. Ecosystem services were identified using De Groot et al. (2002) definitions, separating them into: (a) *provisioning*, originated in the coastal zone (e.g., fishes, macroalgae, mussels) and inland (e.g., wood, bryophytes, grasses), (b) *regulation*, from the coastal ocean (e.g., control harmful algae blooms) and inland (e.g., pests control), (c) *cultural*, e.g., scenic beauty and recreation, rural and ethnic tourism, and (d) *supporting*, e.g., living space, productive systems.

The model was built using bibliographic references (Annex I), and it was validated through key social actors' interviews and a social-ecological survey (Pérez-Orellana 2019). The model starting year (1825) corresponds to the beginning of the process when the island was included as part of the Chilean territory (The Tantauco

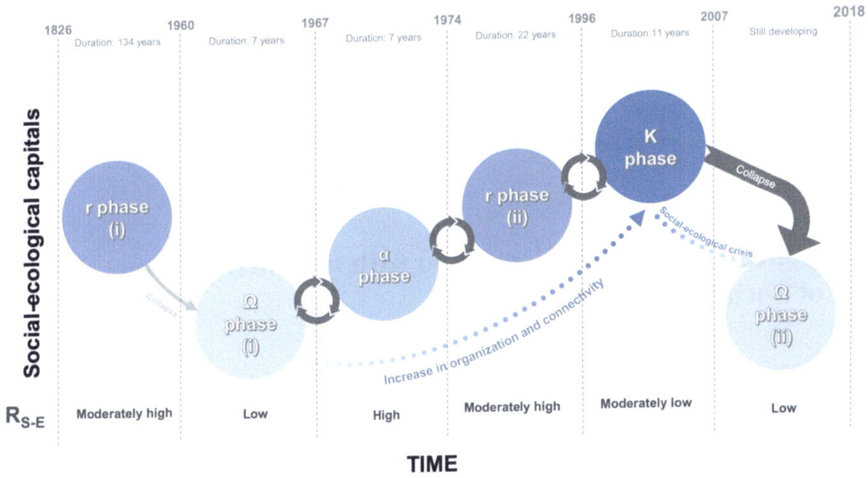

Fig. 3 Conceptual SES-ES adaptive model for Chiloé Island. See Annex I for details on each phase of the model

Treaty, Matte-Varas 1990), and it covers 193 years of island's social-ecological history, up to the year 2018 (Fig. 3).

The model starts with a first growth phase (*r*) with ecosystem services being used for subsistence of local social actors, with low-to-moderate connectivity within social and ecological subsystems. Yet, in the year 1900, the island starts its configuration as a big export territory for potatoes. The phase stops suddenly due to the 1960 mega earthquake, generating a collapse (Ω) phase that lasted 7 years. In the year 1967, a reorganization (α) phase starts, lasting 7 years. During this period (1967–1974), the SES changes, increasing the diversity of economic activities and the use of ecosystem services both for local consumption and export. The next phase is the second growth period of Chiloé SES, lasting 22 years (1974–1996). The system growths rapidly, but its connectivity starts decreasing due to political changes (e.g., military dictatorship) and the economic hegemony of salmon farming. As the growth period was ending, the system started showing several characteristics of a conservation phase (K). One of the most important issues that moved Chiloé to an eleven-year K phase (1996–2007) was the system's rigidity, where social-economic development depended almost entirely on the export of coastal ocean products (mostly salmon and mussels). Finally, the year 2007 marks the second collapse phase (Ω) which, so far, has lasted 11 years without clear signs of ending. However, while the first collapse was due to a natural phenomenon (i.e., earthquake) this one was triggered by an unwanted result of salmon farming: the sanitary crisis resulting from the spread of the Infectious Salmon Anemia (ISA) in Chiloe's coastal waters and recurrent Harmful Algal Blooms (HAB). Both ISA and HAB processes generated social-economic crises including workers' migration from coastal marine ecosystem to land farming, a Chilean State generating subsidies to low-income families and a coastal ocean full of human garbage.

In summary, the conceptual model shows that it is possible to simplify the temporal dynamic of a complex social-ecological system, such Chiloé Island, based on the adaptive-cycling ideas of Holling (2001). It further shows that the study of social–ecological interactions should be done with a transdisciplinary perspective, where the local context may play an important role.

6 Conceptual Models Generated with the Participation of Social Actors

Participative modeling can be traced back to the 1970s when it was used as a tool to understand changes in the scientific knowledge of children (Ampuero et al. 2013). From this perspective, conceptual modeling becomes a strategic communication tool among social actors; a requisite for the sustainability of social-ecological systems according to Ostrom (2009). The advantage, relative to other forms of communication, is that its structure especially when based on friendly graphical software, allows a clear view of the bases, assumptions, and rationales used in its generation. However, its use in ecosystem management requires participation methods that ensure that most social actors have been involved (Yu Iwama and Delgado 2018; Guerrero-Gatica et al. 2019, see also chapter "Social-ecological Systems and Human Well-Being").

Participatory mapping is one of the methods used at the local level, where social actors express their knowledge and perceptions based on the generation of maps (Yu Iwama and Delgado 2018). The creation of maps, according to Sletto (2010), is a process tightly linked with the identity and territorial connections of social actors, where they represent space autonomously. Another strategy for participatory modeling corresponds to those based on brainstorming (Marín et al. 2008). This modeling strategy allows combining qualitative and quantitative information, where social actors define the spatial limits of their social-ecological systems, their key components and the relationships between them (Carpenter et al. 2009).

A Brainstorming Methodology for the Generation of Conceptual Models

The main objective of the brainstorming methodology for conceptual modeling is to obtain the knowledge from all participants, without constraints (Marín et al. 2008). There are several sources to help in the implementation of brainstorming sessions, and the interested reader will find plenty of Internet links. For example, a Google search conducted on April 2019 showed the existence of 80 million websites that use the term "brainstorming."

There are four rules that, in our experience, are central to develop brainstorming sessions:

- Do not judge the ideas of other participants.
- When it comes to ideas, the more, the better.
- Encourage constructive participation among members.
- All participants have the same rights.

There are many ways to develop a brainstorming session (e.g., unstructured or free flow, structured based on a narrator, in circles, etc.). In our experience, if potential participants do not have familiarity with brainstorming sessions, the most convenient methodology is the structured meeting based on a narrator. The process starts with the identification of concerned social actors (see chapter "Social-ecological Systems and Human Well-Being"), organizing them according to their interests and fields of action (e.g., government personnel, NGOs, enterprises personnel). At this point, we should emphasize that when issues are controversial and there may be conflicts within some social actors (e.g., enterprises and NGOs), those in charge should consider running separate sessions for each type of social actor, which will then leave the problem of assembling all ideas in the hands of the organizers (e.g., Delgado et al. 2009).

After potential participants have been identified, organizers should send a series of questions about the social-ecological system to be modeled and the main question to be answered (e.g., what is the carrying capacity of the fjord for salmon farming? Marín et al. 2008: 11,113). The questions to be sent to participants most of the time will be based on historical analyses based on bibliographic references, information available on the Internet and a consultation process with social actors.

Narrators or brainstorming facilitators play a rather specific role within a session: they record answers of all participants without adding his/her own ideas. They should also have working knowledge of iconographic modeling software (e.g., Vensim®, STELLA®) for the next step. When the brainstorming session stops, after all ideas have been recorded and all opinions listened, facilitators should split participants in small groups (not larger than 5 people) who then will build the conceptual model of the social-ecological system, by means of a member of the facilitator's group, based on the information gathered during the first phase. After models have been built, participants may share results or do it later through email. In some cases, all models generated by the small groups will easily converge into a single model; in others, models may be different enough that they will have to stay as different proposals for the problem under analysis. The generation of a participative conceptual model, using the brainstorming strategy described above, may take between 6 and 8 h. In the next subsection, we describe an example of this participatory method.

A Brainstorming-Based Participative Conceptual Model: Forestry–Ecosystem Relationships in Southern Chile

Participative conceptual models are useful tools in the generation of strategic business ideas and for resolution of conflict (Delgado et al. 2009). They help working groups to articulate the necessary knowledge for new actions and also as a window through which people can see each other's ideas. In this specific case, a Chilean forestry company (whose name shall remain anonymous) wanted to analyze the ways in which they could fulfill their auto-imposed commitment to ecosystem-based management. The brainstorming session was conducted by a narrator and included the participation of 12 people from the company all of whom were identified as key actors (see chapter "Social-ecological Systems and Human Well-Being") regarding the development of ecosystem-based management. All participants received seven questions, 2 weeks before the session (Table 1).

The first part of the brainstorming session showed that while all participants understood the concept of ecosystem management, the generated models (three) included different components and interactions. When compared with the opinions of the ecosystem experts, results showed that company experts only identified 24% of the components and interactions of a forestry ecosystem. Thus, each group of experts only identify the components that his/her professional training allows them to see. Interestingly, when all three models were combined, all participants agreed that it was what they wanted to say (Fig. 4). Still, most processes were suggested by the ecosystem experts to company people.

Thus, the mixture of three different models on a single ecosystem provided a consensual view of the relationships between the company's forestry actions and ecosystem changes. The generation of the conceptual model by the people that will later use it would correspond to an agreement between company's personnel, where all opinions were included arriving at a product that was satisfactory to everyone (Pérez-Teruel et al. 2014).

Table 1 Questions sent over an email to the participants of the ecosystem-based forestry management brainstorming session

1. What do you understand by ecosystem management?
2. What is the relationship between the following concepts?
(a) Biodiversity conservation and ecosystems
(b) Watershed water quality and ecosystem
3. What ecological process can you currently monitor?
4. What ecosystem process do you monitor that may be used to prevent catastrophes?
5. Do you apply concepts included in the idea of adaptive management?
6. What improvements would you add to your current initiatives?
7. How often do you interact with people from other divisions within the company?

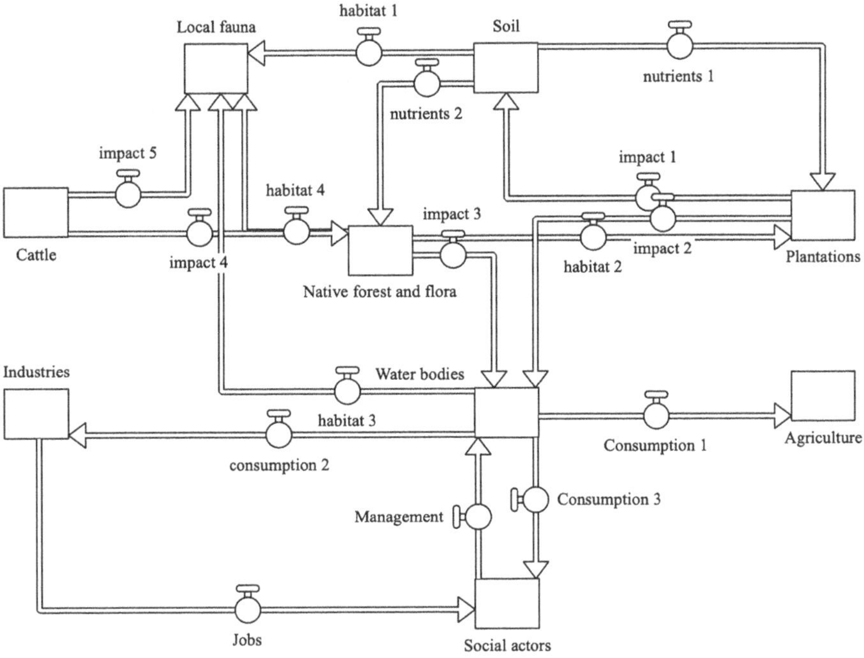

Fig. 4 Forestry–ecosystem conceptual participative model. Symbols are those of the STELLA® Architect Modeling & Simulation modeling Software (Version 1.8.3). The model is the final version that combined three models generated during a brainstorming session

7 Conclusions

The world, as we know it, is changing at an accelerated rate. One of the issues of this change is whether or not we will understand it in time to do something about it. This learning process requires, and it will require, listening to each other in ways that all may comprehend. Conceptual modeling of complex social-ecological systems is one potential way to improve this process.

Acknowledgements The work contained in this chapter was financed by CONICYT-Chile FONDECYT Grant N° 1170532 awarded to L. Delgado.

Annex I

Phases of the social-ecological system (SES)/Ecosystem services (ES) adaptive model for Chiloé Island. All cited bibliographic references are listed at the end of this annex. $R_{S\text{-}E}$ social-ecological resilience, R_S social resilience and R_E ecological resilience.

r phase (i) (1825–1960): rapid growth, the potential of both subsystems increases while their connectivity is low.

$R_{S\text{-}E}$: Moderate R_S and high R_E. Human activities changed after the island was incorporated to the Chilean territory with more options and a growing trend of the connectivity within the SSE. Ecosystem services were used mostly as provisioning for subsistence.

In 1850 agriculture decreases in intensity, being replaced by industrial activities (e.g. sea lion hunting and native wood cutting) in the south of the island (Cárdenas and Villagrán 2005). In 1900 some agricultural activities restart, positioning the island as the main producer of potatoes by the year 1950.

Ω phase (i) (1960–1967): collapse, where exogenous and endogenous perturbations generate crises within the SES.

$R_{S\text{-}E}$: Low. Both subsystems do not show adaptive capabilities after the 1960 mega earthquake. There is an increase in human emigration and the ecological subsystem changes after the sinking of coastal areas, generating an increase in coastal wetlands and a social-ecological crisis (Urbina 1996; Andrade 2017).

α phase (1967–1974): reorganization, after the perturbation (mega earthquake) several innovations start, moving the SES to a new adaptive cycle.

$R_{S\text{-}E}$: high for both subsystems. There is a diversification of economic activities and uses of ecosystem services. One of the main social products is the generation of mutual cooperation networks, cooperatives and new extractive activities both in terrestrial and marine ecosystems (Guajardo 1970; Altieri and Rojas 1999).

r phase (ii) (1974–1996): rapid growth, the potential of both subsystems is increased but with low connectivity.

$R_{S\text{-}E}$: Moderate R_S and high R_E. Human activities adjust to re-organization processes and there is an increase in the use of ecosystem services. This period shows an increase in human capital (professionals and technicians) required by the newly developing industries (Zanlungo et al. 2015). However, union social capital decreases due to the military dictatorship in Chile (1973–1990). One example is the end of the use of fishing pens (corrales de pesca in Spanish), a tradition in Chiloé (Ramírez et al. 2009). Also, during this period, artisan fishermen start organizing in communities and cooperatives as a consequence of the general law of fisheries and aquaculture (Marín and Gelcich 2012). In 1980 there is a sudden increase in economic resources due to the expansion of salmon farming on the island (Fløysand et al. 2010). Later, during the 1990s, diverse pathologies derived from salmon farming started appearing, generating the need to use antibiotics to maintain the fish population.

K phase (1996–2007): conservation, characterized by the accumulation of resources, increase in connectivity but also the monopoly of human activities leading toward an increase in vulnerability.

$R_{S\text{-}E}$: moderate R_E and low R_S. The SES becomes rigid, where the exportation of marine products becomes the social-economic basis of local society. The coastal ecosystem starts showing an increase of harmful algal blooms such as *Alexandrium catenella* during 2002 and 2006 (Molinet et al. 2017). Terrestrial ecosystems start changing due to the modification of the Chilean law regarding forestry (D.L.

701/1998) and the exploitation of mosses, *Sphagnum spp.* (Zegers et al. 2006). Finally, the island starts receiving tourists due to the declaration of the island's churches as world heritage sites by UNESCO (Ortiz et al. 2014).

Ω phase (ii) (2007–?): collapse, due to the outbreak of the Infectious Salmon Anemia (ISA) virus and recurrent harmful algal blooms (NCEI 2018).

$R_{S\text{-}E}$: low. This collapse phase, generated by a sanitary crisis, revealed a low social and ecological resilience, triggering social protests locally known as "mayo Chilote" (Chilote's may; Vargas 2018). The Chilean State resolves to provide subsidies to local people, decreasing, even more, their adaptive capabilities. Finally, during the year 2015 FAO expresses its concern about the low recruitment in mussel's banks in Chiloé coastal waters.

Bibliography

Altieri MA, Rojas A (1999) Ecological impacts of Chile's neoliberal policies, with special emphasis on agroecosystems. Environ Dev Sustain 1:55–72

Andrade RB (2017) El contexto cultural de la papa en Chiloé. Colecciones Digitales, Subdirección de Investigación. Dirección de bibliotecas, archivos y museos (Dibam), 23 pp

Cárdenas R, Villagrán C (2005) Chiloé: Botánica de la cotidianidad. Consejo Nacional del Libro y la Lectura, Consejo. 365 pp

Fløysand A, Haarstad H, Barton J (2010) Global economic imperatives, crisis generation and local spaces of engagement in the Chilean aquaculture industry. Norwegian J Geogr 64:199–210

Guajardo AC (1970) Los asentamientos campesinos y las sociedades agrícolas de reforma agraria. Revista de Derecho Económico 31-32:69–116

Marín A, Gelcich S (2012) Gobernanza y capital social en el co-manejo de recursos bentónicos en Chile: aportes del análisis de redes al estudio de la pesca artesanal de pequeña escala. Cuhso Cultura-Hombre-Sociedad 22:131–153

Molinet C, Díaz M, Marín SL, Astorga MP, Ojeda M, Cares L, Asencio E (2017) Relation of mussel spatfall on natural and artificial substrates: analysis of ecological implications ensuring long-term success and sustainability for mussel farming. Aquaculture 467:211–218

National Centers for Environmental Information (NCEI) (2018) The Southern Oscillation Index (SOI). https://www.ncdc.noaa.gov/teleconnections/enso/indicators/soi/

Ortiz R, Párraga M, Navarrete J, Carrasco I, de la Vega E, Ortiz M, Herrera P, Jurgens JA, Blanchette RA (2014) Investigations of biodeterioration by fungi in historic wooden churches of Chiloé, Chile. Microb Ecol 67:568–575

Ramírez E, Modrego F, Macé JC, Yáñez R (2009) Caracterización de los actores de Chiloé Central. Documento de trabajo/Programa Dinámicas Territoriales Rurales. RIMISP-Centro Latinoamericano para el Desarrollo Rural; no. 55

Urbina R (1996) Castro, castreños y chilotes: 1960-1990. Ediciones Universitarias de Valparaíso de la Universidad Católica de Valparaíso, Valparaíso. 354 pp

Vargas D (2018) Reflexiones en torno al mayo chilote. https://www.lemondediplomatique.cl/Reflexiones-en-torno-al-mayo.html

Zanlungo M, Pablo J, Katz J, Araya G (2015) Servicios intensivos en conocimiento en la industria salmonera chilena. Inter-American Development Bank, Washington, DC. 61 pp

Zegers G, Larraín J, Díaz MF, Armesto JJ (2006) Impacto ecológico y social de la explotación de pomponales y turberas de Sphagnum en la Isla Grande de Chiloé. Revista Ambiente y desarrollo 22:28–34

References

Adger WN (2000) Social and ecological resilience: are they related? Prog Hum Geogr 24:347–364

Ampuero D, Miranda CE, Delgado LE et al (2013) Empathy and critical thinking; primary students solving local environmental problems through outdoor learning. J Advent Educ Outdoor Learn 15:64–78

Anderies JM, Janssen MA, Ostrom E (2004) A framework to analyse the robustness of social-ecological systems from an institutional perspective. Ecol Soc 9:18–35

Arnorld-Cathalifaud M (2003) Fundamento del Constructivismo sociopoiético. Cinta Moebio 18:162–173

Ávila VS, Perevochtchikova M (2018) Sistemas Socio-Ecológicos: Marcos Analíticos y Estudios de caso en Oaxaca. Universidad Nacional Autónoma de México, Instituto de Investigaciones Economicas, México, México

Berger PL, Luhmann T (1968) La construcción social de la realidad. Amorrortu, Buenos Aires

Berkes F, Folke C (2000) Linking social and ecological systems for resilience and sustainability. In: Berkes F, Folke C, Colding J (eds) Linking social and ecological systems: management practices and social mechanisms for building resilience. Cambridge University Press, Cambridge, pp 1–25

Berkes F, Ross H (2013) Community resilience: toward an integrated approach. Soc Nat Resour 26:5–20

Bio-Bio. (2013). Fallo-Celco. http://media.biobiochile.cl/wp-content/uploads/2013/07/Fallo-Celco.pdf. Accessed 9 Jun 2019

Bourdieu P (1986) The forms of capital. In: Richardson JG (ed) Handbook of theory and research for the sociology of education. Greenwood Press, New York, pp 241–258

Bourdieu P (2001) Poder, derecho y clases sociales. Editorial Desclée de Brouwer, Bilbao

Burkhard B, Kroll F, Nedkov S et al (2012) Mapping ecosystem service supply, demand and budgets. Ecol Indic 21:17–29

Carpenter S, Mooney HA, Agard J et al (2009) Science for managing ecosystem services: beyond the millennium ecosystem assessment. Proc Natl Acad Sci U S A 106:1305–1312

Cinner J, Fuentes M, Randriamahazo H (2009) Exploring social resilience in Madagascar's marine protected areas. Ecol Soc 14:41–61

Coleman JS (1990) Foundations of social theory. Harvard University Press, Cambridge

Costanza R, d'Arge R, De Groot R et al (1997) The value of the world's ecosystem services and natural capital. Nature 387:253–260

De Groot RS, Wilson MA, Boumans RMJ (2002) A typology for the classification, description and valuation of ecosystem functions, goods and services. Ecol Econ 41:393–408

Delgado LE, Marín VH, Bachmann PL et al (2009) Conceptual models for ecosystem management through the participation of local social actors: the Río cruces wetland conflict. Ecol Soc 14(1):50. http://www.ecologyandsociety.org/vol14/iss1/art50/. Accessed 9 Jun 2019

Delgado LE, Tironi A, Vila I et al (2014) El Humedal del Río Cruces, Valdivia, Chile: una síntesis ecosistémica. Lat Am J Aquat Res 42:937–949

Delgado LE, Tironi-Silva A, Marín VH (2019) Sistemas socioecológicos y servicios ecosistémicos: modelos conceptuales para el humedal del Río Cruces (Valdivia, Chile). In: Cerda C, Silva-Rodriguez E, Briceño C (eds) Naturaleza en sociedad: Una mirada a la dimensión humana de la conservación de la biodiversidad. Editorial Ocho Libros, Santiago, pp 177–205

Folchi DM (2001) Conflictos de contenido ambiental y ecologismo de los pobres: no siempre pobres, ni siempre ecologistas. Ecología política: cuadernos de debate internacional 22:79–100

Guerrero-Gatica M, Aliste E, Simonetti JA (2019) Shifting gears for the use of the shifting baseline syndrome in ecological restoration. Sustainability 11(5):1458 https://doi.org/10.3390/su11051458 Accessed 9 Jun 2019

Gunderson L, Holling CS (2002) Panarchy: understanding transformations in human and natural systems. Island Press, Washington, DC

Holling CS (1973) Resilience and stability of ecological systems. Annu Rev Ecol Syst 4:1–23

Holling CS (2001) Understanding the complexity of economic, ecological, and social systems. Ecosystems 4:390–405
Jaramillo E, Lagos N, Labra FA et al (2018) Recovery of black-necked swans, macrophytes and water quality in a Ramsar wetland of southern Chile: Assessing resilience following sudden anthropogenic disturbances. Sci Total Environ. 628–629: 291–301 https://doi.org/10.1016/j.scitotenv.2018.01.333 Accessed 9 Jun 2019
Kay JJ, Regier HA, Boyle M et al (1999) An ecosystem approach for sustainability: addressing the challenge of complexity. Futures 31:721–742
Lawrence RJ, Després C (2004) Futures of transdisciplinarity. Futures 36:397–405
López-Angarita J, Sánchez M, del Pilar R et al (2009) A socioecological resilience approach for evaluating management effectiveness of marine protected areas. Final report: NOAA CORAL Grant NA07NOS463002. https://repository.library.noaa.gov/view/noaa/589/noaa_589_DS1.pdf?. Accessed 9 Jun 2019
Maass M (2018) Los sistemas socioecológicos (SSE) desde el enfoque socioecosistémico (SES). In: Ávila VS, Perevochtchikova M (eds) Sistemas Socio-Ecológicos: Marcos Analíticos y Estudios de caso en Oaxaca, México. Universidad Nacional Autónoma de México, Instituto de Investigaciones Económicas, México, pp 19–66
Marín VH, Delgado LE (2008) Conceptual PHES-system models of the Aysén watershed and fjord (southern Chile): a brainstorming strategy. J Environ Manage 88:1109–1118
Marín VH, Delgado LE (2013) From ecology to society and back: the (in) convenient hypothesis syndrome. Int J Sustain Dev 16:46–63
Marín VH, Delgado LE, Bachmann P (2008) Conceptual PHES-system models of the Aysén watershed and fjord (southern Chile): testing a brainstorming strategy. J Environ Manage 88:1109–1118
Marín VH, Delgado LE, Tironi A (2015) Transdisciplina, sistemas y ecosistemas. In: Montecino V, Orlando J (eds) Ciencias ecológicas 1983-2013: treinta años de investigaciones chilenas. Editorial Universitaria, Santiago, pp 283–294
Marín VH, Delgado LE, Tironi-Silva A et al (2018) Exploring social-ecological complexities of wetlands of international importance (Ramsar sites): the Carlos Anwandter Sanctuary (Valdivia, Chile) as a case study. Wetlands 38:1171–1182
Matte-Varas JJ (1990) Tantauco: nobleza, patriotismo y fidelidad. Boletín de la Academia Chilena de la Historia N° 57, Santiago
Millennium Ecosystem Assessment (MEA) (2005) Ecosystems and human well-being: Synthesis report. Millennium Ecosystem Assessment. Island Press, Washington, DC
Müller F, Burkhard B (2012) The indicator side of ecosystem services. Ecosyst Serv 1:26–30
Ostrom E (1990) Governing the commons: the evolution of institutions for collective action. Cambridge University Press, Cambridge
Ostrom E (2009) A general framework for analyzing sustainability of social-ecological systems. Science 325:419–422
Peluffo MB, Catalán Contreras E (2002) Introducción a la gestión del conocimiento y su aplicación al sector público. Series de la CEPAL. Instituto Latinoamericano y del Caribe de Planificación Social, Santiago
Pérez-Orellana DC (2019) Uso histórico de los servicios ecosistémicos en la Isla Grande de Chiloé, Región de Los Lagos, Chile: una aproximación conceptual. Seminario de Título entregado a la Universidad de Chile en cumplimiento parcial de los requisitos para optar al Título de Biólogo Ambiental. Facultad de Ciencias, Universidad de Chile, Santiago
Pérez-Teruel K, Leyva-Vázquez L, Estrada V (2014) Proceso de concenso en modelos mentales y aplicación al desarrollo de software agíl en bioinformatica. Revista Cubana de Informatica en Ciencias de la salud 25(83):318–332
Piccolo JJ (2017) Intrinsic values in nature: objective good or simply half of an unhelpful dichotomy. J Nat Conserv 27:8–11
Piñero L (2008) La teoría de las representaciones sociales y la perspectiva de Pierre Bourdieu: Una Articulación conceptual. Revista de Investigación Educativa 7:1–19

Pomeroy RS, Parks JE, Watson LM (2004) How is your MPA doing? A guidebook of natural and social indicators for evaluating marine protected area management effectiveness. IUCN, Gland
Popper K (1962) La lógica de la investigación científica. Editorial Tecnos, Madrid
Prieto Barboza EA (2013) Resiliencia y panarquía: claves para enfrentar la adversidad en sistemas sociales. Multiciencias 13(1):23–29
Rincón-Ruiz A, Arias-Arévalo P, Núñez JM et al (2019) Applying integrated valuation of ecosystem services in Latin America: insights from 21 case studies. Ecosyst Serv 36:1–12
Rozzi R (2019) Áreas Protegidas y Ética Biocultural. In: Cerda C, Silva-Rodriguez E, Briceño C (eds) Naturaleza en sociedad: Una mirada a la dimensión humana de la conservación de la biodiversidad. Editorial OchoLibros, Santiago, pp 25–69
Sletto BI (2010) Autogestión en representaciones espaciales indígenas y el rol de la capacitación y concientización: el caso del Proyecto Etnocartográfico Inna Kowantok, Sector 5 Pemón (Kavanayén-Mapauri). La Gran Sabana Antropológica 53(113):43–75
Wickson F, Carew AL, Russell A (2006) Transdisciplinary research: characteristics, quandaries and quality. Futures 38(9):1046–1059
Yu Iwama A, Delgado LE (2018) Acción: Participación Comunitaria en Procesos de decisión en la Conservación del territorio. Cuadernos del Pensamiento Crítico Latinoamericano 56:1–3

Social Actors and Participation in Environmental Issues in Latin America

Daniela C. Pérez-Orellana, Ailan Villalón-Cueto, Romina De Ríos, Carolina Velázquez-Mendoza, Marcela Torres-Gómez, David Quiñones-Guerrero, Luisa E. Delgado, Jaime Caro-Vera, and Franco Capriroli

Abstract The social participation of human beings and the term used for participants have changed through time. First, it was civil society, later citizenship, and currently social actor. Each concept has implied larger involvements, as rational and reflexive beings, in environmental issues. Today, social actors have become key figures for the governance of ecological systems. In this chapter we discuss the social actor concept and its relationship with environmental governance.

Keywords Social-ecological systems · Latin America · Complexity · Social actors · Women · Public policies

D. C. Pérez-Orellana (✉) · A. Villalón-Cueto · R. De Ríos · D. Quiñones-Guerrero
J. Caro-Vera · F. Capriroli
Laboratorio de Modelación Ecológica (LME), Departamento de Ciencias Ecológicas, Facultad de Ciencias, Universidad de Chile, Santiago, Chile
e-mail: danielaperezorellana@ug.uchile.cl; ailan.villalon@ug.uchile.cl; romina.derios@ug.uchile.cl; david.quinones@ug.uchile.cl; jaimecaro@ug.uchile.cl; franco.capriroli@ug.uchile.cl

C. Velázquez-Mendoza
Facultad de Ciencias, Universidad Nacional Autónoma de México, Ciudad de México, Mexico
e-mail: caro.vmt7@ciencias.unam.mx

M. Torres-Gómez
Fundación Centro Transdisciplinario de Estudios FES-Sistémicos, Santiago, Chile
e-mail: mtorresg@fen.uchile.cl

L. E. Delgado
Fundación Centro Transdisciplinario de Estudios FES-Sistémicos, Santiago, Chile

Escuela de Ciencias, Facultad de Ciencias, Universidad de Chile, Santiago, Chile
e-mail: ldelgado@uchile.cl

L. E. Delgado, V. H. Marín (eds.), *Social-ecological Systems of Latin America: Complexities and Challenges*, https://doi.org/10.1007/978-3-030-28452-7_3

1 Introduction

A social-ecological system, SES (see chapter "Postnormal Science and Social-ecological Systems") is structured by the interactions between bio-geo-physical and reflexive (human beings) components. The latter is today known as social actors (Delgado et al. 2019). From this perspective, there are no differences between either the intrinsic value of any actor, regardless of its knowledge, origin, or perceptions, or between human beings and nature (Piccolo 2017). Consequently, there may be a plurality of legitimate perspectives about SES structure and dynamics (i.e., postnormal science; chapter "Postnormal Science and Social-ecological Systems") and the need of social participation to manage these complex systems (Berkes and Folke 1998; Holling 2001; Binder et al. 2013; Marín and Delgado 2013).

SES analysis is necessarily contextual since social and ecosystemic interactions occur in a given geographic area wherein perceptions and types of cultural appropriation (including the use of components and their valuation) vary according to the connection between both subsystems (social and ecological) and temporal and spatial scales (Delgado et al. 2019). In fact, increasing the spatial scale of analysis also increases the number of components to be analyzed, their heterogeneity, and their interactions and adaptive responses, thus increasing complexity (Holling 2001; Moser 2014). For example, the dynamic of a watershed, when analyzed as a complex SES from a local perspective, is affected by processes operating at larger scales both social (e.g., national development policies) and ecological (e.g., climate change) which can be conceptualized as external forcing functions. However, if we increase the scale of analysis, these external forcing functions become the components of the system, and the number and variety of social actors also increase. Furthermore, not all social actors have the same possibility of participation even at a local scale (e.g., people with low income, women, and ethnic groups; Calame 2009), which makes them vulnerable since they have to live a life that others decided for them. In this chapter, we analyze the role and concepts associated with social actors as SES components and forms of participation.

2 Social Actors or Stakeholders? Conceptual Implications for Social-ecological Systems

Human beings have different forms to appropriate nature (i.e., cultural appropriation), either building or modifying it, and they have changed through our history. In this chapter, a social actor, either individuals or organizations, is understood as an agent that generates its own reality, affecting social and ecological structures and with a capacity to interact reflexively with other actors about its living environment and the forms of cultural appropriation. Touraine (1997) further adds that the concept refers to a single or group of individuals situated between structural determinism and freedom of action. So, a social actor may act based on its own psychosocial context or representing a collectivity, both of which define its social identity.

Many social-environmental conflicts generate because some social actors are not included within a study or in the application of a public policy (De Castro et al. 2015). Other issues that contribute to the generation of conflicts are the lack of communication between different actors, lack of information, and lack of ability to understand social-ecological processes. Indeed, García Sánchez (2007) questions using the social actor concept in cases where there are difficulties in identifying their acting responsibilities, especially for collective actors, since the original definition (Giddens 1979) considers that actor's interventions have intentions. Thus, the social actor concept incorporates the relationships within SESs and considers human beings as agents of change that determines micro, meso, and macrosocial structures (Merton 1949; Moser 2014) with the capacity to modify natural ecosystems as originally proposed by Tansley (1935).

Stakeholders, or interested parties (Gallie 1956), is another concept that may be found in the literature in relation to environmental issues such as ecological conservation, management, and sustainability. This concept was generated within the domain of economy and management of the enterprise; that is why it originally included both shareholders and any individual and organization that may affect or be affected by the accomplishment of an organization's objectives (Freeman 1984). The term has been used so widely that currently there is a large discussion about its meaning (McGrath and Whitty 2017). For example, Mainardes et al. (2011) have documented 66 different definitions of stakeholder. The authors conclude that all of them relate to organizations without applications to individual behavior. Finally, it appears that stakeholder implies only the existence of interest when a given group has been affected by the objectives of an organization, referring to those responsible for the management or the dynamic of a specific project.

Although the term stakeholder is used to describe social-environmental conflicts and processes such as management and biodiversity conservation, Reed (2008) comments that most conservation initiatives involve only those with participation in their goals, instead of all actors. That is why the term "key stakeholder" is also used to evaluate biodiversity management, valuations, and instruments (Cerda and Bidegain 2018). However, if the issue is environmental governance for the sustainability of ecosystems, the participation of all social actors is vital (see also chapter "A New Environmental Governance"). Indeed, sustainable development considers intergenerational justice (Howarth 1996); this, according to economic theories, is a function of social well-being. Thus, when environmental governance is considered as the main focus, the relationships between human beings change redrafting our collective responsibility toward nature. This, in turn, makes the social actor concept more suitable for this purpose than stakeholder.

3 Identification and Classification of Social Actors

The analysis and governance of social-ecological systems require identifying who are the social actors. And, how do they relate with the ecosystem from which they receive services? Answering those questions is essential to generate, organize, and encourage the necessary participation of all actors. Social subsystems, as their ecological counterparts, are characterized by its diversity. That is why identifying and classifying social actors is vital when designing governmental SES projects and programs.

The Department for International Development of the United Kingdom (DFID 2002) proposes that social actors may be classified into three types: key, primary, and secondary. Key actors refer to those that significantly influence the success of a management project. Primary actors are those directly affected, either positively or negatively, by project development. Finally, secondary actors correspond to all other individuals or groups with an interest or medium role in project management. The main objective of the DFID analysis is to identify actors according to management levels (see chapter "A New Environmental Governance"), which helps in the definition of interactions between social and ecological components, origin of conflicts, and social actors' vulnerabilities.

Mapping social actors through sociograms is another method to identify actors. The main difference with the previously discussed method is that it only involves characterizing groups of actors or institutions affecting or being affected by specific activities. That is actors who have information, resources, experiences, or any form of power that may influence the actions of other groups (EU-FAO 2006). Then, the generated sociogram may be used to identify social relationships that form networks, also pointing to the hierarchical level occupied by different actors (Fig. 1).

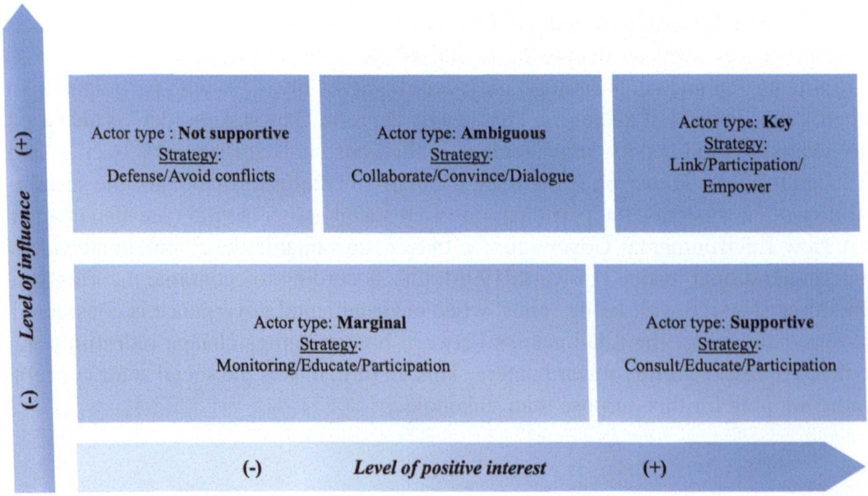

Fig. 1 Levels of power and influence of key social actors affecting the success of a project (modified from Tapella 2007)

Thus, the method can be used to search for the different actors that may participate in a given initiative and to know their potential actions and participation objectives. However, it applies mostly to the most relevant actors, making it closer to the stakeholder concept already discussed. That is why it is generally used when management is top-down or for co-management, two management alternatives still common in Latin America especially when international organizations intervene.[1]

One key issue, in relation to new environmental governance for Latin America (see chapter "A New Environmental Governance"), is the decentralization of decision-making, where the territory becomes highly relevant. A "territory point of view" allows expressing, in a simple way, the large variety of social imaginaries associated with the same space and how points of view and individual interactions generate different conceptions.

Thus, the largest challenge when attempting to implement wide social participation is to represent the diversity of social actors when making decisions, allowing their input during all the phases of the process. Representations and perceptions of local peoples allow incorporating local society's cognitive and cultural elements; consequently, it is inherently subjective and diverse.

Piccolo (2017) remarks that human beings value nature in different ways: instrumental value, intrinsic and relational, depending upon the reality that the ecosystem, its processes, and components represent to them. Therefore, the participation of local people in environmental management is very different from that of none residents, which in turn affects the design and application of public environmental policies. In this regard, social-ecological analyses contribute to a theoretical perspective called participatory research action (Colmenares 2012). This research strategy, derived from the social sciences, use two qualitative tools from psychology and politics when analyzing a concrete reality: (1) a "from within optic," where the studied community is since within its own context, and (2) a "from below approach," integrating all those peoples who have not been studied. This strategy transforms local participants into social actors with rights, which can be co-producers of knowledge, aside from experts. In other words, it allows the generation of postnormal science or science with and for the people (Funtowicz and Ravetz 2000).

4 Social Participation

The concept of "social participation" has received several interpretations, depending upon the application area. A literature analysis showed that its evolution started in the 1970s with Cohen and Uphoff (1977). Schneider and Libercier (1995) and The World Bank (1995) state that participation is a process through which diverse actors dialog and make decisions. Actors' interpretations are linked to their own perspective on

---

[1] http://www.py.undp.org/content/paraguay/es/home/projects/paisajes-de-produccion-verde%2D%2Dcommodities-sustentables.html

development. Notwithstanding, the definition proposed by Cohen and Uphoff (1977) has had a large influence on the definition of citizen's levels of involvement, from simple queries to participative planning and empowerment (FAO 1999).

Korc and Maisonet (1999) state that participation is a gradual process through which citizens are involved, either individually or collectively, in decision-making, auditing, control and implementation of public and private actions affecting political, economic, social, and environmental issues. Later, Jiménez and Mujica (2003) propose that participation is playing active roles in what today is known as the "public sphere."

Despite the diversity of proposals about the concept, there is agreement that it promotes human development (Iyer-Raniga and Treolar 2000), it increases development efficiency, and it stimulates different perspectives for nature's integrated management (Secretariat of the Convention on Biological Diversity 2004), supporting the exchange and feedback of ideas, so alternatives are generated by the different social actors. The need of social participation, in relation to environmental issues in Latin America, was stated explicitly as part of the 1992 Rio Declaration, in its Principle 10[2]: "Environmental issues are best handled with the participation of all concerned citizens, at the relevant level."

From a postnormal-constructivist perspective, the perception of social actors about a given ecosystem is vital when discussing and validating alternatives for its integrated management (Ravetz 1999; Tognetti 1999; Haag and Kaupenjohann 2001; Marín and Delgado 2005). Thus, obtaining those perspectives is a key step in social-ecological research. Social actors' participation, through participative modeling or brainstorming (Marín et al. 2008), can stimulate knowledge integration and governance adaptation in the different contexts necessary for sustainable development. Furthermore, using "learning through doing," participation may promote adaptive management and knowledge acquisition in social actors and governments (Casares and Arca 2002).

From this perspective, economic development, nature conservation, use of ecological components, and social participation are all interactive processes directly influencing the sustainability of social-ecological systems. Thus, the social-economic development of a given region is just one more component of the physical-ecological-social systems (PHES-systems) that actors define, study, conserve, and transform (Marín and Delgado 2005). So, most social characteristics (e.g., poverty, illiteracy, isolation, and lack of participation) will sooner or later affect social-ecological systems and the natural patrimony on which social actors depend.

Social participation, from the perspective of postnormal science, has been discussed at length; however, practical applications in Latin America are few. For example, Bachmann et al. (2007) analyzed the levels of participation for the integrated management of Aysén watershed in southern Chile. Results show that citizen's participation is only symbolic without a real commitment to the idea of local sustainability. Thus, Latin American institutions developing and implementing public policies are characterized by a "top-down" approach, which will normally

[2] http://www.unesco.org/education/pdf/RIO_E.PDF

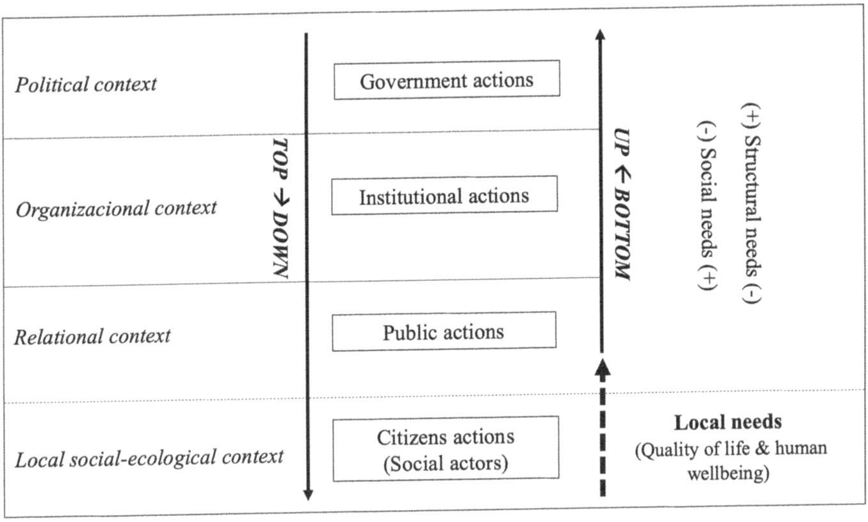

Fig. 2 Schematic model for the application and management of environmental public policies from a social-ecological perspective (modified from Oliver-Mora and Iñiguez-Rueda 2016)

listen only to higher hierarchical levels. On the contrary, when a "bottom-up" approach is used, there are more chances that local social needs may be listened to. However, both of them will be affected by the political and organizational contexts. These three contextual spheres (politics, organizational, and relational) correspond to the three domains for the application of public policies related to environmental issues (Oliver-Mora and Iñiguez-Rueda 2016). We propose that these domains cannot be properly applied unless the local social-ecological context is considered (Fig. 2).

Latin American countries still develop top-down participation (e.g., Barragán et al. 2005), which implies that interested or affected citizens only get involved during project implementation without participating in decision-making (Bachmann 2006). This participation strategy does not allow empowerment by the local society, maintaining it as a passive receptor of the potential benefits. However, local sustainable strategies such as integrated ecosystem management (Christensen et al. 1996) require bottom-up approaches, where local actors get involved in the whole management process (Fig. 2). Bottom-up participation considers:

1. Involvement of all interested actors
2. Transparent and open decision-making
3. Access to all necessary information
4. Adaptive approach (learn from errors and correct them)
5. Multidisciplinary coherency
6. Subsidiarity; decisions should be made at the lowest appropriate level
7. Respect for human rights and quality of life

8. Third-party responsibility; related to economic, social, and environmental management results
9. Transparent and impartial application of legal norms

5 Types of Social Participation

Latin American countries have implemented different social participation strategies in environment-related issues (Annex I). Most of them can be classified in two types: (a) informal participation or self-governance and (b) formal participation (public policies). Next, we discuss both types of social participation.

Informal Participation in Latin America

Informal participation allows evaluating the development context of a given community, its virtues and shortages. This participation is frequently dynamic; that is, local peoples usually will have several degrees of involvement. It can go from passive participation up to complete control of the development process (i.e., self-development actors). Thus, building local governance is possible by strengthening democratic rights (Grindle 2000; PNUD 2004; Dagnino 2006).

This type of participation generates big challenges such as a permanent search of methods and tools to facilitate the actor's involvement and the generation of collective solutions through teamwork. Consequently, its success will depend to a large extent on the degree of organization of the community.

Two Case Studies of Informal Participation

Social cartography of the African descendants' territory of Río Cauca watershed (Colombia)

This case study shows the opportunities and challenges of social cartography when characterizing social-ecological conflicts between inhabitants of three territories of Río Cauca watershed (Torres et al. 2012). Local communities have been affected by monocultures of pines and sugar cane, urban expansion, mining operations, and hydroelectric power plants, generating conflicts and violence within African-descendant communities. The authors, using satellite images and cartographic workshops (including validation of the location of social conflicts), analyzed how social actors perceive and mobilize in relation to these development processes. Results show important social relationships in the region, but minimum relationships with formal institutions. Authors then propose that these informal participative tools should be used in different social and political spheres. However, the information generated with this methodology is only useful as a tool to foment dialogs and to expose points of view; institutions have still to listen in order to make these dialogs fertile for contextualized public policies.

Has the revolution a peasant face? (Páramo, Venezuela)

This study is related to the participation of rural women in decision-making processes through participative mapping and interviews (Caretta et al. 2015). Results show that although women are the major components of the "Communal Council," their participation is mostly passive. Thus, the most important result is that there are gender differences in the perception of the environment and productive activities. Men see themselves as farmers while women defined themselves in relation to home needs and recreational, educational, and environmental activities.

Women's Roles in Sustainability

The social actor concept includes all the heterogeneity that exists in social-ecological systems. This implies that participative processes should emphasize incorporating the largest possible variety of opinions regarding an environmental problem, considering that values and interests will be different among the actor's groups (Quétier et al. 2007). In this chapter we have included the role of women as preponderant social actors in ancestral and contemporary cultures, influencing the sustainability of social-ecological systems.

The International Union for Conservation of Nature (IUCN[3]) has established that the participation of all actors (or sectors) must be guaranteed when discussing decisions about the ecosystem on which they depend. In this way, planification and management, and subsequent monitoring will allow continuous ecosystem observations from the perspective of those concern about it. The result is likely to be a dynamic flow of information that in turn should improve management toward nature's sustainability (Guerrero et al. 2006). Furthermore, integrated ecosystem management involves identifying key social actors from the perspective of development including their organization and participation in decision-making.

The Economic Commission for Latin America and the Caribbean (ECLAC) established "gender equity" among its principles of the 2030 agenda for sustainable development. In this regard, the incorporation of women has shown to be indispensable to deepen democracy, economic growth, and inclusive and sustainable development (Mies and Shiva 1998). The visualization of women as "secondary gender" has ended considering them as a second-category human being; a posture reinforced during a long time by science, history, religion, and cultural dynamics (López Pardina 2002). According to Beauvoir (1965), women have been oppressed, being obliged to live under the shadow of men, and to perpetuate such oppression, preestablished molds have been created where they should fit (even by force) and stay all lifelong.

In Latin America and the Caribbean, it is possible to find the origins of gender binaries when analyzing the concept of "power colonialism," proposed by Quijano

[3] https://www.iucn.org

(2000). This refers to the fact that current power schemes were established during colonization (Segato 2014). This concept naturalized hierarchies by means of races and gender giving them obligations and conditioning pre-Columbian communities in their access to resources such as land, work, and decision-making participation (Lugones 2008).

Although women have had a difficult history, they have also opened roads increasing their importance on activities such as agriculture, especially improving agro-biodiversity, and cultural activities (e.g., traditional medicine), generating strong links with their territories (Arias Toledo 2009; Llaja and Emily 2015; Ortega et al. 2017). Promoting women's rights, fomenting their participation, and incidence on social and political spaces are among the objectives of the gender equality plans proposed by ECLAC (2017). Indeed, there have been advances in gender equality since 1980, where most Latin American countries (Table 1) have modified civil and family codes, land proprietorship rights, improving their participation in decision-making (Lastarria-Cornhiel 2011).

Table 1 Examples of environmental issues from Latin American countries where women have had important roles

Country	Organization	Participative process	Results	References
Bolivia	Central Intercomunal de Mujeres de la Capitanía de Isoso	Women from Isoso remained active demanding recognition of the territories, requesting and demanding their land for decades	Land titles to indigenous farmers. Women remain active in local organizations	Bórquez and Ardito (2009)
Brazil	Women from a diversity of organizations	National political plan for women	Women accomplished their participation in issues such as environmental policies, social-environmental development and economic activities related to sovereignty and food security	Secretaria de Políticas para as Mulheres (2013)
Chile	Aymara indigenous association	Participation in claims about ancestral rights and recuperation of ancestral territories	Judicial recognition of water ancestral rights by The Chilean Supreme Court	Bórquez and Ardito (2009)
	National Association of Rural and Indigenous Women	Second national congress	Identification of causes and potential solutions of land problems to be forwarded to the state	ANAMURI (2014)

(continued)

Table 1 (continued)

Country	Organization	Participative process	Results	References
Colombia	Academia and industrial sector	Generation of industrial women's knowledge network to lead clean production processes and healthy communities	Platform benefiting rural women's rural strategies	Aguilar et al. (2015)
Ecuador	Azuay and Cuenca women's chapter	Analysis of environmental policies and gender equity	2016 equal opportunities plan, including environmental rights and women's participation in environmental management	Ortega (2018)
El Salvador	Women from the Berlin district, LaGeo and FundaGeo	Provide opportunities to women living close to a geothermal plant. Usage of geothermal energy and residue's management	Women have been instructed on cultivating and selling roses and fishing	Aguilar et al. (2015)
Guatemala	National coordination of farmer's organizations	Regional meeting of farmer women to elaborate an integral agrarian reform	The agrarian reform was approved in 2005. The text guarantees women participation	CNOC (2005)
Perú	Life out of plastic	Women create a company to commercialize recycling products and to build awareness of plastic pollution	More than 24,000 people have participated in cleaning activities	Aguilar et al. (2015)

Institutional, Formal, Participation in Latin America

Last decade has witnessed the challenges associated with worldwide environmental deterioration[3]. Citizens' participation in a democracy should ideally sustain governments' decision-making. However, many times decisions do not represent the opinions of civil society, which correspond to a reductionist way for development (Guillen et al. 2008). Thus, governments and civil societies should, from the standpoint of the participation paradigm, stimulate citizen's participation (Rovalo 2015).

Citizen's participation, according to Arnstein (1969), has eight levels that can be grouped into three categories: no participation, simulation degrees, and citizen's power degrees. In the first category, participation spaces are used for public relations by authorities and citizens are listened, but their opinions are not considered. The second category includes bidirectional information transfers with citizens involved as communities. Only the third category includes shared responsibilities, corresponding to direct democracy, where the people continuously participate in decision-making (Prud'Homme 1997). Chapter "A New Environmental Governance" of this book analyzes the mechanisms contributing to participative citizenship.

Consequently, direct citizens' participation changes the classical concept of top-down decision-making, where citizens' interests are excluded. This change goes hand in hand with a postnormal conceptualization of science that incorporates social actors (Yu Iwama and Delgado 2018). Thus, a key issue is the legal development of bottom-up participation schemes (Sol 2012); indeed, they have been considered as a critical step in meeting the objectives of the 2030 Agenda for Sustainable Development (CEPAL 2018).

6 Concluding Remarks

In this chapter, we have described and analyzed several forms, types, and opportunities associated with social participation, related to environmental issues, in Latin America. Our analysis shows that 13 of the 18 analyzed countries (Annex I) have incorporated citizens' participation in their constitutions. Yet, few countries allow direct participation in the exercise of power, and fewer have norms and rights related to its application (Prud'Homme 1997). Accordingly, top-down perspectives have generated many social-ecological conflicts, since specific territorial contexts are not visualized. That is why, if we want to strengthen environmental governance in Latin American countries, there is an urgent need to decentralize and generate multi-scale ways of participation.

The concept of participation, as already stated, has changed through history with multiple interpretations. If we relate it to the environment, social actors' perceptions about ecosystems are key to build and validate alternatives for sustainable management. Furthermore, social actors are reflexive components of nature (Delgado et al. 2019). They relate to its surrounding environment in many ways depending upon philosophical, ethical, political, and social considerations and also conditioned to the type of environment (private, public, and collective) and the scales of perception (micro, meso, and macro). Consequently, social-ecological systems are complex, and unless participation is effective, most social actors will disagree with each other.

Although each social actor may have a role within a social-ecological system, women have historically been relegated from decision-making processes although they have close relationships with their territories in activities such as agriculture and traditional medicine. Latin American countries have implemented, in recent years, women's participation and gender equity, especially on environmental issues. Although it is a slow process, achievements seem to be important, and we anticipated it will keep growing.

Finally, human beings interact with nature modifying it, having their well-being and quality of life as main goals. Therefore, maintaining those processes during the Anthropocene is the main goal of participative, sustainable, environmental governance. In this respect, all social actors now have a new role: their responsibility about the fate of the ecosystems of the biosphere. Consequently, it is vital that all actors shall make informed decisions, considering the environmental, economic, and social consequences of their acts; this requires a cultural change. This change can be accomplished through a learning process focused not only on reflections but on actions, which should be based on a visualization of the historical development of social-ecological systems, their past dynamics that have generated our present conditions and the possibilities for the future. Then, our main goals should be equity, diversity, and benefits of common human beings.

Annex I

Latin American legal frameworks in relation to Arnstein (1969) second and third degree of social participation

Country	Legal framework that incorporates direct social participation	Year	Legal provisions	Territorial scale
Argentina (**)	Constitution (Art. 39)	1995	Citizens have the right to propose law projects to The Congress in diverse issues of their interest	National, provinces and municipalities
	Law N°24.747	1996	Regulates Art. 39	National
	Decree 1172/03 on Access to public information and national strategy for development and land use planning	2013	Public hearings and elaboration of participation norms	National
		2016	Strategic plan as co-management mechanism between civil society and the State at local level. In most cases only legal entities may participate	National, provinces and municipalities
Belice	–		–	–
Bolivia (***)	Constitution (Art. 20, Art. 26, Art. 241, Art. 242)	2009	Participation and control by citizens at all levels	National, departments and municipalities
	Law N°031/(Title VII, Chapters "Postnormal Science and Social-ecological Systems" and "Simplifying the Complexity of Social-ecological Systems with Conceptual Models")	2010	Social participation and control mechanisms to develop laws in autonomous governments	Departments and municipalities
	Law N°341/Social participation and control	2013	Enforcement of Articles 241 and 242 of the Constitution	National, departments and municipalities

(continued)

Brazil (**)	Constitution (Art.14, Art.29)	1988	Establishes participation mechanisms	National, municipalities
	Law N°9.790	1999	Highlights the possibility of cooperation between the State and civil society. It also creates the category of public interest organizations from the civil society, facilitating their access to public resources	National, municipalities
	Law N°12.257/City statute	2001	Municipal participation budget	Municipalities
	Law N°12.187/ National policy on climate change	2009	Stimulate and support civil society's participation in climate change issues	National, municipalities
Chile	Law N°20.500/ Citizens' participation on public management	2011	The State recognizes the right of people to participate in its policies, plans, programs and actions in order to improve their management	National, regional, communities
	Presidential instrument N°007 on citizens' participation	2014	Norm that allows the implementation of Law N° 20.500	National, regional, communities
Colombia (**)	Constitution (Art. 23, Art. 270)	1991	It establishes the citizens' right to claim and intervene in public management	National
	Law N°80	1993	Citizens' control of State contracts	National
	Law N°134/On citizens' participation mechanisms	1994	Establishes the popular legislation and normative initiative	National, departments, municipalities
	Law N°489	1998	Establishes the principles for democratic participation	National, departments, municipalities
Costa Rica (*)	Law N°7.554/Law of the environment	1995	Promotes citizens' participation to elaborate territorial instruments and plans	National, regional, municipalities
	Law N°7.794/ Municipal Code	1998	Establishes three participatory mechanisms: plebiscites, referendum and chapters	Municipalities
	Ley N°8.491/Popular initiative	2006	Citizens may propose law projects	National
	Ley N°8.492/ Referendum law	2006	Citizens may approve or disapprove laws through referendums	National
Ecuador (**)	Constitution (Art. 61, Art. 81, Art.95, Art. 103, Art. 238)	2008	Establishes participation and control rights for citizens	National, provinces
	Law on citizens' participation	2010	Establishment of participation mechanisms	National

(continued)

El Salvador	Decree N°274/ Municipal code	1986	Establishes participation mechanisms	Municipalities
	Decree N°233/Law of the environment	1998	Advisory participation on environmental issues	Zones, departments, municipalities
Guatemala	Peace agreement	1996	Encourages participation to identify, prioritize and solve citizens' needs	National
	Law on development councils	2002	Encourages ethnic participation on development plans	National, regional, departments, municipalities, communities
	General decentralization law	2002	Strengths local power	National, regional, departments, municipalities
Honduras (*)	Constitution (Art. 5)	1982	Participation in public management	National
	Law on citizens' participation	2006	Establishes forms of participation	National
	Law on participation mechanisms	2013	Regulates participation mechanisms	National
México (*)	Constitution (Art. 26)	1917	Participation in democratic planification of national development	National
	Planification law	1983	Establishes the principles for social participation in territorial planification	National
	Participation law (Federal District)	2016	Regulates participation instruments	States
	Federal to promote activities from civil society organizations	2018	Encourages citizens' organizations to participate on environmental protection, restoration of ecological equilibrium and sustainable development	National
Nicaragua	Constitution (Art. 50)	1986	Encourages inhabitants' participation in public issues	National, departments
	Citizen's participation law	2003	Encourages institutional mechanisms to link the State and society	National
Panamá	Constitution (Art. 80, 108, 124)	1972	Recognizes the right to participate in culture development. Encourages the participation of ethnic groups and farmers on economy, society, culture and politics	National
	Law N°6/Access to information	2002	Regulates participation mechanisms	National

(continued)

Paraguay	Constitution (Art. 47, 56, 65, 117)	1992	Promote equal opportunities when using nature's benefits, material goods and culture	National
Perú (∗∗∗)	Constitution (Art. 2, 31)	1993	Establishes mechanisms for the participation in public issues	National
	Law on participation rights and citizens' control	1994	Establishes citizens' rights to participate on public issues and the mechanisms	National
Uruguay (∗∗∗)	Constitution (Art. 331)	1967	Citizens can modify the constitution	National
	Law N°19.272 / Political decentralization and citizens' participation	2014	Establishes municipal rights and encourages citizens 'participation	National
Venezuela	Constitution (Art. 62, 168)	1999	Citizens have the right to freely participate on public affairs	National
	Law on public and popular planning	2010	Its objective is to develop and strengthen peoples' power	National

Asterisks correspond to those countries where the legal framework incorporates the concept of Payment for Environmental Services: (∗) national scale, (∗∗) regional, states or provinces and (∗∗∗) local scale

References

Aguilar L, Granat M, Owren C (2015) Roots for the future: the landscape and way forward on gender and climate change. IUCN & GGCA, Washington, DC

ANAMURI (2014) Segundo Congreso Nacional ANAMURI. Ediciones Tierra mía, Santiago

Arias Toledo B (2009) Diversidad de usos, prácticas de recolección y diferencias según género y edad en el uso de plantas medicinales en Córdoba, Argentina. Bol Latinoam Caribe Plantas Med Aromat 8(5):389–401

Arnstein SR (1969) A ladder of citizen participation. J Am Inst Plann 35(4):216–224

Bachmann P (2006) La participación ciudadana y el manejo integrado: un análisis eco-social de la cuenca del río Aysén (XI región, Chile). Seminario de título carrera de Biología Ambiental. Facultad de Ciencias, Universidad de Chile, Santiago

Bachmann PL, Delgado LE, Marín VH (2007) Analysis of the citizen's participation concept used by local decision makers: the case of the Aysen watershed in southern Chile. Int J Sustain Dev 10(3):251–266

Beauvoir (1965) El segundo sexo. Ediciones Siglo Veinte, Buenos Aires

Berkes F, Folke C (eds) (1998) Linking social and ecological systems: management practices and social mechanisms for building resilience. Cambridge University Press, New York

Binder CR, Hinkel J, Bots PWG, Pahl-Wostl C (2013) Comparison of frameworks for analyzing social ecological systems. Ecol Soc 18(4):26

Bórquez R, Ardito L (2009) Experiencias activas de acceso a la tierra: estrategias de empoderamiento y aseguramiento de derechos desarrolladas por organizaciones de mujeres campesinas e indígenas rurales. Procasur/International Land Coalition, Santiago

Calame P (2009) Hacia una revolución de la gobernanza. LOM Ediciones, Santiago

Caretta MA, Montero C, Cadena Y, Sulbaran L, Sandoval R (2015) ¿La revolución tiene cara de campesina? Un caso de estudio de la participación activa de las mujeres en el riego del páramo venezolano. Rev Lat Am Geogr Gênero 6(2):3–23
Casares JJ, Arca JC (2002) Gestión estratégica de la sostenibilidad en el ámbito local: La Agenda 21 Local. Rev Galega Econ 11(2):1–18
De Castro F, Hogenboom B, Baud M (eds) (2015) Gobernanza ambiental en América Latina. Clacso, Buenoa Aires
Cerda C, Bidegain I (2018) Spectrum of concepts associated with the term "biodiversity": a case study in a biodiversity hotspot in South America. Environ Monit Assess 190(4):207
Christensen NL, Bartuska AM, Brown JH et al (1996) The report of the ecological Society of America Committee on the scientific basis for ecosystem management. Ecol Appl 6:665–691
Cohen J, Uphoff N (1977) Rural development participation: concepts and measures for project design, implementation and evaluation. Rural Development Committee, Cornell University no 2, Ithaca
Colmenares EAM (2012) Investigación-acción participativa: una metodología integradora del conocimiento y la acción. Voces y silencios Revista Latinoamericana de Educación 3(1):102–115
Comisión Económica para América Latina y el Caribe (CEPAL) (2018) Acceso a la información, la participación y la justicia en asuntos ambientales en América Latina y el Caribe: hacia el logro de la Agenda 2030 para el Desarrollo Sostenible. Naciones Unidas, Santiago de Chile
Coordinadora Nacional de Organizaciones Campesinas (CNOC) (2005) Propuesta de Reforma Agraria Integral. Ciudad de Guatemala
Dagnino E (2006) Sociedad civil, participación y ciudadanía: ¿De qué estamos hablando? In: Insunza Vera E, Olvera AJ (eds) Democratización, rendición de cuentas y sociedad civil: participación ciudadana y control social. Universidad Verazruzana, México, pp 223–242
Delgado LE, Tironi-Silva A, Marín VH (2019) Sistemas socioecológicos y servicios ecosistémicos: modelos conceptuales para el humedal del Río Cruces (Valdivia, Chile). In: Cerda C, Silva-Rodríguez E, Briceño C (eds) Naturaleza en sociedad: Una mirada a la dimensión humana de la conservación de la biodiversidad. Editorial Ocho Libros, Santiago, pp 177–205
DFID (2002) Tools for development. A handbook for those engaged in development activity. http://webarchive.nationalarchives.gov.uk/+/http:/www.dfid.gov.uk/Documents/publications/toolsfordevelopment.pdf. Accessed 9 Jun 2019
ECLAC (2017) Gender equality plans in Latin America and the Caribbean. Road maps for development. Gender Equality Observatory for Latin America and the Caribbean, Studies 1. Economic Commission for Latin America and the Caribbean. https://www.cepal.org/sites/default/files/events/files/gender_equality_plans_in_latin_america_and_the_caribbean._road_maps_for_development_0.pdf Accessed 9 Jun 2019
EU-FAO (2006) Stakeholders analysis. European Union and Food and Agriculture Organization. http://www.fao.org/elearning/course/f6/en/pdf/trainerresources/annex0140_01.pdf. Accessed 9 Jun 2019
FAO (1999) Grupo informal de trabajo, enfoques y métodos participativos http://wwwfaoorg/participation/espanol/defaulthtm Accessed 9 Jun 2019
Freeman RE (1984) Strategic management: a stakeholder approach. Pitman, Boston
Funtowicz SO, Ravetz JR (2000) La ciencia posnormal: ciencia con la gente. Icaria editorial, Buenos Aires
Gallie WB (1956) Essentially contested concepts. In: Proceedings of the Aristotelian society. Aristotelian Society, vol 56, pp 167–198
García Sánchez E (2007) El concepto de actor: Reflexiones y propuestas para la ciencia política. Andamios 3(6):199–216
Giddens A (1979) Central problems in social theory: action, structure, and contradiction in social analysis. University of California Press, Berkeley, CA
Grindle MS (2000) Audacious reforms: institutional invention and democracy in Latin America. JHU Press, Baltimore
Guerrero E, De Keizer O, Córdoba R (2006) La aplicación del enfoque ecosistémico en la gestión de los recursos hídricos. UICN, Quito

Guillen A, Badii MH, Blanco M (2008) La participación ciudadana en el contexto del desarrollo sustentable. Innovaciones de Negocios 5(1):131–146

Haag D, Kaupenjohann M (2001) Parameters, prediction, post-normal science and the precautionary principle—a roadmap for modelling for decision-making. Ecol Model 144(1):45–60

Holling CS (2001) Understanding the complexity of economic, ecological, and social systems. Ecosystems 4(1):390–405

Howarth RB (1996) Discount rates and sustainable development. Ecol Model 92(2–3):263–270

Iyer-Raniga U, Treolar G (2000) Forum: a context for participation in sustainable development. Environ Manag 26(4):349–361

Jiménez M, Mujica P (2003) Participación Ciudadana y Cohesión Social. Corporación PARTICIPA, Santiago

Korc M, Maisonet M (1999) Directrices para la elaboración de planes de acción locales para mejorar la calidad del aire. https://www.paho.org/hq/index.php?option=com_docman&view=download&category_slug=mandatos-estrategias-9836&alias=44637-directrices-elaboracion-planes-accion-locales-mejorar-calidad-aire-637&Itemid=270. Accessed 9 Jun 2019

Lastarria-Cornhiel S (2011) Las mujeres y el acceso a la tierra comunal en América Latina. In: Coastas P (ed) Tierra de mujeres: Reflexiones sobre el acceso de las mujeres rurales a la tierra en América Latina. Fundación Tierra, La Paz, pp 19–38

Llaja S, Emily K (2015) Identificación del trabajo de la mujer como usuaria y protectora de la agrobiodiversidad en comunidades de la cuenca del Momón, distrito de Punchana, Región Loreto. Memoria de Ingeniera Agrónoma. Iquitos, Perú, Universidad Nacional de la Amazonia peruana, Facultad de Agronomía, Perú

López Pardina T (2002) Significado de El segundo Sexo en la historia de la Teoría Feminista. In: 50 aniversario de El Segundo Sexo. Tertulia Feminista de Les Comadres, Gijón, pp 51–73

Lugones M (2008) Coloniality and gender. Tabula Rasa 9:73–102

Mainardes EW, Alves H, Raposo M (2011) Stakeholder theory: issues to resolve. Manag Decis 49(2):226–252

Marín VH, Delgado LE (2005) El manejo ecosistémico de los recursos marinos vivos: un desafío eco-social. En Biodiversidad Marina: Valoración, Usos y Perspectivas ¿Hacia dónde va Chile? Editorial Universitaria, Santiago, pp 555–570

Marín VH, Delgado LE (2013) From ecology to society and back: the (in) convenient hypothesis syndrome. Int J Sustain Dev 16:46–63

Marín V, Delgado L, Bachmann P (2008) Conceptual PHES-system models of the Aysén watershed and fjord (Southern Chile): Testing a brainstorming strategy. J Environ Manage 88:1109–1118

McGrath SK, Whitty SJ (2017) Stakeholder defined. Int J Manag Proj Bus 10(4):721–748

Merton R (1949) Social theory and social structure. Free Press, New York

Mies M, Shiva V (1998) La praxis del ecofeminismo: biotecnología, consumo y reproducción. Icaria Editorial, Barcelona

Moser G (2014) Psicología ambiental: aspectos de las relaciones individuo-medioambiente. Ecoe-Ediciones, Bogotá

Oliver-Mora M, Iñiguez-Rueda L (2016) The impetus of experience from the bottom up as a mechanism for contributing to the design of public services. Rev Administr Públ 50(3):377–394

Ortega J (2018) La agenda de Mujeres en Cuenca, el resultado de la articulación entre el Cabildo de las Mujeres Azuay y el Municipio de Cuenca. Agendas Locales de Género 4:18–21

Ortega T, Vázquez García V, Flores Sánchez D et al (2017) Agrobiodiversidad, género y soberanía alimentaria en Tlaxiaco, Oaxaca. Revista mexicana de ciencias agrícolas 8(SPE18):3673–3682

Piccolo JJ (2017) Intrinsic values in nature: objective good or simply half of an unhelpful dichotomy? J Nat Conserv 37(1):8–11

PNUD (2004) La democracia en América Latina: hacia una democracia de ciudadanas y ciudadanos. Programa de las Naciones Unidas para el Desarrollo, Buenos Aires

Prud'Homme JF (1997) Consulta popular y democracia directa. In: IFE (ed) Cuadernos de de Divulgación de la Cultura Democrática, vol 15, p 56

Quétier F, Tapella E, Conti G et al (2007) Servicios Ecosistémicos y actores sociales. Aspectos conceptuales y metodológicos para un estudio interdisciplinario. Instituto Nacional de Ecología México. Gaceta Ecológica 84–85:17–26

Quijano A (2000) Colonialidad del poder, eurocentrismo y América Latina. In: Landwr E (ed) La colonialidad del saber: eurocentrismo y ciencias sociales. Perspectivas Latinoamericanas. CLACSO, Consejo Latinoamericano de Ciencias Sociales, Buenos Aires, p 246

Ravetz IR (1999) What is post-normal science. Futures 31(7):647–654

Reed MS (2008) Stakeholder participation for environmental management: a literature review. Biol Conserv 141(10):2417–2431

Rovalo M (2015) Participación Pública en Materia Ambiental: aspectos del futuro acuerdo regional en América Latina y el Caribe. Derecho Ambiental y Ecología 69:41–44

Schneider H, Libercier M (1995) Concepts, issues and experiences for building up participation. In: Schneider H, Libercier M (eds) Participatory development: from advocacy to action. Organization for Economic Co-operation and Development, Paris, pp 29–64

Secretaria de Políticas para as Mulheres (2013) Plano Nacional de Políticas para as Mulheres 2013–2015. Editora Movimento, Brasilia

Secretariat of the Convention on Biological Diversity (2004). Integrated marine and coastal area management (IMCAM) approaches for implementing the convention on biological diversity. CBD technical series no. 14, Montreal

Segato RL (2014) Colonialidad y patriarcado moderno: expansión del frente estatal, modernización, y la vida de las mujeres. In: Espinoza Y, Gómez D, Ochoa K (eds) Tejiendo de otro modo: feminismo, epistemología, apuestas descoloniales en Abya Yala. Universidad del Cauca, Colombia, pp 75–90

Sol R (2012) El Desafío de la participación ciudadana en el estado democrático de derecho y retos de la participación ciudadana en la gestión de políticas públicas, en espacios institucionales de los estados centroamericanos. FLACSO, Buenos Aires

Tansley AG (1935) The use and abuse of vegetational concepts and terms. Ecology 16(3):284–307

Tapella E (2007) El mapeo de Actores Claves, documento de trabajo del proyecto Efectos de la biodiversidad funcional sobre procesos ecosistémicos, servicios ecosistémicos y sustentabilidad en las Américas: un abordaje interdisciplinario, Universidad Nacional de Córdoba, Inter-American Institute for Global Change Research (IAI)

Tognetti S (1999) Science in a double-bind: Gregory Bateson and the origins of post-normal science. Futures 31:689–698

Torres IV, Gaona SR, Corredor DV (2012) Cartografía social como metodología participativa y colaborativa de investigación en el territorio afrodescendiente de la cuenca alta del río Cauca. Cuadernos de Geografía: Revista Colombiana de Geografía 21(2):59–73

Touraine A (1997) ¿Podremos vivir juntos?: iguales y diferentes. Fondo a la Cultura económica United Nations Programa 21. http://www.un.org/spanish/esa/sustdev/agenda21/agenda21sp-toc.htm#top

World Bank (1995) World Bank participation sourcebook, environment department papers, Participation Series. World Bank, Washington, DC

Yu Iwama A, Delgado LE (2018) Acción: participación comunitaria En procesos de decisión en la Conservación del territorio. Cuadernos del Pensamiento Crítico Latinoamericano 56:1–3

Social-ecological Systems and Human Well-Being

Luisa E. Delgado, Iskra Alejandra Rojo Negrete, Marcela Torres-Gómez, Amanda Alfonso, and Francisco Zorondo-Rodríguez

Abstract Human well-being is a socio-cultural construct, originated within human experiences, that changes through time. The current literature shows a renewed interest to develop new measures and approaches to relate it to the progress of nations. One of the challenges is whether inequality is a key component and how to measure it. Furthermore, are indicators appropriate for the realities and diversity of environments of Latin American countries? In this chapter, we analyze human well-being and its subjective and objective dimensions in social-ecological systems. We propose that one way to study the relationships between both concepts is through an ecosystem services perspective. This includes the social-ecological interactions reflecting human-nature dependencies and their contributions to the well-being of local people and local and national economies. Its application shows a direct dependence of subsistence economies within Latin American rural zones on those services. One consequence is that these zones will be more affected by the degradation of ecosystems than populations from developed countries. Latin American countries still have important groups of ancestral populations whose social, cultural, and economic development has occurred within society-nature interactions for hundreds

---

L. E. Delgado (✉)
Escuela de Ciencias, Facultad de Ciencias, Universidad de Chile, Santiago, Chile

Fundación Centro Transdisciplinario de Estudios FES-Sistémicos, Santiago, Chile
e-mail: ldelgado@uchile.cl

I. A. Rojo Negrete
CEDUA (Centro De Estudios Demográficos, Urbanos y Ambientales),
El Colegio de México A. C, Mexico City, Mexico
e-mail: iarojo@colmex.mx

M. Torres-Gómez
Fundación Centro Transdisciplinario de Estudios FES-Sistémicos, Santiago, Chile
e-mail: mtorresg@fen.uchile.cl

A. Alfonso · F. Zorondo-Rodríguez
Departamento de Gestión Agraria, Facultad Tecnológica, Universidad de Santiago de Chile, Santiago, Chile
e-mail: amanda.alfonso@fundaciontierraaustral.cl; francisco.zorondo@usach.cl

L. E. Delgado, V. H. Marín (eds.), *Social-ecological Systems of Latin America: Complexities and Challenges*, https://doi.org/10.1007/978-3-030-28452-7_4

of years, shaping their well-being. We assert that their co-evolution and adaptations to maintain their interactions are of worldwide interest since they represent learning experiences for contemporary cultures that may help on the generation of co-learning and management structures.

Keywords Social-ecological systems · Latin America · Complexity · Human well-being · Ecosystem services · Traditional ecological knowledge · Shifting baseline syndrome

1 Introduction

The relationships between human well-being and the provision of ecosystem services (ES) can be characterized as an unsettled challenge for science. Their relationship was emphasized by the Millennium Ecosystem Assessment (MEA 2005). Cruz-García et al. (2017) conducted a thorough literature search for studies in countries of Africa, Asia, and Latin America. They state that the exploration of conceptual and methodological frameworks for ecosystem services and human well-being within social-ecological systems has increased in recent years. Still, we should continue the research on them (Costanza et al. 2017). This is especially true for Latin America, where societies depend on the direct use of some ES and where the application of global knowledge to local management is still slow. Among the characteristics that have influenced the speed of this process are the heterogeneity of social-ecological components and their dependency on the scale of the analysis (Maass 2018; McGinnis and Ostrom 2014). We propose that an interesting and contemporaneous framework for understanding social-ecological systems is the ecosystem services framework or frameworks (Sarkki 2017). This perspective is opening new ways of inter- and trans-disciplinary research on this field, where new approaches on ES and well-being can be formulated based on the interactions between social sciences (geography, economy, sociology) and natural sciences (ecology, physics, chemistry).

Delgado and Marín (2016) propose that relationships between society and nature are contextual, an issue also raised by McGinnis and Ostrom (2014). Consequently, their type, intensity, and valuation are unique to each social-ecological system. They are built on the characteristics of each ecosystem and the social practices of people living in its vicinity, which are specific for each of the diverse groups of human beings inhabiting planet earth (Bentacourt and Nahuelhual 2017). Furthermore, they are affected by their worldview (ecocentric/anthropocentric; Binder et al. 2013) and the valuation category (instrumental, intrinsic, relational, symbolic; Piccolo 2017). These characteristics and the local ecological knowledge are incorporated in the valuation schemes of ecosystem services, and as a consequence, they have not produced the results on decision-making originally anticipated (Weyland et al. 2019). On the other hand, the disciplinary perspective (e.g., ecology, economy, sociology) also affects the application of results. Here is where communication between disciplines, managers and social actors has to be improved, especially when dealing with human well-being issues.

The relationships between societies, ecosystems, and human well-being are processes of dynamic systems (social-ecological systems using the original ideas from Berkes and Folke (1998), Ostrom (2007) and McGinnis and Ostrom (2014) co-evolving in time and space. This evolution will depend on the biophysical state of each ecosystem, its components and functions, the associated groups of human beings, their culture, and the ways they manage and govern them (see also chapter "A New Environmental Governance"). Several Latin American societies base their culture and economy on a continuous modification of natural ecosystems, transforming them into novel ecosystems that generate the services they require. This is an ongoing process, happening in rural areas and geographically isolated social-ecological systems of Latin America (Maass 2018). On the other hand, the degradation of the ecosystem affects the quality of life and human well-being on subsistence economies closely related to nature (e.g., the rural environment) where communities take directly from ecosystems, without intermediaries, what they need to live. Furthermore, in Latin America, these ecosystems are related to worldviews, to territories that generate identity. They convey a deep sense of belonging for aboriginal and traditional (rural) people because these are the systems where they have lived and evolved historically (Alvarez and Ther 2016; Ostrom 2009). Thus, for traditional peoples, living in an ecosystem with low modern anthropogenic intervention is an opportunity and a condition for a good quality of life (Delgado and Marín 2016). Here, we offer a review of how several factors would affect ecosystems and their services influencing human well-being. First, we review a general context of human well-being in Latin America and discuss theoretical frameworks to understand the human-nature relationships. Then, we discuss a set of approaches including governance, environmental management instruments, perception of environmental changes (e.g., shifting baseline syndrome) and traditional ecological knowledge.

2 Analyzing Human Well-being in Latin America

Currently, there is a renewed worldwide interest to incorporate new approaches and methods to evaluate human well-being, the social progress, and wealth of different countries (Aguado et al. 2012). This search has included new indicators including happiness and its relationship with well-being. Indeed, ever since Easterlin (1974) posed the question if economic growth does actually increase happiness, the concept was established as fundamental in the research about the relationships between richness and happiness, including proposals that both are not reciprocal (Reyes del Villar 2017).

In the case of Latin America, the interest possibly lies in the regional analysis conducted by the Economic Commission for Latin America and the Caribbean (Galindo et al. 2014), focused on the effects of development models and their negative externalities to nature and human beings. The document exposes that the region requires generating long-term risk strategies to solve several paradoxes in order to transit toward sustainable development, including a more holistic and interdisciplinary view of social, economic, and environmental aspects. Still, although visualizing the non-sustainable social-environmental condition is an opportunity to integrate the realities and interdependencies of local societies, it also generates medium- to long-term challenges regarding public policies related to the economic development and the environment.

One of the issues regarding human well-being is its relationship with the state of ecosystems. In 2010 Raudsepp-Hearne et al. asked the question: Why is human well-being increasing as ecosystem services degrade? Later, Delgado and Marín (2017) contested that the authors did not pay attention to the fact that the index they used to analyze human well-being (i.e., the Human Development Index from United Nations) was not designed to analyze the relationship between humans and the environment. Thus, when analyzing a region in the world where human well-being is tightly related to nature, as in Latin America, there is an urgent need to include components and the services that the human population, especially from the rural environment, uses for their subsistence.

Prevailing academic theories on human well-being centers on (a) psychological or subjective reasons, (b) economic or objective arguments, and (c) sociological or normative ideals (Aguado et al. 2012). However, we propose that there is a need to incorporate the natural components and the available services, especially for rural people. This way of looking at the human well-being implies accepting territorial specificities such as the use of natural resources and the existence of capitals (social, political, symbolic, and economic) as elements to consider in the subsistence of families and their well-being (MEA 2005). For example, Delgado et al. (2013) show that the use of natural elements (e.g., wood from native forests and water) in semi-pristine ecosystems (Aysén watershed in southern Chile) provide monthly economic resources contributing to their well-being. Zorondo-Rodríguez et al. (2019), on the other hand, analyzing forestry watersheds in southern Chile, show that the use of forestry products, including mushrooms and Gevuina's nuts, represents one third of the assets of rural households. Thus, rural populations from Latin America have a better well-being when living within low anthropogenically intervened ecosystems (either historical or novel; see chapter "Social-ecological Complexities and Novel Ecosystems") providing the conditions for a good quality of life. The main implication of this idea is that, at least for Latin America, human well-being should be analyzed from a social-ecological perspective.

3 An Ecosystem Approach Is a Fundamental Element of Human Well-being

The scientific knowledge of nature and its several approaches to identify its contribution to human well-being is relatively recent. Tansley (1935), when describing the ecosystem as the total system, states that this conceptual unit includes: "*not only the organism-complex, but also the whole complex of physical factors forming what we call the environment of the biome- the habitat factors in the widest sense"* (p. 229). He further adds that we should not leave humans out of the study since they affect the ecosystems. However, research on ecosystem ecology was centered for a long time on biotic components (with the exception of human beings), abiotic components, and their interactions (Marín and Delgado 2008). Still, as the effects of human societies on nature grew, a new concept was required; one that would explicitly include nature-society relationships, and as a result, the social-ecological system concept was born (see chapter "Simplifying the Complexity of Social-ecological Systems with Conceptual Models") as well as the ecosystem service concept (MEA 2005).

The ecosystem services are the result of ecosystemic bio-geo-physical processes and include the social-ecological dependency, human-nature interactions, that contributes to the well-being of people and also to the local and national economies (Delgado et al. 2019; IPBES 2019). They are, in turn, the result of complex feedback interconnections between multiple social-ecological systems at several scales (Scholes et al. 2013). Human well-being, in developing countries, has been related to the supply of ecosystem services (Delgado et al. 2013). However, multidisciplinary efforts to study human well-being and natural ecosystems are still a pending work in Latin America (Costanza et al. 2017).

Bachmann et al. (2016) and Delgado and Marín (2016) propose that homologate the ecosystem services to all ecosystems of the biosphere is not possible. The relationships between ecosystems and services will depend on the perspective and visions of the researchers, the state of the ecosystem, and the specificity of human societies. In other words, it is a contextual relationship. The ecosystem services have been classified into four types (CICES 2018): (a) *provisioning*, corresponding to tangible ecosystem products used by human beings for nutrition, raw materials, energy products, etc. These products are normally commercialized or used directly by the local population (De Groot et al. 2002); (b) *regulation*, corresponding to the contribution of ecosystem to human well-being through the regulation of natural processes such as water purification, erosion control, carbon uptake, among others (Kandziora et al. 2013); (c) *cultural*, corresponding to all non-material benefits that people obtain from ecosystems through spiritual, cognitive, reflexive, and recreational activities including knowledge development, social relationships, and esthetic values.

Identifying the relevant ecosystem services for a given community requires incorporating local ecological knowledge (see below) through co-learning processes between formal science and the knowledge generated by the culture and the livelihood of the people who interact with the given ecosystem. Thus, one of the most relevant principles related to human well-being and ecosystem services is the social participation of all actors in a given area (see chapter "A New Environmental Governance"). Participation, in this context, can be defined as the mechanisms and processes through which communities and social actors are present in decision-making and the resulting actions as related to ecosystem management and services (see chapter "Social-ecological Systems and Human Well-Being").

Although social participation may occur at several scales, the local scale is the most appropriate to design, execute, and recognize sustainable participation models. In fact, it is the scale where deliberation processes occur and where actions to solve problems such as sustainability, good living, social inequities, and economic transformations are best designed (Berr and Diemer 2016). Thus, understanding the relationships between ecosystems, human well-being, and traditional ecological knowledge is a process that has to be looked at a local scale since it is contextual. Still, the practical use of that knowledge requires that both scientists and the people generating public policies accept the value of traditional awareness when generating proposals for local management as related to well-being. Understanding local ecological knowledge may "provide an arena for knowledge co-production, trust building, sense making, learning, vertical and horizontal collaboration, and conflict resolution" even impacting political spheres (Berkes 2009: 1695).

4 The Role of Governmental Environmental Institutions on Human Well-being

One of the most important challenges for environmental governance is to encourage state policies that may consider natural processes (e.g., growth of the ecosystem, ecological succession, biodiversity restoration) which in turn influence human well-being. Consequently, there is an urgent need for a renewed vision for environmental governance that may include the participation of social actors. In this regard, the concept of new environmental governance encourages the development of arrangements to advance auto-governance through public policies that may consider human well-being as related to security and the sustainability of ecosystems and their resources. Past ideas on environmental governance (i.e., old governance) were based on breakdowns, separations, and distinctions, for example, separation of competences among state institutions (e.g., ministries and secretariats) each working in a sectorial way and with exclusivity on a single issue (e.g., forestry, water, biodiversity, poverty, and education). Most of the time they worked in a centralized and hierarchical mode. However, today, the perspective is contextual and integrated (Delgado and Marín 2016), with special reference to sociological components that may generate human well-being, especially in rural areas of Latin America.

Human well-being incorporates the sociological or normative component (e.g., through the formulation of laws and norms) that acts as a facilitator. In particular, the relationships between ecosystem services and human well-being are determined by institutions and other forms of anthropological assets (Diaz et al. 2015), including cultural, social, and economic conditions, in a given context of space and time (Fisher et al. 2009; Laterra et al. 2016). Even more, ecosystem services are today equally or more important than economic factors for the well-being satisfaction (Zorondo-Rodriguez et al. 2016). So, institutions and governance systems over ecosystem services are the ways in which people and societies organize themselves, and their interactions with ecosystems, at different scales (Diaz et al. 2015). Institutions encompass all formal (like environmental-related acts, agencies, tools) and informal interactions (see chapter "Social-ecological Systems and Human Well-Being" for informal institutions) among people, stakeholders and social structures in relation with nature. The topic is a current challenge for the humanity, particularly for developing nations within Latin America where the percentage of families living under the poverty line is high, especially in rural areas, and well-being is variable.

Understanding the role of institutions on human well-being brings us key insights on how people, especially from rural areas, can improve it. Institutions would determine the process of making decisions, how power is exercised, and how responsibilities are distributed among all stakeholders and social structures in rural communities (Brondizio et al. 2009; Ostrom 2005). Institutions would also define the people's opportunities of access, control, allocation, and distribution of the benefits from ecosystems (Diaz et al. 2015). However, their effectiveness to ensure sustainable use of ecosystem services depends upon their relationships with people, the existence of de-centralized governments generating contextual, participative, public policies based on local ways of life. Thus, as institutions would influence, positively and/or negatively, the opportunities to satisfy human well-being, their valuation by people will strongly determine the level of effectiveness to manage nature in a sustainable way. Positive perceptions of institutions among rural people would increase their effectiveness, and consequently, it would increase the opportunity to maximize both well-being satisfaction and sustainable uses (Basurto et al. 2013; Ostrom 2005; Sayer et al. 2013).

The next problem is that most of the empirical knowledge on environmental tools (e.g., management procedures) for human well-being have been generated to operate on protected areas (Andam et al. 2010; Ferraro and Hanauer 2014; Ferraro et al. 2011; Ferraro and Pattanayak 2006) rather than on ecosystems located in non-protected land. This is a paradox because much of the interactions between humans and nature occur in lands outside protected areas (Garcia et al. 2007, 2009). The lack of knowledge about how tools in non-protected lands contribute to human well-being limits the opportunities to satisfy it for rural people as well as to implement a successful ecosystem's management. Thus, the challenge is to understand how instruments, applied to lands where people and biodiversity coexist, would regulate the uses of ecosystem services and improve well-being. An effective administration of environmental instruments depends upon the social perception of benefits and costs toward livelihood. Higher satisfaction with the instruments would be associated with better contributions to landholder's well-being. The analysis of the relationship between environmental instruments and people is important in the design and execution of public policies impinging on social and environmental issues since, without local knowledge, adaptive learning is simply not possible. Sadly, in most cases, local knowledge does not reach either public policies or decision makers (Delgado et al. 2007; Saarikoski et al. 2018). On the other hand, local social actors affect their ecosystems when public policies do not generate specific baselines (Guerrero-Gatica et al. 2019). For example, the lack of economic resources in rural areas has stopped the necessary investments to maintain the long-term productivity of natural resources. This, in turn, generates overexploitation and environmental degradation reaching, in some cases, depletion of resources.

Thus, Latin America is facing critical times insofar as human well-being, ecosystems, and governance. We urgently need the local development of strategic policies involving social actors and their relationships with natural resources. However, we should also mention that the conditions and models used today to exploit the ecosystems and its services have modified the historical way in which social actors generated their ecosystem-based well-being. Although ecosystem services are generally valued and used by the people living in a given territory, they are also exploited by extra-territorial actors who generate important disequilibria related to the benefits and accumulated richness. If local historical knowledge is not incorporated when planning the management of ecosystems, the result may be what Soga and Gaston (2018) call the "*shifting baseline syndrome*." That is, in the absence of historical information, "each new generation accept the situation in which they were raised as being normal" (p. 222).

5 The Shifting Baseline Syndrome (SBS) and the Generation of Environmental Public Policies

The SBS refers to the limitation of using human perception to study ecosystem changes, due to lack of people's experience, memory, or knowledge about past conditions. According to this concept, each generation accepts as a baseline the ecosystem state that occurred at the beginning of their life and uses it to evaluate subsequent changes. Over time, and regardless the disturbances the ecosystem may suffer, the altered state will be used as the new baseline condition by the next generation, causing the reference state to drift further away from its original starting point over the decades (Sheppard 1995). As a result, perceptions about the former system state and the causes of their degradation inevitably change with generations (Humphries and Winemiller 2009) and could lead society to gradually accommodate to degraded ecosystems, with an accepted decrease in well-being.

The SBS is also referred to as "environmental generational amnesia," as the process under the population's perception of normality continually updates and the past conditions are forgotten (Soga and Gaston 2018). In general, the authors propose the lack of generational communication as the main driver for this loss of local knowledge, as information of native species and ecosystems from the recent past is not transmitted to new generations (Papworth et al. 2009). This may be explained by shifting patterns of communication between age groups or disuse of some natural resources due to industrialization or rural to urban migration, and therefore, the loss of interest in ecological knowledge (Hanazaki et al. 2013), or because people holding traditional ecological knowledge are at risk (Giday et al. 2010). Also, the loss of interaction and familiarity with the natural environment and changes of livelihoods over time may also explain the degradation of ecological knowledge (Soga and Gaston 2018).

Although the SBS perspective was initially used to describe the limitations in fisheries management, the syndrome would be equally relevant in other disciplines across the natural sciences (Sheppard 1995). Studies have shown that observers from younger generations have unperceived changes in vegetation and forest cover, species composition and abundance, habitat degradation and transformations in agricultural landscapes (Hanazaki et al. 2013). For instance, in the Bolivian Amazonia, which has been deeply affected by deforestation, it was reported that younger respondents perceive a lower number of locally extinct tree and fish species, lower changes on local wildlife composition, and habitat degradation in general, compared with older individuals (Fernández-Llamazares et al. 2015). Another example is the study by Sáenz-Arroyo et al. (2005), where older fishers from Mexico's Gulf of California identified up to five times as many species and fishing sites as once productive, but now depleted. On the other hand, few younger fishers acknowledged that large species were once abundant in common productive sites.

The SBS may translate into differences in ecosystem assessment according to the observer's age. This can carry different consequences, such as an increase in societal tolerance for progressive environmental degradation (Soga and Gaston 2018), less cooperation with conservation programs as younger generations do not acknowledge significant change (Papworth et al. 2009) and an increased social acceptance of non-native species (Clavero 2014). Also, ecological models based on baseline conditions may be programed with erroneous starting points (Sheppard 1995), leading to an underestimation of the ecosystem degradation and failure to recognize the actual long-term societal effects.

Although there is no single correct baseline to use as a starting point (Soga and Gaston 2018), inappropriate reference points for assessing changes may arise under SBS, and therefore, the identification of unsuitable targets for ecosystem management and rehabilitation measures (Pinnegar and Engelhard 2008). The distorted societal perception of environmental degradation can have pervasive effects on ecosystem management and biodiversity conservation, as policymakers may set inappropriate conditions as targets for conservation, restoration, and management of nature (Humphries and Winemiller 2009). As a consequence, policymakers may become compliant with their current conservation efforts and, therefore, become unlikely to effectively address ecosystem degradation (Soga and Gaston 2018). Furthermore, these ramifications may generate positive feedback loops, accelerating, even more, the human impact on ecosystems and accelerating the manifestation of the Shifting Baseline Syndrome (SBS) (Soga and Gaston Op. cit.). This is especially relevant when considering the fact that some ecosystems may change radically over the years and may even shift into a different stable state (e.g., Marín et al. 2009).

6 Rural Territories: Developing Strategies for Human Well-being

Different chapters of this book show that Latin American countries are heterogeneous; each one built through contextual historical relationships between different groups of human beings, ethnic or ancestral peoples, and a large diversity of immigrants shaping the contemporaneous cultures through a fantastic syncretism. In this section, we would like to center on the rural areas of Latin America since they have been neglected when discussing well-being issues, which is counterintuitive, considering that these are the zones where local ecological knowledge and traditions are most visible.

Land use regulations in Latin American countries have been characterized by a lack of orientations and integrations into global policies. Indeed, in several cases, they are simply a set of unconnected sectorial norms that have only increased conflicts of interest (Gastó et al. 2002). These conditions have generated territorial consequences. For example, the occupation of vast surfaces with monocultures has replaced native ecosystems and/or excluded traditional activities in rural areas (Teubal 2001). This transformation of rural areas has been defined by Milton Santos (1993) as "globalization spaces," given the effects from global markets and state policies that promote economic growth at the expenses of cultures.

This change in rural territories has generated a loss of ecological functions impinging on the social environment, triggering in turn emigration from these areas. Furthermore, the negative effects of the "productivity model" is also causing reactions against it, emerging from scientists, politicians, and the civil society, which is reflecting a growing concern on the social characteristics of rural areas as patrimony (Silva 2010).

Currently, there is a dual view of rural areas and their development. The productivity model, from an economic perspective, is still dominant although in some areas there is a growing trend toward multifunctionality of agricultural spaces (Silva 2010). On the other hand, from a social and environmental perspective, the sustainable development concept is opening new orientations toward the resilience (sensu Sánchez-Zamora et al. 2016) of rural territories with the goal of maintaining the social and environmental equilibria (Folke et al. 2002; Sánchez-Zamora et al. 2016; Zuindeau 2007). A resilient system, from a social-ecological perspective, can be associated with several characteristics such as environmental heterogeneity and diversity of opportunities, including multiple economic alternatives, allowing learning and adaptation processes derived from personal and social experiences (Folke et al. 2002). One key component in those processes is to embrace traditional ecological knowledge, which we discuss next.

7 Traditional Ecological Knowledge: The Contribution of Ancestral Cultures

Traditional ecological knowledge (TEK) corresponds to interpretation frameworks to culturally understand ecosystem services, processes, and functions derived from native communities (ancestral cultures) and farmers (traditional cultures) (Berkes et al. 2000). How does it relate to well-being and ecosystem services? Charnley et al. (2007) propose that it relates through their effects on survival and adaptation. The second point of view is that TEK contributes to diversifying management alternatives (Gomez-Baggethun et al. 2009). Finally, TEK can be conceived as a holistic perspective of ecosystems, mixed with spiritual and moral issues, generating diachronic data through experiments with nature, in relation to the ecosystem services, at local and micro-regional scales (Berkes 1993).

The TEK is generated from complex, dynamic, and historical local processes associated with community knowledge arising from continuous social and social-ecological interactions (Barthel et al. 2010). Then, it is validated, adapted, and transmitted through the cycles of social constructions of nature (Davidson-Hunt and O'Flaherty 2007). This, in turn, comes from the continuous and historical systematization of information arising from collective experiences, observations, and learning (Charnley et al. 2007). This systematization represents a guarding, socially shared, memory that generates mental maps capable of defining a complex world with constructive narratives (Barthel et al. 2010).

Several authors (Agrawal 1995; Berkes 1993; Doubleday 1993; Berkes et al. 2000; Charnley et al. 2007; Bonny and Berkes 2008; Barthel et al. 2010; Saarikoski et al. 2018) have identified the main characteristics of TEK, which are appropriate to understand its role in promoting human well-being:

1. It is a way to evaluate and plan the environment.
2. It makes explicit how a resource, ecosystem service, or natural space is conceived, managed, or conserved.
3. High levels of communication and experimentation are explicit.
4. Joint work with local institutions is a key element.
5. It is specific for a given geographic area, although it may shed light over phenomena at larger scales.
6. Its generation process may be formal or informal.
7. The longer a social group stay together, the more complete and complex is the TEK.
8. The knowledge is used many times as a way to prove that it works.
9. It includes different forms of knowledge, participation, compromises, and respect.
10. It includes cultural values, associated with cosmological perspectives, articulated toward survival, adaptation, and well-being where culture and environment are mixed in a single worldview system.

8 TEK in Latin America and Its Relationship with Ecosystem Services and Well-being

Traditional knowledge and the relationships between rural societies and nature were not always recognized. Indeed, under the modern view of the world, they were initially considered archaic when compared with the dominant occidental civilization model (Silvetti 2011). However, in recent years through the perspective of political ecology, among others, there has been a re-evaluation of the diversity of knowledges available in rural societies (ancestral and farmer) that although they are contextual, their main issue is to understand and harmonize with the environment (Silvetti Op. cit.). Latin America is a region with a high diversity and complexity of ancestral ecological knowledge that have survived until now (Leff 2006). For example:

1. In Guava Islands (Orinoco's delta, Venezuela) people have identified the ecosystem services contributing to their well-being including a diversity of food products, types of cultures, sectors that people should use to walk and the temporal changes of nature (Ruddle and Chesterfield 1977).
2. Mexico's "Milpa" has been used by the academy as an example of a cultural framework for the management of resources, corresponding to a social-ecological complex generating provisioning, regulation, cultural, and supporting services. Alcorn and Toledo (1998) have characterized the Milpa as a "cultural script," an internalized plan consisting of a series of steps with alternative subroutines, decision nodes, and space for experimentation. Ecological knowledge is coded within the Milpa as generated by farmers from past generations. Although it was generated in ancient times, people still use it given its multiple benefits and contributions to well-being derived from local ecosystem services.

Recently, the relationships between ecosystem services, human well-being, and TEK have been clarified. One example is the search to incorporate native populations from The Andes in relation to their water cosmovision since without it the management of water resources is limited (Ramsar COP 2008). Thus, in most cases, the TEK shows harmony with nature through the application of ancient-communitarian management schemes tightly linked with ecosystem services that generate not only economic well-being but good living conditions.

9 Living with Traditions: Vulnerability and Strength of Rural Life in Latin America

Ways of life is a useful conceptual approach to understand how people have access and use goods to generate products and to develop survival strategies (Berdegué et al. 2015). This approach has been adopted as a conceptual framework to promote and examine development issues and to fight poverty in rural areas in México and other Latin American countries. Ellis (2000) has emphasized that there is a diversity of strategies that people develop within the rural ways of life. The ways of life approach, unlike others, start with the household as the analytical and social-economic unit which is analyzed through five types of capitals: human, social, financial, natural, and physical. Some studies combine local ecological knowledge and traditional ways of life to study the vulnerability of rural Chilean households in relation to climate change (Delgado et al. 2015; Delgado and Marín 2016). If we consider the impact of the hydric stress on rural populations, social vulnerability becomes one of the dimensions of risk, also including danger, exposure, and uncertainty (Natenzon 2007). A social group, in this case, rural households, is vulnerable when their resources or capitals are not sufficient or strong enough to generate resilient responses to environmental impacts. Quiñones et al. (2017) state that national surveys do not capture the complexity of rural ethnic households and of isolated rural households. Thus, we do not know about their capitals and vulnerabilities, and as a consequence, it is simply not possible to help them to overcome poverty, let alone reach sustainable development. One example is Chiloé Island in Southern Chile, where anthropogenic and natural events have affected the local population given their dependency on local ecosystem services (Alvarez and Ther 2016).

10 Final Reflections

Human well-being has been studied by several national and international agencies through time, changing its meaning as we learn about human beings and the environments. Currently, the concept is considered to be multidimensional, dynamic, culture-specific, person-specific, and nature-specific as we have shown it for the rural people of Latin America.

The available information shows that rural zones are being seriously affected by external (e.g., global economy) and local forcing (e.g., lack of contextualized and interdisciplinary policies). Consequently, we should make more efforts to integrate science, politics, and social actors' knowledge if we want to improve social-ecological systems, especially people's well-being.

It is clear to us, after writing this chapter, that the people living in the countryside (rural areas) in Latin America enjoy and value a peaceful way of living in a place where they can develop psychologically, economically, and socially in contact with nature. Although they value their traditions, they also accept adapting to the changes in historical and novel ecosystems (see chapter "Social-ecological Complexities and Novel Ecosystems"). There are several elements that influence human well-being, as we have shown in this chapter, but their integration requires having contextual baseline information, traditional ecological knowledge, and the political will to generate management schemes promoting sustainability, fight poverty, and the deterioration of rural areas, in other words, to take them out of the invisible condition on which the currently survive.

References

Agrawal A (1995) Dismantling the divide between indigenous and scientific knowledge. Dev Change 26:413–439

Aguado M, Calvo D, Dessal C et al (2012) La necesidad de repensar el bienestar humano en un mundo cambiante. PAPELES de relaciones ecosociales y cambio global 119:49–76

Alcorn JB, Toledo VM (1998) Resilient resource management in Mexico's forest ecosystems: the contribution of property rights. In: Berkes F, Folke C (eds) Linking social and ecological systems: management practices and social mechanisms for building resilience. Cambridge University Press, Cambridge, pp 216–249

Alvarez R, Ther F (2016) Fragmentos de una cosmovisión del Entorno Costero en el Archipielago de Chiloé. Diálogo Andino 49:123–129
Andam KS, Ferraro PJ, Sims KRE et al (2010) Protected areas reduced poverty in Costa Rica and Thailand. Proc Natl Acad Sci 107(22):9996–10001
Bachmann P, de la Barrera F, Tironi A (2016) Recopilación y sistematización de información relativa a estudios de evaluación, mapeo y valorización de servicios ecosistémicos en Chile. Informe final. Ciencia Ambiental, Santiago
Barthel S, Folke C, Colding J (2010) Social–ecological memory in urban gardens—retaining the capacity for management of ecosystem services. Glob Environ Chang 20(2):255–265
Basurto X, Gelcich S, Ostrom E (2013) The social ecological system framework as a knowledge classificatory system for benthic small-scale fisheries. Glob Environ Change Human Policy Dimen 23(6):1366–1380
Bentacourt R, Nahuelhual L (2017) Servicios ecosistémicos y bienestar local: caso de estudio sobre productos de medicina natural en Pannguilli, sur de Chile. Ecol Austral 27:1–14
Berdegué JA, Bebbington A, Escobar J (2015) Conceptualizando la Diversidad espacial en el Desarrollo Rural Latinoamericano: Estructuras, Instituciones y Coaliciones. Serie documentos de trabajo N° 164. Grupo de trabajo Cohesión Territorial para el Desarrollo. Rimisp, Santiago
Berkes F (1993) Traditional ecological knowledge in perspective. Traditional ecological knowledge concepts and cases. In: Traditional ecological knowledge: concepts and cases, vol 1. Canadian Museum of Nature/International Development Research Centre, Ottawa, pp 1–9
Berkes F (2009) Evolution of co-management: role of knowledge generation, bridging organizations and social learning. J Environ Manage 90(5):1692–1702
Berkes F, Folke C (1998) Linking social and ecological systems for resilience and sustainability. In: Berkes F, Folke C (eds) Linking social and ecological systems. Management practices and social mechanisms for building resilience. Cambridge University Press, Cambridge, pp 1–25
Berkes F, Colding J, Folke C (2000) Rediscovery of traditional ecological knowledge as adaptive management. Ecol Appl 10(5):1251–1262
Berr E, Diemer A (2016) De l'écodéveloppement au Buen Vivir, ou comment replacer les savoirs locaux au cœur des processus de coopération décentralisée dans les pays du Sud. Mondes en développement 175(3):23–38
Binder CR, Hinkel J, Bots PWG et al (2013) Comparison of frameworks for analyzing social-ecological systems. Ecol Soc 18(4):26. https://doi.org/10.5751/ES-05551-180426. Accessed 9 Jun 2019
Bonny E, Berkes F (2008) Communicating traditional environmental knowledge: addressing the diversity of knowledge, audiences and media types. Polar Record 44:243–253
Brondizio ES, Ostrom E, Young OR (2009) Connectivity and the governance of multilevel social-ecological systems: the role of social capital. Annu Rev Env Resour 34:253–278
Charnley S, Fischer AP, Jones ET (2007) Integrating traditional and local ecological knowledge into forest biodiversity conservation in the Pacific northwest. For Ecol Manage 246(1):14–28
CICES (2018) Common International Classification of Ecosystem services. European Environment Agency. https://CICES.EU. Accessed 9 Jun 2019
Clavero M (2014) Shifting baselines and the conservation of non-native species. Conserv Biol 28:1434–1436
Costanza R, de Groot R, Braat L et al (2017) Twenty years of ecosystem services: how far have we come and how far do we still need to go? Ecosyst Serv 28:1–16
Cruz-García G, Sachet E, Blundo-Canto G et al (2017) To what extent have the links between ecosystem services and human well-being been researched in Africa, Asia, and Latin America? Ecosyst Serv 25:201–212
Davidson-Hunt IJ, O'Flaherty RM (2007) Researchers, indigenous peoples and place-based learning communities. Soc Nat Resour 20:291–305
De Groot RS, Wilson MA, Boumans RNJ (2002) A typology for the classification, description and valuation of ecosystem functions, goods and services. Ecol Econ 41:393–408
Delgado LE, Marín VH (2016) Well-being and the use of ecosystem services by rural households of the Río Cruces watershed, Southern Chile. Ecosyst Serv 21:81–91

Delgado LE, Marín VH (2017) Human well-being and historical ecosystems: the environmentalist's paradox revisited. Bioscience 67:5–6

Delgado LE, Bachmann PL, Oñate B (2007) Gobernanza ambiental: una estrategia orientada al desarrollo sustentable local a través de la participación ciudadana. Revista Ambiente y Desarrollo de CIPMA 23:68–73

Delgado LE, Sepúlveda MB, Marín VH (2013) Provision of ecosystem services by the Aysén watershed, Chilean Patagonia, to rural households. Ecosyst Serv 5:102–109

Delgado LE, Torres-Gomez M, Tironi A et al (2015) Estrategia de Adaptación Local al cambio Climático para el acceso equitativo al agua en zonas rurales de Chile. Rev Am Latina Hoy 69:113–137

Delgado LE, Tironi-Silva A, Marín VH (2019) Sistemas socioecológicos y servicios ecosistémicos: modelos conceptuales para el humedal del Río Cruces (Valdivia, Chile). In: Cerda C, Silva-Rodriguez E, Briceño C (eds) Naturaleza en sociedad: Una mirada a la dimensión humana de la conservación de la biodiversidad. Editorial OchoLibros, Santiago, pp 177–205

Diaz S, Demissew S, Carabias J et al (2015) The IPBES conceptual framework—connecting nature and people. Curr Opin Environ Sustain 14:1–16

Doubleday NC (1993) Finding common ground: natural law and collective wisdom. In: Inglis JT (ed) Traditional ecological knowledge: concepts and cases. IDRC, Ottawa, pp 41–53

Easterlin RA (1974) Does economic growth improve the human lot? Some empirical evidence. In: David PA, Reder MW (eds) Nations and households in economic growth. Essays in honor of Moses Abramovitz. Academic Press, Cambridge, pp 89–125

Ellis F (2000) Rural livelihoods and diversity in developing countries. Oxford University Press, New York

Fernández-Llamazares Á, Díaz-Reviriego I, Luz AC et al (2015) Rapid ecosystem change challenges the adaptive capacity of local environmental knowledge. Glob Environ Change 31:272–284

Ferraro PJ, Hanauer MM (2014) Quantifying causal mechanisms to determine how protected areas affect poverty through changes in ecosystem services and infrastructure. Proc Natl Acad Sci 111(11):4332–4337

Ferraro PJ, Pattanayak SK (2006) Money for nothing? A call for empirical evaluation of biodiversity conservation investments. PLoS Biol 4(4):482–488

Ferraro PJ, Hanauer MM, Sims KRE (2011) Conditions associated with protected area success in conservation and poverty reduction. Proc Natl Acad Sci 108(34):13913–13918

Fisher B, Turner RK, Morling P (2009) Defining and classifying ecosystem services for decision making. Ecol Econ 68:643–653

Folke C, Carpenter S, Elmqvist T et al (2002) Resilience and sustainable development: building adaptive capacity in a world of transformations. Ambio 31:437–440

Galindo LM, Samaniego JL, Alatorre JE et al. (2014) Paradojas y riesgos del crecimiento económico en América Latina y el Caribe. Una visión ambiental de largo plazo. Serie Medio Ambiente y Desarrollo 156. CEPAL Naciones Unidas

Garcia C, Mane-Vivien D, Kushalappa CG et al (2007) Geographical indications and biodiversity in the Western Ghats, India: can labeling benefit producers and the environment in a mountain agroforestry landscape? Mt Res Dev 27(3):206–210

Garcia CA, Bhagwat SA, Ghazoul J et al (2009) Biodiversity conservation in agricultural landscapes: challenges and opportunities of coffee agroforests in the Western Ghats. India Conserv Biol 24(2):479–488

Gastó J, Rodrigo P, Aránguiz I et al (2002) Ordenación territorial rural en escala comunal: Bases conceptuales y metodología. Facultad de Agronomía e Ingeniería Forestal, Pontifica Universidad Católica, Santiago http://www.cartografia.cl/download/05-Territorio_Rural.pdf. Accessed 9 Jun 2019

Giday M, Asfaw Z, Woldu Z (2010) Ethnomedicinal study of plants used by Sheko ethnic group of Ethiopia. J Ethnopharmacol 132:75–85

Gomez-Baggethun E, Olsson P, Montes C (2009) Learning with crises: building resilience to secure ecosystem services in Mediterranean resource management systems. In: Proceedings of

the 8th International Conference of the European Society for Ecological Economics, University of Ljubljana, Slovenia, 29 June–2 July, 2009
Guerrero-Gatica M, Aliste E, Simonetti JA (2019) Shifting gears for the use of the shifting baseline syndrome in ecological restoration. Sustainability 11(5):1458. https://doi.org/10.3390/su11051458. Accessed 9 Jun 2019
Hanazaki N, Herbst DF, Marques MS et al (2013) Evidence of the shifting baseline syndrome in ethnobotanical research. J Ethnobiol Ethnomed 9:75
Humphries P, Winemiller KO (2009) Historical impacts on river Fauna, shifting baselines, and challenges for restoration. Bioscience 59:673–684
IPBES (2019) The International Science-Policy Platform on Biodiversity and Ecosystem Services (IPBES). Science and policy for people and nature. https://www.ipbes.net/. Accessed 9 Jun 2019
Kandziora M, Burkhard B, Müller F (2013) Interactions of ecosystem properties, ecosystem integrity and ecosystem service indicators. A theoretical matrix exercises. Ecol Indic 28:54–78
Laterra P, Barral P, Carmona A et al (2016) Focusing conservation efforts on ecosystem service supply may increase vulnerability of socio-ecological systems. PLoS One 11:e0155019
Leff E (ed) (2006) Manifiesto por la Vida. Por una Ética para la Sustentabilidad. Red de Formación Ambiental. Programa de las Naciones Unidas para el Medio Ambiente, Oficina Regional para América Latina y el Caribe, México
Maass M (2018) Los Sistemas Socio-Ecologicos (SSE) desde el enfoque socioecosistémico (SES). In: Avila S, Perevochtchikova M (eds) Sistemas Socio-Ecologicos: Marcos Analítico y Estudios de Caso en Oaxaca, México. Editoral Universidad Autonoma de México, Instituto de Investigaciones Económicas, Ciudad de México
Marín VH, Delgado LE (2008) Modelos conceptuales en ecología de ecosistemas: descubriendo al elefante. Rev Chil Hist Nat 81:437–439
Marín VH, Tironi A, Delgado LE et al (2009) On the sudden disappearance of *Egeria densa* from Ramsar wetland site of southern Chile: a climate event trigger model. Ecol Model 220:1752–1763
McGinnis MD, Ostrom E (2014) Social-ecological system framework: initial changes and continuing challenges. Ecol Soc 19(2):30. https://doi.org/10.5751/ES-06387-190230. Accessed 9 June 2019
Millennium Ecosystem Assessment (MEA) (2005) Ecosystems and human well-being: synthesis report. Millennium Ecosystem Assessment. Island Press, Washington, DC
Natenzon C (2007) La vulnerabilidad social como dimensión del Riesgo: Analisis de la Zona Costera del río de la Plata. In: Betocello R, Banco J, Ciccolella P et al (eds) Geografia y Territorios en Transformación. Ed. Noveduc, Buenos Aires
Ostrom E (2005) Understanding institutional diversity. Princeton University Press, Princeton NJ
Ostrom E (2007) A diagnostic approach for going beyond panaceas. PNAS 104:15181–15187
Ostrom E (2009) A general framework for analyzing sustainability of social-ecological systems. Science 325:419–422
Papworth SK, Rist J, Coad L et al (2009) Evidence for shifting baseline syndrome in conservation. Conserv Lett 2:93–100
Piccolo JJ (2017) Intrinsic values in nature: objective good or simply half of an unhelpful dichotomy. J Nat Conserv 27:8–11
Pinnegar JK, Engelhard GH (2008) The ‘shifting baseline’ phenomenon: a global perspective. Rev Fish Biol Fish 18:1–16
Quiñones D, Caro J, Marín VH et al (2017) Manejo resiliente de cuencas forestales de Chile: La cuenca de Cayucupil (Cañete) como caso de estudio. Boletín Nahuelbuta Natura 1:7–30
Ramsar COP (2008) Estrategia Regional de Conservación y Uso Sostenible de los Humedales Altoandinos. Ramsar COP9 DOC.26 http://ramsar.rgis.ch/pdf/cop9/cop9_doc26_s.pdf. Accessed 9 June 2019
Raudsepp-Hearne C, Peterson GD, Tengö M et al (2010) Untangling the environmentalist's paradox: why is human well-being increasing as ecosystem services degrade? Bioscience 60:756–589

Reyes del Villar S (2017) La Felicidad de los Chilenos. Una aproximación a la "Paradoja Latinoamericana". Análisis 24:1–17
Ruddle K, Chesterfield R (1977) Education for Traditional Food Procurement in the Orinoco Delta. Ibero-Americana 53. University of California Press, Berkeley
Saarikoski H, Primmer E, Saarela SR et al (2018) Institutional challenges in putting ecosystem service knowledge in practice. Ecosyst Serv 29:579–598
Sáenz-Arroyo A, Roberts Callum M, Torre J et al (2005) Rapidly shifting environmental baselines among fishers of the Gulf of California. Proc R Soc B Biol Sci 272:1957–1962
Sánchez-Zamora P, Gallardo-Cobos R, Ceña F (2016) La noción de resiliencia en el análisis de las dinámicas territoriales rurales: Una aproximación al concepto mediante un enfoque territorial. Cuadernos de Desarrollo Rural 13(77):93–116
Santos M (1993) Los espacios de la globalización. Anales de Geografía de La Universidad Complutense 13:69–77
Sarkki S (2017) Governance services: co-producing human well-being with ecosystem services. Ecosyst Serv 27:82–91
Sayer J, Sunderland T, Ghazoul J et al (2013) Ten principles for a landscape approach to reconciling agriculture, conservation, and other competing land uses. Proc Natl Acad Sci U S A 110(21):8349–8356
Scholes RJ, Reyers B, Biggs R et al (2013) Multi-scale and cross-scale assessments of social–ecological systems and their ecosystem services. Curr Opin Environ Sustain 5(1):16–25
Sheppard C (1995) The shifting baseline syndrome. Mar Pollut Bull 30:766–767
Silva R (2010) Multifuncionalidad agraria y territorio. Algunas reflexiones y propuestas de análisis. EURE 36(109):5–33
Silvetti F (2011) Una revisión conceptual sobre la relación entre campesinos y servicios ecosistémicos. Cuadernos de Desarrollo Rural 8(66):19–45
Soga M, Gaston KJ (2018) Shifting baseline syndrome: causes, consequences, and implications. Front Ecol Environ 16:222–230
Tansley AG (1935) The use and abuse of vegetational concepts and terms. Ecology 16:284–307
Teubal M (2001) Globalización y nueva ruralidad en América Latina. CLACSO (ed) Una nueva ruralidad en América Latina? http://goo.gl/EVDBBM. Accessed 9 Jun 2019
Weyland F, Mastrangelo ME, Auer AD et al (2019) Ecosystem services approach in Latin America: from theoretical promises to real applications. Ecosyst Serv 35:280–293
Zorondo-Rodríguez F, Grau-Satorras M, Kalla J et al (2016) Contribution of natural and economic capital to subjective well-being: Empirical evidence from a small-scale society in Kodagu (Karnataka), India. Soc Indic Res 127:919–937
Zorondo-Rodriguez F, Carrasco-Oliva G, Alfonso A et al (2019) Vinculando bienestar humano y servicios ecosistémicos: ganancias y pérdidas de bienestar de comunidades rurales por cambios ecosistémicos. In: Cerda C, Silva-Rodríguez E, Briceño C (eds) Naturaleza en Sociedad. Una mirada a la dimensión humana de la conservación de la biodiversidad. Ocholibros, Santiago, pp 207–239
Zuindeau B (2007) Territorial equity and sustainable development. Environ Values 16:253–268. https://doi.org/10.3197/096327107780474564

Part II
Challenges of Latin America Social-ecological Systems

Studying Social-ecological Systems from the Perspective of Social Sciences in Latin America

Ricardo Castro-Díaz, Maria Perevochtchikova, Catherine Roulier, and Christopher B. Anderson

Abstract Latin America can be understood from multiple perspectives, due to its high biological and ecosystemic diversity, intertwined with myriad historical, cultural, social, economic, and political contexts that together condition its social-ecological systems (SES). However, frequently within academic and management agencies, dominant paradigms and models have been imported from the Global North. Consequently, there is a need to recognize and incorporate local and regional (i.e., context-specific) characteristics to understand the SES of territories where there are complex interdependences.

In this chapter, we propose to enhance a Latin American SES perspective by "culturalizing" the ecosystem and the environment, which we perceive as a necessity to understand the interdependence occurring in specific territories. Here, we discuss specific social science contributions to the SES framework by recognizing the influence of Latin American efforts, like the *Modelo Mundial Latinoamericano*. We also conduct a philosophical analysis to compare the SES history and paradigm as a "trialogue" with territorial development, political ecology, and social science disciplines that are well-developed in Latin America. Moreover, we look at how

---

R. Castro-Díaz (✉)
Regional Center of Geomatics, Autonomous University of Entre Rios/National Scientific and Technical Research Council—Argentina, Oro Verde, Argentina
e-mail: ircastrod@unal.edu.co

M. Perevochtchikova
Centro de Estudios Demográficos, Urbanos y Ambientales, El Colegio de México A. C., Ciudad de México, Mexico
e-mail: mperevochtchikova@colmex.mx

C. Roulier · C. B. Anderson
Grupo de Estudios Socio-Ecológicos (Grupo SocioEco), Centro Austral de Investigaciones Científicas (CADIC), Consejo Nacional de Ciencia y tecnología (CONCIET), Ushuaia, Argentina

Grupo de Estudios Socio-Ecológicos (Grupo SocioEco), Instituto de Ciencias Polares, Ambiente y Recursos Naturales (ICPA), Universidad Nacional de Tierra del Fuego, Antártida e Islas del Atlántico Sur (UNTDF), Ushuaia, Argentina
e-mail: croulier@untdf.edu.ar; canderson@untdf.edu.ar

L. E. Delgado, V. H. Marín (eds.), *Social-ecological Systems of Latin America: Complexities and Challenges*, https://doi.org/10.1007/978-3-030-28452-7_5

Ostrom's SES analytical framework has been operationalized in Mexico. Finally, a literature review of SES publications was conducted to determine the state-of-the-art regarding achievements and challenges for social sciences.

Keywords Social-ecological systems · Latin America · Complexity · Social sciences · Ecosystems · Cultures

1 Introduction

It is a common practice to talk about systems in many natural and physical sciences, such as ecology, geology, or physics; however, systems theory and models are also useful concepts to explain social phenomena. In this chapter, we consider why social science disciplines should include or are already including the spatial and temporal scales required to understand complex human-nature relationships that recognize territorial processes as part of integrated social-ecological systems (SES). Indeed, such an assessment is necessary because the issues and problems that were traditionally categorized as "environmental" and approached from the perspective of the biophysical sciences (e.g., climate models of global warming) are increasingly recognized as possessing both social causes *and* consequences (e.g., energy policy and socio-economic impacts of desertification) (IPCC 2018). In turn, traditional "social" concerns (e.g., territorial planning, immigration policies, and social justice) are increasingly understood as being affected by the degradation of biodiversity and ecosystems (e.g., emergent diseases in fragmented landscapes and loss of traditional resource-based livelihoods) (Lira-Noriega and Soberón 2015; IPBES 2018a).

Still, though disciplinary in-breeding and biases are common, even when scientists employ formalized methods to obtain knowledge with the goal of answering questions about these phenomena, and for this reason, some cross-disciplinary marriages—known as interdisciplinarity and transdisciplinarity—have arisen along the lines of political ecology, ecological economics, social ecology, etc., challenging old conceptualizations of separating humans and nature to propose new integrated paradigms as part of ongoing scientific revolutions, sensu Kuhn (1962). Multiple paradigms that can support systems thinking are also well established in the human/social sciences. For example, criticism and phenomenology in geography, functionalism in sociology, relativism in history, and Keynesianism in economics are just a few examples of paradigms used for determining what is "normal social science." But this diversity of theoretical frameworks and core concepts allows us to understand the foundational premises of our scientific communities and cultures. Indeed, systems thinking is a way to abstract reality by organizing it into elements, components, structures, subsystems, and systems, and where we envision humans in these systems is critical to defining the object/subject of study and how to explore or manage it.

In this context, the interaction between social and ecological systems is emerging as a one perspective for understanding the complexity surrounding and embedded in socio-environmental problems, which are being exacerbated by global ecological and climate changes. Even though Latin America has been highly influenced by the paradigms proposed in developed countries, a lack of stability in national scientific policies during last decades has created a crucible that motivates social scientists to re-think their own paradigms by confronting decades of economic fluctuations, poverty, biodiversity loss, climate threats, war and violence, and the multiplicity of situations derived from unattained sustainable development goals. Furthermore, in a world where social sciences are threatened by their political status or governments' economic recipes, they are often marginalized even within academia, and including them in the SES perspective provides a dynamic tool to develop knowledge for studying complexity in the interdependence of communities and the environment (Scholz and Binder 2003, 2004), such as integrated ecosystems and their implications for environmental services (Castro-Díaz 2014), adaptive governance (Folke et al. 2005) and water governance (Pahl-Wostl and Kranz 2010; Pahl-Wostl et al. 2010), social vulnerability and ecosystems services feedback (Castro-Díaz and Natenzon 2018a), and water provision and land cover changes (Castro-Díaz and Natenzon 2018b). Therefore, the study of socio-ecological problems should be considered by paying attention to their complexity and functioning from a systemic (integrated) vision, which changes over space and time (García 2006: 21; Farhad 2012; Castro-Díaz 2017).

Understanding the SES causes and consequences of these relationships allows us to find solutions in a functional (and ethical) way, starting with the origin of the problems and then addressing related actions from the complexity of specific situations. The term complex systems can be considered from the paradigm that reorients scientific thinking toward the complexities, wholes, and open dynamic systems that are present throughout the world (García 2006). The main components of a complex system, as proposed by García (2006), are (1) limits (spatial and temporal), (2) structure (hierarchy of components), (3) scales (of interaction), and (4) processes. In this sense, engaging that which is "complex" allows researchers and managers to emphasize on the composition of an entire system and on the heterogeneous elements in constant interaction with each other, immersed and surrounded by other systems. This perspective, in turn, requires an integrative, inter- or transdisciplinary approach (Castro-Díaz 2017; Merçon et al. 2018).

In recent years, a great interest has been observed in the development of academic studies and public-policy instruments that incorporate the concept of SES, promoting the publication of numerous investigations on analytical frameworks, related concepts, and ways of operationalizing based on causal relationships (Perevochtchikova 2016; Avila-Foucat and Perevochtchikova 2018). The concept of complexity, therefore, contributes to the analysis of these current ecological, social, and economic problems and crises, such as climate change, poverty, injustice, and environmental degradation, among others. It seeks to understand a world in constant transformation and adaptation (or not) to the influence of internal and external stressors or "shocks" through the dynamics of self-organization and self-development (Postigo et al. 2013) with profound epistemological, ontological, and ethical implications.

2 "Culturing" Ecosystems Through the SES Concept

The idea that we live in a time of catastrophe and harm leads humans to be conceived of as aliens to Mother Earth, and this perspective has become well-established in many academic and policy discourses. Indeed, there are major reasons to support it, including climate change (Weart 2008), the modern species extinction crisis (Thomas et al. 2004), the mass extinction of the Australian fauna (Roberts et al. 2001), or the European invasion to the Americas and their devasting ecosystem modifications (Koch et al. 2019). However, this phenomenon can also be considered from the lens of what it means to our cultural relationships to nature.

The idea of "culturing" nature encompasses all actions taken by the human being as an individual (indivisible), community, or society with ecosystems. Its implications, of course, are as diverse as the disciplines of social and human sciences, especially when they are referred to multiple explanations, theories, models, schemes, and all approaches to the human dimension at every possible scale. They are numerous and varied that we should define the socio-ecological relationship as the focus of our attention that, even with constraints, it solves the issue of the metaphor of human actions on the ecosystem as a result of their cultural activities. For understanding the ways of "culturing" nature, we can refer to Nisbet et al. (2009), which states that every aspect of the human life is related to the environment (i.e., natural relatedness). Morin (2009) includes this approach when considering the relationship with the whole to the parts and propose complexity as a feature of the link they hold. This link, in turn, is built into so-called "time-space," a concept that grew from geography and refers to the territorial processes holding the spatial dependence and the temporal causality (Pillet 2004) and including territorial changes between the present and the past for a given location. For example, the way to study a currently flooded valley, a cut forest, the city of Brasilia, and global change are all the result of spatial-temporal dynamics. This "culturing" of nature approach has been developed in several social/human sciences, such as geography or anthropology. However, it can include every such discipline (e.g., sociology, psychology, health, economics, engineering, and others) that can locate their field of study in the diversity of territorial processes.

For comprehending an SES with a social sciences lens, we should, therefore, determine its ecological foundations, but also its expression in human spatial-temporal relationships to understand the dynamic agent causality of human beings. Even though anthropic actions are widely evident in our planet, the study object/subject being investigated with the SES framework needs to recognize that humans go beyond the negative prejudge that many natural and physical sciences have established and instead incorporate a social science perspective that acknowledges societies interacting reciprocally with ecological fluxes. There are multiple examples for identifying these types of socio-ecological relationships: religion, cosmovision, technology, energy production, urbanizations, tourism, agriculture, scientific production, transport, and others such as processes intervening, modifying, and conditioning the ecosystem (see also chapter "Social-ecological Complexities and Novel Ecosystems" in this volume). In short, the social science SES perspective conceives anthropogenic action as part of the ecosystem response, and at the same time, it cannot be studied without appeal to human dimensions that span social, cultural, religious, political, and economic factors.

3 Social Sciences Contributions to SES in Latin America

Globally and regionally, SES scholarship has been based largely on the dominant natural science discourses and paradigms that have influenced a global "brain circulation" of these ideas (Anderson et al. 2015a). However, we seek to contextualize this narrative by putting it into dialogue with social science traditions from Latin America, which to date have been sub-alternate voices. To diagnose the role of Latin American social sciences, in this section, we identify both their achievements, but also their gaps, or what de Sousa Santos (2006) has called a "sociology of the absent," to understand when, where, and why they have been present (or not).

We put forward that Latin America has traditions that can support and enhance regional and global SES research and practice, which is relevant to global efforts to recognize multiple approaches, stakeholders, and worldviews in SES (e.g., Díaz et al. 2015). To test that assertion, we explore here three specific avenues of analysis: (a) a socio-historical perspective of how Latin American scholars have confronted and responded to dominant SES ideas that arise (and often are imposed) from the Global North, the developing of the *Modelo Mundial Latinoamericano* as a reaction to Meadows et al. (1972); (b) a philosophical (epistemological and metaphysical) evaluation of SES in relationship to territorial development (TD) and political ecology (PE), social science fields that are well-developed in Latin America; and (c) an operationalized use of Ostrom's SES framework for understanding multilevel and multiscale interaction, using case studies focused on Mexico's research experiences to see the ways that social sciences have been involved.

The History of the Modelo Mundial Latinoamericano

Understanding Latin America's (potential) contribution to the SES debate requires having a historical perspective. When "environmental" problems were capturing the Western (or "Northern") imagination around the 1960s and 1970s (Estenssoro 2007), Latin American scholars quickly recognized that this environmental crisis was not only ecological, nor merely driven by human demands (e.g., population growth, migration from rural to urban areas, and subsequent urban expansion), but it was also intimately related to human quality of life, social well-being, justice, and equity. In particular, marginal peoples' lives were being greatly affected by these environmental changes. At the same time, though, in the Global North, the world's marginal people were thought of as a "population bomb" (sensu Ehrlich 1975) that was largely considered by developed countries to be the main driver of the environmental and civilization crisis. Arising from this thinking, we see such seminal reports as the Meadows et al. (1972) *Limits of Growth*, which proposed a global model (World3) for rationalizing sustainable human use of resources, based largely on reducing consumption by reducing birth rates (particularly in the Global South).

Based on the treatise *Catastrophe or New Society?* (Herrera et al. 2004), the Latin American World Model (*Modelo Mundial Latinoamericano*, Goñi and Goin 2006) was developed, whose name is important in itself because it purports to be a global model (*modelo mundial*), like Meadows et al. (1972), but, as is frequently the case when the social science perspective is brought to bear, it recognizes its own subjectivity by identifying where it comes from (*Latinoamérica*), rather than being some disembodied idea about a supposedly objective reality. Furthermore, *The Limits to Growth* had a universal conceptualization of humans and undertook a neo-Malthusian approach to the issue, based on resources and reproduction (i.e., a biological approach), where central countries ceased their ever-increasing consumption and peripheral countries ceased their population growth. The Latin American model introduced a novel and integral approach to the issue of sustainability and development by considering the satisfaction of humanity's "basic needs" of food, housing, education, and health for everyone as a way to reach a balance between society, nature, politics, and the economy. During this same time, "Northern" development agencies like the International Monetary Fund and World Bank were only looking at economic variables (Oetiza 2004).

The Latin American approach differed from that which was reflected in the *Limits of Growth* and validated by the countries members of the Club of Rome because it explicitly recognized its own normative character, as well as the power dynamics that are inherent in decisions regarding natural resources and human well-being. While Meadows and colleagues concluded that if their recommendations were not considered, then catastrophe would be imminent, the work derived from the Latin American report reveals that two thirds of humanity already was living a catastrophe, as impoverished and marginalized people. So, in this way, they asserted that the Club of Rome's report did not address the real problem. Of course, as Gallopin (2004) observes, both models have embedded values, but only the *Modelo Mundial Latinoamericano* makes them explicit, which is a hallmark of a social science perspective.

The historical perspective provided by this exemplary case study, however, also illustrates how social science ideas and their impact in the SES debate between the Global South and North are contextual and conditioned by broader societal processes. An inherently unequal power relationship regarding the production and dissemination of its proposal (e.g., it was not fully recognized by the national and international academic community, Goñi and Goin 2006), but the *coup d'état* that installed the Argentine dictatorship in 1976 coincided with the final stage of its work and truncated its continuity and potential influence on regional and global debate and outcomes. Therefore, an entirely external socio-political process in Latin America vitiated the ability of Latin American thinkers, particularly social scientists contribute to these SES issues at the regional and global scales in dialogue with the dominant, natural science-based ideas of Meadows, and others.

A Trialogue Between Social-ecological Systems, Territorial Development, and Political Ecology

SES research shares a common study object/subject (human-nature interface) with territorial development (TD) and political ecology (PE), but each has different philosophical foundations and assumptions that can hinder productive collaboration. While SES arose largely in the context of the ecological sciences striving to integrate a human dimension, TD and PE came from social science traditions to understand the environment. In this way, we would expect them to have both epistemological and metaphysical similarities and differences, which demand attention to put them into constructive "trialogue" and avoid unconstructive arguments in the context of Latin American interdisciplinary socio-ecological research. To test this hypothesis, we analyzed these three fields from historical and philosophical perspectives to see where they complement, contradict, and/or enhance one another to be able to promote interdisciplinary (or integrated) studies in Latin America of complex human-nature dynamics in the Anthropocene (Table 1).

SES, TD, and EP are relatively young academic fields that have emerged in the last 50 years. Early notions of SES can be found within the realm of ecology and natural resource management in the Global North (Holling 1973; Odum 1953, 1973; Berkes and Folke 1998). Only more recently has an explicit SES approach get visibility in Latin America (e.g., Delgado and Marín 2005; Maass et al. 2010; Castro-Díaz 2017; Easdale et al. 2016). While TD came into its own in the 2000s, its roots could be found in the 1970s (and even earlier to the 1950s) with different efforts at local economic development in Europe and Latin America. Also beginning in the 1970s, PE consolidated in Europe and North America, becoming firmly established in the 1990s (Martínez Alier 2005). By the 2000s, though, PE also came to have globally influential scholars from Latin America, where authors like Colombian A. Escobar (2000), Mexican E. Leff (2004), and Argentine H. Alimonda (2004) began to work with this perspective on local problems, such as peasant and indigenous social movements and environmental conflicts in defense of natural resources (land, water, mining).

Regarding their objects/subjects of study, SES uses systems-thinking and complexity and networking theories to attempt to study the whole. For its part, although TD began with a focus on economics, it has since expanded to a more encompassing

Table 1 Summary of analytical axes for the conceptualization of perspectives in socio-ecological topics for Latin America

Analytical axes	Characteristics	Political ecology	Territorial development	Socio-ecological systems
History	*Origins (temporal and spatial)*	1970s in Global North; 1990s in Latin America	1980s in Global North and Latin America	Beginning in the 1970s and consolidating in the 1990s in Global North; 2000s in Latin America
	Key figures and institutions (countries)	Martínez Alier (Spain), Alimonda (Argentina), Escobar (Colombia, USA), Leff (Mexico)	Alburquerque (Spain) Costamagna (Argentina), Instituto Praxis (Argentina)	Holling (Canada), Odum brothers (USA), Berkes (Netherlands, Canada), Folke (Sweden)
Epistemology	*Predominant research types and foci*	Basic, social science-based research, largely academic and theoretical Started with the "environmental crisis" and added political and power dimensions	Applied, social science-based analysis of local productive systems Started from an economic perspective, but moved toward holistic understanding	Basic, but often applied to real-world problems, natural science-based studies that often use complexity and network theories and resilience concepts Started as ecology integrating humans, but expanding to toward the social domain
	Methods	Qualitative	Increasingly participatory research action to co-construct knowledge(s)	Mixed, but primarily quantitative
	Confirmation and validation	Qualitative methods	Statistics are used, but also the putting into practice of information based on transferability, viability, and credibility. Data for decision-making	Uses statistics and modeling

(continued)

Table 1 (continued)

Analytical axes	Characteristics	Political ecology	Territorial development	Socio-ecological systems
	Study object	Process and dynamics of power that determine the distribution (access and benefits sharing) of natural resources	Productive processes/ microprocesses that lead toward improvement of quality of life of the people living in a territory	Feedback processes and thresholds or parts of the system
Metaphysics	*Objectivity*	Assumes subjectivity	Assumes subjectivity and the co-construction with stakeholders. It criticizes supposed objectivity in science and territorial construction. It assumes subjectivity from the political posture that conditions processes being studied	Tends toward objectivity
	Reductionism versus holism	Emergent approach	Emergent approach	Systems approach that recognizes both holism and reductionism
	Conceptualization of nature	Nature as an element of power and dispute between social actors	Nature is part of the "scenario" of the territory, but not central to it. Nature is mostly conceived of as natural resources for production. In recent years, the perspective of "sustainable development" gave nature a greater role, but it continues to be one of various dimensions	Nature conditions social practices and should be reconciled with human uses. Nature is recognized as being a source of benefits for humans and also the recipient of their actions (reciprocity)

understanding of human well-being and the needs to innovate in production systems to fulfill these needs. Meanwhile, PE looks more specifically at the political dimensions of natural resources, while not denying other elements. These foci condition these disciplines' methodologies, for example, SES studies frequently concern system resilience with thresholds and feedbacks (Adger 2000; Cumming 2011;

Castro-Díaz 2013), which use quantitative methods from basic science disciplines that use statistics and models. TD also addresses systems, analyzing productive processes (or micro-processes) that tend to be applied to a specific territory (e.g., local food systems), but its experiences are systematized, and case studies are often used in a specific spatio-temporal context. Additionally, in TD, research-action techniques are often used, and validation is conducted by putting findings into practice via transferability, viability, and credibility with local stakeholders (e.g., the Argentine city of Rafaela, see also Karlsen and Larrea 2015). Finally, PE is mostly concerned with basic research and uses qualitative methods that are not always subject to statistical validation techniques. Some PE studies, though, concern such topics as environmental justice and use participatory approaches, like TD, to not only study but transform or affect reality (e.g., avoid conflicts related to natural resource industries, such as mining). However, in Latin America, most PE continues to be academic, but often related to social and environmental justice movements, and therefore also has an applied intention, even if not an applied approach. Indeed, throughout Latin America, PE observatories have arisen to maintain monitoring of these human-nature power conflicts (e.g., *Observatorio de Conflictos por Recursos Naturales*, https://ocrn.info/; *Observatorio Latinoamericano de Conflictos Ambientales*, http://olca.cl/oca/index.htm).

While each of these disciplines approaches human-nature relationships from different epistemological positions, histories, and interests in socio-scientific questions, they share many points in common, and these similarities and synergies provide ample space for collaboration and dialogue. Latin American social science traditions, such as TD and PE, have consolidated bodies of literature and knowledge communities and, therefore, should be further considered by natural scientists and interdisciplinary scholars as the new paradigm of SES is implemented. Many natural and applied scientists may have an inherent affinity to the SES approach, given its history and philosophical orientation (e.g., systems modeling and quantitative methods), but they should also be aware of these other traditions that have been developing from Latin America for several decades, and whose research production is often made in Spanish or in local and regional journals, which many not be part of the "global brain circulation" (sensu Anderson et al. 2015a).

Applying Ostrom's SES Analytical Framework to Latin America

The SES analytical framework proposed by Ostrom (2007, 2009) considers social-ecological interactions at macro and micro levels (multilevel), from local to regional spatial scales (multi-scale), and applicable to specific case studies (Perevochtchikova 2018). It is an integrative framework developed from a bibliographic review of more than 40 years of different approaches and theories analyzing the relationship between society and ecosystems, adaptable to other theories and diverse SES. The framework takes up and integrates big groups of ecological variables, conceived of as resource systems and units, which include biodiversity and ecosystems, and

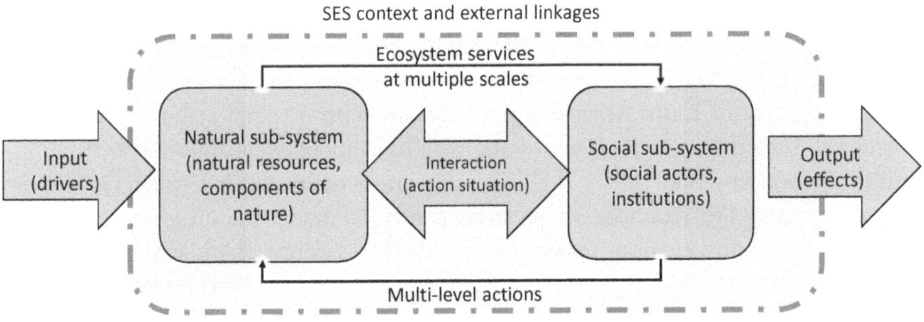

Fig. 1 Summary of Ostrom's (2007, 2009) framework for conceptualizing socio-ecological systems (SES)

social variables, considered of as governance systems and users, which encompass stakeholders and institutions. In turn, these sub-systems are interconnected through the action situation that has inputs (drivers) and leads to certain outputs (effects) on the operation processes of a hypothetical SES. Each SES has connections with other, external systems, and it has certain characteristics of the political, social, and economic context that is present at different territorial and temporal scales (Fig. 1).

After selecting the analytical framework for a specific study, which depends on objectives, goals, and resources, the next stage of formalization refers to the selection of variables that later allow to arrive at the filling and breakdown of information into each group and even to formulate indicators. These variables can be analyzed qualitatively or quantitatively, taking information from a documentary analysis of existing sources (official and academic) and/or constructing it from fieldwork. Each group of these variables is desegregated in corresponding levels, and interdisciplinary, inter-sectoral, and inter-institutional collaborations are required for integrated analysis (Perevochtchikova 2018).

A practical contribution in terms of formalizing the framework and presenting quantitative relationships can be found in Schlüter et al. (2014), which used a format of equations and mathematical language to model the case of a fisheries SES. Bennett and Gosnell (2015) pre-select some second-level variables from Ostrom's framework (2009) and adjust them to the needs and context of particular cases. Some interesting exercises can be highlighted as the adaptation of the general framework for forestry and fishing systems in Hinkel et al. (2014), and other case studies dealing with the process of formalizing the SES framework at a local scale and ranging from the conceptual determination to the definition (Hinkel et al. 2015).

In Latin America, there are still few examples of formalizing the SES framework and even less of its operationalization (Perevochtchikova 2018). Among the almost absent publications on the operationalization of the framework (which refers to the analysis of variables based on obtained information), the study by Leslie et al. (2015) developed a regional-scale analysis in several fishing communities from Baja California, Mexico, which was more focused on determining economic benefits related to different ways of fishing. However, these cases confirm that this SES analytical framework is adaptive to a variety of SES contexts and can be modified to practically any study case, with better results at local or regional scales being integrative and considering the great potential for use in scientific research and public policy.

The majority of Latin American publications with a social science focus have objectives linked to the analysis of the relationship between human well-being, multilevel actions, and policy-making with multi-scale ecosystem services over time and space. For example, these studies seek to detect the effects of applying governmental conservation programs in Mexico (Perevochtchikova, 2019), to understand the well-being and the use of ecosystem services by rural households in Chile (Delgado and Marín 2016), to analyze the vulnerability of SES in Colombia (Berrouet et al. 2018), to determine livelihood strategies in complex SES in Nicaragua (Williams and Kramer 2019), and to study resilience and dynamic use of biodiversity in Costa Rica (Rodríguez and Davidson-Hunt 2018).

The social science works have explicitly incorporated historical analysis to reconstruct the trajectory of change of SES, with the use of geographical information systems (GIS) for space analysis of physical and biological variables and land use changes. On the other hand, it is very common to find interviews with key stakeholders and survey applications for governance studies. Also, ethnographic data collection is seen through participant observation, assisting different social actors' labors, and during fieldwork and workshops. Ethnoecological approaches also include transect walks and life story interviews. The analytical techniques found in these social studies combine the quantitative and quantitative approaches, but in each of these cases interdisciplinarity is a necessity, where social sciences and scientists can play a strong role for analysis of interactions, related to actors, institutions, and governance aspects ranging from inputs (e.g., public policy) to outputs (e.g., benefits) of SES.

4 Literature Review of SES Publications

Many proposals are being developed in the global scientific literature (1) to define SES (Haberl et al. 2006), (2) to operationalize models to study SES (Collins et al. 2011), (3) to apply SES knowledge to public policies and private decisions (Carpenter et al. 2009), and (4) to develop research and governance models that include divers social actors (e.g., the Inter-Governmental Platform on Biodiversity and Ecosystem Services (IPBES): Díaz et al. 2015; Anderson et al. 2019). Anderson et al. (2015b) found an exponential growth in the use of SES-related terms in ecology and sustainability journals beginning in the 2000s. However, there is also an indication that the conceptualization of SES in this body of literature has been influenced mostly by quantitative social sciences (e.g., studies of institutions, economic incentives, land use, population, social networks, and social learning), with less emphasis on those approaches with interpretative tradition (Stojanovic et al. 2016).

We carried out a systematic literature review to identify, organize, and analyze the scientific production regarding the use of SES in the world and in Latin

America. The review was based on the proposal by Grant and Booth (2009), Booth et al. (2012), and Perevochtchikova et al. (2019). This review used standards and procedures established in the declaration of Preferred Reporting Items for Systematic Reviews and Meta-Analyzes (PRISMA) (Moher et al. 2009) and applied the Search, Appraisal, Synthesis, Analysis (SALSA) framework, which is shared by many studies that perform systematic reviews of scientific literature (Grant and Booth 2009; Codina 2015). Taking as a reference the search strategies used in ecosystem services literature reviews (Perevochtchikova and Oggioni 2014; Martinez-Harms et al. 2015; Ezzine-de-Blas et al. 2016; Locatelli et al. 2017; Himes-Cornell et al. 2018; Perevochtchikova et al. 2019), the decision was made to perform an advanced search in the Scopus database, which gathers information on publications of high scientific rigor at an international level.

The first phase of the analysis considered the construction of syntaxes with the use of keywords in English and Spanish linked to the concept of "socio-ecological systems." The search for the selected terms was carried out in titles, abstracts, and keywords, and then we applied a filter to select only those publications in journals related to social sciences. In the second phase of evaluation, a sub-search focused on the geographic delimitation of SES studies referring to Latin America and/or to the countries of the region, also with a social science filter, to assess the importance of this focus for SES.

Globally, a total of 7300 records were obtained, of which 2198 were open access. Publications are mostly made in an article format (74.3%) and in English (99.8%); much fewer contributions to SES are published as reviews (9.4%), book chapters (6.9%), conference proceedings (4.2%), and books (1.1%). The earliest work appeared in 1970, and there has been an exponential growth since 2003, which closely correlates to the Millennium Ecosystem Assessment (MEA 2005), with more than 1000 publications in 2018.

Within this literature, the disciplinary approaches were diverse, and for the purposes of calculating percentages, one study could pertain to more than one research domain. Globally, interdisciplinary (i.e., environmental sciences, 55.7%), social science (37.5%), and agricultural and biological sciences (22.5%) approaches dominated. Lesser contributions came from medicine (13.8%), earth and planetary sciences (8.3%), physiology (6.3), and economics (7%).

Among the 158 countries involved in these publications, the dominant political entities were (in decreasing order) the USA, Australia, Canada, the United Kingdom, Germany, Switzerland, the Netherlands, France, and Spain. From Latin America, Brazil (in 14th place with 168 publications), Mexico (in 17th place with 137 publications), and Chile (in 20th place with 108 publications). Regarding study sites, the largest percentage were located in North America (15%), but 11% were from Latin America. This situation represents an opportunity to develop works on this subject in our region and potentially means that, despite the relatively low scientific production, Latin America has study areas that are of interest to the international community due to the high biological and cultural diversity.

For Latin America, a total of 556 publications (with 197 as open access) were found until April 2019. This constitutes only about 7.6% of the international SES

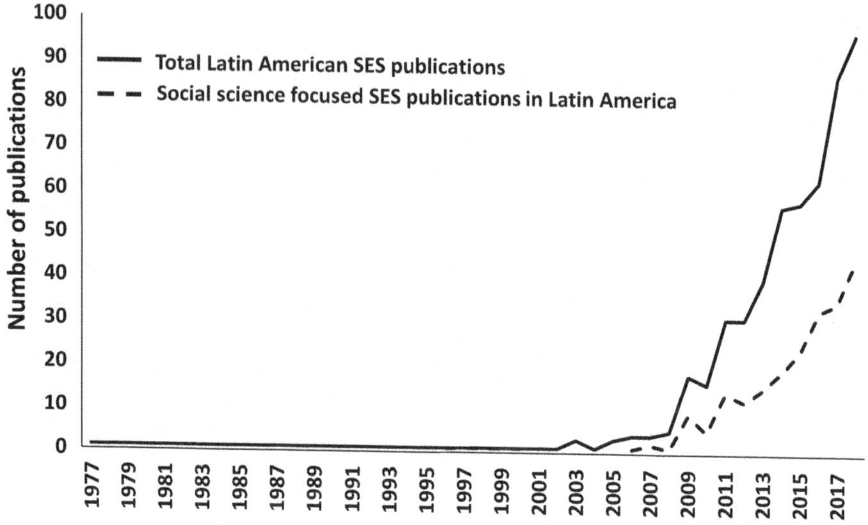

Fig. 2 Total number of socio-ecological (SES) publications in Latin America from the Scopus database (solid line) and those SES studies that had a specific social science orientation (dashed line)

research production. While globally the first publication was in 1970, in Latin America it was in 1977, and a growth trend is observed since 2006, increasing markedly since 2016 (Fig. 2). The sharp increase in Latin American SES research coincides with the publication of the works of importance and international reference in the topic of SES, such as MEA (2005) and especially other influential papers like McGinnis and Ostrom (2014), Schlüter et al. (2014), Hinkel et al. (2014), among others.

Almost 85% of the Latin American SES studies are published as scientific articles (with only 5% as book chapters, 4% as reviews, and 3% as conference papers). The research disciplinary domains are similar to the results found at an international level: environmental sciences (64%), social sciences (42%), agricultural and biological sciences (32%), earth and planetary sciences (9.5%), medicine (6.7%), economics (5.9%), energy (4.9%), arts and humanities (4.7%), and engineering (4.1%). While the social sciences are found at second place in both the global and regional scales, Latin American also had SES studies published in arts and humanities journals.

Among the 71 countries mentioned in the Latin American SES publications, the following trends were observed for their contribution to overall research productivity: USA (38%), Brazil (17%), Mexico (15%), Chile (11%), UK (10%), Canada (10%), among other countries in Europe, Latin America, also New Zealand and China, and even with a few works from Morocco and the Russian Federation. Regarding the affiliation of the corresponding authors of the publications, we found the greatest representation from academic institutions in Brazil, Mexico, and Chile, followed by Argentina, Colombia, Ecuador, and Peru. Also, financial support for these projects mainly came from national science and technology commissions of the mentioned countries, and then also from different entities and agencies in the USA.

Focusing explicitly on Latin American SES studies that were undertaken with a social science orientation, a total of 235 publications were found (60 as open access). Among these publications, articles clearly dominate (87%), followed by book chapters (4.3%) and reviews (3.4%). The publications are also made in international journals, which are predominantly in English and use the ISI Impact Factor. *Interciencia* is the only Spanish language journal that was found among the list of the top ten sources of Latin American SES articles. This journal has also provided space to present theoretical-conceptual discussions and case studies, which many international journals are reticent to accept.

Among the 51 countries mentioned in these publications, most of the studies came from the USA, Brazil, Chile, Mexico, Canada, the UK, the Netherlands, Germany, Spain, Argentina, and Colombia (Fig. 3). The same trend was found regarding the affiliation of the corresponding author and with the sources of financial support. In this sense, among the funders, there are 159 sources mentioned in acknowledgments, including government agencies, international agencies, foundations, as well as national science and technology councils, or academic and national universities. Important support particularly comes from the USA, Mexico, Brazil, Canada, Germany, and the European Union. From Latin America, countries such as Mexico, Brazil, and Argentina resulted as the most important in support of SES research.

In general, Latin American SES studies consider case studies at local or regional scales, from multi- and, to a lesser extent, interdisciplinary perspectives with the use

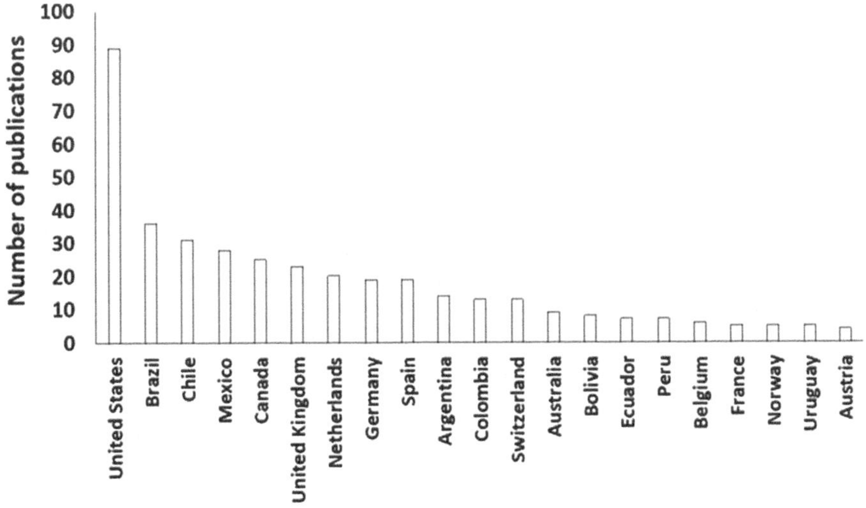

Fig. 3 Countries with the most research productivity in Scopus regarding Latin American socio-ecological publications with a social science orientation

of mixed or quantitative methods. The purposes of the studies include improvement of the human vulnerability and adaptability to external stressors (e.g., climate change or other risk situations), improvement of mechanisms for public management of natural resources (e.g., incorporating integrated vision into the proposals), or improvement in productive systems (e.g., agroforestry). Among the challenges that remain to be addressed, we can mention areas of opportunity in the development of practices: interdisciplinary analysis (with not only the construction of physical models of ecological functioning or conceptual models of interaction between social-ecological variables); transdisciplinary studies (from different sectors of society); impulse to modeling (e.g., based on dynamic systems); and even application of techniques such as social networks.

5 Final Reflections

Latin America has much to offer the world regarding the human dimensions of SES, given our high cultural and ecological diversity (IPBES 2018b). In addition, a pantheon of important social and environmental scholars come from Latin America, including social ecology in Uruguay (Gudynas and Evia 1991); environmental sociology in Argentina (Svampa 2008), the implications of political and social movements in the face of intense economic production during armed conflict in Colombia (Escobar 2000), ecological economics that questions orthodoxy by questioning underlying rationalities of modernity, based on such pillars as technology, monetary cost-benefit analyses, and science in Mexico (Leff 2010) and the need to consider the human-face of development in Chile (Max-Neef 1994). However, promoting knowledge dialogue between disciplines is inherently difficult, and even more so when much of this SES-related scientific production has been made in regional journals or in Spanish.

Therefore, we should take into account intrinsic factors that have limited social science contribution of Latin America to the global debate in SES. For example, in many cases, these disciplines are still young with less than 20 years of development in post-dictatorial societies. At the same time, there are external relationships that condition local dynamics, such as North/South power dynamics that are inherent in the “global brain circulation” (Anderson et al. 2015a). Here, though, we find hope that knowledge dialogue can be improved between North/South and disciplines, which is evidenced in experiences such as the Inter-Governmental Science-Policy Platform on Biodiversity and Ecosystem Services (IPBES), that is working to bring to light these previously underrepresented voices in the assessment and governance of SES.

In this chapter, we have taken a social science perspective to SES and human-nature relationships, which has identified and legitimized the study of social variables beyond merely economic considerations (Anderson et al. 2019). The depth and breadth of these contributions is only clear, however, if SES researchers and practitioners take the time to learn the history of these traditions in our continent and also make the effort to understand how to dialogue with these complementary and sometimes contradictory proposals, which requires understanding not only the history but also the philosophical underpinnings of our and other's disciplines. The "disciplinary" perspectives of the environment without question can be integrated into an interdisciplinary approach, but it is important to recognize how humans are being conceived of as part of nature (e.g., defending it, destroying it, using it, and living in it). All of these approaches can advance the study of processes related to society-nature, but it is important to distinguish different elements to then be able to integrate and articulate those concepts and knowledge(s). The ongoing challenge to attain mutual recognition among these fields and improve dialogue (or trialogue: territorial development and political ecology) can be aided by finding complementary perspectives that provide different tool sets to more fully address the complexity of socio-ecological study topics.

To date, the historical and current contributions that Latin American social science and humanist traditions have made to SES research have been limited, but there are reasons to hope that it is possible to improve this deficit. On the one hand, clearly, these academic traditions are advancing. Therefore, it would be important for them to become self-aware and strategic to engage in the global debate. Latin American countries are contributing strongly to the conceptual framework and operationalization of such initiatives as IPBES (Díaz et al. 2015; IPBES 2018b; Anderson et al. 2019), including the recognition and incorporation of diverse knowledge sources(s) into decision-making. Also the Latin American Social Sciences Council (*CLACSO*) develops many efforts for support existing research networks by its Work Team Program, including a Network for Transdisciplinary Studies of Ecosystems and Society (*Red de Estudios transDisciplinarios sobre el Ecosistema y la Sociedad*), which involves social scientists throughout Latin America striving to understand the complexity beyond the territories in the region. Finally, further engagement in these efforts by social scientists from Latin America is one way to take their voice (including ideas, concepts, methods, and paradigms) into a global platform that empowers and seeks to integrate plural values and perspectives, augmenting participation through publications and helping integrate the social science into solving real-world problems.

References

Adger WN (2000) Social and ecological resilience: are they related? Progr Hum Geogr 24(3):347–364

Alimonda H (2004) Una introducción a la Ecología Política latinoamericana (pasando por la historia ambiental). Documento curso de posgrado en Ciencias Sociales (Universidad Federal Rio de Janeiro)

Anderson CB, Monjeau A, Rau J (2015a) Knowledge dialogue to attain global scientific excellence and broader social relevance. Bioscience 65:709–717

Anderson CB, Pizarro JC, Estevez R et al (2015b) ¿Estamos avanzando hacía una socio-ecología? Reflexiones sobre la integración de las dimensiones "humanas" en la ecología en el sur de América. Ecol Austral 25:263–272

Anderson CB, Seixas CS, Barbosa O et al (2019) Determining nature's contributions to achieve the sustainable development goals. Sust Sci 14:543–547

Avila-Foucat VS, Perevochtchikova M (eds) (2018) Sistemas socio-ecológicos: marcos analíticos y estudios de caso en Oaxaca, México

Bennett DE, Gosnell H (2015) Integrating multiple perspectives on payments for ecosystem services through a social–ecological systems framework. Ecol Econ 116:172–181

Berkes F, Folke C (1998) Linking social and ecological systems: management practices and social mechanisms for building resilience. Cambridge University Press, Cambridge

Berrouet LM, Machado J, Villegas-Palacio C (2018) Vulnerability of socio—Ecological systems: a conceptual framework. Ecol Indic 84:632–647

Booth TH, Williams KJ, Belbin L (2012) Developing biodiverse plantings suitable for changing climatic conditions 2: using the Atlas of Living Australia. Ecol Manage Restor 13(3):274–281

Carpenter SR, Mooney HA, Agard J et al (2009) Science for managing ecosystem services: beyond the Millennium Ecosystem Assessment. Proceedings of the National Academies of Science of the United States of America 106:1305–1312

Castro-Díaz R (2013) Implicancias de la resiliencia espacial en la prestación de servicios ambientales en cuencas norandinas. Contribuciones Científicas. GAEA 25:71–87

Castro-Díaz R (2014) Implicancias territoriales de los esquemas de pago por servicios ambientales (PSA) en cuencas norandinas. Cuadernos de Geografía-Revista Colombiana de Geografía 23(1):61–74

Castro-Díaz R (2017) Epistemología y pragmatismo en el análisis de los sistemas complejos. Revista Latinoamericana de Metodología de las Ciencias Sociales 7(2):e026. https://doi.org/10.24215/18537863e026. Accesed 8 June 2019

Castro-Díaz R, Natenzon CE (2018a) The social vulnerability and ecosystem services feedback: approaching social-ecological analysis in water supply for Andean communities. World Social Science Forum "Security and Equality for Sustainable Futures", Fukuoka

Castro-Díaz R, Natenzon CE (2018b) Análisis de las transformaciones espaciales del suelo y sus implicancias para la provisión de agua en la Laguna de Fúquene, Colombia. In: Díaz I (ed) Servicios Ecosistémicos en Humedales. SEDEMA, Veracruz, pp 181–217

Codina L (2015) No lo llame Análisis Bibliográfico, llámelo Revisión Sistematizada. Y cómo llevarla a cabo con garantías: Systematized Reviews + SALSA Framework. https://www.lluiscodina.com/revision-sistematica-salsa-framework. Accesed 8 Jun 2019

Collins SL, Carpenter SR, Swinton SM et al (2011) An integrated conceptual framework for long-term social–ecological research. Front Ecol Environ 9:351–357

Cumming GS (2011) Conceptual background on social-ecological systems and resilience. In: Cumming GS (ed) Spatial resilience in social-ecological systems. Springer, Dordrecht, pp 7–33

De Sousa Santos B (2006) La sociología de las ausencias y la sociología de las emergencias: para una ecología de saberes. In: De Sousa Santos B (ed) Renovar la teoría crítica y reinventar la emancipación social (encuentros en Buenos Aires). CLACSO, Buenos Aires, pp 13–41

Delgado LE, Marín VH (2005) FES-sistema: un concepto para la incorporación de las sociedades humanas en el análisis medioambiental en Chile. Revista Ambiente y Desarrollo 21(3):18–22

Delgado LE, Marín VH (2016) Well-being and the use of ecosystem services by rural households of the Río Cruces watershed, southern Chile. Ecosyst Serv 21:81–91

Díaz S, Demissew S, Carabias J et al (2015) The IPBES conceptual framework: connecting nature and people. Curr Opin Environ Sustain 14:1–16

Easdale MH, Aguiar MR, Paz R (2016) A social–ecological network analysis of Argentinean Andes transhumant pastoralism. Reg Environ Change 6:2243–2252

Ehrlich PR (1975) The population bomb. Rivercity Press, Rivercity

Escobar A (2000) El lugar de la naturaleza y la naturaleza del lugar: ¿globalización o postdesarrollo? In: Lander E (ed) La colonialidad del saber: eurocentrismo y ciencias sociales. Perspectivas latinoamericanas. Editorial Consejo Latinoamericano de Ciencias Sociales, Buenos Aires, pp 68–87

Estenssoro JF (2007) Antecedentes para una historia del debate político en torno al medio ambiente: la primera socialización de la idea de crisis ambiental. Rev Universum 22:92–111

Ezzine-de-Blas D, Wunder S, Ruiz-Pérez M et al (2016) Global patterns in the implementation of payments for environmental services. PLoS One 11(3):e0149847
Farhad S (2012) Los sistemas socioecológicos. Una aproximación conceptual y metodológica. XIII Jornadas de Economía Crítica, España
Folke C, Han T, Olsson P et al (2005) Adaptive governance of social-ecological systems. Annu Rev Env Resour 30:441–473
Gallopin GC (2004) El Modelo Mundial Latinoamericano ("Modelo Bariloche"): Tres décadas atrás. In: Herrera AO, Sconick HD, Chichilnisky G et al (eds) ¿Catastrofe o Nueva Sociedad? Modelo Mundial Latinoamericano. Centro Internacional de Investigaciones para el Desarrollo, Ottawa, pp 13–26
García R (2006) Sistemas complejos: conceptos, métodos y fundamentación epistemológica de la investigación interdisciplinaria. Ed. GEDISA, Barcelona
Goñi R, Goin F (2006) Marco Conceptual para la Definición del Desarrollo Sustentable. Salud Colectiva 2(2):191–198
Grant MJ, Booth A (2009) A typology of reviews: an analysis of 14 review types and associated methodologies. Health Info Libr J 26:91–108
Gudynas E, Evia G (1991) La Praxis por la Vida—Introducción a las metodologías de la Ecología Social. CIPFE—CLAES—NORDAN, Montevideo
Haberl H, Winiwarter V, Andersson K et al (2006) From LTER to LTSER: conceptualizing the socioeconomic dimension of long-term socioecological research. Ecol Soc 11(2):13. http://www.ecologyandsociety.org/vol11/iss2/art13/. Accesed 8 Jun 2019
Herrera AO, Sconick HD, Chichilnisky G et al (eds) (2004) ¿Catastrofe o Nueva Sociedad? Modelo Mundial Latinoamericano: 30 años después, 2nd edn. Centro Internacional de Investigaciones para el Desarrollo, Ottawa
Himes-Cornell A, Pendleton L, Atiyah P (2018) Valuing ecosystem services from blue forests: a systematic review of the valuation of salt marshes, sea grass beds and mangrove forests. Ecosyst Serv 30:36–48
Hinkel J, Bots PWG, Schlüter M (2014) Enhancing the Ostrom social-ecological system framework through formalization. Ecol Soc 19(3):51
Hinkel J, Cox ME, Schlüter M et al (2015) A diagnostic procedure for applying the social-ecological systems framework in diverse cases. Ecol Soc 20(1):32
Holling CS (1973) Resilience and stability of ecological systems. Annu Rev Ecol Syst 4:1–23
IPBES (2018a) Summary for Policymakers of the Assessment report on Land Degradation and Restoration of the Intergovernmental Science-Policy Platform on Biodiversity and Ecosystem Services. IPBES Secretariat, Bonn
IPBES (2018b) Summary for Policymakers of the Assessment Report on the Intergovernmental Science-Policy Platform on Biodiversity and Ecosystem Services on Regional Assessment for the Americas. Secretariat of the Intergovernmental Science-Platform on Biodiversity and Ecosystem Services, Bonn
IPCC (2018) Summary for policymakers. In: Masson-Delmotte V, Zhai P, Pörtner HO et al (eds) Global warming of 1.5°C. An IPCC special report on the impacts of global warming of 1.5°C above pre-industrial levels and related global greenhouse gas emission pathways, in the context of strengthening the global response to the threat of climate change, sustainable development, and efforts to eradicate poverty. World Meteorological Organization, Geneva, p 1–24
Karlsen J, Larrea M (2015) Desarrollo territorial e investigación acción. Innovación a través del dialogo. Instituto Vasco de competitividad, Serie Desarrollo territorial. Universidad de Deusto
Koch A, Brierley C, Maslin MM et al (2019) Earth system impacts of the European arrival and Great Dying in the Americas after 1492. Quater Sci Rev 207:13–36
Kuhn T (1962) The Structure of Scientific Revolutions. The University of Chicago, Chicago
Leff E (2004) Racionalidad Ambiental. La Reapropiación Social de la Naturaleza. Ed. Siglo XXI, Mexico
Leff E (2010) Economía ecológica, racionalidad y sustentabilidad. Sustentabilidades 2:106–119
Leslie HM, Basurto X, Nenadovic M et al (2015) Operationalizing the social-ecological systems framework to assess sustainability. PNAS 112(19):5979–5984

Lira-Noriega A, Soberón J (2015) The relationship among biodiversity, governance, wealth, and scientific capacity at a country level: disaggregation and prioritization. Ambio 44(5):391–400

Locatelli B, Aldunce P, Fallot A et al (2017) Research on climate change policies and rural development in Latin America: scope and gaps. Sustainability 9(10):1831. http://www.mdpi.com/2071-1050/9/10/1831. Accessed 8 Jun 2019

Maass M, Díaz-Delgado R, Balvanera P et al (2010) Redes de Investigación Ecológica y Socio-Ecológica a Largo Plazo (LTER y LTSER) en Iberoamérica: Los casos de México y España. Revista Chilena de Historia Natural 83(1):171–184

Martínez Alier J (2005) El Ecologismo de los Pobres. Conflictos Ambientales y Lenguajes de Valoración. Icaria, Barcelona

Martinez-Harms MJ, Bryan A, Balvanera P et al (2015) Making decisions for managing ecosystem services. Biol Conserv 184:229–238

Max-Neef MA (1994) Desarrollo a Escala Humana: Conceptos, Aplicaciones y Algunas Reflexiones. lcaria Editorial SA, Barcelona

McGinnis MD, Ostrom E (2014) Social-ecological system framework: initial changes and continuing challenges. Ecol Soc 19(2):30

MEA (2005) Ecosystems and human well-being: scenarios. Island Press, Washington, DC

Meadows DH, Meadows DL, Randers J et al (1972) The limits to growth: a report for the club of Rome's project on the predicament of mankind. Universe Books, New York

Merçon J, Ayala-Orozco B, Rosell JA (2018) Experiencias de colaboración transdisciplinaria para la sustentabilidad: Construyendo lo común. CopIt-arXives, México

Moher D, Liberati A, Tetzlaff J et al (2009) Preferred reporting items for systematic reviews and meta analyses: the PRISMA statement. PLoS Med 6(7):e1000097. https://doi.org/10.1371/journal.pmed1000097

Morin E (2009) Introducción al pensamiento complejo. Gedisa, México

Nisbet EK, Zelenski JM, Murphy SA (2009) The nature relatedness scale: linking individuals' connection with nature to environmental concern and behavior. Environ Behav 41(5):715–740

Odum EP (1953) Fundamentals of ecology, 2nd edn. W. B. Saunders, Philadelphia

Odum HT (1973) Energy, ecology and economics. Ambio 2:220–227

Ostrom E (2007) A diagnostic approach for going beyond panaceas. Proc Natl Acad Sci U S A 104:15181–15187

Ostrom E (2009) A general framework for analyzing sustainability of social-ecological systems. Science 325:419–422

Pahl-Wostl C, Kranz N (2010) Editorial to special issue: water governance in times of change. Environ Sci Policy 13:567–570

Pahl-Wostl C, Holtz G, Kastens B et al (2010) Analyzing complex water governance regimes: the management and transition framework. Environ Sci Policy 13:571–581

Perevochtchikova M (2016) Estudio de los efectos del Programa de Pago por Servicios Ambientales: Experiencia en Ajusco, México. Editorial El Colegio de México, México

Perevochtchikova M (2018) Formalización de un sistema socio-ecológico forestal mexicano. In: Ávila Foucat VS, Perevochtchikova M (eds) Sistemas socio-ecológicos: marcos analíticos y estudios de caso en Oaxaca, México. IIEc-UNAM, México, pp 129–146

Perevochtchikova M, Oggioni J (2014) Global and Mexican analytical review of the state of art on ecosystem and environmental services: a geographical approach. Investig Geogr 85:47–65

Perevochtchikova M, De la Mora G, Hernández Flores JA et al (2019) Systematic review of integrated studies on functional and thematic ecosystem services in Latin America, 1992–2017. Ecosyst Serv 36:1–13. https://doi.org/10.1016/j.ecoser.2019.100900. Accessed 9 June 2019

Pillet F (2004) La geografía y las distintas acepciones del espacio geográfico. Investig Geogr 34:141–154

Postigo JC, Blanco Wells G, Chacón Cancino P (2013) Las ciencias sociales en la encrucijada: el cambio ambiental global en América Latina y el Caribe. In: Informe mundial sobre ciencias sociales, 2013: cambios ambientales globales. UNESCO, Paris, pp 166–177

Roberts RG, Flannery TF, Ayliffe LK et al (2001) New ages for the last Australian megafauna: continent-wide extinction about 46,000 years ago. Science 292(5523):1888–1892

Rodríguez M, Davidson-Hunt IJ (2018) Resilience and the dynamic use of biodiversity in a Bribri Community of Costa Rica. Hum Ecol 46(6):923–931

Schlüter M, Hinkel J, Bots PWG et al (2014) Application of the SES framework for model-based analysis of the dynamics of social-ecological systems. Ecol Soc 19(1):36

Scholz RW, Binder CR (2003) The paradigm of human-environment systems. Working Paper 37. Natural and Social Science Interface. Swiss Federal Institute of Technology, Zürich

Scholz RW, Binder CR (2004) Principles of human-environment systems research. In: Pahl C, Schmidt S, Jakeman T (eds) iEMSs 2004 International Congress: Complexity and Integrated Resources Management. International Environmental Modelling and Software Society, Osnabrueck, pp 791–796

Stojanovic T, McNae H, Tett P et al (2016) The "social" aspect of social-ecological systems: a critique of analytical frameworks and findings from a multisite study of coastal sustainability. Ecol Soc 21(3):15. https://doi.org/10.5751/ES-08633-210315

Svampa M (2008) La disputa por el desarrollo: territorio, movimientos de carácter socio-ambiental y discursos dominantes. In: Svampa M (ed) Cambio de época. Movimientos sociales y poder político. Siglo XXI, Buenos Aires, pp 1–31

Thomas JA, Telfer MG, Roy DB et al (2004) Comparative losses of British butterflies, birds, and plants and the global extinction crisis. Science 303(5665):1879–1881

Weart SR (2008) The discovery of global warming. Harvard University Press, Harvard

Williams NE, Kramer DB (2019) Agricultural biodiversity maintenance in a coastal socio-ecological system: the pearl lagoon basin, Nicaragua. Hum Ecol 47(1):111–120

Environmentalism of the Poor: Environmental Conflicts and Environmental Justice

Mauricio Folchi

Abstract This chapter examines the theory of the 'environmentalism of the poor' proposed by Catalan ecological economist Joan Martínez-Alier and Indian historian Ramachandra Guha. The authors identify two types of environmentalism: one ideological and based on values, found in countries of the global North, and the other materialistic and based on interests, found in the poor countries of the global South. The chapter also offers a critical analysis of the concept of 'environmental conflict', a key element in the theory of the environmentalism of the poor, and finally an assessment of the relationship between the environmentalism of the poor, environmental conflict and the broad concept of 'environmental justice'.

Keywords Political ecology · Enviromentalism · Social-ecological · Environmental conflicts · Environmenta justice

1 Two Differing Environmentalisms

The environmentalist movement achieved international recognition during the late 1960s and early 1970s with the rise of organisations like *Friends of the Earth* (USA 1969), *Les Amis de la Terre* (France 1970), and *Greenpeace* (Canada 1971). A few years later, the movement gained strength with the formation of the first green political parties: *Die Grünen* (Germany 1979), *Ecolo* (Belgium 1980), *Miljöpartiet de Gröna* (Sweden 1981), *Les Verts* (France 1982) and *Os Verdes* (Portugal 1982). These parties and civil society organisations worked towards the inclusion of environmental issues—such as air pollution, acid rain, the nuclear threat, waste management, noise mitigation and wildlife conservation—in the public agenda, and towards

M. Folchi (✉)
Departamento de Ciencias Históricas, Facultad de Filosofía, Universidad de Chile, Santiago, Chile
e-mail: mfolchi@u.uchile.cl

L. E. Delgado, V. H. Marín (eds.), *Social-ecological Systems of Latin America: Complexities and Challenges*, https://doi.org/10.1007/978-3-030-28452-7_6

the formulation of the first environmental policies in their countries (Dryzek et al. 2003).

The environmentalist movement was fuelled by a large number of diverse, but convergent, intellectual ideas and traditions. Some have their roots in the nineteenth century, such as the idea of nature conservation, which led to the creation of the first protected wilderness areas in the USA (Yosemite in 1864 and Yellowstone in 1872). Others are more recent, such as those related to radioactivity and the nuclear threat. These ideas came together to form an ideology which gave the movement a clear political objective: to combat environmental deterioration or, euphemistically, to 'save the planet'. The environmentalism of the 1970s and 1980s was seen as a movement of educated citizens from wealthy countries, concerned with issues far more sophisticated and complex than the traditional clamour of the popular classes over employment, prices and personal security.

In 1977, British sociologist Ronald Inglehart put forward scientific support for this image in the form of his theory of value change in developed societies. Based on the idea of the 'hierarchy of needs' propounded by the North American psychologist Abraham Maslow (1908–1970), Inglehart proposed that the level of material development achieved by the advanced societies of Europe was driving a change in values, leading to an evolution of the political attitudes of citizens. He observed that what he called 'materialist' values, which he defined as those having to do with people's economic and physical security, were increasingly thought to be less important than 'post-materialist' values, the latter being associated with Maslow's idea of self-actualisation, which he had situated at the top of the pyramid of needs. He identified two indicators for the phenomenon: the aesthetic valuation of surroundings ('to make our cities and countryside more beautiful') and the aspirations of participation ('people have more say in how things get decided at work and in their communities') (Inglehart 1977: 40).[1]

Inglehart observed that concern over urban and industrial environmental issues was evident in developed countries, but unevenly applied. In fact, he concludes that 'a concern for 'beauty' is intimately involved only in those countries where economic development and urbanization are relatively far advanced—sufficiently advanced that the public is relatively sensitive to the lack of beauty in the environment' (1977: 48). In the poorer and less-urbanised societies of Europe, such as Ireland and Italy, economic growth was valued relatively highly, with only:

> a weak tendency to feel that it may be detrimental to the beauty of the environment, which is given relatively low priority in any case. The Irish and Italians [...] rank 'Beautiful cities' lower than any public [...]. In Ireland and Italy an anti-Industrial dimension is present, but a concern with environmental beauty does not play a significant part (Inglehart 1977: 48).

According to this theory, *environmentalism*—although Inglehart did not address it in these terms, nor was it the object of his studies—constituted a form of intellectual

[1] This change in the value system begins in the most developed societies, specifically within the best-situated groups socio-economically speaking. It is also clearly seen among the younger generations of these countries, as those people socialised under conditions of peace and relative prosperity would be the most likely to have post-materialist values (Inglehart 1977: 28).

sophistication peculiar to developed countries and, in the words of Guha (1994: 138) and Martínez-Alier (1994: 13) 'a social phenomenon of people with full stomachs, a new luxury fashion for leisure time'. As a consequence, there was little hope that this ideology would spread to countries of the Third World, which had yet to reach the well-being threshold required to embrace post-materialist values. A similar argument was put forward around the same time by economist Lester Thurow (1938–2016):

> If you look at the countries that are interested in environmentalism, or at the individuals who support environmentalism within each country, one is struck by the extent to which environmentalism is an interest of the upper middle class. Poor countries and poor individuals simply aren't interested […]. Environmentalism is a demand for more goods and services (clean air, water, and so forth) that does not differ from other consumption demands […]. From this perspective, environmentalism is a natural product of a rising real standard of living (Thurow 1980: 104–105).

Many opponents made this same Maslowian criticism of environmentalism, as did a number of third world leaders who feared that the new ideology would serve as justification for new forms of intervention in their territories, giving way to the imposition of new mechanisms of control over their natural resources—resources which they had only recently succeeded in nationalising.[2]

The theory of the *environmentalism of the poor* developed by Indian historian Ramachandra Guha and Catalan ecological economist Joan Martínez-Alier in the early 1990s was posed as a rebuttal to that of Inglehart.[3] These authors rejected the idea that the rich are more environmentalist than the poor, or that the poor are too poor to be environmentalists. They proposed the existence of a second type of environmentalism, different to that of the developed world, and found among poor people in poor countries who in both the past and present have striven to 'obtain the ecological requirements for life: energy, water and a place to live' (Martínez-Alier 1994: 239).

According to the environmentalism of the poor, defence of the environment, far from being motivated by abstract ideals or sentiments towards the environment or nature, constitutes a response on the part of the poor—primarily the indigenous and peasant populations of the global South—to a wholly tangible and entirely materialistic situation: the deterioration of the environment in which they live and the consequent impossibility of subsistence. In this sense, the 'environmentalist' component is implicit in their actions. A good example of this was the Brazilian environmentalist Chico Mendes (1944–1988), the renowned rubber tapper union leader

[2] This explains why the declaration made following the first United Nations Conference on the Human Environment, held in Stockholm in 1972, included among its principles the sovereign right of States 'to exploit their own resources pursuant to their own environmental policies', and the consideration that the "extent of the applicability" of environmental standards should be set in accordance with the means of each country, given that "standards which are valid for the most advanced countries [...] may be inappropriate and of unwarranted social cost for the developing countries" (cf. principles 21 and 23).

[3] The allusion to Inglehart is explicit in Martínez-Alier (1994, 1995, 2002), Guha (1994), Gadgil and Guha (1995) and Guha and Martínez-Alier (1997).

from the state of Acre in Brazil, who 'realised only a few years before his murder that he was an environmentalist, even though his life-long opposition to the privatisation and pillage of the Amazon had made him one many years previously' (Martínez-Alier and Roca 2000: 21).

Inglehart considered concern for the environment to be an 'aesthetic' issue situated at the highest point of Maslow's pyramid of needs and, as such, a post-materialist value orientation. By contrast, concern over environmental deterioration according to the environmentalism of the poor is linked to people's basic needs, belonging therefore to the materialist sphere. The health and subsistence of poor rural populations depend on the quality of the environment in which they live, and on their level of access to environmental resources and services. 'Indeed, a clean and safe environment is a need for all humans rather than a luxury good' (Martínez-Alier 2014: 240).

It is ironic to think that the two dimensions identified by Inglehart as materialist values were *physical and economic security* given that they both are so obviously dependent on the quality of the environment, as observed by the theory of the environmentalism of the poor. Also surprising is the fact that Inglehart was apparently unaware of the (materialist) social movements against air pollution and other environmental problems that had taken place in his own United Kingdom and other European countries since the nineteenth century—far before those countries achieved their 'developed' status—in reaction to increasing industrialisation and urbanisation. During the 1830s and 1840s, the English sanitary movement emerged in response to the deplorable environmental health conditions that gripped many working-class neighbourhoods, whose populations had for years been forced to live with inadequate provision of drinking water and sewerage, air polluted by smoke from factories, and the 'miasmas' that rose from stagnant water and the mounting piles of waste (Rosen 1993 [1958]; Flinn 1968). During the same period, the numerous copper foundries of Wales were the target of outcry from local farmers who had witnessed the ruin of their crops by the sulphur dioxide that spewed from industrial chimneys (Rees 1993; Newell 1997). Similarly, the industrial and domestic burning of coal in English cities made atmospheric pollution a perennial problem that was difficult to solve, especially in London. During the winter of 1952, particularly unfavourable wind conditions engulfed the city in a blanket of smoke that did not disperse for a week, causing the deaths of four thousand people and sparking social and political outcry (Wise 2001 [1968]; Brimblecombe 2012 [1987]).

Some years later, as Inglehart extended his study area to the whole world, he was surprised to find that in the countries of Latin America and the former Soviet bloc, much greater value was put on 'beautiful cities'. Rather than revise his theory, Inglehart concluded that in these countries, the basis of such concerns was materialistic and not associated with the aesthetic values found in the West. His explanation for the anomaly was that:

> In these societies, environmental pollution has become a massive and life-threatening problem, far more severe than in the West. In these countries, pollution is not perceived primarily as an aesthetic problem, but one that is directly life-threatening (Abramson and Inglehart 1995: 116).

Besides the underlying methodological problems with this argument, there is undeniable theoretical inconsistency. There is no reason to assume that sulphur dioxide or breathable particulate matter would have a different effect on an organism in the global North compared to one in the South. The same logic is applicable to soil or water contamination, or to nuclear radiation. The effects would be the same all over the world, and those afflicted would be equally interested in securing a solution to their situation wherever they are. It is possible that perceptions of these problems and thresholds of tolerance vary from place to place, as do forms of expression of discontent. However, differences should be explained on a case by case basis, taking into consideration a diversity of factors unrelated to national or personal income levels.

One such factor that exerts an influence over the political attitudes of the citizenry towards environmental problems is, fairly obviously, their value orientation, and this is a theme central to the theory of the environmentalism of the poor. Ramachandra Guha asserts that the main difference between first world environmentalism and the environmentalism of the poor is that first world environmentalists defend 'pure nature' for its intrinsic biological and aesthetic value, while the environmentalism of the poor is a practice of people who depend almost exclusively on the natural resources found in their locality, inspiring them to protect forests, meadows, fishing grounds and other resources key to their survival (Guha 1994; Gadgil and Guha 1995). Guha and Madhav Gadgil dub these groups 'ecosystem people'. Martínez-Alier maintains that there are two approaches that run through the environmentalist movement: the cult of wilderness, and the environmentalism of the poor.[4] The former is based on a love of beautiful landscapes and bestows profound non-material values on nature. This is the root of first world environmentalism, which is characterised by commitments or pledges towards nature. By contrast, 'a material interest in the environmental resources and services provided by Nature for human livelihood characterizes the environmentalism of the poor' (Martínez-Alier 2002: 253).

Upon closer inspection of the issue, we are able to identify three factors that differentiate between the two types of environmentalism: values, the concept of environment, and political action. The first has to do with the long-established dispute within environmentalism between anthropocentric and ecocentric positions. There has been a good deal of interest in this issue from within the field of Environmental Psychology, and it has been proposed that, while both doctrines promote environmental protection, the motivations and behaviours of individuals that subscribe to each one are different (Gagnon and Barton 1994; De Young 2000; Suárez et al. 2007). Anthropocentric individuals consider that the environment should be

[4] Joan Martínez-Alier mentions a third form of environmentalism: 'the gospel of eco-efficiency'. However, the phrase tends to be used more to refer to an environmental school of thought than to a social movement, and has thus been omitted here. The idea has rationalist roots and a utilitarian moral foundation, and stems from early environmental sciences such as modern silviculture. This was the chosen discipline of Gifford Pinchot, one of the first and most distinguished representatives of the school of thought, which today is expressed in concepts such as 'ecological modernisation', 'natural capital', 'ecosystem services' and the controversial 'sustainable development'.

protected for the sake of human well-being and the quality of life of those people who depend on environmental quality and the health of ecosystems. By contrast, ecocentric individuals maintain that nature deserves protection due simply to its own inherent value. For them, environmental preservation is worthwhile regardless of the economic implications or effects on the lifestyle of human beings. For ecocentrists, 'nature has a spiritual dimension and intrinsic value that is reflected in their experiences in nature and feelings about natural settings' (Gagnon and Barton 1994: 150). According to these concepts, first world environmentalism can be taken as a primarily ecocentric movement, with the environmentalism of the poor being more anthropocentric.

Secondly, the concepts of 'environment' as held by the two forms of environmentalism are very different. According to the environmentalism of the poor, the environment is a place in which people live and where they access resources and services that they need to conduct their lives. By contrast, first world environmentalism views the environment as separate from the individual and believes that its conservation is not vital to the life or well-being of an individual or group (although it is insisted that 'the survival of the human race' is under threat). Rather than thinking in terms of the environment of people—which includes artificialised environments in which peasants and the majority of indigenous groups live—these environmentalists are concerned with Nature, and if they defend it, they are driven by their ideology (their aesthetic and ethical post-materialist values) rather than by personal or collective interest. First world environmentalists dream that rivers will once again flow crystal clear, and strives for renewed air purity and the conservation of wild flora and fauna, regardless of the fact that none of these issues threatens or affects them directly. By contrast, in the case of the environmentalism of the poor, people defend not 'the environment', 'nature' or 'planet Earth', but their own habitat, that is, the specific place from which they obtain sustenance and shelter.

The authors mentioned above lament that this elitist and ecocentric view of environmental struggle has become so widespread both within and outside environmentalist circles; it would also be fair to add that this has become the case among academics. More than one author has cited the words of Hugo Blanco, a prominent Peruvian indigenous and peasant leader whose extensive contribution to public life in Peru eventually saw him elected as a Senator, and who wrote in 1991 that:

> environmentalists or conservationists are nice, slightly crazy guys whose main purpose in life is to prevent the disappearance of blue whales and pandas. The common people have more important things to think about, for instance how to get their daily bread [...]. However, there are in Peru a very large number of people who are environmentalists [...]. Isn't the village of Bambamarca truly environmentalist, which has time and again fought valiantly against the pollution of its water from mining? Are not the town of Ilo and the surrounding villages which are being polluted by the Southern Peru Copper Corporation truly environmentalist? Is not the village of Tambo Grande in Piura environmentalist when it rises like a closed fist and is ready to die in order to prevent strip-mining in its valley?.[5]

[5] Article published in *La República*, Lima, 6 April 1991. Cited by (Martínez-Alier 1994: 11; Guha and Martínez-Alier 1997: 24; Martínez-Alier 1998: 26; Guha 2000: 104; Martínez-Alier 2002: 264).

A third difference between the two forms of environmentalism is the political orientation of each. Political Science reminds us that all subjects have values and interests which, in the normal course of events, they share with other people, forming collectives or organisations in an attempt to influence political power. These groups—known as 'pressure groups' or 'interest groups'—may be of a 'promotional' or 'protective' nature. They are open to the whole of society, and promote a cause, a set of ideals, or certain values. They pursue objectives that do not necessarily benefit their own members, but rather society as a whole (Kuper and Kuper 2005: 1010; Jerez 1997). It is clear that first world environmentalism falls into this category.

Meanwhile, protective groups defend the material interests of a certain group of people. Their membership is therefore more limited, although their demands may cover an extensive range of public policy issues. They tend to involve people who carry out similar functions within the socio-economic system (and are in fact often referred to as 'functional groups'), such as workers, business people or peasants, who organise themselves into unions or other associations in order to protect their interests. Formation of these protective groups may also be based on territorial interests. So-called 'geographic groups' emerge when the shared interests of people living in the same location are threatened by projects such as a new highway or rail link. Unlike functional organisations, which are permanent by nature, geographic groups are formed when lives, families and neighbours are threatened; situations like this tend to be temporary (Hague et al. 2016: 306: Briggs 2015: 116). It is clear that the environmentalism of the poor involves this second type of political actor.

2 Environmental Conflicts in the Theory of the Environmentalism of the Poor

Environmental conflicts are a central component of the theory of the environmentalism of the poor. The latter is expressed in the struggles through which people of the global South defend their environment from external aggression. Ramachandra Guha developed the theory based on his research on conflicts sparked by commercial forestry expansion in the upper Ganges basin (the Chipko movement), in the state of Karnataka and other parts of India (Guha and Gadgil 1989; Guha 1994, 2000 [1989]; Guha and Martínez-Alier 1997). Martínez-Alier, for his part, founded the idea on his knowledge of peasant struggles in the Andean regions of South America.[6]

The works of Guha and Martínez-Alier—and of many other authors that subscribe to the theory—offer numerous examples of conflicts of this nature from

[6]Concerning the origin of this society, Martínez-Alier commented that 'Environmentalism of the Poor' was a term that he and Guha began using shortly after they met for the first time in August 1988 in Bangalore. 'We noticed, then, how his work on the Chipko movement fitted with the work I was starting to do on Latin American environmental movements' (Martínez-Alier 2014: 241).

around the world. For the most part they involve the struggles of peasants and indigenous groups, but also of urban dwellers against water, air or soil pollution caused by mining or industrial activity, endeavours to halt the loss of native flora and fauna, and efforts to regain access to water, forests, fishing banks and diverse other resources. However, these abundant conflict cases are not simply a manifestation of the environmentalism of the poor; they point to a bigger issue. If this were not the case, 'this book would merely become an entertaining catalogue of environmental struggles, with a tendency to select anecdotal evidence showing a black-and-white picture of the good guys (and girls) against the bad guys' states Martínez-Alier (2002: 317). At the heart of the theory of the environmentalism of the poor lies the fact that the innumerable local conflicts that occur around the world, both historically and in the present, are an expression of a far greater conflict: the conflict between capitalist economics (or market economics) and the environment.[7]

Continuous growth in economic activity drives economic agents, be they large, private, often transnational companies or State agencies, to occupy new spaces—'the frontiers of extraction advance into new territories as old sources become depleted or too expensive' (Martínez-Alier 2002: 54)—in order to exploit the resources found there and to 'press them into the service of the growing commercial and industrial economy' (Guha 1994: 139) or, to put it another way, to incorporate them into the social metabolism of industrial economies (Martínez-Alier 2002: 54, 2014: 240).

The spatial expansion of the capitalist economy invades rural spaces in which poor people live, usurps their resources and inflicts environmental degradation against which those people rebel.

> The conflict between economy and environment does not manifest itself only in the attacks on remaining pristine Nature but also in the increasing demands for raw materials and for sinks for residues in the large parts of the planet inhabited by humans, and the planet as a whole [...]. The case for a general 'win–win' solution (better environment with economic growth) is far from proven (Martínez-Alier 2002: 317).

The theory of the environmentalism of the poor was strongly influenced by the work of British historian E. P. Thompson (1924–1993), and in particular by his theory of the 'moral economy' (Guha and Gadgil 1989; Guha 2000 [1989]; Martínez-Alier 1990, 1994, 2002; Goebel 2010). Thompson questioned whether the frequent riots and popular uprisings that took place in England during the eighteenth century were simply the consequence of the starvation that periodically beset the population, or whether they arose in response to something more profound. He argued that these protests held certain legitimacy, in that the men and women that became involved did so with the conviction that they were 'defending traditional rights or customs; and, in general, that they were supported by the wider consensus of the community' (Thompson 1971: 78). Up until the eighteenth century, economic practices—in

[7] For the most part, Martínez-Alier uses the name 'economy' as a synonym for various related expressions, such as 'capitalist economy', 'capitalism', 'industrial economy', 'market economy', 'modern economy' and 'industrialising economy'. It would perhaps be more precise to refer to the 'growth economy', or to 'economic growth'.

particular those to do with the circulation of cereals, which comprised the basis of the diet—were guided by a set of moral precepts aimed at ensuring the provision of food for the poor. Once the rules of the 'free market' began to take effect, the masses rebelled. The same occurred as capitalist property relations were imposed in the countryside. Traditionally, poor peasants were entitled to glean, collect firewood and graze their livestock along roads and on stubble, as these resources came under a common regime. The 'enclosures' redefined property, changing the rules of access and destroying the fragile subsistence economy of the poor (Thompson 2012 [1980]: 243–247). This led Thompson to suggest that there was large-scale conflict in eighteenth-century England, and that this stoked riots among the poor.

The theory of the environmentalism of the poor offers a similar assessment. Rural communities attempt to preserve their way of life and forms of occupation of space in the face of advancing capitalism. They fight not only to conserve and maintain access to the ecosystem goods and services upon which they subsist, but they also defend the traditional systems of management of these resources—such as common access—which are also thrown into conflict with expanding market economics, as was the case in eighteenth-century England.

> In the environmentalism of the poor, the most common form of action consists of a rejection of the inclusion of environmental resources in the general market system, in order to keep them within or return them to the non-commercial sphere of the 'moral economy' (Martínez-Alier 1994: 185).

The structural conflict that frames environmental disputes occurs not only between the growth economy and the environment, but between two systems of thought: neoclassical economics and ecological economics. According to Martínez-Alier (2002: 317), 'ecological economics provides the theory on the structural conflict between the economy and the environment'.

This explains why Martínez-Alier refers to environmental conflicts as 'ecological distribution conflicts', or conflicts between 'languages of valuation'. For him, environmental conflict opposes the capitalist economic system (or free market system) and, at the same time, subjects neoclassical economic theory to scrutiny.

A system of production cannot be understood without also considering the distribution of those assets that facilitate that production. The system is organised according to the norms or customs of distribution. Ecological economics sees distribution not as an 'economic' problem (of the distribution of assets and incomes), but as an 'ecological' problem (of distribution of resources and waste). It is the 'agreements or customary arrangements on how to get the natural resources and what to do with the waste' that make certain productive systems possible. The decision to produce cars depends on the possibility of releasing CO_2 into the atmosphere and of dumping old vehicles into landfill. The decision to produce wood pulp depends on the possibility of planting trees and pouring waste into rivers or the sea. Conflicts that occur in response to these production decisions are ecological distribution conflicts, because they stem from a resource and waste distribution framework which is damaging to certain actors—usually the poor (Martínez-Alier 2002: 23–25).[8] As a result, the struggle of the poor constitutes forcing the economic system to internalise the costs, thus correcting the unequal ecological distribution upon which it functions.

However, this leads to a further problem: that of economic valuation. Regardless of the mechanism adopted to address the environmental costs of a productive activity (e.g. ecosystem restoration, compensation of those affected, investment in mitigation technologies, green taxes and so forth), a value must first be put on these costs. Conventional economics knows only one way of doing this: to assign a monetary value to environmental damage. Numerous methods have been created for doing so. Ecological economics has devised other methods of valuation, besides monetary, based on biophysical and social indicators that attempt to respect the numeraires specific to nature and to people.

Valuation of externalities is an epistemological and political problem. The social groups engaged in the conflict defend their rights and interests, resorting to one or other language of valuation. Sometimes they accept the economic language of monetary value, but in other instances they adopt their own languages of valuation which may be unfathomable to the conventional economy, but not so to the ecological economy.

> Environmental struggles resort sometimes to the language of economic valuation, for instance when compensation for externalities is asked for. This is the case in forensic contexts claiming damages. In such a case the bottom line is money. In many other cases, the poor and indigenous have tried to stop degradation of the environment by arguing not in terms of economic costs but in terms of rights (territorial rights, human rights), or in terms of sacredness. Valuation languages are often not translatable into one another. There is no common currency. Commensuration would be an act of power (Martínez-Alier 2014: 241).

3 Environmental Conflicts Beyond the Environmentalism of the Poor

Distinguishing between a materialist environmentalism of the poor in the countries of the global South and a contrasting post-materialist ideological environmentalism in countries of the North may constitute an oversimplification of a complex and heterogeneous social phenomenon. Similarly, a direct association between environmental conflicts and the environmentalism of the poor conceals all of those environmental conflicts that do not conform to the theory, but which nevertheless are relevant.

The theory of the environmentalism of the poor defines environmental conflict (or an ecological distribution conflict) as one which originates from the action of a

[8] Other authors define the 'distributive' nature of ecological conflicts in slightly different terms. They frame it not as an economic problem, but as one of social justice: the unequal distribution of the gifts of nature and of environmental damage.

powerful economic actor who appropriates or degrades a space or resource upon which a poor (Southern) community depends, and which that community resists by defending the environment. Accordingly, environmental conflicts are fought between predefined and asymmetric roles, and are environmentally orientated. The theory of the environmentalism of the poor maintains that the poor have an 'ecological conscience' and commitment to the protection of the environment that may be comparable to that of an ideological environmentalist. Thus, we may conclude that the poor are agents of an economic project different from that of capitalism, focused on sustainability. While this is all possible, it is not necessarily the case.

Firstly, there is no reason to consider that environmental disputes of a 'materialist' nature are exclusive to the poor. Poverty is not a prerequisite for involvement in a conflict of this nature. Given their material origin, these struggles may involve any group of people that feel threatened or are in some way harmed by an intervention in their habitat that affects their material conditions of reproduction (Folchi 2001). Paraphrasing Gadgil and Guha, it is not only the poor who are 'ecosystem people', but all those whose economy and health are dependent upon their habitat. Some of these people live on the limits of subsistence; others have a higher level of income. However, all of them would equally resent—and eventually resist—interventions that may damage or modify their habitat.

Equally, these materialist struggles that the theory of the environmentalism of the poor attributes to the global South could occur in any country of the North. This was the case historically, from the struggle of E. P. Thompson's peasants against the eighteenth-century enclosures, to the urban and rural inhabitants who fought against the externalities of the Industrial Revolution of the nineteenth century and the urban growth of the twentieth. The current situation remains the same, although the threats of today are not the same as those of the past, and the citizens affected (and who move in protest) are no longer poor like their forebears. If we accept that environmental struggles are motivated by materialist factors, there is no reason to think that the poor of the South are alone: the middle classes and the wealthy of both North and South may share such motivations.

Secondly, it should be emphasised that people involved in a 'materialist' environmental struggle are protecting a way of life, and at the same time defending a particular way of relating to their environment or managing resources. These ways of life are not necessarily environmentally sustainable (whatever the definition of this concept may be), nor are they necessarily conducted in complete harmony with the natural world. In fact, we are talking about rural ways of life that involve changes in soil use and the modification of natural balances, changes which—unintentionally—have more than once triggered processes of degradation of the resource base (Blaikie 1985). The claimed—and granted—legitimacy of these struggles lies first and foremost in tradition and custom, not in sustainability.

There are certain environmental conflicts that serve very well to illustrate this lack of environmental orientation, namely those triggered by the creation of protected wild areas within inhabited spaces. These prohibit traditional resource management practices, impose restrictions to access, and may even involve the eviction of the inhabitants themselves (West et al. 2006). In the context of these conflicts, the environmentalism of the poor is unwavering in its alignment with poor peasants whose ways of life are undermined by these conservation initiatives. Ramachandra Guha has hard words for this type of project, and for their ideological foundations (Guha 1989, 1994). It seems clear that the poor are not on the side of environmental preservation when faced with the rulings of conservation biology, opting instead—perfectly reasonably—to oppose environmental initiatives of this kind.[9]

It should also be borne in mind that environmental conflicts are not always the result of environmental pillage or overexploitation of resources. To put it another way, they are not necessarily a response to a state of affairs deemed negative by environmentalist ideology or from the point of view of ecology. If we cast our gaze over the complete history of environmental conflict, we will summarise that the phenomenon is the result of any non-consensual transformation of the environment. The 'positive' or 'negative' nature of a given change to the environment is decided by people (Folchi 2001). Thus, the actions of those involved in an environmental conflict cannot be considered in advance as a form of 'environmentalism'. Thus, people that enter into an environmental conflict do so not to defend the environment or the natural world, but to protect themselves from a threat or from harm to their living conditions. The interests of an affected actor may sometimes coincide with the principles of sustainability, and sometimes not.

That being said, it seems relevant to seek a broader and more inclusive definition of environmental conflict, and to understand it as a situation of confrontation between two or more actors that arises when the action or decision of one threatens the material and symbolic ties established by the other with their environment, threatening their material well-being, security or identity.[10]

4 Environmental Conflicts and Environmental Justice

The theory of the environmentalism of the poor presents three theoretical dilemmas that need to be examined. The first is the contrast between an environmentalism based on values (albeit often controversial)[11] and a material environmentalism dictated by interests (legitimate and noble interests maybe but interests nonetheless).

[9] Joan Martínez-Alier holds an ambiguous position on this. On one hand, he criticises initiatives such as the reintroduction of brown bears native to the Balkans into the Catalan Pyrenees, a move which threatens rural inhabitants engaged in sheep farming (2002: 255). On the other, he is in favour of seeking alliances 'between the interests (and the values) of poor people and the disinterested 'wilderness' values of 'deep ecologists'' (2002: 26).

[10] In a previous work I used the expression "conflicts with environmental content" to refer to environmental conflicts that are not ideologically environmentalist in nature. Other authors (Soto et al. 2007) have proposed similar distinctions. However, given the broad (and ambiguous) use of the expression 'environmental conflict', it seems more appropriate to define the concept clearly rather than create new definitions.

[11] See Guha (1989, 1994) and Guha and Martínez-Alier (1997).

Aside from establishing this dichotomy, the theory suggests that the poor embody the rejection of the industrial economy (anti-environmental, by definition) and, consequently, sees in them a commitment to the preservation of the environment and of the natural world. It also highlights that in their struggles, the poor mobilise 'languages of valuation' of nature and the environment that differ from those of conventional economics. If this is the case, it can only be concluded that the environmental struggles of the poor have an epistemological and value-based dimension that transcends 'material interest'. When Martínez-Alier covers this point, he acknowledges that the environmental struggles of the poor may be based on values, and that this may even be a requirement. He offers the example of the struggles of the U'Wa people in Colombia, who in the late 1990s moved in opposition to oil drilling in their territory. The people consider their lands—including the part below the ground—to be sacred and would not suffer them to be defiled. Once the U'Wa had brought the language of the sacred into the discussion, any use of the language of finance became futile, including, for example, monetary compensation (Martínez-Alier 2002: 253).

Indigenous peoples of Latin America are not limited to values based on 'sacredness'. For the most part they are peasants who, understandably, have values associated with rural life and their productive practices, just the same as other rural inhabitants such as shepherds, small-scale farmers, fishermen and shellfish gatherers. For all of these people, soil, water, animals (domestic and wild), flora (wild and cultivated), rain, tides and everything else hold meaning and value. All these elements are integrated into their culture and ways of life. As a result, every time these ways of life are threatened by interventions in their territory, people will tend to fight them. They do this for material reasons, yes. However, they also act in defence of the values that make sense of these practices and their threatened ways of life. Thus, interests and values are interconnected.

In order to understand how environmental conflicts arise and what fuels them, taking into consideration both interests and values, we may refer to the theory of the 'circle of conflict' proposed by the North American sociologist and politician Christopher Moore (1996). According to his model, there are five underlying causes or 'drivers' which may create or fuel a conflict: relationship, data, value, structural and interest.[12]

Inspired by these concepts, we may imagine an environmental conflict as an iceberg floating in the ocean, with only a small part of its mass protruding above the surface. This part of the conflict visible to the eye is the 'conflict of interest': two actors with opposing and incompatible interests in the environment. Rejection by local people of a power plant that threatens their health; disapproval by peasants of a mining project that will pollute a fertile valley; resistance by a neighbourhood community to a development project that will deny them access to natural

[12] The model makes it clear that diagnosis of these causes is essential, as each one requires a different form of intervention in order to achieve its resolution. For example, if the conflict is rooted simply in a discrepancy of information, a suitable intervention would be to put in place a process of information gathering deemed appropriate by both parties.

surroundings; a dispute between two groups of artisanal fishers over the delimitation of their fishing grounds. These are all examples of 'conflicts of interest'.

Below the waterline we find the 'structural problems' of a country which drive an actor to adopt a course of action that may violate the rights of others. An unclear or lacking definition of property rights, and an overlap (or contradiction) between different legislative bodies that oversee the management of natural resources are common causes of environmental conflict, as too are the unfair laws of economics, urbanism or the environment which deny certain rights to certain groups. This combination of gaps, contradictions and denial drives parties to confront one another, each claiming one legitimate right or other. In cases such as these, the struggle, more than being directed at the circumstantial opponent, points to structural unfairness. If we develop the examples above, the true adversary is not the power company, but the country's power grid; not the mining company, but the system that awards mining concessions; not the real estate developer, but the city's planning regulations; not the fishermen from the next cove, but the fisheries legislation.

In fact, a conflict may be more deeply rooted, concerning, for example, a dispute over the principles that govern social order or, in other words, the values or 'world views' of the opposing actors. This occurs when the action that triggers the conflict is justified by a system of values or a world view that is not shared by the group which will ultimately suffer from that action. In such cases, the conflict transcends the specific interests that gave rise to it, becoming an ideological dispute over questions such as the development model, the organisation of land, or the notion of social justice. Controversy arises around the definition of these concepts and others involved in the discussion, such as 'nature', 'natural resource', 'forest', 'glacier' and 'wetland' (Folchi and Godoy 2016). Reverting once again to the examples used previously, conflicts may at this level constitute a dispute between a sustainable power grid versus an efficient one; a subsistence economy versus a primary and export economy; a high-density urbanisation model versus a low-density model; and the economic function of marine resources versus their social function.

In summary, the difference between material motivations (interests) and ideological motivations (values) makes little contribution to the study of environmental conflicts. Any given conflict may also include structural and ideological components, although often only material interests are explicitly considered from the start. One characteristic of these conflicts is that during the course of events, people involved may recognise or become aware of those deeper layers of the conflict.

A second theoretical dilemma of the theory of the environmentalism of the poor is the inclusion of urban environmental struggles. All conceptualisation behind the theory refers to the rural poor, that is, 'ecosystem people'. Consequently, the majority of conflicts used as examples involve indigenous groups, peasant communities, artisanal fishers, shepherds, and rural inhabitants in general. However, the 'poor of the city' are almost always added to the list of representatives of the environmentalism of the poor, although examples of this type are fewer in number and generally not taken into consideration.

Martínez-Alier proposes that the presence of the environmentalism of the poor in the global South is due to the fact that, in general, people of the North have lost the idea of the environment as their source of livelihood, while the poor and largely rural populations of the South 'are more connected to the environment, and thus have a more intimate understanding of what is at stake by not managing it carefully' (Martínez-Alier 2014: 240). If this is true, the disconnection of a habitat that supplies sustenance is something which affects all urban inhabitants in equal measure, poor or otherwise. In his own words, 'hence the proverbial response of urban children to the question of where does the milk or do the eggs come from —the supermarket' (Martínez-Alier 2002: 26). It is entirely likely that, posed with the question of where water comes from, these children would respond 'from the tap', and electricity 'from power poles'.

So, in what way can the urban poor be compared to the poor found in rural spaces? The poor that live in cities are not sustained by the services of the 'urban ecosystem' (with the exception of those who use waste as fuel and are recyclers, although these are not always poor people). The relationship of the urban poor with the environment is no different to that of other city dwellers, or indeed to that of any other human being. All require a set of satisfactory conditions in order to live, such as air, water, shelter, transportation, recreation and so on. If somebody does something to worsen these conditions, a conflict may arise; however, this act would occur independently of the social condition of those affected, and independently of their place of residence.

Conflicts on the part of the urban poor do not have the same origin or meaning as that involving poor rural inhabitants—Guha and Gadgil's 'ecosystem people'—whose habitats are invaded by economic activities that degrade the environment or usurp their resources. For the most part, the urban poor have never known a healthy habitat which has subsequently suffered degradation as a result of invasion by an external agent. That is not their story. In fact, it is usually the poor who are forced to colonise previously degraded spaces (Been 1994; Mohai and Saha 2015).

The relationship of the urban poor with their environment is functionally no different to that of the rest of the urban population and cannot be compared to the intimate relationship of the rural poor with their own environment. However, this is not to say that the urban poor enjoy the same environmental conditions as the rest of the urban population. It is a well-known fact that environmental problems are distributed unequally in cities, often concentrated in neighbourhoods with an industrial past, or on the margins of the urban zone where land is cheaper and public investment is scarce. For this same reason, it is the poor who tend to live in these neighbourhoods, in the same way that the poor quality of the environment that dominates in those areas becomes an expression of the social injustice with which the poor must contend. This phenomenon is known as 'environmental injustice' and is the third dilemma of the theory of the environmentalism of the poor when it claims that the urban notion of environmental justice and the environmentalism of the poor 'can be understood as one single current' (Martínez-Alier 2002: 13).

In general terms, environmental injustice can be defined as the unequal distribution of environmental costs and benefits between different segments of society. It can also be defined as spatial inequality of environmental quality. The key effect of this concept is to turn the environmental question into one of social justice, by claiming the right to live in an environment free of contamination and with equal access to the environmental goods and services necessary for a decent quality of life.

In some countries, such as the USA, this inequality may also involve racial aspects. Neighbourhoods with greater environmental degradation tend to be home predominantly to black or Latino populations, which explains why the environmental justice movement that originated in this country has strong roots in the civil rights movement.[13] Originally, and for many years, the environmental justice movement in the USA had no links to the country's environmental movement, which was dominated by white, educated people with medium to high income. The latter movement had launched the fight for the environment under the auspices of the 'cult of wilderness' which, as mentioned previously, places the 'environment'—or 'nature'—at the centre of the issue, rather than the habitat of people. For this reason, the movement failed to appreciate the link between environmental deterioration and questions such as race or poverty, until the advent of the environmental justice movement in the early 1980s (Rhodes 2003).

In the cities of Latin America, there is considerable evidence of these spatial inequalities. It is likely that in many cases this has to do with racism, although the suggestion is rarely made. A similar phenomenon to that which gave rise to the environmental justice movement in the USA is the location of pollution sources. There are population centres—such as La Oroya or Ilo (both in Peru), or Ventanas in Chile—which for many years have been affected by atmospheric pollution, primarily sulphur dioxide generated by copper foundries. Inhabitants of these cities have fought—unsuccessfully—for decades to free themselves from this contamination (Dore 2000; Folchi 2006; Li 2015). In other cases, these *pobladores* have been forced to live with the pollution released by petrochemical plants and industrial or domestic refuse sites, such as in Buenos Aires, Tamaulipas and Temuco (Auyero and Swistun 2007; Hernández-Rejón 2014; Castillo 2018). This discomfort—or suffering—has not always given way to open conflict.

However, perhaps the most obvious and large-scale expression of environmental injustice in Latin America is the existence of informal settlements, spawned by urban and rural poverty and by forced displacement. The inhabitants of these settlements are forced to unite and fight for decent living conditions, including the provision of drinking water, sewerage and energy, and access to transport and waste removal services.[14] Furthermore, these settlements are often situated on land considered non-buildable, such as ravines, hillsides and river terraces which may be susceptible to flooding, landslides, forest fires and other disasters. As a result, the residents of these settlements suffer a second form of environmental injustice, that

[13]In 1979, residents of Warren County, North Carolina—the majority of whom being black—unsuccessfully contested the opening of an industrial landfill site. This event is considered to mark the beginning of the environmental justice movement (Taylor 2014).

[14]Cities in which these struggles have been studied include Lima, Maracaibo, Cochabamba, Medellín and Quito (Meneses 2008; Petzold 2010; Linsalata 2014; Zibechi 2015; Gómez and Cuvi 2016).

of having to live in conditions of extreme vulnerability. This situation is also true in the USA, particularly in the wake of Hurricane Katrina (Cigler 2007; Bullard and Wright 2009).[15]

There are clear similarities between the environmentalism of the poor and environmental justice. However, a careful analysis of the two concepts reveals their extent and the differences between them. The struggles of rural inhabitants—the 'ecosystem people'—and the urban poor to improve or retain their standard of living differ both in terms of the relationship that people establish with their environment and the social specifics of that relationship. However, both types may be considered struggles for environmental justice, which is a sufficiently broad category to encompass the environmentalism of the poor, but not the other way around (Anguelovski and Martínez-Alier 2014).

It is important to bear in mind that the concept of environmental justice has grown considerably in recent decades, both geographically and socially, and the same is occurring with its associated movement. It is not only in the USA that references to environmental justice are heard, but all over the world (Pellow 2007; Reed and George 2011; Martínez-Alier et al. 2016; Allen et al. 2017). The concept is no longer applied exclusively to those societies in which environmentally disadvantaged ethnic minorities exist, but also to societies where the population discriminated against for their ethnicity constitutes the majority, such as in South Africa (Bond 2000; McDonald 2002). It is also applied to those societies in which inequality has nothing to do with race, referencing groups defined in non-socio-economic terms (Boone 2008; Anguelovski 2013). Mentions are beginning to be made of water justice, energy justice, and even climate justice (Boelens et al. 2011; Zwarteveen and Boelens 2014; Hall et al. 2013; Borras and Franco 2018). All the struggles of environmentally disadvantaged social groups—urban and rural, from North and South—come together in the environmental justice movement. Each and every one claims the right to live in an environment compatible with their ways of life and complete with the requirements for a decent existence.

Environmental justice has also become broader in theoretical terms. A number of authors have proposed taking the concept beyond the issue of distribution, and directing it towards the analysis of the conditions or mechanisms that produce environmental injustice; in other words, using it to understand the concept of injustice as a 'process' rather than as a 'result' (Pellow 2000, 2001; Schlosberg 2004, 2007; Boone 2008). Schlosberg (2004), for example, argues that justice is a balance of three key interlinked elements: distribution, recognition and participation. The author pinpoints the procedural dimension of environmental justice, as well as its political nature: environmental injustice is the result of a lack of recognition and participation in the decision-making process. This assessment calls for the further broadening of the scope of the environmental justice movement to include groups that are not socially marginalised (Bustos et al. 2017).

[15] Cities studied in this context include San Salvador, Morelia, Guatemala City, São Paulo and Valparaíso (Lungo and Baires 1995; Hernández and Vieyra 2010; Sánchez del Valle 2014; Jacobi et al. 2015; Muñoz et al. 2018).

The approach offers an explanation of why the range of citizen movements that may be seen as expressions of the environmental justice movement tend to make political demands for acknowledgement and participation in decision-making processes to do with the environment (or in those involving environmental factors) and with territorial issues in general. Furthermore, these groups politicise issues of structure and value involved in the specific conflict that motivates their mobilisation: they question environmental institutionalism, territorial and economic organisation, the energy grid and even the distribution of power. During these political disputes, they mobilise and debate ideas and concepts that stem from the different environmental schools of thought and from other ideological sources. Struggles for environmental justice become scenarios in which these ideas can be applied, discussing issues fundamental to environmentalism such as 'sustainability', 'economic development' or the state of the 'natural world'. By sparking these political moments, environmental conflicts bring about the formation of a new type of citizenship: an 'environmentalised' citizenship with socio-environmental or territorial political conscience. These are citizens aware of their relationship with their surroundings and of the territorial rights involved, of environmental problems on different scales, and of the economic or political origins of these problems. If we put together all of these elements, we may conclude that these struggles for environmental justice are really attempts to democratise the environment.

Acknowledgements This chapter was made possible by the support of CONICYT (Chilean National Commission for Scientific and Technological Research) through projects: PIA-Anillo Soc 1404 and FONDECYT 1150770.

References

Abramson P, Inglehart R (1995) Value change in global perspective. University of Michigan Press, Ann Arbor, MI

Allen A, Griffin L, Johnson C (eds) (2017) Environmental justice and urban resilience in the global south. Palgrave Macmillan, New York

Anguelovski I (2013) New directions in urban environmental justice: rebuilding community, addressing trauma, and remaking place. J Plan Educ Res 33(2):160–175

Anguelovski I, Martínez-Alier J (2014) The 'Environmentalism of the Poor' revisited: territory and place in disconnected glocal struggles. Ecol Econ 102:167–176

Auyero J, Swistun D (2007) Expuestos y confundidos. Un relato etnográfico sobre sufrimiento ambiental. Iconos 2(2):137–152

Been V (1994) Locally undesirable land uses in minority neighborhoods: disproportionate siting or market dynamics? Yale Law Rev 103(6):1383–1422

Blaikie P (1985) The political economy of soil erosion in developing countries. Longman Scientific & Technical, London

Boelens R, Cremers L, Zwarteveen M (eds) (2011) Justicia Hídrica. Acumulación, conflicto y acción social. Instituto de Estudios Peruanos, Lima

Bond P (2000) Economic growth, ecological modernization or environmental justice? Conflicting discourses in post-apartheid South Africa. Capital Nat Social 11(1):33–61

Boone CG (2008) Environmental justice as process and new avenues for research. Environ Justice 1(3):149–153
Borras S Jr, Franco J (2018) The challenge of locating land-based climate change mitigation and adaptation politics within a social justice perspective: towards an idea of agrarian climate justice. Third World Q 39(7):1308–1325
Briggs J (2015) Doing politics. Routledge, Abingdon
Brimblecombe P (2012) [1987] the big smoke: a history of air pollution in London since medieval times, 2nd edn. Routledge, Cambridge
Bullard R, Wright B (eds) (2009) Race, Place, and Environmental Justice after Hurricane Katrina. Struggles to Reclaim, Rebuild, and Revitalize New Orleans and The Gulf Coast. Westview Press, Boulder
Bustos B, Folchi M, Fragkou M (2017) Coal mining on pastureland in southern Chile; challenging recognition and participation as guarantees for environmental justice. Geoforum 84:292–304
Castillo M (2018) Pueblo Mapuche y sufrimiento ambiental en el caso de Boyeco. La dimensión socioecológica de la desigualdad en Chile contemporáneo. Antropologías del Sur 5(9):29–43
Cigler B (2007) The "Big Questions" of Katrina and the 2005 great flood of New Orleans. Public Adm Rev 67(1):64–76
De Young R (2000) Expanding and evaluating motives for environmentally responsible behavior. J Soc Issues 56(3):509–526
Dore E (2000) Environment and society: long-term trends in Latin American mining. Environ Hist 6(1):1–29
Dryzek JS, Downes D, Hernes HK et al (2003) Green States and Social Movements. Environmentalism in the United States, United Kingdom, Germany, and Norway. Oxford University Press, Oxford
Flinn MW (1968) Public health reform in Britain. Macmillan, London
Folchi M (2001) Conflictos de contenido ambiental y ecologismo de los pobres: no siempre pobres, ni siempre ecologistas. Ecol Política 22:79–100
Folchi M (2006) Historia Ambiental de las Labores de Beneficio en la Minería del Cobre en Chile, siglos XIX y XX. Dissertation, Universitat Autònoma de Barcelona
Folchi M, Godoy F (2016) La disputa de significados en torno al Proyecto Hidroeléctrico Alto Maipo (Chile, 2007-2015). HALAC 6(1):86–104
Gadgil M, Guha R (1995) Ecology and equity. The use and abuse of nature in contemporary India. Routledge, Abingdon
Gagnon SC, Barton MA (1994) Ecocentric and anthropocentric attitudes toward the environment. J Environ Psychol 14(2):149–157
Goebel A (2010) Ecologismo de los pobres y marginalidad social: vehículos de complementariedad y puentes dialógicos. Reflexiones 89(1):127–142
Gómez A, Cuvi N (2016) Asentamientos informales y medio ambiente en Quito. AREAS. Revista Internacional de. Ciencias Sociales 35:101–119
Guha R (1989) Radical American environmentalism and wilderness perservation. A third world critique. Environ Ethics 11(1):71–83
Guha R (1994) El ecologismo de los pobres. Ecol Política 8:137–151
Guha R (2000) [1989] the unquiet woods: ecological change and peasant resistance in the Himalaya, 2nd edn. University of California Press, Berkeley
Guha R, Gadgil M (1989) State forestry and social conflict in British India. Past Present 123:141–177
Guha R, Martínez-Alier J (1997) Varieties of environmentalism. Essays North and South. Earthscan, London
Hague R, Harrop M, McCormick J (2016) Political science: a comparative introduction, 8th edn. Palgrave Macmillan, New York
Hall SM, Hards S, Bulkeley H (2013) New approaches to energy: equity, justice and vulnerability. Introduction to the special issue. Local Environ 18(4):413–421

Hernández J, Vieyra A (2010) Riesgo por inundaciones en asentamientos precarios del periurbano. Morelia, una ciudad media mexicana. ¿El desastre nace o se hace? Revista de Geografía Norte Grande (47):45–62
Hernández-Rejón EM (2014) Sustentabilidad y calidad de vida urbana. Revista de Comunicación de la SEECI (Special Issue):159–169
Inglehart R (1977) The silent revolution. Changing values and political styles among Western publics. Princeton University Press, Princeton, NJ
Jacobi PR, Silva-Sánchez S, Fracalanza AP (2015) Gobernanza del agua en la ciudad de São Paulo: actores sociales, degradación socioambiental y acciones públicas innovadoras. Am Latina Hoy 69:35–51
Jerez MJ (1997) Los grupos de presión. In: del Aguila R (coord.), Manual de Ciencia Política, Trotta, Madrid, p 291–318
Kuper A, Kuper J (eds) (2005) The social science encyclopedia. Routledge, New York
Li F (2015) Unearthing conflict: corporate mining, activism, and expertise in Peru. Duke University Press, Durham and London
Linsalata L (2014) Ni público, ni privado: común. Prácticas y sentidos de la gestión comunitaria del agua en la zona sur de Cochabamba en Bolivia. In: Composto C, Navarro ML (eds) Territorios en disputa. Despojo capitalista, luchas en defensa de los bienes comunes naturales y alternativas emancipatorias para América Latina. Bajo Tierra Ediciones, México, DF, pp 249–266
Lungo M, Baires S (1995) San Salvador: Crecimiento Urbano, Riesgos Ambientales y Desastres. Alternativas para el Desarrollo 29:4–12
Martínez-Alier J (1990) La interpretación ecologista de la historia socio-económica. Hist Soc 7:137–162
Martínez-Alier J (1994) De la Economía Ecológica al ecologismo popular, 2nd edn. Icaria, Barcelona
Martínez-Alier J (1995) The environment as a luxury good or "too poor to be green"? Ecol Econ 13(1):1–10
Martínez-Alier J (1998) La economía ecológica como ecología humana, Fundación César Manrique, Lanzarote, Islas Canarias
Martínez-Alier J (2002) The environmentalism of the poor a study of ecological conflicts and valuation. Edward Elgar, Northampton, MA
Martínez-Alier J (2014) The environmentalism of the poor. Geoforum 54:239–241
Martínez-Alier J, Roca J (2000) Economía ecológica y política ambiental. Fondo de Cultura Económica, México, DF
Martínez-Alier J, Temper L, Del Bene D, Scheidel A (2016) Is there a global environmental justice movement? J Peasant Stud 43:731–755
McDonald D (2002) Environmental justice in South Africa. Ohio University Press, Athens
Meneses M (2008) El problema del agua y el saneamiento en los asentamientos humanos de Lima Sur 2008. Scientia 12(12):161–191
Mohai P, Saha R (2015) Which came first, people or pollution? Assessing the disparate siting and post-siting demographic change hypotheses of environmental injustice. Environ Res Lett 10(11):1–17
Moore CW (1996) The mediation process: practical strategies for resolving conflict, 2nd edn. Jossey-Bass, San Francisco
Muñoz P, Gascón F, de Armas T (2018) El riesgo socio-ambiental en los procesos de urbanización de América Latina. El caso de Vista al Mar, un campamento emplazado en los cerros de Valparaíso. Quidditas 16(8):25–51
Newell E (1997) Atmospheric pollution and the British copper industry, 1690-1920. Technol Cult 38(3):655–689
Pellow DN (2000) Environmental inequality formation: toward a theory of environmental injustice. Am Behav Sci 43(4):581–601
Pellow DN (2001) Environmental justice and the political process: movements, corporations, and the state. Sociol Q 42(1):47–67

Pellow D (2007) Resisting global toxics: transnational movements for environmental justice. MIT Press, Cambridge
Petzold A (2010) Acceso al agua en asentamientos irregulares de Maracaibo: variantes de un servicio. Estudios del Hábitat 11:7–21
Reed M, George C (2011) Where in the world is environmental justice? Prog Hum Geogr 35(6):835–842
Rees R (1993) The great copper trials. Hist Today 43(12):38–44
Rhodes EL (2003) Environmental justice in America. A new paradigm. Indiana University Press, Bloomington
Rosen G (1993) [1958] a history of public health, 2nd edn. Johns Hopkins University Press, Baltimore
Sánchez del Valle R (2014) Percepción del riesgo a desastres en asentamientos precarios urbanos de la ciudad de Guatemala: apuntes para su abordaje. Avance 4(1):25–30
Schlosberg D (2004) Reconceiving environmental justice: global movements and political theories. Environ Politics 13(3):517–540
Schlosberg D (2007) Defining environmental justice. Oxford University Press, Oxford
Soto S, Herrera A, González-de-Molina M, Ortega A (2007) La protesta campesina como protesta ambiental. Siglos XVIII-XX. Hist Agraria 42:277–302
Suárez E, Salazar ME, Hernández B et al (2007) Qué motiva la valoración del medio ambiente? La relación del ecocentrismo y del antropocentrismo con la motivación interna y externa. Rev Psicol Soc 22(3):235–243
Taylor D (2014) Toxic communities. Environmental racism, industrial pollution, and residential mobility. New York University Press, New York and London
Thompson EP (1971) The moral economy of the English crowd in the 18th century. Past Present 50:76–136
Thompson EP (1993) Customs in common. Penguin Books, London
Thompson EP (2012) [1980] La formación de la clase obrera en Inglaterra. Capitan Swing, Madrid
Thurow L (1980) The zero-sum society. Distribution and the possibilities for economic change. Basic Books, New York
West P, Igoe J, Brockington D (2006) Parks and peoples: the social impact of protected areas. Ann Rev Anthropol 35(1):251–277
Wise W (2001) [1968] killer smog: the World's worst air pollution disaster, 2nd edn. Universe, Lincoln
Zibechi R (2015) Medellín. La ladera grita, resiste y construye. Kavilando 7(1):39–46
Zwarteveen M, Boelens R (2014) Defining, researching and struggling for water justice: some conceptual building blocks for research and action. Water Int 39(2):143–158

A New Environmental Governance

Luisa E. Delgado, Francisco Zorondo-Rodríguez, Pamela Bachmann-Vargas, Carmiña Soto, Veronique S. Avila Foucat, Ricardo A. Gutierrez, Andrea Muñoz-Barriga, and Oscar E. Ferreiro

Abstract At present, there is no unified theoretical framework to deal with environmental governance issues. Consequently, there is a diversity of interpretations of the concept at the public-political arena both nationally and internationally. Recent

L. E. Delgado (✉)
Escuela de Ciencias, Facultad de Ciencias, Universidad de Chile, Santiago, Chile

Fundación Centro Transdisciplinario de Estudios FES-Sistémicos, Santiago, Chile
e-mail: ldelgado@uchile.cl

F. Zorondo-Rodríguez
Departamento de Gestión Agraria, Facultad Tecnológica, Universidad de Santiago de Chile, Santiago, Chile
e-mail: francisco.zorondo@usach.cl

P. Bachmann-Vargas
Environmental Policy Group, Wageningen University, Wageningen, The Netherlands
e-mail: pamela.bachmannvargas@wur.nl

C. Soto
Dirección General de Postgrado y Relaciones Internacionales, Universidad Nacional de Asunción, San Lorenzo, Paraguay
e-mail: csoto@rec.una.py

V. S. Avila Foucat
Instituto de Investigaciones Económicas, Universidad Nacional Autónoma de México, Ciudad de México, Mexico
e-mail: savila@iiec.unam.mx

R. A. Gutierrez
School of Politics and Government, Universidad Nacional de San Martín and CONICET, San Martín, Argentina
e-mail: ricardo.gutierrez@unsam.edu.ar

A. Muñoz-Barriga
Research Group Socio-ecological systems, Faculty of Human Sciences, Pontificia Universidad Católica del Ecuador, Quito, Ecuador
e-mail: amunoz@puce.edu.ec

O. E. Ferreiro
Facultad de Ciencias Agrarias, Universidad Nacional de Asunción, San Lorenzo, Paraguay
e-mail: ferreiro@rieder.net.py

L. E. Delgado, V. H. Marín (eds.), *Social-ecological Systems of Latin America: Complexities and Challenges*, https://doi.org/10.1007/978-3-030-28452-7_7

Latin American efforts have given a step forward conceptualizing environmental governance from the South and systematizing experiences to illustrate a diverse contemporaneous reality. At a regional scale, within the last decades, discursive turns in national policies such as the introduction of the sustainable development concept have triggered an increase in studies and applications of environmental governance (e.g., forest's governance, climate change, marine coastal zones) including the use of the ecosystem services concept. The instrumentation of public actions in relation to environmental governance derives from the states. However, if analyzed with a beyond-the-States view, governance can be understood as a process involving the participation of governmental and non-governmental actors reaching decisions, for mutual benefits, through negotiation processes. However, there is not, still, within the countries of the region, inclusive and participative governance oriented toward the sustainable use of natural resources. Although there are many challenges, in this chapter we discuss two of them: (1) to build an analytical framework to understand the environmental governance modes currently available in Latin America and (2) to generate a new sociopolitical interdisciplinary framework involving both natural and sociopolitical systems as a contribution to a new analytical framework for environmental governance. In other words, new environmental governance for Latin America.

Keywords Social-ecological systems · Latin America · Complexity · Environmental governance · Public policies · Adaptation

1 Introduction

The Dawn of Environmental Concerns in Latin America

The concept of environmental governance has increased its relevance in the twenty-first century, associated with the need for the sustainable management of social-ecological systems. Indeed, a literature search (conducted on February 2019) by means of the Web of Science between the years 2005 and 2019 and using the terms "environmental governance" and "Latin America" as keywords, generated 75 articles. Articles included countries such as Paraguay, Argentina, Peru, Mexico, Costa Rica, Bolivia, Chile, and Guatemala among others. This development has been perceived as a new democratic, participative, and collaborative challenge among social, economic, and political actors of the region (Castro et al. 2015). Environmental governance is today used for the management of social, political, economic, and ecological problems and to deconcentrate power, implementing more efficient and transparent public actions as key elements for equity and wellbeing (Calame 2009; Arnouts et al. 2012). In Latin America, it has also been associated with local-territorial movements related to environmental, social, and ecological problems affecting local populations and, in some cases, tightly related to historical and novel ecosystem's goods and services used for economic subsistence and at times playing important cultural roles (Álvarez and Ther 2016).

Since the end of the 1980s and beginning of the 1990s, there has been a shift in environmental political discourses. These changes have incorporated modifications in the definition of what constitutes a complex social-ecological problem, its definition and structure, and the way to approach it. Examples are the ever-growing use of concepts such as sustainability, biodiversity, integrated evaluation, environmental quality among others. Environmental problems now are defined as associated with social, economic, and technological issues and, therefore, their solution involves fields such as public policies, agriculture, infrastructure, and technology. The main result of these changes is that solving these problems is not the exclusive resort of institutions and agencies related to environmental policies (Leroy and Arts 2014).

This change in public discourses was spread and popularized after the Rio de Janeiro Earth Summit of 1992 when the world generated the bases for a new vision of regional development: sustainable development. This concept can be conceived as a new paradigm that put human beings in the center of modernity, considering economic development as a mean and not as the end in itself, protecting the life of present-day and future generations and recognizing that the integrity of the natural systems is the basis of life on planet Earth. This event opened, both at national and regional levels, discussions on the likelihood of compatibility between development models and the sustainability of social and ecological systems (Calix 2016). Thus, starting in the 1990s the environmental legislation gradually became a multisectoral field, appealing to shared responsibilities among different domains of public policies and posing questions about their coordination and integration. It also represented an opportunity for the civil society to start questioning the role of the public institutions regarding the ecological systems as a debt to be solved.

Environmental governance issues acquired relevance in Latin America only at the beginning of the twenty-first century, mostly due to the advent of social-environmental conflicts. Also, science starts playing an important role in openly discussing environmental issues reaching society at large through reports such as those generated by the Millennium Ecosystem Assessment (MEA 2005).

Castro et al. (2015) conducted a detailed analysis of the different action spheres of public policies and social organizations in Latin America in relation to natural resources and other social issues (e.g., dictatorships). The book describes how civil society develops an environmental awareness through processes of self-empowerment, addressing issues such as culture, life, and endangered environments.

Thus, in the dawn of the new millennium, an inflection point is reached; environmental territorial demands become citizen's concern, including social manifestations on Latin American cities with environmental problems appearing on mass media. One example is the "social-environmental crisis of the Rio Cruces wetland" in southern Chile. This crisis mobilized local-national and international civil society, academia, political and judicial actors participating in environmental governance issues (Delgado et al. 2009; Marín et al. 2018; Delgado et al. 2019).

Environmental Governance

Environmental governance research focuses on the ways in which society organizes to solve dilemmas and to create new opportunities, analyzing the conditions and capacities involved, as well as the intervening social actors and their interactions (Calame 2009). Lately, Latin American countries have occupied key roles in global debates over the causes and solutions of environmental problems such as climate change, biodiversity conservation, and others (Castro et al. 2015). Our region has transformed into an innovation space searching for new alternatives for environmental governance where social movements, governments, and firms may have agreements and disagreements. Inter- and transdisciplinary research, as applied to environmental governance, offers a perspective that connects social and environmental changes with governance issues involving public policies and civil learning (Lemos and Agrawal 2006). Furthermore, even disciplinary science has changed its ontology embracing systemic worldviews and postnormal approaches (see chapter "Postnormal Science and Social-ecological Systems").

Although the contemporary environmental governance concept may have emerged as a neoliberal proposal for non-state management of natural resources and environmental issues, it has been shaped by several disciplines to incorporate new perspectives (Hogenboom et al. 2012). Swyngedouw (2005) points out that changes in government, from closed to open (modern), is associated with the use of new technologies and a re-structuring of democracy parameters. Still, he argues that governance-beyond-the-state may be Janus-faced.

Currently, environmental governance takes into consideration the capacities within each country, its constitutional structure, the type of political regime and government, market conditions, science and civil society with the goal of understanding collective problems and to provide solutions that can even be modified through time (Kooiman 2003). Thus, environmental governance is understood as the establishment, confirmation, or change of institutions to solve conflicts of environmental issues (Paavola 2007; Eakin and Lemos 2006). This perspective is related with the environmental justice concept, that put in the center of the debate the distribution of environmental costs and benefits, and the empowerment of the people that depends only on ecosystem's goods and services (Delgado and Marín 2016).

Castro et al. (2015) define environmental governance as the process of formulating, designing, and executing procedures and practices to configure the access, use and control of natural resources among several actors. Lemos and Agrawal (2006) define it as the set of regulatory procedures, mechanisms, and organizations through which actors influence the actions related to environmental issues.

Sarkki (2017) proposes that environmental governance should consider all structures and processes, political and social, of a given country with sustainable development as its main common goal. McGinnis and Ostrom (2014) and Delgado et al. (2019) further propose that complex and nested social-ecological systems are affected by many forms of governance, that may develop on different scales of time and space, and where those larger constrain the responses at smaller scales. Calame

(2009) mentions that one of the important requirements to achieve a common objective (e.g., sustainability) is to consider the participation of all social actors, also sharing the responsibilities of the negative effects of human actions over the environment. Under these conditions, multi-scale environmental management becomes relevant. From the local scale, with collective and territorial learning, up to the national scale where government responses get coupled to develop strategies for the people.

Thus, the new environmental governance describes and defines a desired social future, representing values of human coexistence as the main objectives of social action (social agenda) such as environmental sustainability (Aguilar 2007). Still, accepting the social valuation of ecosystems implies that social actors may perceive and judge negatively their social condition if they are conscious of their local and global environmental risks. Hence, several experts define our present environmental governance condition as a problem to solve; we may arrive at a solution soon, but we are not there yet (Aguilar 2007; Castro et al. 2015).

2 Structures and Typologies of the Environmental Governance

There are few studies on the typologies and/or structures of environmental governance, including the processes and relationships between social and political actors. The most frequent analysis deals with political changes and the essence of governability in itself.

Arts et al. (2000: 54), proposed the concept of policy arrangement, defined as "the temporary stabilization of the content and organization of a particular policy domain." Represented as a flexible and interlinked tetrahedron, the arrangement is composed of four dimensions: actors and their coalitions, power and resources, the rules of the game, and discourses (Arts et al. 2006). Empirical applications have shown that the policy arrangement framework provides a suitable tool to analyze environmental policy change (Contesse et al. 2018). Afterward, Arnoust et al. (2012) proposed four fundamental elements influencing the ways in which environmental governance will develop in a given country: (1) sociopolitical trends, (2) shock events, (3) adjacent arrangements, and (4) policy entrepreneurs (i.e., those with the capability to generate changes). If we then add elements for good governance, as proposed by the United Nations (Córdova Montúfar 2018), we arrive at a conceptual framework that describes the complexity associated with environmental governance (Fig. 1).

Arnoust et al. (2012) proposed two analytical categories: (a) hierarchical governance and (b) closed co-governance. These two governance typologies consider traditional governmental structures and they belong to the first steps of modernity in our region, during the decades of the 1970s and the 1980s (i.e., old governance). The other stage (new governance) can be characterized by investment in the

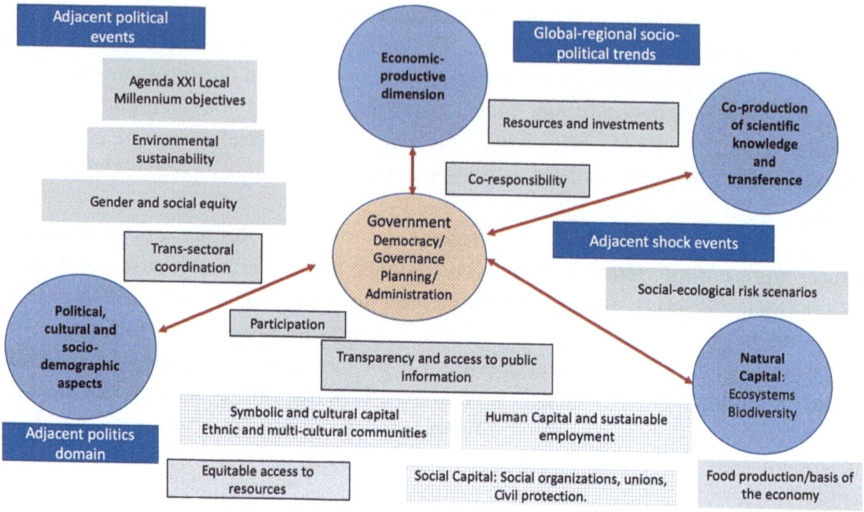

Fig. 1 Pillars, fundamental principles, and capitals for the development of an environmental governance (modified from Arnouts et al. 2012)

innovation of public policies, with a renewed interest in environmental issues in the region at large. This new perspective, that started near the end of the 1980s, is known as "good governance." It propitiates the development of self-government to overcome poverty based on (a) small states, (b) market incentives (e.g., privatization and liberation of resources), and (c) participation (decentralization and NGOs).

Thus, changes in governance can be understood as a historical-relational revolution, both inside and outside a given country, where the necessary elements for a contemporaneous, territorial, environmental governance have been generated. A contemporaneous typology of environmental governance has been proposed by Primmer et al. (2015), where they classified it into four types: (a) *hierarchical*, corresponding to a structure where ideas are transferred from higher to lower political levels; (b) *scientific-technical*, that emphasizes the transference of knowledge from scientists to local social actors, with the associated uncertainty; (c) *adaptive-collaborative*, where the main emphasis is on the participation of local social actors; and (d) *strategic*, with self-organized networks within the civil society with the common good as main objective.

The currently dominant governance type in Latin America is hierarchical, where most of the control lies on state actors. As a result, it has been difficult to incorporate the participation of non-state actors and the civil society, given its top-down dynamic. Yet, states seem to be losing control over the effectiveness of public policies oriented to the management of natural systems, where vulnerable people that depend on the quality of those systems fight to "to keep the resources they need for livelihood" (Martínez-Alier 2014: 241).

Thus, environmental governance modes or types imply understanding public policies beyond pragmatism, since they are the result of the type of institutional system (Córdova Montúfar 2018), the balance of forces among the different social actors, and the capacity of the state to generate contingent policies oriented to the contemporaneous reality.

3 Environmental Governance in Latin America: A Democratic Construction Process

The environmental governance from governmental institutions and their procedures (e.g., law systems, policies, programs, and competencies) are currently highly discussed issues in Latin America. However, the main issues still relate to hierarchical modes of governance (see also chapter "Social-ecological Systems and Human Well-Being"). In the case studies analyzed in this section, we show that there already exist the mechanisms and structures, as well as the necessary social processes to check, from time to time, the governance modes within the region. Still, we propose that Latin America should move toward adaptive strategies including continuous improvements and co-learning (Córdova-Montúfar 2018; Perevchtchikova 2014).

Case Studies

The National Environmental System of Paraguay: An Attempt to Move Toward a New Environmental Governance

Governance, according to Calame (2009), can be understood as a legitimation mechanism, not only of institutions and their rules but also of the actions oriented to common benefits. Thus, a segmented view of reality is no longer possible. The author suggests that it is necessary to change the focus trying to see the interrelationships between the different components since the main challenge is to articulate competencies through networks with a systemic approach (Calame 2009: 37). This is called new governance, governance revolution, and active and/or open governance.

In the year 2000, Paraguay, through the Law N° 1561, reorganizes the public institutions in charge of the environmental management, creating the National Environmental System (SISNAM), The National Council for the Environment (CONAM), and the Secretariat for the Environment (SEAM). The main goal was to manage environmental issues more operationally at the different government levels: central, departments, municipalities. SEAM, an autonomous institution, was created in 2005 with the objectives of formulating, coordinating, executing, and supervising the national environmental policy. CONAM was integrated by several actors from the public domain, businesses and civil society being open to participative governance. Thus, it became the main defining group for the national environmental policy of Paraguay, the operational structure of SISNAM. However, although it was conceived with a systemic view, CONAM operated more like a machine, where environmental issues were attended as linear cause-effect chain structures. Yet, the systemic view (related to the idea of new governance) appeals rather to an "organismic" behavior, including growth, cyclical information guidelines, non-linear interconnections, self-correction, and renovation (Duffield 2001).

Despite its overall behavior, CONAM represented an attempt to include most social actors in environmental issues. Thus, its suppression in 2018 when SEAM acquired a ministry status (Law N° 6123) represents a retrogression in Paraguayan environmental governance.

Environmental Governance in Argentina: A Multi-level Design

The 1994 Argentinian constitutional reform introduced a series of new rights to the original (1853) constitution, including the right to live in a healthy environment. The new Article 41 establishes, among other things, that the Nation is in charge of dictating the norms to provide the minimum protection on issues such as the rational utilization of natural resources, environmental conservation, preservation of natural and cultural patrimony, and the necessary information for environmental education.

Therefore, although the reformed constitution maintains the original domain on the natural resources in the hands of provinces, it also gives power to the National State to dictate the rules on minimum protection throughout the territory. Under these circumstances, it obliges both national and provincial governments to guarantee a healthy environment. The General Law for the Environment (Law N° 25.675 from 2002) defined the concept of minimum environmental protection. This includes political and management instruments such as spatial planning, environmental impact assessment, a control system for the development of anthropic activities, and so on. In particular, it establishes mechanisms for the participation of citizens in environmental impact assessment and territorial ordering procedures.

Finally, the law generates the Federal Environmental System, appointing the preexisting Federal Council for the Environment (COFEMA) as its maximum authority. In this way, the law for the environment generated a complex network of public organizations in charge of protecting the right for a healthy environment and the implementation of environmental policies.

Among the many specific environmental laws, the Law for the Native Forests is the only minimum protection law that applies and combines territorial ordering and citizen's participation mechanisms. Furthermore, it incorporates the ecosystem services concept as criteria for territorial ordering and schemes for the payment of those services (Laterra et al. 2017). Thus, it can be stated that it is the first and only law that establishes environmental governance in the whole country (Aguiar et al. 2018).

When analyzed in terms of institutional governance, the Native Forests Law and its regulation distribute competences and responsibilities between COFEMA, the National State and the provinces. This governance scheme is then combined with three levels for the participation of social organizations, economic actors, and experts generating the conditions for true environmental governance. If we now center on the processes for the formulation of the Territorial Orderings of Native Forests (OTBNs), the law, in practical terms, has worked as a space for the participation and confrontation of social and community organizations, small and large economic actors, and experts of different disciplines and institutions.

Studies conducted in some provinces show that approved OTBN contents and their subsequent regulations can be explained as the result of struggles between production and protection-oriented actors within participation spaces (commissions, fora, etc.) and also within the legislatures (Figueroa and Gutiérrez 2018). OTBNs in those provinces do not fully satisfy social actors and they are even resisted by production-oriented groups. Moreover, people in some provinces have mobilized against OTBN that they did not consider adequate from their points of view. All these issues show that institutional environmental governance does not leave aside the contentious participation of civil society; quite the contrary, it seems to strengthen it and complement it.

Hence, we can state that the Law for the Native Forests suffers from a series of limitations that go beyond citizen's participation mechanisms. Still, official data show that the deforestation rate has been reduced notoriously in Argentina after the implementation, and as a consequence, of this law (Figueroa and Gutiérrez 2018). If that is the case, it means that the established environmental governance arrangement has generated positive results despite its limitations.

Forest Institutions in Chile

In this case study, we analyze the institutional roles related to the forest's management of rural zones. Institutions would determine the process of making decisions, how power is exercised, and how responsibilities are distributed among all stakeholders and social structures in rural communities (Brondizio et al. 2009; Ostrom 2005). Institutions would also define the people's opportunities of access, control, allocation, and distribution of the benefits from ecosystems (Diaz et al. 2015). However, their effectiveness to ensure the sustainable use of ecosystem services depends upon their relationships with rural people, the existence of a decentralized government, and the local ways of life (see chapter "Studying Social-ecological Systems from the Perspective of Social Sciences in Latin America"). As institutions would influence positively and negatively the opportunities to satisfy human wellbeing, the valuation of people over institutions strongly determines the level of effectiveness to manage nature in a sustainable way. Positive perceptions of institutions among rural people would increase their effectiveness, and consequently it would increase the opportunity to maximize both wellbeing satisfaction and sustainable uses (Basurto et al. 2013; Brondizio et al. 2009; Ostrom 2005; Sayer et al. 2013).

Let us consider the valuation of a forest institution in Chile as a study case. The Chilean Forestry Department (CONAF) is the Chilean environmental institution whose duties include overseeing the Native Forest Act obedience, protection of forest ecosystems, and managing the national system of protected areas. The Native Forest Act (MINAGRI 2009) regulates landowners' native forest management (Pellet et al. 2005), and it was created to protect the natural forest that had been replaced and degraded by agricultural use, cattle raising, fires, forestry, urbanization, and an increasing demand for firewood (Pellet et al. 2005). This act includes three types of management plans: (1) Forest Management Plans, (2) Forest Management Plans under order criteria; and (3) Preservation Management Plans. While the act considers economic benefits, these management plans and their benefits are difficult to obtain due to complex application processes, where the amount of money that the owners can receive is smaller than other benefits of the forestry sector, such as for planting exotic species (Reyes et al. 2014). The act also sets monetary penalties depending on the extent of the damage and which species have been affected by an unauthorized cut of native forest. Also, if a landowner does not follow the management plan, they will be fined for noncompliance, losing the benefits obtained and being obliged to return the subsidy received (Reyes et al. 2014).

Forest fragments under management plans would keep a greater natural capital and would offer a larger range of ecosystem services than those forests without plans (Nahuelhual et al. 2007). Therefore, greater social support of forest institutions would facilitate the implementation of management plans in more fragments located outside protected areas or private lands. The establishment of management plans in more forest fragments is a key issue to assure the conservation and protection of highly threatened forest ecosystems. Nevertheless, one of the big gaps that still remain in our knowledge about forest management is identifying the biophysical limits for extraction of forest products in order to achieve the sustainable use of ecosystem services. The biophysical limits should be also defined as a function of the characteristics of the socio-ecological system, where social expectations, perceived costs and benefits, and social beliefs in the institutions should be considered. Unfortunately, the costs, benefits, and social support for forest institutions still do not appear to be reflected in decision-making regarding forest management and conservation in Chile (Nahuelhual et al. 2007).

4 The Multiple Levels of the Environmental Governance

Several sustainability initiatives are generated at multiple levels, from global (e.g., conferences, intergovernmental agreements and actions) to regional and local (Fig. 1). If we add the millennium objectives,[1] wanted in many regions, and national/local initiatives then it becomes a rather complex social-ecological process (see also chapter "Social Actors and Participation in Environmental Issues in Latin America"). These initiatives, according to Ostrom (2009), should be constantly evaluated to monitor their long-term actions and to gradually integrate changes in dominant

[1] https://www.who.int/topics/millennium_development_goals/about/en/

management paradigms and the types of natural resources governance under an integrated transdisciplinary approach (Perevchtchikova 2014).

We should also remember that there are several conditions that should be met for reaching sustainability objectives, related to the characteristics of the time and space levels (or dimensions) where contemporaneous environmental governance operates (top-down, bottom-up, inside-out). On the one hand, global institution's initiatives utilize a unifying approach, that at times resembles interventionism, where the objective is to palliate sustainability problems that, according to them, countries cannot face. On the other, several initiatives are not implemented due to the lack of local capacities and proficiencies (e.g., Aguilar 2007). Thus, except for some virtuous examples, it has not been possible to complement and generate synergies between these two levels (global and local). Therefore, the challenge for Latin America is to overcome the inter-level conflicts in relation to sustainability and its implementation, always paying due consideration to national realities. In other words, implement global objectives with a contextual approach (see also chapter "Studying Social-ecological Systems from the Perspective of Social Sciences in Latin America").

Environmental governance deals with global environmental problems and their local expressions, where risks are distributed between and within the countries of the region (Martinez Alier 2014). Some of them will be more or less vulnerable depending on the state of the ecosystems, the causal structure of the human population, and how much their wellbeing depends on the direct use of nature's goods and services (Natenzon and Ríos 2015; Delgado et al. 2015). This global risk scenario is where environmental governance for Latin America should act, coordinating actions, agents, and actors in several spatial and temporal scales through integrated and sustainable measures.

If social-ecological models are applied only at some spatial scales, the result for environmental governance modes is of partial views with erroneous perspectives of the complex problems of our contemporary world (Delgado et al. 2019). This approach does not allow seeing social-ecological systems as interconnected structures affecting each other. Environmental management should consider social-ecological analyses at several scales so the relationships individual–environment may fit into one another like Russian dolls. In the words of Latour (2005: 180): "There are two different ways of envisaging the macro-micro relationship: the first one builds a series of Russian Matryoshka dolls – the small is being enclosed, the big is enclosing; and the second deploy connections – the small is being unconnected, the big one is to be attached."

In each level (global, national, regional, local), individuals relate with its environment in its multiple dimensions (i.e., biophysical, social, economic, and political; Delgado et al. 2019). The practical application of this concept is known as multi-level governance, currently used in the design of policies in the European Union (Calame 2009). Applying this governance design, which includes indicators and qualitative/quantitative measurements of action, could shed light on how to deal with missing links (i.e., the connection) between levels. Furthermore, this design is supported by the nested, hierarchical, condition of social-ecological systems, where processes operating at large scales restrict those operating at smaller scales (Delgado et al. 2019).

One example is the perception of different groups of social actors in different governance levels in relation to environmental catastrophes and the associated social-ecological resilience. A social group may show good skills to cope with changes, adapting to the point of reaching wellbeing if operating only with social dimensions at a local scale. Meanwhile, regional governmental responses operating at different levels and speeds may still be discussing programs and approaches while the local social-ecological system has already adapted (Delgado et al. 2019). Still, the partial functionality of each level, if others are not considered, may generate tight conclusions and environmental unsustainability (Cruz-Garcia et al. 2017) or generate partial explanations for problems containing several scales.

Dietz et al. (2003) describe multi-level adaptive governance as a strategy to mediate social management conflicts. The author relates governance with the resilience of social-ecological systems mentioning that learning, knowledge generation, learning to organize, cooperation, power, participation, a flexible organization, trust, leadership, social memory, and the formation of groups are key elements to generate adaptive co-management. Brunner et al. (2005) base their adaptive governance proposal in co-management experiences where the local is coordinated and organized with larger scale governance so a desirable social-ecological state may be reached through adaptation or transformation (Chaffin et al. 2014). In summary, when it comes to environmental governance, understanding it from multiple levels or scales is vital.

5 Polycentric and Adaptive Governance: Examples of Connections Between Levels

Polycentric governance can be used as an example of multi-level governance, especially if we refer to resources and ecosystems commonly used by several people such as coastal zones. The main idea beyond this governance scheme is the implementation of multiple decision centers at different scales over a single resource in a given territory (Schöder 2018). However, its polycentrism can be defined in terms of structures and processes, the autonomy of the decision centers, the diversity of organizations, scales and the overlap of functions.

The literature on polycentric governance and the conservation of natural resources cover several systems such as water governance (Baldwin et al. 2018), land–ocean interaction (Pittman and Armitage 2019) and fisheries (Carlisle and Gruby 2018). Baldwin et al. (2018) argue that this type of governance promotes collective actions between different scales. However, collective actions within each scale generated by social capitals allow communities the capacity to organize themselves generating structures for decision-making (Buciega and Esparcia 2013).

On the other hand, polycentric governance is considered as an element that promotes resilience within social-ecological systems, associated with overlapping territorial areas and their interdependent decision-making (Garrick et al 2018; Biggs et al. 2012). The main proposed basis for this resilience is the diversity of social actors who generate multiple responses to deal with an adverse event (either social or ecological). If, on top of these characteristics, we add redundancy and participation then conditions are met to generate co-learning allowing the implementation of adaptive governance.

Chaffin et al. (2014) define adaptive governance as the interactions between actors, networks, organizations, and institutions arising in the common search for a desirable state of a social-ecological system. Such state will then depend upon the social actors and their working mechanisms and strategies (e.g., participative modeling, identification of local sustainability objectives). This bottom-up strategy is built from understanding and knowing the perceptions and valuations of local social actors and their ecosystem knowledge. Furthermore, social-ecological adaptation capacities related, for example, to situations such as fishing closures, the extinction of natural components, or the scarcity of provisioning ecosystem services play an important role.

Adaptive governance, as mentioned by Schultz et al. (2015), is based on flexible collaborations, based, in turn, on knowledge and the decision-making processes involving governmental and non-governmental actors with the objective of negotiating and coordinating the management of social-ecological systems. This strategy provides the opportunity to incorporate traditional ways of life to public programs and policies, incorporating local learning to contribute to the sustainability of these ecosystems (Yu Iwama and Delgado 2018; Álvarez and Ther 2016).

There is a consensus that the success of community resource's management depends on several factors such as the institutional environment, the social cohesion of the involved communities (i.e., social capital), local ecological knowledge, and the degree of interactions between communities and the local markets (e.g., chain values) among others. In the case of complex fisheries in particular (e.g., multiple species, multiple users and communities), the argument is that it is very difficult that an institution based on self-governance and managed exclusively by the users may be a real option. This is mostly due to the market's pressures and potential lack of integration of local social actors with the rest of society that may mine collective management. Still, geographically isolated areas, such as Isla Grande de Chiloé in southern Chile, may be ideal for this type of management processes (Paredes 2019).

6 New Environmental Governance for the South: A Proposal

Latin America, as we have discussed at length in this chapter, seems to be in an inflection point, where the possibility of advancing toward a renewed environmental governance is clearer. In this process, it is important that research may be action-oriented; that is, not only analyzing problems conceptually or from a disciplinary scientific perspective but also considering virtuous practices as gears for adaptive and strategic learning. One virtuous example is the project "Integrating biodiversity conservation and land sustainable management in all bio-regions and biomes from Paraguay- Green commodities" led by the United Nations Development Program-Paraguay.[2] We are convinced that before proposing types of interactions, communication channels and indicators or multiple-levels actions it is necessary to characterize the diversity of governance modes (structures and processes) in relation to the social-ecological system to be managed and the components of each subsystem (social and ecological) targeted for sustainability. This is a way to make visible the different levels if they exist, and the potential strategies (see chapter "Environmental Governance for the Coastal Marine Ecosystem Services of Chiloé Island (Southern Chile)").

We propose that a new or revitalized environmental governance in the region will require giving due consideration to the following, minimal, aspects with the purpose of consolidating an adaptive, strategic, and participative social-ecological process:

1. *A world in constant change*. It should be clear to all of us that we live in a changing world where technology, communication, and real-time learning are tools that can be used for cooperation (i.e., globalization). On the other hand, ecosystems and their functions that maintain life on Earth are also changing, being transformed in some cases into novel ecosystems (see chapter "Social-ecological Complexities and Novel Ecosystems" with a clear adaptation necessity. Under these conditions (i.e., high uncertainty, self-generated threats, global risk), governance should adopt an adaptive strategy to cope with social and natural context at a local-regional level.
2. *Changes in the relationships between actors and their structures*. It is rather important when developing environmental governance that interacting social actors or agents may change their roles. The market should be an agent of change; the government should incorporate interdisciplinary visions; science should accept the co-production of knowledge with other actors; NGOs should be more inclusive; and social actors should incorporate learning, cooperation, and solidarity. In other words, environmental governance means redefining values and objectives. Improving communication, among all proposed changes, is the most important by far. Multi-level environmental governance (i.e., strategic) requires the plural participation of actors, each bringing several value dimensions (e.g., teleological and factual, anthropocentric and ecocentric,) including ways to value ecosystem's components such as intrinsic, relational, or instrumental among others (Piccolo 2017). This improved communication should trigger changes in the actor's roles, generating a co-responsibility view regarding environmental damages and how to face them together.
3. *Considering different spatial and temporal scales of the social and ecological processes*. Social-ecological systems are hierarchical and nested where processes occurring at smaller scales are constrained by those at larger scales

[2] https://greencomoditiesparaguay.org

(Fig. 2). Ecological changes, visible to social actors (e.g., decrease in biodiversity), occur when previous un-noticed damages had already occurred (Delgado et al. 2014). Furthermore, since ecosystems are historical systems and their responses to triggers depend on previous conditions or states (i.e., hysteresis), the development and continuation of systemic, long-term, studies and monitoring is vital. Furthermore, incorporating spatial analyses may help to identify the heterogeneity on which social-ecological processes operate in their different scales.

4. *Decentralizing decision-making*. If we asked social actors about environmental problems, their first responses will most of the time refer to their local systems.

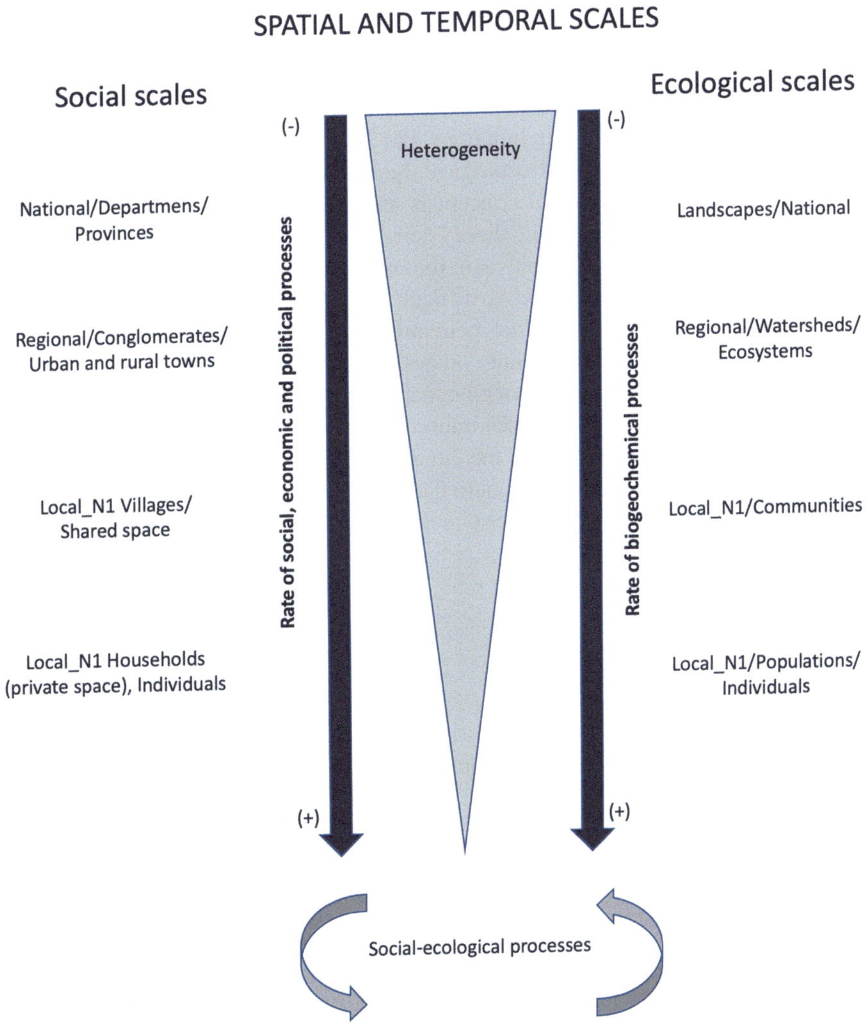

Fig. 2 A scale-dependent conceptual diagram of social-ecological processes

That is, a contextual-territorial response. Therefore, environmental governance should incorporate this territorial dimension and its fundamental principle, active subsidiarity (Calame 2009). In this way, local social actors will gain protagonism both during learning as in decision-making.

7 Final Reflections

The more knowledge advances, based on new theories and experiences, the more our perception of the real world evolves. In the case of environmental governance, advances have been almost revolutionary (sensu Kuhn 1962). The environmental awareness generated by the ecological crises at the end of the twentieth century made mankind shake and re-evaluate our self-generated risks and threats. It also generated a need to change our perspectives regarding the analysis of public policies and their implementation, their functions and actors, and the need to incorporate contextual, and complex, social-ecological dynamics.

The sustainable development concept is associated with intergenerational justice, which according to economic theory corresponds to a social good function that describes social transactions between the wellbeing of different social actors. Beyond the fact that social good is difficult to define (depending on time-space scales), the consideration of future generations requires expanding the focus to include issues such as the uncertainty on desirable conditions for development and the environment. The environmental governance organizes the relationships between humans with sustainability as its common end, reshaping collective responsibility and impact of human actions over the environment. Thus, governance for sustainability generates social empowerment to the local communities together with public and economic actors where the latter two do not have exclusivity over the speech.

References

Aguilar M (2007) Reseña de "Gobernanza y gestión pública". Foro Internacional, El Colegio de México, AC 47(2):443–447

Aguiar A, Mastrangelo ME, García Collazo MA et al (2018) ¿Cuál es la situación de la Ley de Bosques en la Región Chaqueña a diez años de su sanción? Revisar su pasado para discutir su futuro. Ecología Austral 28:400–417

Álvarez R, Ther F (2016) Fragmentos de una Cosmovisión Mesriza Asociada al acceso y uso del entorno costero en el Archipielago de Chiloé. Diálogo Andino 49:123–129

Arnouts R, van der Zouwen M, Arts B (2012) Analysing governance modes and shifts-Governace arragements in Dutch nature policy. Forest Policy Econ 16:43–50

Arts B, Van Tatenhove J, Leroy P (2000) Conclusions and research agenda: political modernisation and the dynamics of environmental policy arrangements. In: Tatenhove J, Arts B, Leroy P (eds) Political modernization and the environment. The renewal of environmental Policy Arrangements. Van Kluwer Academic, Dordrecht, pp 199–215

Arts B, Leroy P, Van Tatenhove J (2006) Political modernization and policy arrangements: a framework for understanding environmental policy change. Publ Organ Rev 6:93–106
Baldwin E, McCord P, Dell'Angelo J et al (2018) Collective action in a polycentric water gobernanza system. Environ Policy Gov 28:212–222
Basurto X, Gelcich S, Ostrom E (2013) The social ecological system framework as a knowledge classificatory system for benthic small-scale fisheries. Global Environ Change Hum Policy Dimen 23:1366–1380
Biggs R, Schlüter M, Biggs D et al (2012) Toward principles for enhancing the resilience of ecosystem services. Annu Rev Env Resour 37(1):421–448
Brondizio ES, Ostrom E, Young OR (2009) Connectivity and the governance of multilevel social-ecological systems: the role of social capital. Annu Rev Env Resour 34:253–278
Brunner R, Steelman T, Coe-Juell L et al (2005) Adaptive governance: integrating science, policy and decision making. Columbia University Press, New York
Buciega A, Esparcia J (2013) Desarrollo, Territorio y Capital Social. Un análisis a partir de dinámicas relacionales en el desarrollo rural. Redes 24(1):1–33
Calame P (2009) Hacia una revolución de la gobernanza: reinventar la democracia. LOM Ediciones, Santiago
Calix JA (2016) Enfoques de Desarrollo en América Latina-Hacia una transformación Social-Ecológica. Análisis 1:1–36
Carlisle KM, Gruby RL (2018) Why the path to polycentricity matters: evidence from fisheries governance in Palau. Environ Policy Gov 28:223–235
Castro F, Hogenboom B, Boud M (eds) (2015) Gobernanza Ambiental en América Latina. Ciudad Autonoma de Buenos Aires. CLACSO, Buenos Aires
Chaffin BC, Gosnell H, Cosens BA (2014) A decade of adaptive governance scholarship. Ecol Soc 19(3):56. https://doi.org/10.5751/ES-06824-190356. Accessed 9 Jun 2019
Contesse M, Van Vliet BJM, Lenhart J (2018) Is urban agriculture urban green space? A comparison of policy arrangements for urban green space and urban agriculture in Santiago de Chile. Land Use Policy 71:566–577
Córdova Montúfar M (2018) Gobernanza y políticas públicas. La seguridad ciudadana de Bogotá y Quito. Editorial Universidad del Rosario/Flacso
Cruz-Garcia GS, Sachet E, Blundo-Canto G et al (2017) To what extent have the links between ecosystem services and human well-being been researched in Africa, Asia and Latin America? Ecosyst Serv 25:201–212
Delgado LE, Marin VH, Bachmann P et al (2009) Conceptual models for ecosystem management through the participation of local social actors: the río cruces wetland conflict. Ecol Soc 14(1):50
Delgado LE, Tironi A, Vila I et al (2014) El humedal del Río Cruces, Valdivia Chile: una síntesis ecosistémica. Lat Am J Aquat Res 42:937–949
Delgado LE, Torres-Gomez M, Tironi A et al (2015) Estrategia de Adaptación Local al cambio Climático para el acceso equitativo al agua en zonas rurales de Chile. Rev Am Latina Hoy 69:113–137
Delgado LE, Marín VH (2016) Well-being and the use of ecosystem services by rural households of the Río Cruces watershed, southern Chile. Ecosyst Serv 21:81–91
Delgado LE, Tironi-Silva A, Marín VH (2019) Sistemas socioecológicos y servicios ecosistémicos: modelos conceptuales para el humedal del Río Cruces (Valdivia, Chile). In: Cerda C, Silva-Rodriguez E, Briceño C (eds) Naturaleza en sociedad: Una mirada a la dimensión humana de la conservación de ls biodiversidad. Editorial OchoLibros, Santiago, pp 177–205
Díaz, S Demissew S, Carabias J et al (2015) The IPBES Conceptual Framework — connecting nature and people. Curr Opin Environ Sustain 14:1–16
Dietz T, Ostrom E, Stern PC (2003) The struggle to govern the commons. Science 302:1907–1912
Duffield M (2001) Las nuevas guerras y el mundo global. La convergencia entre desarrollo y seguridad. Cataratas, Madrid

Eakin H, Lemos MC (2006) Adaptation and the state: Latin America and the challenge of capacity-building under globalization. Glob Environ Chang 16:7–8

Figueroa LM, Gutiérrez RA (2018) Enfrentados por el ambiente: incidencia de las coaliciones sociedad-estado en la protección de bosques nativos. In: Gutiérrez RA (ed) Construir el ambiente: sociedad, estado y políticas ambientales en Argentina. Teseo, Buenos Aires, pp 103–166

Garrick D, Heikkila T, Villamayor-Tomas S et al (2018) The lens of polycentricity: Identifying polycentric governance systems illustrated through examples from the field of water governance. Environ Policy Gov 28:236–251

Hogenboom B, Baud M, de Castro F (2012) Gobernanza ambiental en América Latina: hacia agenda de investigación integradora. Revista del Centro Andino de Estudios Internacionales 12:51–71

Kooiman J (2003) Governing as governance. Sage Publications, London

Kuhn T (1962) The structure of scientific revolutions. University of Chicago Press, Chicago

Laterra P, Martín-López B, Mastrangelo M et al (2017) Servicios ecosistémicos en Latinoamérica: de la investigación a la acción. Ecol Austral 27:94–98

Latour B (2005) Re-assembling the social. An introduction to actor-network theory. Oxford University Press, Oxford

Lemos MC, Agrawal A (2006) Environmental governance. Annu Rev Env Resour 31:297–325

Leroy P, Arts B (2014) Institutional dynamics in environmental governance. Springer, Berlín

Marín VH, Delgado LE, Tironi-Silva A et al (2018) Exploring social-ecological complexities of wetlands of international importance (Ramsar sites): the Carlos Anwandter Sanctuary (Valdivia, Chile) as a case study. Wetlands 38:1171–1182

Martínez Alier J (2014) El ecologismo de los pobres. Editorial Quimantú, Santiago de Chile

McGinnis MD, Ostrom E (2014) Social-ecological system framework: initial changes and continuing challenges. Ecol Soc 19(2):30. https://doi.org/10.5751/ES-06387-190230

Millennium Ecosystem Assessment (MEA) (2005) Ecosystems and human well-being: synthesis report. Millennium Ecosystem Assessment. Island Press, Washington, DC

MINAGRI (2009) Reglamento general de la ley sobre recuperación del bosque nativo y fomento forestal Nro 20.283. Ministerio de Agricultura, Santiago

Nahuelhual L, Donoso P, Lara A et al (2007) Valuing ecosystem services of Chilean temperate rainforests. Environ Dev Sustain 9:481–499

Natenzon C, Ríos D (2015) Una revisión sobre catástrofes, riesgo y ciencias sociales. In: Natenzon C, Ríos D (eds) Riesgos, catástrofes y vulnerabilidades: aportes desde la geografía y otras ciencias sociales para casos argentinos. Ediciones Imago Mundi, Buenos Aires, pp 1–27

Ostrom E (2005) Understanding institutional diversity. Princeton University Press, Princeton NJ

Ostrom E (2009) A general framework for analyzing sustainability of social-ecological systems. Science 325:419–422

Paavola J (2007) Institutions and environmental governance: a reconceptualization. Ecol Econ 63:93–103

Paredes L (2019) Desarrollo de un modelo conceptual para el manejo de servicios ecosistémicos costeros: Isla Grande de Chiloé. Tesis para optar al cargo de Magíster en Gestión y Planificación Ambiental. Facultad de Ciencias Forestales y de la Conservación de la Naturaleza. Universidad de Chile, Santiago

Pellet PF, Ugarte E, Osorio E et al (2005) Biodiversity conservation in Chile, legally enough? The need for mapping the law before deciding. Rev Chil Hist Nat 78:125–141

Perevchtchikova M (ed) (2014) Pago por servicios ambientales en México. Un acercamiento para su estudio. El Colegio de México AC, México DF

Piccolo JJ (2017) Intrinsic values in nature: objective good or simply half of an unhelpful dichotomy? J Nat Conserv 37(1):8–11

Pittman J, Armitage D (2019) Network governance of land-sea social-ecological systems in the Lesser Antilles. Ecol Econ 157:61–70

Primmer E, Jokinen P, Blicharska M et al (2015) Governance of ecosystem services: a framework for empirical analysis. Ecosyst Serv 16:158–166

Reyes R, Sepulveda C, Astorga L (2014) Gobernanza del Sector Forestal Chileno: Tensiones y Conflictos entre las Fuerzas de Mercado y las Demandas de la Ciudadania. In: González DCM, Lara A (eds) Ecología forestal. Bases para el manejo sustentable y conservación de los bosques nativos de Chile. Ediciones Universidad Austral de Chile, Valdivia, pp 693–720

Sarkki S (2017) Governance services: co-producing human well-being with ecosystem services. Ecosyst Serv 27:82–91

Sayer J, Sunderland T, Ghazoul J et al (2013) Ten principles for a landscape approach to reconciling agriculture, conservation, and other competing land uses. Proc Natl Acad Sci U S A 110:8349–8356

Schultz L, Folke C, Österblom H et al (2015) Adaptive governance, ecosystem management, and natural capital: Fig. 1. Proc Natl Acad Sci 112:7369–7374

Schöder NJS (2018) The lens of polycentricity: identifying polycentric governance systems illustrated through examples from the field of water governance. Environ Policy Gov 28:236–251

Swyngedouw E (2005) Governance innovation and the citizen: the Janus face of governance-beyond-the-state. Urban Stud 42:1991–2006

Yu Iwama A, Delgado LE (2018) Acción: Participación Comunitaria en Procesos de decisión en la Conservación del territorio. Cuadernos del Pensamiento Crítico Latinoamericano 56:1–3

A Hierarchical Approach for the Evaluation of Multiple Ecosystem Services

Antonio Tironi-Silva, Marcela Torres-Gomez, and Francisco De la Barrera

Abstract The study of ecosystem services in Latin America has increased exponentially in the last 10 years. During this period, different methodologies have been applied for the identification and valuation of ecosystem services, following experiences designed for developed countries, sometimes ignoring the social and environmental context of each territory where the methods were applied. The objective of this chapter is to review the methodologies applied in Latin America, with particular emphasis on the identification of the providing ecosystems and the particular services that should be considered at each scale of analysis. Finally, based on our findings and the hierarchical theory, we propose a theoretical-methodological framework for the identification and valuation of multiple ecosystem services at different scales.

Keywords Social-ecological systems · Latin America · Complexity · Ecosystem services · Ecosystem delimitation · Multi-level model

1 The Relevance of the Ecosystem Services Approach

Ecosystem services, or the benefits our society obtains from nature, is a recent concept that permeated very quickly through many fields, including academia, governments, international organizations, and private companies. In academia, during the last 15 years, the number of articles written on the subject of ecosystem services

A. Tironi-Silva (✉)
Cienciambiental Consultores, Santiago, Chile
e-mail: atironi@cienciambiental.cl

M. Torres-Gomez
Fundación CTF, Santiago, Chile
e-mail: mtorresg@fen.uchile.cl

F. De la Barrera
Facultad de Arquitectura, Centro de Desarrollo Urbano sustentable (CEDEUS), Universidad de Concepción, Concepción, Chile
e-mail: fdelabarrera@udec.cl

L. E. Delgado, V. H. Marín (eds.), *Social-ecological Systems of Latin America: Complexities and Challenges*, https://doi.org/10.1007/978-3-030-28452-7_8

grew exponentially worldwide, especially after the publication of reports from the Millennium Ecosystem Assessment (Fisher et al. 2009; Balvanera et al. 2012; Delgado and Marín 2015; De la Barrera et al. 2015). This situation is replicated in Latin America, but at lower rates and limited to few services, mostly those related to global benefits such as carbon sequestration or services of regional extent as water regulation (Balvanera et al. 2012; Malinga et al. 2015; Dobbs et al. 2019). Latin America has also been especially active on assessing the link between ecosystem services supply and policies, but big challenges remain for its effective implementation on the national laws and official tools for planning and management (Balvanera et al. 2012). International private multinational companies like Enel, Unilever, and Natura have also included the consideration of the ecosystem services approach on investment plans to assess the "natural" value of their assets (Natural Capital Protocol, WBCSD). Last but not least, even international, supranational organizations are developing a huge global assessment of our biodiversity using an ecosystem services approach for improving biological conservation, environmental management, and sustainability (Mahanty et al. 2013; TEEB 2010; MEA 2005; IPBES 2019). In spite of the great advances in the incorporation of the concept in policies, its practical scope has not reached its maximum potential (Costanza et al. 2017; Daily et al. 2009; Ruckelshaus et al. 2015). Some of the biggest challenges are linked to inconsistencies of the methodological approaches, especially for evaluation, mapping and modeling, the elevated cost of implementing sophisticated methods, the lack of proper institutional frameworks, and the distrust or low understanding of the theoretical approaches behind the concept (Costanza et al. 2017).

All through the world, the ecosystem service approach (ESA) is gaining momentum as a good proxy indicator or conceptual tool to comprehend, plan, and manage the relationships between society and nature, thus becoming the new paradigm for conservation, management, and the assessment of nature's value. This is not trivial since our current development model is structurally failed and needs re-invention in the near future, with a multi- and transdisciplinary, systemic vision, hoping to replace the mechanistic idea of the last century (Capra 1996). In terms of public decisions, this will be a completely new way of managing our relationship with nature.

2 Current Visions of Ecosystem Services Production

The most accepted conceptual model for the provision of ecosystem services is a "cascade" with four fundamental steps (MEA 2005; Potschin and Haines-Young 2011; Spangenberg et al. 2014):

1. *Ecological processes and functions:* There must be an "operative" ecosystem composed of several interacting elements (plant, soil, sun, etc.) providing a function (e.g., wood production).

2. *Human perception:* Humans must have some type of access to these services and the knowledge to use them, with the exception of some services that are provided "automatically" (e.g., regulating services).
3. *Production technology, technique, and knowledge*: Many ecosystem services require at least a basic technology or technique for its effective use (e.g., a pipe for water, an axe for lumber, etc.). Cultural services do not require any "technology" but knowledge about the meaning for a given service, which is usually culturally inherited. Regulating services do not require this step either.
4. *Benefits and values*: Ecosystem services must represent a benefit to the human population that perceives and uses them (Barkmann et al. 2008). Without benefits, there is no value, according to the cascade model.

A complete evaluation of these steps requires knowledge from different disciplines, including natural sciences (ecology, geography, hydrology, etc.), social sciences (economy, sociology, anthropology, etc.), and many others such as history, ethnography, and engineering. Indeed, the interdisciplinarity of the ecosystem services science has been measured and documented (Costanza and Kubiszewski 2012).

The application of these steps, principles, and methodologies of the ESA works globally, but Latin America cannot be analyzed like just another region in the world. The differences are originated mainly on the highly productive and diverse biodiversity still found in the global south—including in Latin America, where foreign investment, mainly in over-exploited areas, stimulates the ever-growing extraction of natural resources, despite high social inequalities persisting for decades and even centuries, in contrast with its highly rich environment, biological diversity, and cultural heritage (Balvanera et al. 2012; Dobbs et al. 2019).

The lack of qualified and updated secondary data in Latin America hinders the chances to evaluate multiple ecosystem services and only few services are evaluated (Martínez-Harms and Balvanera 2012; De la Barrera et al. 2015).

This chapter intends to describe a clear, simple, and replicable conceptual and methodological approach to solve some of the problems of evaluating multiple ecosystem services within a Latin American context that requires rapid responses with scarce secondary data. The proposal is based on our experience on several ecosystem services assessments at local, regional, and national scales. It is our hope to reach a non-technical, global audience of academics, government officials, and private actors.

3 Approaches to Ecosystem Services Studies Worldwide

Among the different ecosystem services approaches, there is the purest, "econometric" line, associated to the Payment for Ecosystem Services (PES), which assumes that the services have a finite and assessable economic value, whose sum corresponds to the value of a delimited extent of nature. Under this scope, every service should have a monetary areal value, and by simply multiplying this unitary value by the assessed area the benefit can be calculated, for any natural land on Earth. This approach is commonly used for some provisioning services (e.g., food from agriculture, wood from forestry, and water from rivers or even from native forests) and it has received considerable attention in Latin America (Balvanera et al. 2012). Its utility is grounded on the simplicity of opposing these monetary benefits against the costs of a given change or project to decide whether to fund/develop/implement the project. There are many limitations to this approach, and one of the most important critic is that it misses half of the real value of nature, since many services do not have an economic value (De Groot et al. 2012).

Another problem with this view is the evident reductionism and anthropocentrism, contrasting with the conceptualization of the symbolic, cultural, infinite and irreplaceable value of nature, found in many places of Latin America, especially for indigenous people (see also Chapter "Studying Social-ecological Systems from the Perspective of Social Sciences in Latin America"). The long-term rooting of different understandings and cosmovisions of nature in Latin America can sometimes limit the application of PES and other approaches coming from classical economic theory.

The mapping of ecosystem services is one of the most frequent methodologies for assessing multiple services, using a matrix that has as its first component the potential provision of services provided by nature elements. These natural elements can be any attribute of biodiversity able to inform about the complex biophysical structures and processes that provide ecosystem services. As the variety of attributes is very high, land cover/uses are frequently employed as proxy variables to represent the most relevant ecosystem processes and services' provision (Harrison et al. 2014). These proxy variables are qualitatively or quantitatively calculated, then estimated by expert knowledge for the ecosystems present in the region under evaluation. The use of secondary information is dominant, especially land cover cadasters, where land cover/use is used as a proxy for ecosystems at large spatial scales (Martínez-Harms and Balvanera 2012; Malinga et al. 2015). Latin America has replicated this methodological pattern but with low development of studies about the spatial distribution of ecosystem services (Malinga et al. 2015; De la Barrera et al. 2015). The diversity of services that can be mapped will depend upon the capacity to generate new primary data (e.g., field measurements, spectral vegetation indices) or the quality of available secondary data (e.g., cadasters, topographical information, databases).

The use of an ecosystem services matrix facilitates the evaluation of multiple services, but requires geodatabases (e.g., land cover or ecosystem maps) of adequate spatial resolution, the participation of experts, common bases of agreement, time to complete large matrices, and a deep understanding of the causal relationships between the biophysical structure and processes of ecosystems and the provision of benefits for the people. Considering all these features, we propose to consider three basic aspects: (1) the delimitation of ecosystems at an adequate scale, (2) the identification of ecosystem services, and (3) the social and economic valuation.

4 Delimitation of Ecosystems for Ecosystem Services Evaluation

The basis of systems theory delivers concepts that are very useful for the delimitation of ecosystems, and it is important to understand them from a practical perspective when it is necessary to define appropriate spatial units in a study of ecosystem services. Ecosystems are complex and hierarchical systems (Jorgensen and Müller 2000; Allen and Star 1982). Complexity is an intrinsic characteristic to any ecosystem and defines how the existence of multiple components and diverse visions of a system cannot be reduced to each other because they represent different aspects of the same ecosystem, associated with different interests and different historical moments (Munda 2004). On the other hand, the hierarchical organization of nature, as a basic philosophical principle of systems theory, establishes that the universe is composed of subsystems (or Holos). They represent functional units stratified in various levels, with a complex structure of vertical and horizontal relationships, including emergent and system-wide properties.

The identification of ecosystems, as the first step of ecosystem services evaluation, needs to deal with these theoretical aspects and the discussion on their arbitrary delimitation based on spatial discontinuities or homogeneities (Tansley 1935; Jørgensen 1992; Marín 1997). This debate generates the responsibility to maintain coherence and be explicit in the description of the ecosystems being addressed, the proper delimitation and identification of the components that are considered part of them, including human groups, and the objectives of the delimitation.

The starting principle for the delimitation and classification of an ecosystem is the identification of spatial units based on the density of relationships and interactions between biotic (biodiversity) and abiotic (climate, hydrology, soil) components. The recognition of interactions as a fundamental principle for the identification of ecosystems acknowledges that homogeneities (or discontinuities) are not absolute but depend on the scale to which it is observed (Klijn and de Haes 1994). Therefore, depending on the scale and the specificity of the ecosystem service to evaluate, a different identification of the ecosystem could be needed.

The homogeneities recognized in the landscape are defined by the observation scale, so the first obstacle to delimit ecosystems is establishing the criteria allowing to identify these homogeneities at different scales (Blasi et al. 2000). The establishment of these criteria allows for the recognition of persistent patterns within a homogeneous unit. However, the factors that establish these criteria are not always suitable for mapping (Blasi et al. 2000). Therefore, the identified factors or variables that determine a homogeneity should facilitate the spatialization of ecosystems. These already spatialized ecosystems should be hierarchized through a classification defined from their spatial and temporal scales (Klijn and de Haes 1994; Blasi et al. 2000). Time considerations will facilitate comparisons of the change in the provision of ecosystem services when ecosystem distribution changes over time (e.g., Montoya-Tangarife et al. 2017 for a Chilean case).

In practical terms, the hierarchical theory provides the basis for decomposing ecosystems in subsystems and thus to simplify their study, considering only the necessary components to delimit and articulate the system to the spatial scale of interest, allowing to define the basic characteristics of the system: (1) the structure or principle of organization of the elements, (2) the external appearance or manifestation of the structure, and (3) the flows of each ecosystem. It is very important to emphasize that only those elements necessary to fulfill the objectives of an assessment should be considered, a decision in which the level of details should necessarily be sacrificed in order to have a more global vision of the system (O'Neill et al. 1989; Wu and David 2002).

In general, the relevant attributes for classifying terrestrial ecosystems are climate, lithology, geomorphology, human activities, soil type, vegetation, and fauna, among others. Klijn and de Haes (1994) have proposed a widely used methodology, where a set of biotic and abiotic variables can be used for the delimitation of ecosystems according to the spatial scale under consideration. For example, at a supraregional or supranational scale, climatic differences are relevant for the delimitation of ecosystems. However, on a smaller scale, such as watershed-scale, the type of climate can be homogeneous, making it unviable to identify heterogeneities based on this variable. At this scale, other variables should be used, for example, geology, soil type, or species communities, to identify heterogeneities or discontinuities for the delimitation of ecosystems. Therefore, an operational delimitation of ecosystems should be able to differentiate homogeneous units that differ between them or that allow finding discontinuities at a spatial or temporal scale relevant to the research objective. Precisely this definition of ecosystem and conceptualization for its delimitation was the one used by the evaluation of the Millennium Ecosystems (MEA 2003).

A functional classification of ecosystems to evaluate ecosystem services requires a great capacity to delimitate an ecosystem using base maps, automatic or manual classification of satellite imagery, or the use of secondary data. The term "functional" refers to its usefulness and suitability for the incorporation of the ESA into environmental and territorial management (Burkhard and Maes 2017), so the data need to be accurate in spatial (and even temporal) resolution, depending on the scale of the analysis. Our experience shows that secondary data of land cover/use, frequently employed as a proxy of ecosystems, are not always useful to evaluate ecosystem services since they were built up for other purposes (e.g., natural resources inventories, biodiversity conservation assessments, etc.). Therefore, the use of existing geodatabases requires an extra effort of adapting the classification schemes in order to evaluate ecosystem services at the proper scale, considering the quality limitation of available data (Burkhard and Maes 2017) and even modifying the spatial resolution by combining databases, performing reclassifications, or adding information. In Europe, the geodatabase of CORINE (Coordination of Information on the Environment) Land Cover (http://land.copernicus.eu/pan-european/corine-land-cover) is frequently used for mapping ecosystem services (e.g., Maes et al. 2012; Burkhard et al. 2012; Koschke et al. 2012). Latin America does not have a similar input data; thus, studies on the ESA need to use or adapt existing schemes or create new classifications (e.g., Montoya-Tangarife et al. 2017; Laterra et al. 2011) based on land cover as input, adding other ecosystems such as wetlands or streams, and other biophysical data as elevation, soil type, geomorphology, etc. (Burkhard and Maes 2017). As it will be seen in the following sections, the classification includes principles and guides of different national and international efforts linked to the delimitation and classification of ecosystems (e.g., MEA (2003) and FEGS-CS (Landers and Nahlik 2013)), and adapts them to obtain a classification suitable for the purposes of this chapter.

5 Identification and Classification of Ecosystem Services

There are different classifications of ecosystem services; three of the most used are MEA (2005), TEEB (2010), and CICES (Haines-Young and Potschin 2011). It is important to note that CICES is considered by many countries and regions as their official classification, even in Latin America (e.g., Chile). This classification includes some services that are provided by very specific biodiversity attributes as steers, bees, palms, or mountain rocks. If the approach followed is rooted in an ecosystem classification, only services provided by the specific ecosystems should be considered. Then, there are services relevant for global, regional, or local scales. Depending on the purpose of the evaluation, different services should be included in the final model. This proposal works over multiple ecosystem services. If only few ecosystem services need to be evaluated or mapped (e.g., less than 5), the definition of the most proper classification is less relevant and other methodological approaches should work better, especially those based on quantifying direct linkages between biophysical structures or processes and provisions of benefits to people.

After having a final classification, delimitation, and mapping of the ecosystems, and at least an initial list of the ecosystem services to valuate, you can start the next step in our proposal, assessing the rate of provision of services by each ecosystem. There are many different methodologies for this, including qualitative and quantitative approaches. After estimating a rate of provision of an ecosystem service in production units by area, calculating the economic value should be straightforward, using one of the many techniques available, including specific studies on the subject for a variety of ecosystems (Himes-Cornell et al. 2018; Burkhard et al. 2012; Cerda and Melo 2019; Vo et al. 2012).

As the use of primary data limits the chances of evaluating multiple ecosystem services, expert knowledge is frequently used for the task of evaluating the capacity of ecosystems to provide services. This means that experts are able to integrate all the complex processes and functions an ecosystem has to perform to provide the services, and even being able to make a semiquantitative estimation in a Likert scale. Despite the weakness of this assumption, expert-based methods are frequently used in the study of ecosystem services (e.g., Montoya-Tangarife et al. 2017; Burkhard et al. 2012; Kopperoinen et al. 2014; Koschke et al. 2012; Vihervaara et al. 2010). Under this technique, the use of apps or online surveys are very helpful to reach a higher number of experts, and facilitate the completion of larger questionnaires, allowing the accumulation of larger amounts of data.

To define the proper scale and facilitate the delimitation of providing ecosystems and choosing the proper valuation method for each ecosystem services, we propose a conceptual model of ecosystem services' provision, aiming to give a blueprint for any ecosystem service assessment.

6 A General Model for Ecosystem Services Provision and Assessment (Fig. 1)

According to our general model, there are three main paths for ecosystem services' provision:

- A non-additive, emergent and "automatic" flow of regulating and supporting ecosystem services, as long as the providing ecosystem maintains its structures and functions. This group of services is at a highest level in the hierarchy of

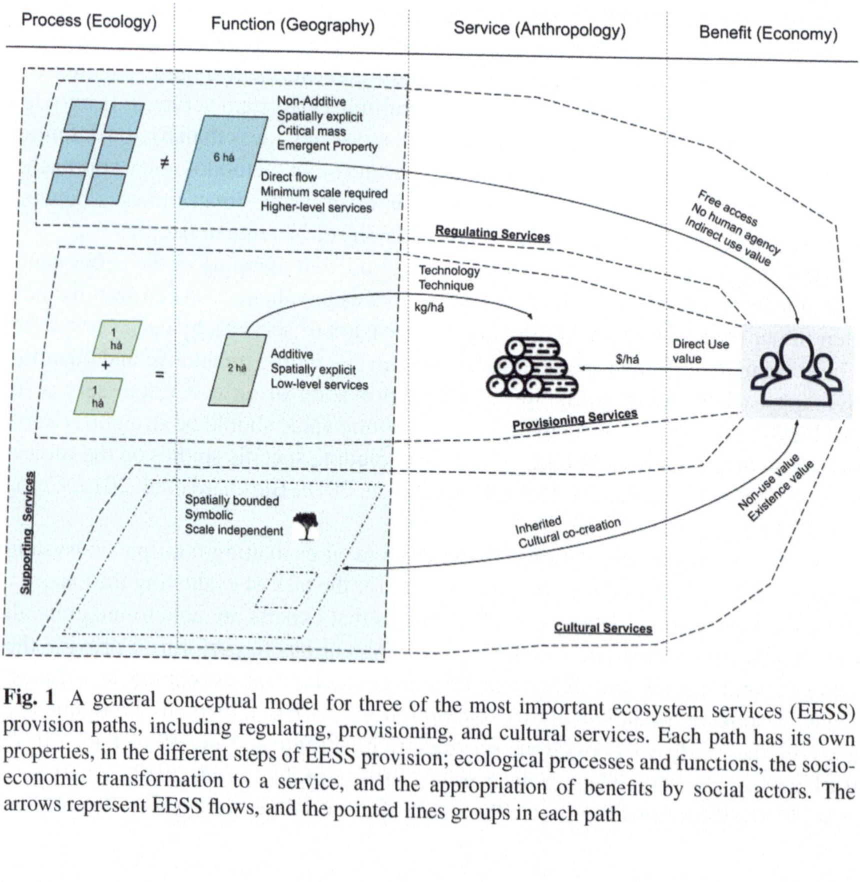

Fig. 1 A general conceptual model for three of the most important ecosystem services (EESS) provision paths, including regulating, provisioning, and cultural services. Each path has its own properties, in the different steps of EESS provision; ecological processes and functions, the socio-economic transformation to a service, and the appropriation of benefits by social actors. The arrows represent EESS flows, and the pointed lines groups in each path

ecosystem services, since they are enabling services for lower level EESS (e.g., the water regulation service is a requirement for the "drinking water provisioning" service). These services only provided at large scales, are spatially explicit, but non-additive, and emerge as regional and global processes. This means that regulating services are undividable into areal values. They can be valued using indirect methods, generally based on the cost of replacement by technology.

- An additive, spatially explicit, linear path of ecosystem service provision, typically observed in provisioning services valuation. These services deliver goods and products historically appreciated by human populations, and provision rates can easily be calculated. These EESS usually have a formal or informal market already established, thus the economic valuation is trivial.
- A symbolic, scale independent, non-spatial flow of cultural ecosystem services. They are only perceived by humans with the required knowledge of the symbolic meaning of the service, a knowledge usually passed by generations through cultural inheritance.

These paths coincide with a characteristic classification of EESS: regulating, provisioning, and cultural services. These three categories are a general rule, so there are many exceptions to this path classification, and there are many hybrid or mixed ecosystem services (e.g., cultural services produced like regulating services by some particular kind of ecosystems).

Ecosystem services' assessment and valuation are one of the best tools available to estimate the real societal value of nature. Our proposal is to include all the dimensions of value, including the cultural, regulating, life-enabling values of many ecosystems. It is urgent to improve the simplistic approach based solely on economic valuation and include in our EESS assessment other sources of value, like culture, tradition, science, and knowledge. It is also important to include the perspective of all relevant social actors, accepting even their different cosmovision, avoiding excessive scientific and cultural impositions. We must understand and accept the limitations of a reductionist approach for ESA, and multi- and transdisciplinary methods must be used to evaluate all three paths of ecosystem services' provisions.

References

Allen TFH, Star TB (1982) Hierarchy: perspectives for complexity. Chicago University Press, Chicago

Balvanera P, Uriarte M, Almeida-Leñero L et al (2012) Ecosystem services research in Latin America: the state of the art. Ecosyst Serv 2:56–70

Barkmann J, Glenk K, Keil A et al (2008) Confronting unfamiliarity with ecosystem functions: the case for an ecosystem service approach to environmental valuation with stated preference methods. Ecol Econ 65(1):48–62

Blasi C, Carranza ML, Frondoni R, Rosati L (2000) Ecosystem classification and mapping: a proposal for Italian landscapes. Appl Veg Sci 3:233–242

Burkhard B, Maes J (2017) Mapping ecosystem services. Adv Books 1:e12837

Burkhard B, Kroll F, Nedkov S et al (2012) Mapping ecosystem service supply, demand and budgets. Ecol Indic 21:17–29
Capra F (1996) He web of life: a new synthesis of mind and matter. HarperCollins
Cerda C, Melo OA (2019) Valoración económica de servicios ecosistémicos y biodiversidad en Chile: marcos conceptuales y experiencias en investigación. In: Cerda C, Silva-Rodríguez E, Briceño C (eds) Naturaleza en sociedad una mirada a la dimensión humana de la conservación de la biodiversidad. Editorial Ocho Libros
Costanza R, Kubiszewski I (2012) The authorship structure of "ecosystem services" as a transdisciplinary field of scholarship. Ecosyst Serv 1(1):16–25
Costanza R, de Groot R, Braat L et al (2017) Twenty years of ecosystem services: how far have we come and how far do we still need to go? Ecosyst Serv 28:1–16
Daily GC, Polasky S, Goldstein J et al (2009) Ecosystem services in decision making: time to deliver. Front Ecol Environ 7(1):21–28
De Groot R, Brander L, Van Der Ploeg S et al (2012) Global estimates of the value of ecosystems and their services in monetary units. Ecosyst Serv 1(1):50–61
De la Barrera F, Bachman-Vargas P, Tironi A (2015) La investigación de servicios ecosistémicos en Chile: una revisión sistemática. Investigaciones Geográficas, Universidad de Chile 50:3–18
Delgado LE, Marín VH (2015) Ecosystem services: where on earth? Ecosyst Serv 14:24–26
Dobbs C, Escobedo FJ, Clerici N et al (2019) Urban ecosystem Services in Latin America: mismatch between global concepts and regional realities? Urban Ecosyst 22(1):173–187
Fisher B, Turner RK, Morling P (2009) Defining and classifying ecosystem services for decision making. Ecol Econ 68(3):643–653
Haines-Young R, Potschin M (2011) Common international classification of ecosystem services (CICES): 2011 update. Report to the European Environmental Agency, Nottingham
Harrison PA, Berry PM, Simpson G et al (2014) Linkages between biodiversity attributes and ecosystem services: a systematic review. Ecosyst Serv 9:191–203
Himes-Cornell AH, Grose SO, Pendleton L (2018) Mangrove ecosystem service values and methodological approaches to valuation: Where do we stand?. Front Mar Sci 5
IPBES (2019) Summary for policymakers of the global assessment report on biodiversity and ecosystem services of the Intergovernmental Science-Policy Platform on Biodiversity and Ecosystem Services. https://www.ipbes.net/document-library-catalogue/summary-policymakers-global-assessment-report-biodiversity-ecosystem. Accessed 12 June 2019
Jørgensen SE (1992) Parameters, ecological constraints and exergy. Ecol Model 62(1–3):163–170
Jorgensen S, Müller F (2000) Ecosystem as a complex system. In: Jorgensen S, Müller F (eds) Handbook of ecosystem theories and management. CRC, Boca Raton, FL, pp 5–19
Klijn F, de Haes HAU (1994) A hierarchical approach to ecosystems and its implications for ecological land classification. Landsc Ecol 9(2):89–104
Kopperoinen L, Itkonen P, Niemelä J (2014) Using expert knowledge in combining green infrastructure and ecosystem services in land use planning: an insight into a new place-based methodology. Landsc Ecol 29(8):1361–1375
Koschke L, Fuerst C, Frank S et al (2012) A multi-criteria approach for an integrated land-cover-based assessment of ecosystem services provision to support landscape planning. Ecol Indic 21:54–66
Landers DH, Nahlik AM (2013) Final Ecosystem Goods and Services Classification System (FEGS-CS). U.S. Environmental Protection Agency, Washington, DC, EPA/600/R-13/ORD-004914
Laterra P, Castellarini F, Orúe ME (2011) ECOSER: Un protocolo para la evaluación biofísica de servicios ecosistémicos y la integración con su valor social. In: Laterra P, Jobbágy EG, Pauelo JM (eds) Valoración de Servicios Ecosistémicos: conceptos, herramientas y aplicaciones para el ordenamiento territorial. Instituto Nacional de Tecnología Agropecuaria, Buenos Aires, pp 359–389
Maes J, Egoh B, Willemen L et al (2012) Mapping ecosystem services for policy support and decision making in the European Union. Ecosyst Serv 1(1):31–39

Mahanty S, Suich H, Tacconi L (2013) Access and benefits in payments for environmental services and implications for REDD+: lessons from seven PES schemes. Land Use Policy 31:38–47
Malinga R, Gordon LJ, Jewitt G et al (2015) Mapping ecosystem services across scales and continents—a review. Ecosyst Serv 13:57–63
Marín VH (1997) General system theory and the ecosystem concept. Bull Ecol Soc Am 78:102–104
Martínez-Harms MJ, Balvanera P (2012) Methods for mapping ecosystem service supply: a review. Int J Biodivers Sci Ecosyst Services Manage 8(1–2):17–25
Millenium Ecosystem Assesment (MEA) (2003) Ecosystem and human wellbeing: a framework for assessment. Millenium Ecosystem Assessment. http://pdf.wri.org/ecosystems_human_wellbeing.pdf. Accessed 10 Jun 2019
Millenium Ecosystem Assesment (MEA) (2005) Ecosystems and human well-being: synthesis. https://www.millenniumassessment.org/documents/document.356.aspx.pdf. Accessed 10 Jun 2019
Montoya-Tangarife C, De La Barrera F, Salazar A et al (2017) Monitoring the effects of land cover change on the supply of ecosystem services in an urban region: a study of Santiago-Valparaíso, Chile. PLoS One 12(11):e0188117
Munda G (2004) Social multi-criteria evaluation: methodological foundations and operational consequences. Eur J Oper Res 158(3):662–677
O'Neill R, DeAngelis D, Waide J et al (1986) A hierarchical concept of ecosystem. Princeton University Press, Princeton, NJ
O'Neill RV, Johnson AR, King AW (1989) A hierarchical framework for the analysis of scale. Landsc Ecol 3(3–4):193–205
Potschin MB, Haines-Young RH (2011) Ecosystem services. Exploring a geographical perspective. Progr Phys Geogr 35(5):575–594
Ruckelshaus M, McKenzie E, Tallis H et al (2015) Notes from the field: lessons learned from using ecosystem service approaches to inform real-world decisions. Ecol Econ 115:11–21
Spangenberg JH, von Haaren C, Settele J (2014) The ecosystem service cascade: further developing the metaphor. Integrating societal processes to accommodate social processes and planning, and the case of bioenergy. Ecol Econ 104:22–32
Tansley AG (1935) The use and abuse of vegetational concepts and terms. Ecology 16(3):284–307
TEEB (2010) The economics of ecosystem services and biodiversity. Earthscan, London
Vihervaara P, Rönkä M, Walls M (2010) Trends in ecosystem service research: early steps and current drivers. AMBIO 39(4):314–324
Vo QT, Kuenzer C, Vo QM, Moder F, Oppelt N (2012) Review of valuation methods for mangrove ecosystem services. Ecol Indic 23:431–446
Wu J, David JL (2002) A spatially explicit hierarchical approach to modeling complex ecological systems: theory and applications. Ecol Model 153(1–2):7–26

Social-ecological Complexities and Novel Ecosystems

Víctor H. Marín and Colin Maxwell Finlayson

Abstract Earth's ecosystems are no longer pristine areas (i.e., historical ecosystems); rather, they have been altered by human beings who have used them as sources of goods and services. While doing this, they have modified their ecological structure and introduced non-native species resulting in emerging or novel ecosystems. In this chapter, we discuss the social-ecological complexities associated with Latin American ecosystems and discuss the need to embrace their novelty.

Keywords Social-ecological systems · Latin America · Complexity · Novel ecosystems · Ecocentrism · Anthropocentrism

1 Ecocentrism v/s Anthropocentrism: A Complex Dichotomy

Nature or Mother Nature, if you consider it "*as a force that affects the world and humans*,"[1] is today the result of what humans have done to our planet over the last 3000 years or even longer (Ellis et al. 2013; Western 2001). This fact was used by Crutzen (2002) to propose that planet earth is in a new geological era: The Anthropocene, a "*human-dominated, geological epoch.*" The planetary dominance of *Homo sapiens* has had several consequences. One of them is shown in the results of the Millennium Ecosystem Assessment (MEA 2005: 1), a planetary analysis based on the concepts of ecosystem services and human well-being, whose findings were summarized as: "*Over the past 50 years, humans have changed ecosystems more rapidly and extensively than in any comparable period of time in human history,…*"

[1] https://www.oxfordlearnersdictionaries.com/definition/english/mother-nature?q=Mother+Nature

V. H. Marín (✉)
Laboratorio de Modelación Ecológica, Departamento de Ciencias Ecológicas, Facultad de Ciencias, Universidad de Chile, Santiago, Chile
e-mail: vmarin@uchile.cl

C. M. Finlayson
Institute for Land, Water and Society, Charles Sturt University, Albury, NSW, Australia
e-mail: mfinlayson@csu.edu.au

L. E. Delgado, V. H. Marín (eds.), *Social-ecological Systems of Latin America: Complexities and Challenges*, https://doi.org/10.1007/978-3-030-28452-7_9

Furthermore, reversing ecosystem degradation will not be easy and would need to involve social and political changes worldwide. The problem, according to Western (2001), is that most changes that humanity has made to our ecosystems have been intentional with many beneficial outcomes for an expanding population, nonetheless they have had important side effects which now have adverse consequence for people (Gordon et al. 2010). Of course, one of the questions engendered by these statements is: should we return to a more natural, pristine, condition or should we accept that we have modified nature beyond the possibilities of returning to what existed before human domination? Kopf et al. (2015) suggest that for aquatic ecosystems, this is necessary given the impossibility of returning to historical baselines.

If we all had the same answer to this question, issues associated with biological conservation and the creation of nature sanctuaries and national parks would be much simpler than in fact they are today. For example, Protected Planet is an information source on protected areas globally, managed by the United Nations Environment World Conservation Monitoring Centre (UNEP-WCMC) with support from the International Union for Conservation of Nature (IUCN) and its World Commission on Protected Areas (WCPA). We obtained the latest available dataset (UNEP-WCMC and IUCN 2017) from its online platform and analyzed the information for Latin America (Table 1).

Table 1 Protected areas, marine and terrestrial, in Latin American countries as percentages of total territorial areas (Source: UNEP-WCMC and IUCN 2017)

Country	% of total territorial areas
Argentina	7.41
Bolivia	30.87
Brazil	28.58
Chile	27.04
Colombia	15.69
Costa Rica	3.04
Ecuador	14.96
El Salvador	2.15
Guatemala	10.12
Guyana	5.30
Honduras	10.86
México	19.06
Nicaragua	15.51
Panamá	5.23
Paraguay	14.31
Perú	13.14
Suriname	8.48
Uruguay	2.30
Venezuela	36.9
Average	**14.26**
World average	**14.49**

Protected areas for Latin American countries range from 2.15% of total territorial area for El Salvador to 30.87% for Bolivia. The average value of 14.26% is very close to the world average of 14.49%. One way to explain the diversity of values in Table 1 is to invoke the dichotomy between ecocentric and anthropocentric approaches to nature. Ecocentrism considers nature for its intrinsic value, regardless of its use by humans, while anthropocentrism values it instrumentally in terms of its utility to humans (Binder et al. 2013). Thus, societies and/or governments closer to an ecocentric perspective may want to increase protected areas, while those closer to an anthropocentric point of view may not necessarily see the urgency of increasing such areas. Others have proposed a third way to understand human/nature values beyond the dichotomy (Piccolo 2017), with some reflecting on justice to humans and non-humans (Kopnina 2016). Finally, others, such as Sessions (1999), have proposed that the origin of the dichotomy can be found at the beginning of human developments on Earth. According to Sessions (p. 140), ancient hunting-gathering cultures "*were permeated with Nature-oriented religions that expressed the ecocentric perspective*" that were replaced by anthropocentric perspectives when agriculture started being dominant.

In Latin America there still are practitioners of ancient, nature-related, ceremonies, mostly from ethnic communities, that are living examples of ecocentrism. For example, the Mapuche culture (Argentina and Chile) celebrates the "Nguillatún,"[2] a ceremony where a leader prays for abundant crops and food. Another example is "The flowering of the llamas,"[3] practiced by Aymaras and Quechuas (Bolivia, Chile, Perú). Cattle for these ancient societies, especially llamas, are part of the family; thus, once a year, women build flower arrangements putting them in llamas' heads and celebrate all day long in order to create a bond between humans and non-humans. In Brazil, more specifically in Salvador da Bahia, they celebrate the festivity of Iemanjá (goddess of the sea), an afro-brazilian (Yoruba) festivity related to the productivity of rivers and estuarine systems.[4]

Nevertheless, as stated on chapter "Postnormal Science and Social-ecological Systems", 80% of Latin American people live and were most likely born in urban areas. Thus, contact with nature in the last 50 years, and even longer, has been scanty, except perhaps for weekends on urban green areas. However, this activity could in fact be perceived as an anthropocentric view of nature, as a provider of recreation ecosystem services. In this situation, the social-ecological concept may come in handy. Kopnina (2016), referring to ecological conservation, states in her conclusion (p. 425) that: "*the issue of conservation cannot be seen as purely ecological because it always involves people,*" and Piccolo (2017), discussing intrinsic values of nature, proposes that we need to pay more attention to relational values. The social-ecological concept was indeed proposed to handle those issues. However, we have already explained (see chapter "Postnormal Science and Social-ecological Systems") that the concept is now 21 years old and its incorporation in Latin

[2] http://www.profesorenlinea.cl/chilehistoria/MapucheGuillatun.htm

[3] https://soniariveracea.blogspot.com/2005/11/el-floreo-tradicion-aymara.html

[4] https://en.wikipedia.org/wiki/Yemoja#Brazil

America has been rather slow, even in terms of scientific research. One proposition to understand this problem is that human societies often take a long time to incorporate what science recommends. For example, when Tansley (1935) proposed the ecosystem concept, it incorporated humans as components. Still, 66 years later O'Neill (2001: 3282) needed to remind us that: "*Homo sapiens is not an external disturbance, it is a keystone species within the system.*" Why? Back in 1935 the total human population was close to 2.19 billion people, growing to 7.63 billion in 2018.[5] Furthermore, back then nature was seen as constant, in an equilibrium condition (O'Neill 2001). Now we know that nature may behave non-linearly, jumping from one equilibrium state to another even when triggered by small disturbances (Scheffer et al. 2001).

Thus, we face the following problem: there seems to be two ways of valuing nature (ecocentric and anthropocentric); the first more related with actions such as conservation of biodiversity and restoration and the second with other activities such as increasing the flux of ecosystem services. And, there may be a third that goes beyond the dichotomy, emphasizing relational values, where the social-ecological concept may be a key element. Can we wait the same 66 years that took ecologists to re-discover the ecosystem concept? How much of nature have we already changed? Potapov et al. (2017) using remote sensing methods (Landsat imagery) show that forests, in the year 2000, covered 44% of Earth ice-free land area, but that only 22% of them (9.68% of Earth land area) corresponded to intact forested landscapes (IFL). For South America, their information (Potapov et al. 2017, Table 1) shows a total forest area of 15.11×10^6 km^2 with 36.99% corresponding to IFL. The information for wetlands in South America, based on the global assessment of inland wetland conservation status of Reis et al. (2017, Table 2), shows that only 17.82% of inland wetland areas can be considered protected.

In summary, if we think in terms of conservation of biodiversity in Latin America, 14.26% of terrestrial and marine areas can be considered protected. When forest and wetland ecosystems are considered as examples, the figures are 36.99% for the former and 17.82% for the latter. What does this tell us about Latin America, in terms of the above-mentioned dichotomy? One answer is that most ecosystems (63% as a minimum up to 85%) have been modified by humans, to serve anthropocentric purposes (i.e., increase in well-being and the supply of ecosystem services, with many changes to the ecological structure of these ecosystems). Thus, waiting for societies to embrace the social-ecological concept so that human/nature relationships are incorporated into nature conservation may take a long time. As a consequence, scientists (social, ecological, and political) need to explore all forms of communication (see also chapter "Social Actors and Participation in Environmental Issues in Latin America") with social actors and governments so that human/non-human relationships and values may be discussed openly in order to identify and establish a new bond with nature.

[5] Sources: https://ourworldindata.org/world-population-growth and http://www.worldometers.info/world-population/

Now, let's suppose that Latin American scientists renew and enlarge their effort to communicate about social-ecological systems with social actors and governments. Furthermore, also suppose that we go beyond the ecocentric/anthropocentric dichotomy and embrace the relational values between humans and non-humans. How do we generate this new way of relating with nature? Do we concentrate on the 14.26% protected ecosystems, trying to increase their percentage? Or, do we somehow incorporate the remaining 85.74%? This is where the "novel ecosystem" concept, which we discuss in the following section, may be of interest.

2 Historical v/s Novel Ecosystems: Adding Elements to Social-ecological Complexities

Chapin III and Starfield (1997) used for the first time the term "novel ecosystem," referring to the alterations that climate change could generate in the Arctic tundra. One year later, Ott (1998) used the same term, but this time to explain human behavior in unknown ecosystems (that is, novel for humans). Later on, Johnson (2002) used the term "novel habitats," in a sense similar to the current use of "novel ecosystems," but only referring to freshwater systems.

The current use of the novel ecosystem concept is attributed to Hobbs et al. (2006: 2), who defined it as ecosystems that: "*have species composition and relative abundances that have not occurred previously within a given biome.*" The article, since its publication, has been cited 874 times, with an average of 95 citations per year between 2012 and 2017.[6] We propose, and we develop our proposal in this section, that the concept is in the process of generating a paradigm shift (in the sense of Kuhn 1962) in relation to ecological conservation and restoration.

One of the topics discussed by Hobbs et al. (2006) is that the management of novel ecosystems should benefit human societies, a perspective that could be defined as anthropocentric. However, Lindenmayer et al. (2008) stressed that we may not have enough knowledge on ecological changes to manage those ecosystems, something that could complicate policy making and resource management. Hobbs et al. (2009), in an opinion article, proposed that current targets for ecological restoration are for ecosystems they defined (p. 600) as "historical," those that: "*retain the biota and ecosystem properties that were prevalent in the past,*" and that there are hybrid (between novel and historical) ecosystems that may cause difficulties for conservation experts. After this article, and Hobbs et al.'s (2013) book on novel ecosystems, conservation and restoration experts started addressing how the concept may change current perspectives; a process that could be defined as the beginning of a change in current conservation paradigms; and a model drift using Kuhn's 1962 terminology where some "conservation/restoration anomalies" (e.g., those related to novel ecosystems) cannot be fully explained by the current paradigm. For Moreno Mateos

[6] Source: https://webofknowledge.com/

(2013), the main issue is threshold crossing, which is one of the characteristics of ecosystems when conceptualized as systems with multiple stable states. For example, slow increases in the phosphorus concentration or in water levels for aquatic ecosystems (i.e., threshold crossings) may change their structure (Marín et al. 2009). Moreno Mateus (2013: 459) proposed that conservation biology and restoration ecology "*need more time to mature*…." Murcia et al. (2014) also proposed that the novel ecosystem concept was a way forward for changing traditional approaches to conservation and restoration. They also added that recovery time scales for ecosystems such as wetlands may be in the order of 50 to 100 years, which brings the topic of whether we should accept novel ecosystems, given the low likelihood of moving them back to historical conditions, or insist in managing them for recovery regardless of the time involved. Hobbs et al. (2014) then discussed the issue of irreversibility. What if the removal of alien species is impossible? We would also like to add a different question, further contributing to Kuhnian conservation anomalies: What if the removal of alien species is undesirable? We illustrate this issue with an example: the Rio Cruces wetland in southern Chile.

The social-ecological issues related to changes in the ecological structure of the Río Cruces wetland, located near the city of Valdivia in southern Chile, have been described and analyzed by several authors (see Jaramillo et al. 2018, Marín et al. 2018 and chapter "Postnormal Science and Social-ecological Systems" of this book). The important details, for the purpose of this discussion, are these: early in its formation after an earthquake in 1962, the wetland was invaded by an alien macrophyte (*Egeria densa*), originally from Brazil. However, it became the main food item of an endemic, and highly symbolic bird species, the black-necked swan (*Cygnus melancoryphus*). The wetland, as stated in chapter "Postnormal Science and Social-ecological Systems", is a Sanctuary of Nature under Chilean law and a wetland on international importance under the Ramsar Convention (Ramsar 1998). A sudden decline in the local population of the swan during 2004 was associated with simultaneous changes in the biomass of *Egeria densa* (Jaramillo et al. 2018). Thus, if we think in classic terms, regarding biodiversity conservation (e.g., Iriarte et al. 2005), we should have been happy to watch the disappearance of the invasive species (i.e., the macrophyte). However, its effects on the endemic bird species made the abrupt decrease in its abundance undesirable. In fact, Jaramillo et al. (2018) describe current recovery of *E. densa* as part of the wetland's resilience. We agree with Hobbs et al. (2014) that the structure of these types of ecosystem (i.e., novel) may generate conundrums if we maintain a classic perspective on restoration and conservation. However, if we accept that the Rio Cruces wetland is a novel ecosystem, we could indeed manage with the invasive species *E. densa* as an ecosystem engineer (Yarrow et al. 2009) whose removal would likely cause undesirable regime shifts, as has occurred in the past (Marín et al. 2008).

We have used the Rio Cruces example as a way of showing that ecological restoration and conservation are going through a Kuhnian paradigm change. However, we could cite many other examples. Morse et al. (2014), proposed a new definition for novel ecosystems, to facilitate its use in practical applications. Collier (2014) emphasized the potential role of novel ecosystems in the provision of ecosystem services. Francis (2014) proposed to embrace ecological novelty in urban rivers and to use ecological engineering techniques. Trueman et al. (2014) showed an example of the use of the concept in an emblematic area: the Galapagos Islands. Collier (2015) stated that societal values are still unknown in relation to novel ecosystems and Lennon (2015: 287) pointed out that policy is still silent about the concept. Finally, Truitt et al. (2015) proposed that novel ecosystems should be managed using adaptive approaches. What it is interesting from this last citation is that Lee (1993) was describing in detail the meanings and approaches for adaptive management some 22 years ago. Thus, it again seems that scientists are not learning quickly and, if we indeed are in the nebulous zone between paradigms for conservation and restoration, somewhere in model drift, model crisis, or model revolution of Kuhn's cycle, speed of change may be even slower. Wherever we are, one thing is clear, ecological restoration and conservation is no longer "normal science" since the novel ecosystem concept was introduced. Also, social issues related to novel ecosystems (including their management) are just being discussed. For example, 51 articles were written on issues related to novel ecosystems during 2018 (Source: WoS[6]) with only 20% of them discussing social issues. Thus, we still have a long way to go. The question is, will we have enough time, given current speed of environmental changes? We do not have the answer to that question.

3 The Challenges of Embracing Novelty

When we looked for information on novel ecosystems in Latin America[6], we only found one reference (Gardener et al. 2012). The authors presented three main findings (p. 226), two of which relate to the issue of this chapter: (1) eradication of invasive plants is inefficient and (2) there is a need to accept the concept of novel ecosystems. If we consider that classical biological conservation and restoration pay special attention to biodiversity, the scientific literature is showing, worldwide, that novel ecosystems can in fact contribute to its conservation. As an example, we looked for references on forestry systems around the world using two subjects in our search: forestry and biodiversity[6]. The result provided a total of 2442 articles, mostly written after 2007. When we confined results to Latin America, there have been articles written for several countries (e.g., Argentina, Bolivia, Chile, Ecuador México, Panama), but their total number was low (14), with an average between 1 and 2 publications per year. Another way to look for information about novel ecosystems in Latin America is to conduct a web search (by means of Google). The search (using both terms in Spanish) generated only 23 web sites, mostly from academic organizations and none from governmental institutions. When the search was conducted in English for web sites around the world, the result was over two million, still mostly from academic societies.

Thus, we conclude from this literature search that: (1) the novel ecosystem concept is gaining momentum worldwide, but still limited to academic institutions/journals with small practical applications within governments, (2) the concept is not being used in Latin America, even within science. Why should this be a problem? The main reason is that we can no longer restrict our concerns about the future ecological conservation of Latin America to 14.26% of our ecosystems (Table 1). Even if we assume that those ecosystems are maintaining, intact, historical biodiversity and that the effects of human beings are kept to an acceptable minimum, we are living in the Anthropocene! Consequently, we should accept that the other 85.74% of the ecosystems are heavily used and managed, or perhaps not, by and for humans. These ecosystems, as discussed above, include invaders from other regions, some of which are supporting healthy populations of endemic species. If, on top of that, we accept the conclusions of Gardener et al. (2012) regarding the difficulty to eradicate them, or that ecosystem changes involving invaders are often irreversible (Hobbs et al. 2014), then we have to accept and manage novel Latin American ecosystems for issues not only dealing with the provision of ecosystem services, but also for biodiversity conservation. This implies accepting the keystone condition of *Homo sapiens*, as Tansley stated more than 80 years ago, but going beyond the dichotomies mentioned above (Piccolo 2017) and with the active involvement of social actors, as shown in several chapters (Parts II and III) of this book.

Governance is an important issue in social-ecological systems (see chapter "A New Environmental Governance"), including those, or rather, especially those built around novel ecosystems. However, governance proposals for novel ecosystems are almost absent from the literature, with the exception of the review work by Clement and Standish (2018), where they analyzed 15 articles that included "*governance and social aspects of novel ecosystems*" (p. 38). They concluded that the speed of policy adaptation is well below that of environmental changes. That leaves us with a sizable problem in Latin America, since we have not even started discussing the novel ecosystem concept within academia, let along with policy makers. Thus, we still do not know if we will find the level of resistance from biodiversity and conservation experts that has occurred in other regions of the planet. Furthermore, many novel ecosystems in Latin America correspond to privately own areas (e.g., private forestry areas where companies grow pine trees; Quiñones et al. 2017). How do we manage privately owned areas if we want them to increase their contribution to biodiversity conservation? One solution is to generate political changes so governments create incentives for such purposes. In other cases, such as novel urban wetlands in Argentina (Schmidt 2018), there are several conflicting actors (e.g., environmental NGOs, local and national governmental structures, local social actors) and insufficient resources to deal with their management.

In summary, the challenges of embracing ecosystem novelty are multiple. For Latin America, it should start with studying them in terms of their biodiversity, ecosystem services, and social-ecological complexities generated mostly from factors such as: (1) the multitude of social actors, (2) the extent of private v/s public ownership of land, and (3) ecocentric/anthropocentric valuations. If the concept of novel ecosystems is incorporated into society at the speed of many other concepts (e.g., that of ecosystems), we may have to wait some 60 years before we see the first results, which in itself may be too late to formulate society-wide responses given the rates of environmental degradation that are currently evident in Latin America.

References

Binder CR, Hinkel J, Bots PWG et al (2013) Comparison of frameworks for analyzing social-ecological systems. Ecol Soc 18(4):26
Chapin FS III, Starfield AM (1997) Time lags and novel ecosystems in response to transient climate change in arctic Alaska. Clim Chang 35:449–461
Clement S, Standish RJ (2018) Novel ecosystems: governance and conservation in the age of the Anthropocene. J Environ Manage 208:36–45
Collier MJ (2014) Novel ecosystems and the emergence of cultural ecosystem services. Ecosyst Serv 9:166–169
Collier MJ (2015) Novel ecosystems and social-ecological resilience. Landsc Ecol 30:1363–1369
Crutzen PJ (2002) Geology of mankind. Nature 415:23
Ellis EC, Kaplan JO, Fuller DQ et al (2013) Used planet: a global history. PNAS 110:7978–7985
Francis RA (2014) Urban rivers: novel ecosystems, new challenges. WIREs Water 1:19–29
Gardener MR, Bustamante RO, Herrera I et al (2012) Plant invasions research in Latin America: fast track to a more focused agenda. Plant Ecology & Diversity 5:225–232
Gordon L, Finlayson CM, Falkenmark M (2010) Managing water in agriculture to deal with trade-offs and find synergies among food production and other ecosystem services. Agric Water Manag 97:512–519
Hobbs RJ, Arico S, Aronson J et al (2006) Novel ecosystems: theoretical and management aspects of the new world order. Glob Ecol Biogeogr 15:1–7
Hobbs RJ, Higgs E, Harris JA (2009) Novel ecosystems: implications for conservation and restoration. Trends Ecol Evol 24:599–605
Hobbs RJ, Higgs ES, Hall CM (eds) (2013) Novel ecosystems. Intervening in the new ecological world order. Wiley, Chichester
Hobbs RJ, Higgs ES, Harris JA (2014) Novel ecosystems: concept or inconvenient reality? A response to Murcia et al. Trends Ecol Evol 29(12):645–646
Iriarte JA, Lobos GA, Jacksic FM (2005) Especies de vertebrados invadores en Chile y su control y monitoreo por agencias gubernamentales. Rev Chil Hist Nat 78:143–154
Jaramillo E, Lagos NA, Labra FA et al (2018) Recovery of black-necked swans, macrophytes and water quality in a Ramsar wetland of southern Chile: assessing resilience following sudden anthropogenic disturbances. Sci Total Environ 628-629:291–301
Johnson WC (2002) Riparian vegetation diversity along regulated rivers: contribution of novel and relic habitats. Freshw Biol 47:749–759
Kopf RK, Finlayson CM, Humphries P et al (2015) Anthropocene baselines: assessing change and managing biodiversity in human-dominated aquatic ecosystems. Bioscience 65:798–811
Kopnina H (2016) Nobody likes dichotomies (But sometimes you need them). Anthropol Forum 26:415–429
Kuhn T (1962) The Structure of Scientific Revolutions. University of Chicago Press, Chicago
Lee K (1993) Compass and Gyroscope. Integrating science and politics for the environment. Island Press, Washington, DC
Lindenmayer DB, Fischer J, Felton A et al (2008) Novel ecosystems resulting from landscape transformation create dilemmas for modern conservation practice. Conserv Lett 1:129–135
Marín VH, Delgado LE, Bachmann P (2008) Conceptual PHES-system models of the Aysén watershed and fjord (Southern Chile): testing a brainstorming strategy. J Environ Manage 88:1109–1118
Marín VH, Tironi A, Delgado LE et al (2009) On the sudden disappearance of Egeria densa from a Ramsar wetland site of Southern Chile: a climatic event trigger model. Ecol Model 220:1752–1763
Marín VH, Delgado LE, Tironi-Silva A et al (2018) Exploring social-ecological complexities of wetlands of international importance (Ramsar sites): the Carlos Anwandter Sanctuary (Valdivia, Chile) as a case study. Wetlands 38:1171–1182

MEA (Millennium Ecosystem Assessment) (2005) Ecosystems and Human Well-being: Synthesis. Island Press, Washington, DC
Moreno Mateos D (2013) Is embracing change our best bet? Science 341:458–459
Morse NB, Pellissier PA, Cianciola EN et al (2014) Novel ecosystems in the Anthropocene: a revision of the novel ecosystem concept for pragmatic applications. Ecol Soc 19(2): 12 https://doi.org/10.5751/ES-06192-190212. Accessed 10 Jun 2019
Murcia C, Aronson J, Kattan GH et al (2014) A critique of the 'novel ecosystem' concept. Trends Ecol Evol 29:548–553
O'Neill RV (2001) Is it time to bury the ecosystem concept? (With full military honours, of course). Ecology 82:3275–3284
Ott J (1998) The Delphic bee: bees and toxic honeys as pointers to psychoactive and other medicinal plants. Econ Bot 52:230–266
Piccolo JJ (2017) Intrinsic values in nature: Objective good or simply half of an unhelpful dichotomy? J Nat Conserv 37:8–11
Potapov P, Hansen MC, Laestadius L et al (2017) The last frontiers of wilderness: tracking loss of intact forest landscapes from 2000 to 2013. Sci Adv 3(1):e1600821
Quiñones D, Caro J, Marín VH et al (2017) Manejo resiliente de cuencas forestales de Chile: La cuenca de Cayucupil (Cañete) como caso de estudio. Boletín Nahuelbuta Natural 1:7–30
Ramsar (1998) Ficha Informativa de los Humedales de Ramsar. Santuario de la Naturaleza e Investigación Científica "Carlos Anwandter" del río Cruces, Valdivia. https://rsis.ramsar.org/ris/222. Accessed 8 Jun 2019
Reis V, Hermoso V, Hamilton SK et al (2017) A global assessment of inland wetland conservation status. BioScience 67:523–533
Scheffer M, Carpenter S, Foley JA et al (2001) Catastrophic shifts in ecosystems. Nature 413:591–596
Schmidt MA (2018) Conflictos por la valoración de humedales en ámbitos urbanos. La cuenca Matanza Riachuelo, Argentina. Bitácora Urbano Territorial 28:89–98
Sessions G (1999) Ecocentrism and the anthropocentric detour. In: Merchant C (ed) Ecology (key concepts in critical theory). Humanities Press International, London, pp 140–151
Tansley AG (1935) The use and abuse of vegetational concepts and terms. Ecology 16:284–307
Trueman M, Standish RJ, Hobbs RJ (2014) Identifying management options for modified vegetation: Application of the novel ecosystem's framework to a case study in the Galapagos Islands. Biol Conserv 172:37–48
Truitt AM, Granek EF, Duveneck MJ et al (2015) What is Novel About Novel Ecosystems: Managing Change in an Ever-Changing World. Environ Manag 55:1217–1226
UNEP-WCMC and IUCN (2017) Protected Planet: Terrestrial and marine protected areas (% of total territorial area). The World Database on Protected Areas (WDPA)/The Global Database on Protected Areas Management Effectiveness (GD-PAME)] [On-line], [January/2019], UNEP-WCMC and IUCN, Cambridge. www.protectedplanet.net. Accessed 10 Jun 2019
Western D (2001) Human-modified ecosystems and future evolution. PNAS 98:5458–5465
Yarrow M, Marín VH, Finlayson M et al (2009) The ecology of Egeria densa Planchón (Liliopsida: Alismatales): a wetland ecosystem engineer? Rev Chil Hist Nat 82:299–313

Social-ecological Systems and the Economics of Nature: A Latin American Perspective

Eugenio Figueroa B.

Abstract Today, based on the best scientific knowledge available, the main concerns regarding nature are the rampant ongoing destruction of Earth's ecosystems and the increasing global warming, which threaten the very survival of the human species. The main reason is that the current functioning of the complex global socioeconomic-ecological systems of the world is driving the planet, with high certainty, towards a completely unsustainable deterioration of its natural capital, biodiversity, and atmosphere. This is the result of the manipulation of the power mechanisms within the socioeconomic-ecological systems of every country by the small powerful elites with the purpose of appropriating for themselves the enormous economic rents generated by the overexploitation of the planet's natural resources and ecosystems. The monumental challenges of climate change, global warming, and nature deterioration are just the physical symptoms of the underlying unsustainable operation of the world's socioeconomic-ecological systems (or social-ecological systems as they are generally known today). On the other side, its social and economic symptoms are the billions of people living in poverty, without water, proper education, and health services, and the prevailing enormous inequities in the distribution of the world's wealth and income. In this chapter, using the Latin American experience, we analyze why this is happening and how enormously difficult is to change the current functioning of the system. We conclude with a rather pessimistic view about the possibility of meeting the challenges that all of us, as the human species, face today. Our only hope seems to be to use the opportunity offered by the modern communication technologies to inform the people of every country about the large risks and dangers we are all exposed to, and mobilize them to take effective control of the decision-making mechanisms that today are in the hands of the small powerful elites. This is obviously a monumental task.

E. Figueroa B. (✉)
Department of Economics, and Center of Environmental and Natural Resource Economics (CENRE); School of Economics and Business, University of Chile, Santiago, Chile
e-mail: efiguero@fen.uchile.cl

L. E. Delgado, V. H. Marín (eds.), *Social-ecological Systems of Latin America: Complexities and Challenges*, https://doi.org/10.1007/978-3-030-28452-7_10

Keywords Social-ecological systems · Latin America · Complexity · Nature economy · Nature deterioration · Climate change

1 Introduction

In 2006, worried about the alarming rate of deterioration of the planet's nature, ecosystems, and biodiversity, and about the resulting destruction of the flow of ecosystem services that support the life on Earth, we wrote a paper (Figueroa and Aronson 2006). Our purpose was to contribute with some theoretical insights to improving the ongoing efforts to conserve and restore ecosystems, protected areas (PAs), and nature in general. In that work, we argued that, because of the increasingly human-dominated planet we are living in, it is necessary to create more realistic, holistic, and durable linkages for nature, ecosystems, and PAs. These linkages are required for the long-term conservation of species and habitats and for securing their provision of the continuous flow of ecosystem services that make life possible in the planet (Cairns 1995; Costanza et al. 1997; Folke 1991; de Groot et al. 2002). Additionally, we explained that there was a need to extending the geographical, physical, and biological concepts of linkages to embrace socioeconomic factors. For this, we used a conceptual approach using the then nascent notions of "emerging ecosystems" (SEEs), "natural capital restoration" (NKR), and "social-ecological systems" (SESs). We also employed the value notion generally used in economic science, i.e., monetary value, as a conceptual tool in the unavoidable assessment process that modern societies must undertake to decide about how much, and how, to conserve and/or restore nature, PAs, or any ecosystem in particular.

Three years later, in 2009, Elinor Ostrom wrote a seminal work (Ostrom 2009), motivated by the worldwide problems related to the considerable damage to or losses of many natural resources, including fisheries, forests, and water resources, the major reductions in biodiversity and the threat of massive climatic change. She indicated that scientific disciplines use different concepts and languages to describe and explain complex social-ecological systems, and as a result of this our understanding of the processes that led to deterioration of (or improvements in) natural resources was limited. Without a doubt, this conclusion remains valid today. Moreover, Ostrom contradicted the then generally accepted theory that assumed that resource users never self-organize to maintaining their resources and thus governments must impose solutions. For this, she developed a general framework for analyzing the sustainability of SESs. She also used this framework to identify ten subsystem variables that affect the likelihood of self-organization in efforts to achieve a sustainable SES.

Now, 13 years after our paper and almost a decade after Ostrom's manuscript, we write this chapter attempting to explain why the SES approach has become even more relevant today to understanding the threats and challenges the world faces due to nature's destruction and climate change, and how urgent it is to realize this nowadays. The reason is, in the first place, that climate change is the largest threat humanity has ever faced because, according to the best scientific knowledge available, even the very survival of our species is at a high risk if correcting measures to efficiently control and stop human-induced global warming are not taken within the next two decades. Additionally, in the second place, in almost every ecosystem of the globe, biodiversity is under a massive threat, degrading and reducing at alarmingly high rates across the planet (CBD 2014; Vitousek et al. 1997). The recent 2018 Living Planet Report of World Wildlife Fund (WWF 2018) provides alarming evidence of the latter: the population abundance of the globally monitored species of vertebrates (birds, mammals, fish, reptiles, and amphibians) has declined by 60% on average between 1970 and 2014.

In this chapter, our focus is two-fold. First, it is multidisciplinary, in the sense that we discuss and highlight the relevance and usefulness of the analytical tools of both social and economic sciences on the one hand, and of physical sciences, on the other hand, to use effectively the SES approach in dealing with the pressing current world challenges. Second, our focus is also Latin American because we employ our region's experience to illustrate the large difficulties that current global challenges imply, as well as to provide some ideas about possible ways to face these challenges.

2 The Enormous Challenges the World Faces

Climate Change and Global Warming

According to the best scientific knowledge and evidence currently available, human influence on the climate system is clear and growing, with impacts observed across all continents and oceans, and many of the observed changes since the 1950s are unprecedented over decades to millennia. The United Nations Intergovernmental Panel on Climate Change (IPCC) estimates, with 95% certainty, that humans are the main cause of current global warming (IPCC 2014). Additionally, human activities are estimated to have caused approximately 1.0 °C of global warming above preindustrial levels (within a range of 0.8 °C to 1.2 °C, estimated with 66–100% certainty), and global warming is likely to reach 1.5 °C between 2030 and 2052 if it continues to increase at the current rate (90–100% confidence) (IPCC 2018). In other words, in its most recent report, the IPCC released an alarming warning indicating that if emissions continue to rise at their present rate, by 2040, the atmosphere will surely warm up by as much as 1.5 °C above the preindustrial level. It also indicated that these global warming will provoke flooding of coastlines, the killing of coral reefs worldwide, and more catastrophic droughts and wildfires (IPCC 2018). To avoid this, greenhouse gas emissions would need to fall by nearly half from 2010 levels in the next 12 years and reach a net of zero by 2050.

The recent predictions of the IPCC's climate models project robust differences in regional climate characteristics between present-day and global warming of 1.5 °C, and between 1.5 and 2 °C[1] and the differences in the expected impacts and costs from these two global warming scenarios are quite relevant (IPCC 2018). For example, by 2100, global mean sea level rise is projected to be around 0.1 m lower with global warming of 1.5 °C compared to 2 °C, which would provide greater opportunities for adaptation in the human and ecological systems. On land, impacts on biodiversity and ecosystems, including species loss and extinction, would be lower at 1.5 °C of global warming compared to 2 °C, which would imply lower impacts on terrestrial, freshwater, and coastal ecosystems and would allow retaining more of their ecosystem services to humans. The increases in ocean temperature, as well as associated increases in ocean acidity and decreases in ocean oxygen levels, would be lower in the global warming scenario of up to 1.5 °C compared to the one of 1,5–2 °C. For the former scenario, this would also imply lower risks to marine biodiversity, fisheries, and ecosystems, and their functions and services to humans, compared to the latter scenario. Climate-related risks to health, livelihoods, food security, water supply, human security, and economic growth are projected to increase less with global warming of 1.5 °C than with global warming of 2 °C. On the other hand, most adaptation needs are estimated to be lower for the first than for the second warming scenario.

Using a less technical language in translating these IPCC's projections into their expected future impacts, one can say that if high world's fossil fuel emissions continue unabated and consequently global CO_2 concentrations continue to increase, there will be predominantly negative impacts for humanity, especially for young people. For example, regarding the increase in sea level, continued high fossil fuel emissions would make coastal cities dysfunctional, with incalculable and very costly consequences in human and economic terms. With respect to biodiversity, the shifting of climate zones, with other stresses, may force many species into extinction, leaving a more desolate planet. Finally, regarding regional climate, if high fossil fuel emissions continue, subtropics and tropics will become dangerously hot, and the resulting emigration chaos may threaten global governance (Hansen 2018).

Because of these findings, in October 2018, the IPCC urged the world to set at 1.5 ° C its definite target to limiting future global warming (IPCC 2018). It has been amply recognized that urgent and deep societal transformations will be needed to actually constrain global climate change to the limit of 1.5–2 °C or of 1.5 ° C global warming. To attaining any of these targets, intentional trajectories of decarbonization should be traveled, and the interconnected technical, economic, social, and political changes that this entails must be implemented soon. In fact, as Rockström et al. (2017) explain, carbon emissions need to be reduced by 50% every decade between now and 2050. Moreover, in the energy sector alone, in order to transit to new and cleaner energy systems, technological changes and adaptations as well as quite important modifications of current political regulations, tax codes, and pricing regimes, and of the behavior of users and adopters will be required (Sovacool 2016; Sovacool and Hirsh 2009; Painuly 2001). In the transportation and the industrial sec-

[1]Robust differences imply that at least two thirds of climate models show the same sign of changes at the grid point scale, and that differences in large regions are statistically significant.

tors of most countries, the needed transition towards a less intensive carbon mode has to be tackled urgently also. Many authors and institutions have indicated that if the transition towards a decarbonized world economic system does not occur soon and quickly, it will be too late (IEA 2012; Giddens 2009). According to Patterson et al. (2018) and McSweeny and Pearce (2017), the world must act within the next 5–17 years to have a 33–66% chance of achieving the goal of limiting the global warming to 1.5–2 °C. All of this imposes unprecedented demands on national and subnational political systems and requires large and profound transformations in many subsystems of the social-ecological systems of most countries, which had made many to question the political feasibility of effectively addressing these challenges.

Leahy (2018) summarized the current situation reporting that the challenge to stay below 2 °C degrees is immense, requiring fossil fuel infrastructure to be phased out, non-fossil energy sources phased in, and large-scale removal of carbon from the atmosphere; and to stay below 1.5 °C simply requires the needed transformation be much faster and deeper than for 2 °C. On top of that, the world has been going in the wrong direction with global emissions increasing above the growth rates required to meet the current global warming challenges. It is clear now that without the full involvement and alignment of our technical, social, and political dimensions, 1.5 °C and even 2 °C will not be possible. Indeed, every possible test has been done and the news is not good. The doctors, in this case, the IPCC and the world scientific community, have explained possible treatment options to ensure our future health. Now, we (the public) have to decide which option to follow.

In fact, the situation is quite appropriately depicted by Leahy (2018). However, and unfortunately, as we will explain below, the complexity of the SES in which we (the public) are living and making our decisions, is immense too. So much that, for instance, the information produced for us by the doctors (the IPCC and the world scientific community) does never reach us, or most of the time we receive it adulterated by the vested interests of the messengers that deliver it to us. Because of this, being uninformed or informed in a purposely biased-way, we make no decision.

Deterioration of Nature, Ecosystems and Biodiversity

Something similar happens regarding the destruction of the world's nature, landscapes, and ecosystems, and the deterioration of global biodiversity. The consequences of the erosion of nature are so large, that Sir Robert Watson, the Chair of the Intergovernmental Platform on Biodiversity and Ecosystem Services (IPBES),[2] stated in May 2019, "The overwhelming evidence of the IPBES Global Assessment, from a wide range of different fields of knowledge, presents an ominous picture. The health of ecosystems on which we and all other species depend is deteriorating more rapidly than ever. We are eroding the very foundations of our economies,

[2] The IPBES is an independent intergovernmental body placed under the auspices of four United Nations entities and administered by United Nations Environment Program (UNEP).

livelihoods, food security, health and quality of life worldwide. The loss of species, ecosystems and genetic diversity is already a global and generational threat to human well-being" (SCOS 2019).

Compared with the average "normal" species extinction rate indicated in the fossil records, the known recent extinction rate appears to be some 100 or 200 times higher.[3] The currently existing estimations indicate that around 300–350 vertebrates and nearly 400 invertebrates have become extinct during the past 400 years. Because mammals and birds tend to be relatively well recorded and leave recognizable skeletal remains, in these two groups the known rate of extinction over the past 400 years averages around 20–25 species per 100 years (CBD 2001).

The major drivers of biodiversity deterioration and the ensuing accelerating loss rate are the habitat loss and degradation, land-use intensification and degradation, climate change (through heat and drought stresses), pollution, overexploitation and unsustainable use, invasive alien species and armed conflicts. Additionally, the drivers linked to agriculture account for 70% of the projected loss of terrestrial biodiversity (CBD 2014). The estimated economic cost of biodiversity and ecosystem services lost, only because of land degradation, is more than 10% of the annual global gross product (CBD 2001).

From an economic point of view, it is quite clear that the main problem that Earth's biodiversity is facing is the bias valuation societies are doing of its contribution to people's well-being and future survival. The available scientific evidence indicates that global biodiversity is being destroyed at a rate that is clearly too high and risky from the social point of view. The underlying explanation is that we, the human species, have not revealed the true value of biodiversity, ecosystems, and ecosystem services for the well-being of all people, not for the well-being of only those who, through the functioning power mechanisms currently in place, are making the decisions regarding biodiversity use, and its conservation or destruction. The key conceptual and practical issue is in this case, very much the same as the one regarding the lack of adequate protection of the planet's atmosphere and the resulting global warming. The failure of current complex socio-ecological systems to manage in a socially efficient way nature's "commons", those resources that belong to all and which are inefficiently allocated by the markets and, therefore, they are affected by the so-called unresolved "government failures".

3 The Complex Functioning of Social-ecological Systems Allowing Nature Deterioration

The purpose of this section is to show that the seemingly insurmountable hurdles to solve the urgent world problems regarding climate and nature come mostly from the large complexity of the social-ecological systems in which human

[3] To achieve greater precision in such estimates is difficult because of the inherent bias in the fossil record, but the general direction of the trend is well supported (WCMC 2002).

societies currently live. This complexity results from the institutions that structure these social-ecological systems and govern their economic, social, and political functioning, through the established institutional arrangements, which are crucially determined by the historical and current power structures designing and legitimizing the prevailing institutional architecture of such social-ecological systems.

The economy and its institutions (markets, property rights, contracts, regulations, banks, etc.) and functioning (gross domestic product, people incomes, labor wages, pensions, economic growth, inflation, pollution, etc.) are key elements of the social-ecological systems. The reason is that people and societies, in their endless pursuit of happiness, spend a very significant part of their time, energy and lives, working, doing business and developing a myriad of different economic activities. The latter should imply no problem at all if Adam Smith's "invisible hand" operated in the real-world economies and markets. In such a case, and in a world of "perfect competition," no harm would arise if individuals pursuing their own individual happiness make their individual decisions caring only about their own personal welfare and security and intending only their own gains, without regard of others' happiness, security, or gains. The reason is that, at the end, these individuals, as guided by an "invisible hand," would end up attaining the largest possible happiness for all of them together, even though this end was no part of any of them individually. In spite of the fact that there is no formal and generally accepted demonstration of this powerful insight of Adam Smith, it has gained the rank of a fundamental truth in mainstream economics, even though it is also generally accepted that "perfect competition" simply does not exist. Additionally, in economic science, it is amply recognized that in the real-world social-ecological systems, interdependencies between individuals and between subsystems of any SES always exist. Any of these two latter conditions by itself has the same effect on Smith's "invisible hand" insight that has the absence of "perfect competition" in the economy: it renders Smith's insight theoretically invalid and useless as a normative concept to apply to the real-world social-ecological systems. In other words, in the real world, markets and economic institutions more often than not "fail", and that grand unintended end so magically attained by the mysterious action of the "invisible hand," in reality, is never attained.

The implication of this is that in the current complex modern world, the economy cannot operate efficiently without norms and regulations imposed by society to the markets, the economy, and the institutional arrangements that allow them to operate without distortions. However, society does not exist as a real entity, but it is a conceptual social construct. Therefore, through the democratic system, individuals have to transfer part of their personal sovereignty to the government, the parliament, and the judiciary system, to design, enact, and enforce norms, regulations, and laws. Thus, politicians and public officials and employees are supposed to use the power transferred to them by the individual citizens to make the economy, markets, and all legal and administrative institutions to function honestly and efficiently to attaining the highest welfare and happiness for everyone and for all people in the society from the use of their always-scare resources.

Then, unfortunately, the usual principal-agent problem arises because politicians and public servants who must serve the interest of the society are not angels.

They are normal human beings, with their own personal circumstances, interests, motivations, and problems. As a result, very often, and in many countries, most often than not, they do not serve society's interests but they serve their own personal or their group's interests, affecting and diminishing the welfare of those who they should protect and care for their well-being. This is why the so-called "government failures" and the rampant corruption of those institutions that should serve society end up transforming some real-world social-ecological systems into highly inefficient socioeconomic machines serving the vested interests of some small groups at the expense of the vast majority of society.

As economic science demonstrated it time ago, nature, ecosystems, biodiversity, and climate are all prone to be affected by the lack of functional "invisible hands," and by the lack of efficiently operating parliaments, governments, courts, and regulatory public agencies, and by the destructive erosion of the social and economic systems provoked by corruption. This is because nature, ecosystems, biodiversity, and climate generally are not "private goods" in an economic sense, due to the fact that, most of the time, they are affected by "non-rivalry," and/or by "non-exclusion" (or high costs of exclusion).[4] The implication of the latter is that, in every country, the powerful Adam Smith's "invisible hand" simply does not operate in the allocation of nature's resources by the economic subsystems of their social-ecological systems. Therefore, to attain an adequate use of nature's resources from the point of view of society, it is necessary to design and implement an adequate system of public regulations. This means that commanded by their populations, governments of all countries should correct the ubiquitous market failures in allocating nature resources by their economic systems. Thus, governments must enact, implement, and enforce appropriate regulations to correct the malfunctioning of their economic systems of every country to realign them with the interests of their populations. The irrefutable evidence that this is not happening is the rampant ongoing degradation of global nature and the existing hard scientific evidence about the worrisome occurrence of global climate change.

The reason why the public regulatory systems are not correcting the highly deleterious effects of the lack of the "invisible hand" in the markets that allocate nature's resources is that the overuse and overexploitation of natural resources, ecosystems, biodiversity, and ecosystem services generate enormous economic rents that are appropriated by those overusing and overexploiting them. Moreover, nature overexploiters are able to appropriate these rents because the legal, economic, and regulatory institutional structure in every country is highly determined by the large political power controlled by those who possess a large wealth and who, very often, also possess a large share of the total wealth and income of their countries. In other words, the large amounts of economic rents from overexploiting nature's resources provide to overexploiters the economic wealth and power to establish and perpetuate the institutional arrangements keeping this vicious circle in place. As explained by several economists, economic rents generated by exploiting nature, natural resources

[4] For explanations of "non-rivalry" and "non-exclusion," see any basic economics textbook, such as Samuelson and Nordhaus (2010), Mankiw (2012), or CORE TEAM (2019).

(fisheries, forests, mines, wetlands, land, water, etc.), ecosystems, biodiversity, and climate provide immense economic power to the exploiters of these nature's resources, which they usually use to capture political power. Political power, in turn, allows them to create, maintain, and perpetuate the economic institutions (taxes, tax codes, fiscal policies, trade policies, financial regulations, etc.) that are also instrumental to perpetuating the mechanisms allowing the appropriation of economic rents (Acemoglu 2008; Collier 2007; Acemoglu and Robinson 2008, 2012; López 2019; López and Figueroa 2016; Figueroa et al. 2013).

These are the socioeconomic and political mechanisms determining economic and political power behind the real-world social-ecological systems operation provoking the ongoing alarming destruction of nature, natural resources, ecosystems, and biodiversity. These mechanisms also explain why climate change, in spite of being the largest challenge the human species is currently facing regarding its future survival, is almost absent of most serious discussions reaching the general public through newspapers, magazines, television, radio, etc. Instead of this, the general public receives every day an amazingly large and detailed information and discussion about football, soccer, tennis, golf, and other sports news, and a multitude of other news, comments, gossips, and discussions about show business, entertainment world, and a variety of other inconsequential issues.

The power mechanisms shaping current world social-ecological systems also explain facts such as why the US government allows, permits, and subsidizes fossil fuel reserves so that the fossil fuels are processed, transported, and burned with little or no control on emissions. In spite of the fact that, as explained by James Hansen, one of the world most renowned climate scientists,[5] the US government has possessed extensive knowledge about the threat posed by fossil fuel-driven climate change, for several decades. Thus, the US government is allowing the atmosphere to be treated as a free dumping ground for waste CO_2 (Hansen 2018). Additionally, the Republican and the Democratic political parties of the United States receive very large amounts of money from the fossil fuel industry. Moreover, in 2011, the Obama Administration opened up hundreds of millions of tons of coal on public lands to new lease sales; and the sales were at prices far below market value, continuing a practice of federal subsidy of coal titans amounting, through those sales alone, to tens of billions of dollars (Hansen 2018). Certainly, all these billions of dollars flowing through the political system lubricate the economic-political-corruption mechanism we are analyzing here.

Moreover, the Trump Administration's astounding recent efforts to accelerate fossil fuel use are pressing the world rapidly towards the climate precipice. As Hansen (2018) argues, the current US government blatantly misrepresents the facts about climate change and specifically about the US contribution to climate change.

---

[5] James Hansen was for 17 years Director of the Goddard Institute for Space Studies (of the US National Aeronautics Space Administration), and currently is Adjunct Professor at Columbia University's Earth Institute, where he directs a program in Climate Science, Awareness, and Solutions

All of this is happening in the United States, and the American governments have been taking these inexplicable decisions even though they are perfectly aware of the consequences of such policies in terms of nature destruction and global warming. These policies of the American governments, as well as their vicious deference with the fossil fuel industry and their sycophantic relationship with this industry, violate the rights of young people and future generations, a problem that cannot be solved at the ballot box (Hansen 2018). The underlying reason for the latter is the ongoing corruption in the economic and political subsystems of the American SES.

If this is the situation in the United States, one cannot expect to find any better picture in the rest of the countries of the American continent located to the south of the US borders, whose social-ecological systems exhibit, in general, less developed democratic political subsystems and less transparent administrative and legal subsystems than the US social-ecological system. Unfortunately, as is shown below, this is the case in Latin American countries.

4 Latin American Experience in Dealing with Nature's Degradation and Climate Change[6]

The Latin America and the Caribbean (LAC) region of the American continent contains over 50% of the world's biodiversity, as well as over the 25% of its forests (UNEP 2010). It includes some of the most species-rich biomes on Earth, such as lowland forests, coral reefs, mangroves, and wetlands, making the region one of the most endowed in terms of natural capital wealth: trees, water, minerals, and fisheries. Six of the world's most biodiverse countries are within the Latin American region,[7] which is also home to the world's most biodiverse habitat, the Amazon rainforest (UNEP 2012). Over 40% of the Earth's biodiversity is held within the South American continent, as well as over 25% of its forests (UNEP 2010). Tropical forests, savannahs, grasslands, and xeric communities originally covered vast areas of the LAC region (Olson et al. 2001), but there has been considerable loss of some habitats. Habitat loss is mostly provoked by land conversion to agriculture and pasture for livestock and is the most important threat to biodiversity in the region. Moreover, the total area transformed per year remains high (Aguiar et al. 2016).

Due to the richness of its natural capital in general, and of its biodiversity in particular, the Latin American region could make a significant difference in facing the enormous world's challenges of the accelerating global warming and the biodiversity rampant degradation, in either direction. Towards successfully meeting and overcoming these challenges, safeguarding the planet and protecting the Earth for human's future survival. Alternatively, to accelerate the encounter of the human

---

[6] A large part of the data and empirical evidence provided in this section comes from IPBES (2018) which is the most recent assessment on the nature and biodiversity of the Americas and the Caribbean continent.

[7] They are Brazil, Colombia, Ecuador, Mexico, Peru, and Venezuela.

species with its final Armageddon. Our analysis here is highly relevant and timely because the existing statistics and empirical evidence show a rather discouraging performance of the Latin American region in the recent decades.

In fact, the three Americas—North, Meso, and South—are endowed with a much greater capacity for nature to contribute to people's quality of life than the global average. The Americas contain 40% of the world ecosystems' capacity to produce nature-based materials consumed by people and to assimilate by-products from their consumption; but they contain only 13% of the total global human population. Such capacity results in three times more resources provided by nature per capita in the Americas than are available to an average global citizen (IPBES 2018). On the other hand, the majority of countries in the Americas are using nature more intensively than the global average and exceeding nature's ability to renew the contributions it makes to people's welfare. As a result, the 13% of the global human population living in the Americas produces 22.8% of the global ecological footprint.

Approximately 25% of the 14,000 species comprehensively assessed in the Americas by the International Union for Conservation of Nature (IUCN) are classified as being at high risk of extinction. Moreover, the risk of populations or species is increasing in almost every type of habitat. Of the threatened endemic species, more than 50% of the species in the Caribbean, more than 40% in Mesoamerica and almost a 25% in North and South America are exposed to a high risk (IPBES 2018).

Although there are in the Americas in general, and in the LAC region specifically, public environmental policies and governance approaches aiming to reduce pressure on nature and on nature's contributions to people,[8] they often fail to achieve their objectives. In fact, in most of the Americas, the functioning of the complex and non-inclusive socio-ecological systems and the power mechanisms operating within these social-ecological systems generally end up provoking two results that are key to understand the current regional (and global) problems regarding nature, biodiversity, climate, and global warming. First, the decision-making processes within the established social-ecological systems subordinate the world's "common resources"—nature, environment, ecosystems, atmosphere, and climate–to a set of narrowly defined economic and productive oriented objectives. Second, the practical operation of those policies and governance approaches determine, at the end of the day, highly inequitable distributions of the benefits provided by these "common resources" to people in all the sub-regions of the continent (and the world). The implication of these two worrisome results can be directly linked to a conclusion, also worrisome, of the most recent IPBES assessment of the current status of biodiversity and ecosystem services in the Americas: on average, biodiversity, and nature's contributions to people have been diminishing under the current governance systems in the Americas (IPBES 2018).

[8] Public environmental policies and governance approaches implemented in the LAC region specifically are, among some others, the following reported by UNEP-WCMC (2016). (1) A range of low carbon sustainable development approaches. (2) Efforts to control illegal trade in wildlife. (3) A significant expansion of protected area coverage in recent years. (4) An increase in regional support for conserving migratory species. (5) The implementation of targeted species management and recovery programs.

Moreover, this 2018 IPBES assessment concludes that the disproportionate and unsustainable use of its "biocapacity" in the Americas has increased steadily in recent decades. Evidence of this are facts such as the following three. First, since the 1960s, renewable freshwater available per person has decreased by 50%, and land devoted to agriculture has increased by 13%. Second, since 1990, forest areas have continued to be lost in South America (9.5%) and Mesoamerica (25%), although there have been net gains in North America (0.4%) and the Caribbean (43.4%). Finally, since the 1960s, the ecological footprint of the Americas has increased two- to threefold in each sub-region. The consequence of these facts and a large list of others that could be added is that the aggregate ecological footprint of the extensive continent of the three Americas and the Caribbean remains unsustainable and continues to grow (IPBES 2018).

Unfortunately, this discouraging current picture of nature, ecosystems, and ecosystem services in the American continent is expected to worsen in the future because of the continuously increasing global demands for food, water, and energy which are served in a significant proportion by the "common resources" from the rich natural endowment of the Americas.[9] In fact, the Americas are currently the largest global exporters of food and one of the largest traders in bioenergy. The unsustainable practices underlying the use and exploitation of their "common resources" to supply the increasing global demands have had negative consequences for nature, with adverse implications for nature's contributions to people and quality of life, and for availability of future options (IPBES 2018). The exploitation of fisheries has peaked in all sub-regions of the Americas and the Caribbean and fish and seafood harvests are decreasing as stocks decline. Moreover, the commerce of timber and fiber from plants and animals is large in the Americas, and timber and fiber production have increased significantly over the last several decades. In recent years, the rates of production increase have begun to slow down, as new technologies and production substitutes emerge and supplies of timber continue to decrease. However, in some cases the overall reduction in hardwood harvest has not reduced pressure on some valuable species. For example, in South America the coniferous production has increased since 2000 (IPBES 2018).

The threats to or declines in all the nature-based securities in the Americas reflect the ongoing reduction of nature's ability to contribute to human quality of life. Past rates of loss are high, and the loss continues, with some biomes under particular pressure. For example, in the biennial of 2014–2015, approximately 1.5 million hectares of the Great Plains were lost to land conversion or reconversion; between 2003 and 2013, the north-east agricultural frontier in Brazil more than doubled from 1.2 to 2.5 million hectares, with 74% of new croplands taken from intact Cerrado biome in that specific region. Even relatively well-conserved high elevation habitats

[9] Grau and Aide (2008) explain that increasing global food demand, particularly from Southeast Asia, accelerates deforestation in areas suitable for modern agriculture (e.g., soybean), severely threatening ecosystems, such as Amazonian rain forests, dry forests, and subtropical grasslands. They also argue that the demand for biofuels may become a much larger threat in the future.

have been degraded. For example, the Peruvian Jalca was converted at an annual rate of 1.5% over a 20-year period starting from 1987.

Brazil, in addition to be the fifth largest country in the world with a total area of 8,515,767 km^2, is indeed a megadiverse country in terms of biological resources, possessing 6 terrestrial biomes—Amazonia, Atlantic Forest, Caatinga, Cerrado, Pantanal, and Pampa. The Brazilian Amazonia biome is referred to as the "lungs of the planet" and is considered one of the most important areas on Earth because, among other reasons, it represents 60% of the Amazon rainforest and about 33% of world's rainforest. It is home to one third of the Earth's species and is a critical global storehouse of carbon (a typical forest of the region contains, on average, 460 tons of biomass per hectare, which corresponds to 230 tons of carbon fixed in plant tissues) (Wurdig et al. 2009).

However, since 1978, over 750,000 square kilometers (289,000 square miles; an area larger than the area of Spain and the United Kingdom together) of Amazon rainforest have been destroyed across Brazil, Peru, Colombia, Bolivia, Venezuela, Suriname, Guyana, and French Guiana. Historically, deforestation in the Amazon was provoked mainly by subsistence farmers who cut down trees to produce crops for their families and local consumption. However, in the later part of the twentieth century, that changed, and a large proportion of deforestation became to be driven by industrial activities and large-scale agriculture. By the 2000s, more than 75% of forest clearing in the Amazon was for cattle-ranching, and deforestation rate in the Amazon was faster than ever before. However, in 2004, this trend was reverted and annual forest loss in the country declined about 80%, due to several reasons: (1) increased law enforcement; (2) satellite monitoring; (3) pressure from environmentalists; (4) private and public sector initiatives; (5) new protected areas and (6) macroeconomic trends. Unfortunately, the encouraging reversion of Amazon forest destruction occurred in Brazil since 2004 did not occur in other Amazon countries, in which their rates of deforestation accelerated since 2000 (Butler 2019).

This implies that, in the future, Brazil will be an even more important factor for climate change and global warming than before. Aguiar et al. (2016) proposed three updated qualitative and quantitative land-use scenarios for the Brazilian Amazon and used computational models to estimate net deforestation-driven carbon emissions for these scenarios. They show that the Brazilian Amazon region could become a sink of carbon after 2020 in a scenario of residual deforestation (~1000 km^2/year) and a change in the current dynamics of the secondary vegetation—in a forest transition scenario. Unfortunately, their estimations showed that the continuation of the current situation of relatively low deforestation rates and short life cycle of the secondary vegetation would maintain the region as *a source* of CO_2—even if a large portion of the deforested area were covered by secondary vegetation. This implies that current perspectives for limiting Amazon deforestation in the future look rather gloomy. Because, the Brazilian Forest Code, the law that regulates how much land in key areas like the Amazon rainforest can be farmed and deforested, and how much of the previously deforested land has to be restored has undergone some big changes in recent years to limiting its power. For example, a new article was introduced in 2012, relaxing the norms that previously mandate private landowners to protect a percentage of the forests on their lands. Only this change in the Forest Code has opened up for legal deforestation an additional 6.5–15.4 million hectares of private land (Betuel 2019). Moreover, and a matter of a larger concern, the Brazilian president, Jair Bolsonaro, who took office just the first day of this year 2019, has declared his intention of converting forests into farmland to produce meat and soy for the world markets. In addition, he has also proposed to build an 870 km paved highway through the Amazon and has indicated that he will limit the powers of Brazil's environmental agencies (Betuel 2019).

In a recent study, Prevedello et al. (2019) has shown that tropical forests are particularly sensitive to the climate effects of forest change, with forest cover losses of ~50% associated with increased land surface temperature (LST) of 1.08 ± 0.25 °C, whereas similar forest cover gains decreased LST by −1.11 ± 0.26 °C. Additionally, these authors show that these changes on LST were largely mediated by changes in albedo and evapotranspiration. They also showed that predicted forest changes in Brazil associated with a business-as-usual land-use scenario through 2050 might increase LST up to 1.45 °C. Although these results refer to local temperature changes, it is quite possible that, given the large relevance of the Amazon region for maintaining the chemical as well as the physical balances within the global atmosphere system, future increases in deforestation rates in the Amazon will end up accelerating climate change and the speed of global warming. This could have serious implications in terms of the possibility of containing future global temperature increases to less than 2 °C by 2100.

5 Future World Challenges from a Latin American Perspective

As it is well known, the world's plan and instruments to facing the big challenges of the current deterioration of global biodiversity are determined in the Strategic Plan for Biodiversity 2011–2020 and its 20 Aichi Biodiversity Targets defined in Nagoya, Japan, in 2010. The vision of this Strategic Plan is a world "living in harmony with nature" and in which, "by 2050, biodiversity is valued, conserved, restored and wisely used, maintaining ecosystem services, sustaining a healthy planet and delivering benefits essential for all people." From this vision, the international community set the explicit mission of taking "effective and urgent action to halt the loss of biodiversity in order to ensure that, by 2020, ecosystems are resilient and continue to provide essential services, thereby securing the planet's variety of life, and contributing to human well-being, and poverty eradication."

Moreover, in September 2015, at the United Nations Sustainable Development Summit in New York, the so-called Transforming our World: The 2030 Agenda for Sustainable Development was adopted, including 17 Sustainable Development Goals (SDG). These instruments jointly set out "a supremely ambitious and transformational vision," of "a world free of poverty, hunger, disease and want, where all life can thrive… a world free of fear and violence. A world with universal literacy … A world where we reaffirm our commitments regarding the human right to safe drinking water and sanitation and where there is improved hygiene; and where food is sufficient, safe, affordable and nutritious. A world where human habitats are safe, resilient and sustainable and where there is universal access to affordable, reliable and sustainable energy … a world of universal respect for human rights and human dignity, the rule of law, justice, equality and non-discrimination; of respect for race, ethnicity and cultural diversity; and of equal opportunity permitting the full realization of human potential and contributing to shared prosperity. A world which invests in its children and in which every child grows up free from violence and exploitation. A world in which every woman and girl enjoys full gender equality and all legal, social and economic barriers to their empowerment have been removed … a world in which every country enjoys sustained, inclusive and sustainable economic growth and decent work for all. A world in which consumption and production patterns and use of all-natural resources—from air to land, from rivers, lakes and aquifers to oceans and seas—are sustainable. One in which democracy, good governance and the rule of law as well as an enabling environment at national and international levels, are essential for sustainable development, including sustained and inclusive economic growth, social development, environmental protection and the eradication of poverty and hunger. One in which development and the application of technology are climate-sensitive, respect biodiversity and are resilient. One in which humanity lives in harmony with nature and in which wildlife and other living species are protected."

The 20 Aichi targets are directed to globally attaining the three objectives of the United Nations Convention on Biological Diversity (CDB): (1) the conservation of biological diversity; (2) the sustainable use of its components and (3) the fair and equitable sharing of the benefits arising out of the utilization of genetic resources, including by appropriate access to genetic resources and by appropriate transfer of relevant technologies, taking into account all rights over those resources and to technologies, and by appropriate funding. These Aichi targets are currently incorporated within the 17 SDG of the 2030 Agenda for Sustainable Development. As we analyzed in the previous section, unfortunately, the LAC region's performance regarding nature protection and biodiversity conservation in the last decades is not encouraging and, in fact, is quite disappointing because it is not contributing to meet the challenges.

Moreover, as concluded in the IPBES Assessment of 2018, it seems that only a few of the Aichi Targets will be met by the 2020 deadline for most countries in the Americas, in part because of policy choices and trade-offs with negative impacts on aspects of biodiversity. Thus, continued loss of biodiversity could undermine the achievement of some of the Sustainable Development Goals, as well as some international climate-related goals, targets, and aspirations (IPBES 2018).

A large number of studies across taxonomic groups in temperate and tropical forests, grasslands, and marine systems support links between biodiversity and productivity, stability, and resilience of ecosystems (Isbell et al. 2015; Oliver et al. 2015; Bommarco et al. 2013; Steneck et al. 2002; Tilman and Downing 1994). Thus, projections of further loss of biodiversity pose significant risks to society because future ecosystems will be less resilient (IPBES 2018). Moreover, ecosystems are expected to face an even wider array of drivers that have been the primary causes of degradation in the past. Some environmental and social thresholds (or tipping points) are being approached or passed. A clear example of this is the combined effect of warming temperatures and pollution that is increasing the vulnerability of coral reefs in the Caribbean (Zaneveld et al. 2016; Ban et al. 2014). Even with a warming scenario of less than a 4 °C, a large coral reef mortality is expected, which will have significant impacts on coral reef ecosystems (IPBES 2018).

As noticed by IPBES (2018), the rapidly increasing urban/rural population ratio in the Americas implies that decision-making power is likely to shift increasingly towards those who have a less direct relationship to nature's contributions to people for their livelihoods. This implies that such power inequalities can strongly influence the outcomes of discussions about trade-offs among nature's contributions to people or between biodiversity protection or use. In fact, the latter points towards even more heavily biased decision-making processes within the social-ecological systems of the Americas in the future than what we mentioned before. This will imply, in turn, a stronger and deeper subordination of nature and ecosystems within the operating social-ecological systems of the continent to those interests focused more exclusively on productive objectives and narrowly defined economic benefits. The expected result will surely be larger future disregard of more inclusive objectives and broadly defined social objectives encompassing a larger well-being for the most, as well as a socioeconomic development path that is sustainable in the long run, and for the current as well as for the future generations.

As a concluding remark, it is possible to affirm that, in the Americas in general, and in the Latin America and the Caribbean region specifically, nature's "common resources" are currently managed in a clearly unsustainable way and most of the evidence shows that the efforts made in the last decades are clearly insufficient to revert the current trend. Moreover, there are additional worrying sings indicating that the situation could even be worse in the future.

6 Final, and Gloomier, Comments About the Future

In 2011, Frank Uekoetter, a German historian and political scientist, wrote an essay on the history of environmentalism entitled "Consigning Environmentalism to History? Remarks on the Place of the Environmental Movement in Modern History." He started his manuscript with the following opening phrase: "One of the more striking features of environmentalism is its remarkable resilience to obituaries." He immediately continued with the following thoughts: "Authors of obituaries usually fall into two categories, those who feel that environmentalism was futile and thus deserved to die, and those who seek an end because they have a different agenda to push. Both types of obituaries tend to flourish in times of crisis: when the German environmental movement was losing steam towards the end of the 1980s, authors as diverse as Herbert Gruhl and Hoimar von Ditfurth abandoned all hopes for environmental reform, thus providing examples for the first type of obituaries. More recently, Michael Shellenberger and Ted Nordhaus showcased the second type when they proclaimed the imminent 'death of environmentalism' their assumption being that a new post-environmental movement would emerge from the ruins. In both cases, the misrepresentation is obvious: what looked like an ending turned out to be, quite simply, 'change'" (Uekoetter 2011).

Because of its remarkable current validity, it is worth mentioning another paragraph of Uekoetter's essay, in spite of the fact that he wrote it a decade ago. "Moreover, it is hard to say where environmentalism is currently heading: the recent Copenhagen summit witnessed an unprecedented number of activists pushing the issue of global warming—but it also had the world's leaders disappointing these aspirations with a weak memorandum (Uekoetter 2011)".[10] In fact, the same is true regarding the most recent international conferences to deal with the global life-threatening problems of climate change and global warming, such as the Paris Agreement on Climate Change of December 2015 (COP15) of the United Nations Framework Convention on Climate Change, or its last COP 24, held in Katowice, Poland, last December 2018. On the one side, large numbers of people from all over the world demanding to the world leaders to make responsible decisions to meet the urgent challenges. On the other side, those interpellated world leaders producing disappointing and bland agreements, which clearly lead nowhere close to what is needed.

Again, the reason behind this is the one already explained above. The non-inclusive economic and political power mechanisms within the global socio-ecological systems currently operating in the world are simply serving a small powerful minority. This 0.1%, 0.01%, or an even smaller percentage of the world population is profiting from the current world's social-ecological systems and is indifferent to the life-threatening risks the world is facing. Because they erroneously believe, they will have enough resources to solve their problems when the time comes, or they simply disregard the existing scientific evidence.[11] These elites have the economic and political power and they used it, managing the public opinion through the manipulation of the information and disinformation they provide to the people through the public communication and press media they control. Thus, the environmental movements all over the world have a hard time to push their agenda to face the pressing world challenges. Moreover, those who control the economic

[10] The so-called Copenhagen Climate Summit was the 2009 United Nations Climate Change Conference, held in the capital of Denmark in December of that year, and corresponded to the 15th Conference of the Parties (COP 15) to the United Nations Framework Convention on Climate Change (UNFCCC) and the 5th Meeting of the Parties (MOP 5) to the Kyoto Protocol

[11] A clear example of this is the way that, on December 26, 2018, Donald Trump dismissed the "Fourth National Climate Assessment." This study, produced by his own administration, involving 13 federal agencies and more than 300 leading climate scientists, warns about the potentially catastrophic impact of climate change and reports that climate change will cost the United States hundreds of billions of dollars annually and damage health. Interviewed by reporters outside the White House about this report, Trump simply said, "I've seen it. I've read some of it, and it's fine," saying later, "I don't believe it" (BBC 2018; Oprysko 2018).

and political power within the world's social-ecological systems set the limits for how far these environmental movements can push their agendas. In the Latin American socioeconomic-political context, this is dramatically clear, because in several countries of the region the political systems, as well as the legal institutions in place, are much weaker than in North America, Europe, or the Scandinavian countries, for example. Therefore, in those Latin American countries when environmental activists go beyond the limits set by those holding the power, they often pay a high price.

In fact, the same phrase with which Uekoetter (2011) started his essay, mentioned above, can perfectly characterize the risky lives the environmentalists live in some Latin American countries. Unfortunately, in this case, because a much more macabre reason than the one Uekoetter had in mind when he wrote, "One of the more striking features of environmentalism is its remarkable resilience to obituaries." The reason this time is that Latin America is by far the world's most dangerous place to be an environmental activist. In fact, almost 60% of the environmental killings recorded in 2017 took place in the region. Only in Honduras, in the 9-year period of 2009–2017, 109 people were killed for taking a stand against dams, mines, logging, and agricultural projects. Two Goldman Environmental prizewinners were shot to death in two Latin American countries, in 2017. One in Mexico, Isidro Baldenegro, an indigenous leader and opponent of illegal logging. The other in Honduras, Berta Cáceres, an active opponent to a rash of development projects and concessions handed out, often illegally, to private companies for dams, mines, and other projects in her country (Pearce 2017; Global Witness 2017).

For decades, a large number of environmentalists have been killed in Latin America. One of the most publicized was the assassination of Francisco "Chico" Alves Mendes, more than 30 years ago, in Xapuri, Brazil. "Chico" Medes was a trade union leader and environmentalist of a poor background. He fought to preserve the Amazon rainforest and defended the Brazilian peasant and indigenous peoples' human rights. He received several national and international awards, including the United Nations Environmental Program Global 500 Roll of Honor Award, in 1987, and the National Wildlife Federation's National Conservation Achievement Award in 1988 (Devine 1999). He was murdered by a single shotgun blast outside his house. In those days, Chico Mendes used to mention "At first, I thought I was fighting to save rubber trees, and then I thought I was fighting to save the Amazon rainforest. Now I realize I am fighting for humanity." That year of 1988 of his assassination, "Chico" Mendes was the 19th environmental activist to be murdered in Brazil (Lallanilla 2018).

This is a menacing and shaming Latin American record. It is also the reflection of the way that the powers behind the large economic interests react when they defend the status quo to be able to continue profiting from the overexploitation of nature's "commons" in the region. These powers react brutally to the clamor of people demanding their rights to the resources provided by nature for the well-being of all and not of only few assassinating and murdering those brave enough to raise their voices. The complexity of the socio-ecological systems we live in today allows these powers to act from the shadows of the institutional arrangements they have created and maintain. From those shadows, these powers use the enormous economic resources they have acquired, through decades and centuries, and in a large proportion, from the economic rents appropriated from the destruction of nature's "commons." In fact, as Acemoglu and Robinson (2012) have explained it, it is through the mechanisms of power embedded in the world's socioeconomic systems along decades and centuries that the functioning of the economies, the legal structures, and the societies of the different nations ends up producing the results they produce. More or less freedom, more or less justice, more or less wealth, more or less equality, more or less order, and more or less happiness. In other words, Acemoglu and Robinson have elucidated the channels by which the elites dispose and arrange the rules that govern the socioeconomic systems of their countries to benefit themselves at the expenses of the many. The degree of the control of the power mechanisms of their socioeconomic systems attained by the elites explains the differences in the socioeconomic development of the different nations of the world. Those poorer and less developed nations are those that have been ruled by small elites that have organized society for their own benefit at the expense of the vast mass of people. In these nations, the political power has been narrowly concentrated and has been used to create great wealth for those who possess it, which results in socioeconomic systems that are "exclusive" because they marginalized the vast majority of their citizens (Acemoglu and Robinson 2012).

Nations of more developed countries, such as England and the United States were able to develop and became rich because their citizens overthrew the elites who controlled power, and created societies in which political rights were much more broadly distributed than before. In these newly created societies, the government was accountable and responsive to citizens, and the great mass of people could take advantage of economic opportunities (Acemoglu and Robinson 2012). England and the United States established more "inclusive" socioeconomic systems than those previously existing in the world, in which the benefits of their nations' collective efforts were much broader and equitably distributed to their peoples. These authors show that the reason why England is today much richer than Egypt and a large number of the less developed countries of the world is that, in 1688, England had a revolution that transformed the politics and thus the economics of its nation.[12] People fought for and won new political rights and they used them to expand their economic opportunities. The result was a fundamentally different political and economic trajectory, culminating in the Industrial Revolution. Acemoglu and

[12] This revolution, known as the "Glorious Revolution," overthrew King James II of England by a union of English politicians with the Dutch stadtholder William III, Prince of Orange. The successful invasion of England by William with a Dutch fleet and army led to his ascension of the former to the throne as William III of England, after the Declaration of Right, leading to the Bill of Rights 1689. This English Bill of Rights, set out certain basic civil rights, laid down limits on the powers of the king, and set out the rights of Parliament, including the requirement for regular parliaments, free elections, and freedom of speech in Parliament. It additionally set out certain rights of individuals including the prohibition of cruel punishment and reestablished the right of Protestants to have arms for their defense within the rule of law. Moreover, the bill also included no right of taxation without Parliament's agreement.

Robison extend their work to show that the large inequality in the world today is also due to the historical structuring and development of the power mechanisms operating within the socioeconomic systems of counties. What the peoples of England and the United States were able to acquire were newly structured socioeconomic systems that allowed the "invisible hand" to operate again in their economic subsystems. This made it possible, in turn, more productive and faster-growing economies.

As we explained before, nature's "commons," such as natural resources (water, minerals, lands, etc.), atmosphere (air, oxygen, carbon dioxide, ozone, etc.), climate (temperature, humidity, precipitations, winds, etc.), ecosystems (fisheries, forests, wetlands, deserts, etc.), and ecosystem services (oxygen generation and sequestration, water purification and regulation, biomass, nutrients cycle, erosion control, food and fibers, etc.) generate enormous amounts of economic rents—pure and new wealth—from the economic activities of societies. These enormous economic rents belong to the owners of these "commons," who are each one and all the inhabitants of the planet and, therefore, these economic rents should be received by every one of them. However, the appropriation of these economic rents has been historically determined, and is currently determined, by the power subsystems operating within the socio-ecological system of every country. These power subsystems are controlled by the socioeconomic-political elites of every country, who use them to appropriate the largest share of these enormous economic rents for themselves. This provides these elites even more power to control the social-ecological systems in their own benefit. In the economic realm, the clearest manifestation of the latter is the large economic inequality existing in the world.

As pointed out by Piketty (2014), the history of the distribution of wealth has always been deeply political; it cannot be reduced to purely economic mechanisms. Moreover, the history of the high economic inequalities is shaped by the moral views of economic, social, and political actors regarding justice and inequity, as well as by their relative powers and the collective choices that result. It is, indeed, the joint product of all relevant actors combined.

Obviously, the distribution of income—the wealth generated every year that can be consumed without diminishing the previously existing wealth—is also deeply political and cannot either be reduced to purely economic mechanisms. Because the distribution of income is also governed by the power mechanisms embedded in the socioeconomic system of every country. With regard of the income inequality in the United States, Piketty shows that, during the 100-year period of 1910–2010, the top decile of the income distribution claimed 45–50% of the US national income in the 1910s–1920s, before dropping to 30–35% by the end of the 1940s. From 1950 to 1970, income inequality stabilized at that level. Then, in the 1980s, inequality increased rapidly until 2000, when it returned to a level of 45–50% of national income.

From the analysis of the US and other countries' historical paths of income inequality, Piketty arrives to the key conclusion of his work. He states that the fundamental inequality in income distribution is explained basically by the fact that the average annual rate of return on capital (r) has been higher than the rate of growth of the economy (g). When r is larger than g by a significant margin, as it occurred through much of history until the nineteenth century and as is likely to be the case again in the twenty-first century, then it logically follows that inherited wealth grows faster than output and income. Thus, Piketty explains, people with inherited wealth need to save only a portion of their income from capital to see that capital grows more quickly than the economy as a whole. Under such conditions, it is almost inevitable that inherited wealth will dominate wealth amassed from a lifetime's labor by a wide margin. Thus, the concentration of capital will attain extremely high levels. These levels are potentially incompatible with the meritocratic values and principles of social justice fundamental to modern democratic societies (Piketty 2014).

It is crucial to understand that Piketty's "r", the average annual rate of return on capital, includes profits, dividends, interest, economic rents, and other income from capital, expressed as a percentage of its total value. Without any doubt, since the industrial revolution and the subsequent inception of the capitalist economic system, economic rents constitute a very significant part of "r". Moreover, as we explained above, economic rents generated by nature's "commons" have been, and are, enormous in the modern economy. Therefore, the appropriation of these economic rents by the elites constitutes a large part of the reason why r has been systematically larger than g. Because the capitalist elites, the owners of capital, which is "physical" capital—machinery, equipment, buildings, plants, tools, vehicles, etc.—have been able to appropriate for themselves a large proportion of the rents of nature's "commons" as if they were part of the wealth generated by their "physical" capital.[13] In fact, these physical capital-elites have used their control over the power mechanisms of the socio-ecological systems of their countries to rig the laws and regulations of the economic subsystems to appropriate for themselves the largest proportion of the economic rents generated by nature's "commons." In this way, these elites have been able to appropriate for themselves the lion's share of nature's generated-economic rents and have perpetuated the high inequality of the distribution of income and wealth in their countries and the world.

Therefore, the destruction of nature, ecosystems, ecosystem services, as well as the increasing human-induced deterioration of the planet's atmosphere and the cur-

[13] Capitalism is generally defined as "an economic system where private entities own the factors of production. The four factors are entrepreneurship, capital goods, natural resources, and labor." (Amadeo 2018). In a capitalist economy, capital assets—such as factories, mines, and railroads—can be privately owned and controlled, labor is purchased for money wages, capital gains accrue to private owners, and prices allocate capital and labor between competing uses (Jahan and Mahmud 2015). The owners of capital goods, natural resources, and entrepreneurship exercise control through companies. The individual owns his or her labor (Amadeo 2018). It is interesting to notice how the ownership of natural resources is implicitly attributed to capitalists because it is simply accepted that companies exercise control over them, and companies are mainly owned by capitalists in the capitalist system. The same is done, when mines are defined as "capital assets," jointly with man-made assets such as factories and railroads, without specifying if the mines include or not the related mineral ores or the fossil fuel resources. Obviously, these views differ from the notion that every inhabitant of planet Earth owns nature's resources (natural resources or nature's "commons").

rent climate change and global warming are, primarily and fundamentally, reflections of the use by the capitalist elites of the power mechanisms imbedded in the socio-ecological systems of their countries. Unfortunately, these elites live alienated by the illusion that the same power that has allowed them until these days to appropriate their enormous current wealth will also allow them in the future to get through any possible problem. It seems that they ignore completely what the final Armageddon of the planet Earth will imply if some of the unpleasant current scientific projections become true.

As a corollary of this chapter, it seems worthy to add a couple of closing suggestions. The first is that many of the arguments presented in this work may seem very alien to experts working on some of the many different dimensions of the ongoing nature destruction, biodiversity deterioration, climate change, or global warming, or to anybody concerned about these crucial problems. However, I am convinced that the more foreign our arguments in this chapter appear to the reader, the more relevant this first suggestion could be for her or him. For that reason, I state this suggestion in a very simple and direct way. I believe that much more holistic and down to the Earth approaches are needed today to meet the big challenges currently threatening our human species. A multidisciplinary approach is the only one that could allow scientists and technical experts to change their (our) narrow specific paradigms in order to be able to define jointly common new ways of dealing with the multiple dilemmas of how to make the highly complex current social-ecological systems to move in the right directions, quickly and effectively. This requires a big effort from physical sciences experts to try to understand the highly complex and foreign to them lessons provided from the social sciences. It also requires a big effort from social sciences experts to try to understand the highly complex and foreign to them lessons provided from the physical sciences. Moreover, once they are able to come up with commonly designed and agreed propositions, they must work with politicians, government officials, public servants, social leaders, and private entrepreneurs in order to mobilize the world socioeconomic-ecological systems to implement those propositions, quickly and effectively.

My second suggestion arises directly from the previous discussion regarding the way the elites use the power mechanisms of the world social-ecological systems to their own benefit. It is absolutely necessary to focus as many efforts as possible in the future to counterbalancing the decisions made by these elites in order to finally revert those decisions and eliminate their disastrous effects on the planet nature, the Earth ecosystems, the global climate, and the vast majority of people. To that purpose, it is necessary to use every space open to peaceful and effective civil and political action, which nowadays imply using all the possibilities provided by the new communication technologies for social interaction and mobilization.

It is necessary to communicate to the people of every country the real risks and the alarming implications of the current and growing world challenges regarding nature destruction and global warming and to transmit effectively to them the urgency of meeting these challenges within the next decades. We must get out of our comfort zones to make the people alarmed of the large risks their comfort zones and their lives are facing today because of the rampant current destruction of the planet nature and ecosystems and the rapidly increasing global warming. We have to convey the messages from our sciences in simple and easily understandable forms to the public. We must do it, and a sufficient reason to do it is that, if we do not do it, we are really doing what those powerful elites want us to do because that is precisely their strategy to manipulate us. Our inaction is our best move for them.

Unfortunately, the efforts that my suggestions demand are immense; it would be much easier otherwise. However, we must be able to meet the current world challenges to assure the survival of our sons and daughters and of our grandchildren. Otherwise, probably they will never have a chance to see grandchildren of their own. I believe that a terrifying conviction like this one might have had in his mind Jay Inslee, the Governor of Washington State when he said, "We are the first generation to feel the impact of climate change, and the last generation that can actually do something about it" (AD 2018; Buzz Feed News 2019).

References

Acemoglu D (2008) Oligarchic versus democratic societies. J Eur Econ Assoc 6(1):1–44

Acemoglu D, Robinson JA (2008) Persistence of power, elites, and institutions. Am Econ Rev 98(1):267–293

Acemoglu D, Robinson JA (2012) Why nations fail. Crown Publishers, Random House, New York

AD (2018) Art Basel Hong Kong 2018: artist Vivek Vilasini seeks to capture climate change atmospherics. https://wwwarchitecturaldigestin/content/art-basel-hong-kong-2018-vivek-vilasini-sakshi-gallery/#s-cust0. Accessed 10 Jun 2019

Aguiar AP, Vieira IC, Assis TO et al (2016) Land use change emission scenarios: anticipating a forest transition process in the Brazilian Amazon. Glob Chang Biol 22(5):1821–1840

Amadeo K (2018) Capitalism, its characteristics, with pros and cons. How it works compared to socialism and communism. The balance. https://www.thebalance.com/capitalism-characteristics-examples-pros-cons-3305588 Accessed 10 Jun 2019

Ban SS, Graham NAJ, Connolly SR (2014) Evidence for multiple stressor interactions and effects on coral reefs. Glob Chang Biol 20(3):681–697

BBC (2018) News. US & Canada. Trump on climate change report: 'I don't believe it'. https://www.bbc.com/news/world-us-canada-46351940 Accessed 10 Jun 2019

Betuel E (2019) Deforestation in Brazil's Amazon rainforest will cause major warming by 2050. https://wwwinversecom/article/54212-climate-change-deforestation-amazon-regions. Accessed 10 Jun 2019

Bommarco R, Kleijn D, Potts SJ (2013) Ecological intensification: harnessing ecosystem services for food security. Trends Ecol Evol 28(4:230–238

Butler RA (2019) Amazon destruction. https://rainforestsmongabaycom/amazon/amazon_destructionhtml Accessed 10 June 2019

Buzz Feed News (2019 Jay Inslee is running for president as the climate candidate. https://www-buzzfeednewscom/article/zahrahirji/jay-inslee-2020-president-climate-change. Accessed 10 Jun 2019

Cairns J Jr (1995) Ecosystem services: an essential component of sustainable use. Environ Health Perspect 103(6):534

CBD (2001) Global biodiversity outlook 1. Secretariat of the Convention on Biological Diversity, Montréal

CBD (2014) Global biodiversity outlook 4. Secretariat of the Convention on Biological Diversity, Montréal
Collier P (2007) The bottom billion. Oxford University Press, New York
CORE TEAM (2019) The Economy. https://core-econ.org/the-economy/?lang=en. Accessed 10 Jun 2019
Costanza R, d'Arge R, de Groot R et al (1997) The value of the world's ecosystem services and natural capital. Nature 387:253–260
de Groot RS, Wilson MA, Boumans RMJ (2002) A typology for the classification, description and valuation of ecosystem functions, goods and services. Ecol Econ 41:393–408
Devine C (1999) Mendes, Chico. In: Devine C, Poole H (eds) Human rights: the essential reference. The Oryx Press, Phoenix, AZ, p 202
Figueroa E, Aronson J (2006) New linkages for protected areas: making them worth conserving and restoring. J Nat Conserv 14:225–232
Figueroa E, López RE, Gutierrez P (2013) Structural barriers to long-term productivity growth and sustainable development: lessons from Chile's 'implicit industrial policy'. Annual meeting of the Latin American studies association; Washington, DC, May 29–June 1, 2013
Folke C (1991) Socio-economic dependence on the life-supporting environment. In: Folke C, Kåberger T (eds) Linking the natural environment and the economy: essays from the eco-eco group. Ecology, economy & environment, vol 1. Springer, Dordrecht, pp 77–95
Giddens A (2009) The politics of climate change. Polity, New York
Global Witness (2017) Environmental defenders. Annual report 2017: spotlight on corruption. Global Witness, London, pp 23–26
Grau HR, Aide M (2008) Globalization and land-use transitions in Latin America. Ecol Soc 13(2):16. http://www.ecologyandsociety.org/vol13/iss2/art16/. Accessed 10 Jun 2019
Hansen J (2018) Climate change in a nutshell: the gathering storm. http://www.columbia.edu/~jeh1/mailings/2018/20181206_Nutshell.pdf. Accessed 10 Jun 2019
IEA (2012) World energy outlook. International Energy Agency, OECD, Paris
IPBES (2018) The IPBES regional assessment report on biodiversity and ecosystem services for the Americas. Secretariat of the Intergovernmental Science-Policy Platform on Biodiversity and Ecosystem Services, Bonn
IPCC (2014) Climate Change 2014: Synthesis Report. Contribution of Working Groups I, II and III to the Fifth Assessment Report of the Intergovernmental Panel on Climate Change IPCC, Geneva, https://www.ipcc.ch/site/assets/uploads/2018/02/SYR_AR5_FINAL_full.pdf. Accessed 10 Jun 2019
IPCC (2018) Global Warming of 1.5°C. An IPCC Special Report on the impacts of global warming of 1.5°C above pre-industrial levels and related global greenhouse gas emission pathways, in the context of strengthening the global response to the threat of climate change, sustainable development, and efforts to eradicate poverty: summary for Policymakers. World Meteorological Organization, Geneva
Isbell F, Craven D et al (2015) Biodiversity increases the resistance of ecosystem productivity to climate extremes. Nature 526(7574):574–577
Jahan S, Mahmud AS (2015) What is capitalism? International Monetary Fund. Finance Dev 52(2)
Lallanilla M (2018) Learn about the life of and death of Rainforest Activist Chico Mendes. ThoughtCo. https://www.thoughtco.com/who-killed-chico-mendes-1708804. Accessed 10 Jun 2019
Leahy S (2018) Climate change impacts worse than expected, global reports warns. National Geographic. https://www.nationalgeographic.com/environment/2018/10/ipcc-report-climate-change-impacts-forests-emissions/. Accessed 10 Jun 2019
López RE (2019) Power in economics: growth, inequality and politics. Santiago, Department of Economics, University of Chile
López RE, Figueroa E (2016) On the nexus between fiscal policy and sustainable development. Sustain Dev 24:201–219
Mankiw G (2012) Principles of economics, 6th edn. South-Western Cengage Learning, Mason

McSweeny R, Pearce R (2017) Analysis: only five years left before 1.5C carbon budget is blown. Carbon Brief. https://www.carbonbrief.org/analysis-only-five-years-left-before-one-point-five-c-budget-is-blown
Oliver TH, Heard MS, Isaac NJB et al (2015) Biodiversity and resilience ecosystem functions. Trends Ecol Evol 30(11):673–684
Olson DM, Dinerstein E, Wikramanayake ED et al (2001) Terrestrial ecoregions of the world: a new map of life on earth. BioScience 51(11):933–938
Oprysko C (2018) 'I don't believe it': Trump dismisses grim government report on climate change. https://www.politico.com/story/2018/11/26/trump-climate-change-report-1016494. Accessed 10 Jun 2019
Ostrom E (2009) A general framework for analyzing sustainability of social-ecological systems. Science 325:419–421
Painuly JP (2001) Barriers to renewable energy penetration; a framework for analysis. Renew Energy 24:73–89
Patterson JJ, Thaler T, Hoffmann M et al (2018) Political feasibility of 1.5° C societal transformations: the role of social justice. Curr Opin Environ Sustain 31:1–9
Pearce F (2017) Honduras, where defending nature is a deadly business. The Guardian, 22 March. https://www.theguardian.com/environment/2017/mar/22/honduras-berta-caceres-activism-defending-nature-deadly-business. Accessed 10 June 2019
Piketty T (2014) Capital in the Twenty-first Century. The Belknap Press of Harvard University Press, Cambridge
Prevedello JA, Winck GR, Weber MM et al (2019) Impacts of forestation and deforestation on local temperature across the globe. PLoS One 14(3):e0213368
Rockström J, Gaffney O, Rogeli J et al (2017) A road map for rapid decarbonization. Science 355:1269–1271
Samuelson P, Nordhaus W (2010) Economics. MacGraw-Hill, New York
SCOS (2019) Story: New IPBES Global assessment Informs Ocean Biodiversity and Ecosystem services Goals. Stanford Center for Ocean Solutions, Stanford Woods Institute for the Environment, and Hopkins Marine Station of Stanford University. https://oceansolutions.stanford.edu/news-stories/new-ipbes-global-assessment-informs-ocean-biodiversity-and-ecosystem-services-goals. Accessed 10 Jun 2019
Sovacool BK (2016) How long will it take? Conceptualizing the temporal dynamics of energy transitios. Energy Res Soc Sci 13:202–215
Sovacool BK, Hirsh RF (2009) Beyond batteries: an examination of the benefits and barriers to plug-in hybrid electric vehicles (PHEVs) and a vehicle-to-grid (V2G) transition. Energy Policy 37(3):1095–1103
Steneck RS, Graham MH, Bourque BJ et al (2002) Kelp forest ecosystems: biodiversity, stability, resilience and future. Environ Conserv 29(4):436–459
Tilman D, Downing JA (1994) Biodiversity and stability in grasslands. Nature 367:363–365
Uekoetter F (2011) Consigning Environmentalism to History? Remarks on the Place of the Environmental Movement in Modern History. Rachel Carson Center. RRC Perspectives, N° 7 https://www.jstor.org/stable/10.2307/26240343. Accessed 12 Jun 2019
UNEP (2010) State of biodiversity in Latin America and the Caribbean. United Nations environment program. https://www.cbd.int/gbo/gbo3/doc/StateOfBiodiversity-LatinAmerica.pdf. Accessed 10 Jun 2019
UNEP (2012) Chapter 12: Latin America and the Caribbean. In: Global Environment Outlook-5: environment for the future we want. United Nations Environment Program Valetta, Malta
UNEP-WCMC (2016) The state of biodiversity in Latin America and the Caribbean: a mid-term review of progress towards the Aichi Biodiversity Targets. UNEP-World Conservation Monitoring Centre, Cambridge
Vitousek PM, Mooney HA, Lubchenco J et al (1997) Human domination of Earth's ecosystems. Science 277(5352):494–449

WCMC (2002) World atlas of biodiversity: Earth's living resources in the 21th century. UN Environment World Conservation Monitoring CENTRE. http://www.international-food-safety.com/pdf/World%20Atlas%20Of%20Biodiversity.pdf. Accessed 10 Jun 2019

Wurdig LF, Beber FC, Alves V et al (2009) The Brazilian Pampa: a fragile biome Luiz Fernando. Diversity 1:182–198

WWF (2018) Living planet report 2018 technical supplement: living planet index. World Wildlife Fund http://awsassetspandaorg/downloads/lpr2018_technical_supplement_for_lpipdf. Accessed 10 Jun 2019

Zaneveld JR, Burkepile DE, Shantz AA et al (2016) Overfishing and nutrient pollution interact with temperature to disrupt coral reefs down to microbial scales. Nat Commun 7(1):1–12

Part III
Case Studies

Spatial Modeling of Social-ecological Systems of Hydrological Environmental Services in Las Conchas Creek Basin, Argentina

Ricardo Castro-Díaz, Walter Sione, Brián Ferrero, Virginia Piani, Griselda Urich, and Pablo Aceñolaza

Abstract Analyzing social-ecological systems should include both temporal and spatial dimensions as components of their complexity. Four conditionalities integrate the response of the territory to external and internal impacts: biophysical determinants, social structures, economic forces, and political reconfigurations. The causality behind the dynamics of systemic elements represents multiple relationships causing transformations in the structure and identity of the system.

In this chapter, we describe a spatial model to understand the processes that support the hydric regulation as an environmental service in the Arroyo Las Conchas watershed in Entre Rios (Argentina). We build the contextual conditions of biological diversity and ecosystem structures in the watershed, followed by the identification of the social structures which interacts with the agricultural production and the

---

R. Castro-Díaz (✉)
Centro Regional de Geomática, Universidad Autónoma de Entre Rios/CONICET, Oro Verde, Argentina
e-mail: ircastrod@unal.edu.co

W. Sione
Facultad de Ciencia y Tecnología, Centro Regional de Geomática, Universidad Autónoma de Entre Ríos, Diamante, Argentina
e-mail: sione.walter@uader.edu.ar

B. Ferrero
CIT-Rafaela, CONICET, Santa Fe, Argentina
e-mail: brianferrero@conicet.gov.ar

V. Piani
Centro Regional de Geomática-Universidad Autónoma de Entre Ríos/Centro de Investigaciones Científicas y Transferencia de Tecnología a la Producción-CONICET, Parana, Argentina

G. Urich
Centro Regional de Geomatica, UADER, Parana, Argentina

P. Aceñolaza
CICyTTP-CONICET and FCA-UNER, Diamante, Argentina
e-mail: pablo.acenolaza@fca.uner.edu.ar

L. E. Delgado, V. H. Marín (eds.), *Social-ecological Systems of Latin America: Complexities and Challenges*, https://doi.org/10.1007/978-3-030-28452-7_11

political reconfiguration focused on the capitalization of nature. Finally, we generate a waterlogging probability model in two scenarios (current state and reforestation scenario) using remote sensing and GIS data for integrating the interdependence of these conditioners and establishing the state of this environmental service of hydrological regulation in the basin.

Keywords Social-ecological systems · Latin America · Complexity · Argentina · Hydrology · Spatial models

1 Introduction

The understanding of environmental or ecosystem services is linked to the study of biophysical variables and land use (Nelson et al. 2009), global dynamics (Boumans et al. 2002), knowledge-based (Turner et al. 2016), and the inclusion of agent-based models for sustainable behaviors (Alonso-Betanzos et al. 2017). These perspectives imply the so-called territory resignification as a complex system wherein coexist dynamic relationships among the social, economic, historic-politic, and biophysical dimensions and focusing in permeable spaces like the core of the landscape concept (Christensen et al. 2016).

The landscape dynamics include disturbances, interactions, relationships, and feedbacks creating spatiotemporal drivers for heterogeneity, scale, and thresholds (Turner 2010). This complexity is mainly based on the existence, modification, dependence, and sustainability related to social actors and their endogenous and exogenous impacts which could modify their own structure and identity in social-ecological systems (SES) (Cumming and Collier 2005).

In this context, of understanding complex processes in a territory, spatial modeling emerges as a way to configure the patterns nurturing the behavior of the SES (Cumming 2011). Some of these tools have been developed in Latin America for modeling ecosystem services (e.g., the Collaborative Protocol ECO-SER oriented to analyze the dynamics and behavior of several ecological fluxes, Laterra et al. 2011). Accordingly, the mapping ecosystem functions and services is held by the need of understanding the behavior of biophysical factors affecting the spatial and temporal patterns which support the hydrological regulation as an environmental service (Dawson et al. 2010).

The spatial data are not only a requirement for modeling SES, but also to the procedures to study thresholds, feedbacks, and alternative states, which are directly linked to the society (Castro-Díaz 2017). Consequently, the connections of spatial modeling fall inside the dialectic between "The tragedy of the commons" (Hardin 1968) and "The commons governance" (Ostrom 1990) as elements for the primary discussion of our reality versus the ideal. These spatial dynamics between the territory as a social construction and the ecosystem as a biotic/abiotic interactive network are revealed as the integrative conception of spatial resilience, involving the quantification of systemic elements.

Notwithstanding, it is primordial to accomplish the social-spatial analysis in the establishment of different territorial manifestations influenced by the actors on the permanence/decadence of ecosystem services (Castro-Díaz and Natenzon 2018b). Nahuelhual et al. (2015) refer to a weak relationship between the purpose of "the map" and their methods because the study of ecosystem services is based on economic values or biophysical transferences, leaving behind the spatial-temporal component and the scaling perspective. This leads to the necessity to find ways to "spatialize society and their effects" by explaining the intervening and defining human factors of the ecosystem services at multiple scales.

In this chapter, we establish the systematic elements which could help to find the implications between the changes generated by the communities and the ecosystem responses. We developed the analysis of several social elements which influence—and are influenced—by ecological fluxes of the environmental services of the hydrological regulation in a case study of Las Conchas Creek (LCC) basin in Entre Ríos (Argentina).

The Basin as a Social-ecological System

Drainage basins are basic units for water management and planning, including the hydrogeological system that supplies aquifers (Molina Garate 2008). Water management systems and their catchment basins are created to avoid conflicts between human beings and their environment. Within these conflicts, it can be mentioned the greater demand for water, pollution, natural disasters, and deterioration in basins. Conflicts are more pronounced in the so-called urban basins, which are those where populations settle or supply urban areas (Dourojeanni and Jouravlev 1999; Gencer et al. 2018).

Indeed, basin management appeared as an activity linking conservation of the elements and natural resources with the specific management of water; combining aspects of protection, conservation, and use of those resources. It has been considered an instrument of environmental management that must adapt to the geomorphological complexity and the eco-hydrological dynamics, as well as to the different ways of land appropriation (Benegas and Faustino 2008).

Ramsar's manuals (2010) state "the need to plan at the catchment or basin level, which implies integrating the management of water resources and the conservation of wetlands." The main challenge that currently exists in pursuit of sustainable management is the need to balance the requirements and availability of water, in order to assure future generations, the same qualitative and quantitative levels that are currently available (Ruiz de Galarreta et al. 2010).

In Argentina, different studies on basin management have been carried out, most of them highlighting the problems that arise from poor management or lack of planning. The most common examples found are those related to environmental pollution and changes in land use by the agro-industry sector. Pochat (2005) stressed that the analysis of Argentina's experience in water management at basin and basin organizations level, both provincial and interjurisdictional, may be of interest due to the variety of nuances that it presents. Resulting from climatic and hydrological diversity, but, above all, its political-institutional structure, with features that differentiate it from other countries of the region.

An important level of historical intervention, at the landscape scale, is presented at LCC basin, where changes have taken place both in the natural and anthropogenic systems. Therefore, repercussions on the functioning of the system are expected (Bortoluzzi et al. 2008). Provincial Law N° 9757, of the province of Entre Ríos, has the purpose of creating Basin Committees and Water Consortiums of the province, which will generate conditions and projects to ensure regional and provincial integration, rational exploitation of hydraulic works and the sustainable use of water in the public domain (Entre Ríos 2006). Through this law, a participatory and democratic system is established in the region for the management of the basin. However, despite having a legal framework, there has not been enough progress in terms of environmental management of the basin. This is worrisome, especially considering the pace at which changes in land use are advancing, with a tendency towards urbanization on the one hand, and agricultural expansion on the other.

The promotion of participation and sustainable local development are essential to implement more efficient environmental policies in the LCC (Castro-Díaz et al. 2018). Local actors or "communities" are usually conceived as homogeneous entities, ignoring inequalities and internal power relations, which are often conflicting and complex (Stoll-Kleemann et al. 2006). Ignoring this social reality brings serious consequences when a group of external actors tries to establish a new regime of environmental governance. Although local actors, as collective actors, tend to develop internal social organization schemes to regulate the use and access to natural resources, the (forced) modification of arrangements by external actors tends to politicize the context and generates new ways of interacting with the environment. Therefore, local actors must be seen as a heterogeneous social group, subject to tensions and conflicts of social, political, economic, and cultural nature. This establishes, voluntarily or necessarily, relations with other actors at different spatial levels. The idea is to ensure the use and access to the natural resources of a given territory when generating strategies for new conservation spaces. This, in turn, is necessary to devise mechanisms that guarantee the control of the territory by local actors and limit the use and access to external actors (Brenner 2012).

2 Methods

Following the Ostrom's conceptual framework (2009), we characterized SES elements as relationships and components of the whole set involved in the supply and restoration of diverse ecological fluxes (Bruckmeier 2016). They were selected by representing a clear relationship between the biophysical determinants, economic forces, political reconfigurations, and social structures stated as conditioners in Castro-Díaz (2013, 2017) and Castro-Díaz and Natenzon (2018a).

For understanding the biophysical determinants, we collected information on the changes in the landscape since the eighteenth century and other secondary information about patterns in phytography dynamics. The economic forces showing the environmental stressors emerged by the agricultural intensification and their effects on the natural landscape, and they were studied using satellite imagery and secondary information.

The political reconfiguration was centered in the understanding of the perspective of the payment for ecosystem services (PES) and the capitalization of nature propelled as a response of the ecosystem degradation of the LCC. Finally, the social structures are aimed at explaining the current socio-environmental conflicts caused by the intensive agriculture production and derived implications on water pollution that affect multiple rural communities as stated in mobilizations and sues against the Argentinian State.

We focused in these four conditioners to provide relevance to the land use and land cover (LULC), agriculture and economical production, and the socio-environmental conflicts as components of the current disturbance in the hydrological environmental services.

Satellite Imagery and Field Work

The LCC basin is located in the south of the Department of Paraná (Entre Ríos, Argentina) on an area of 2.184 km^2 and a maximum length on the order of 50 km (Aceñolaza and Sayago 1980; Bortoluzzi et al. 2008). Bortoluzzi et al. (2008) delimited 13 sub-basins related to streams and other 8 headwaters directly associated with the Las Conchas Creek main channel. Derived from canonical classification analysis (Fig. 1), this study defined seven main land use classes showing a high predominance of agriculture, followed by hay, forest, and urban (including industrial and residential in diverse densities) (Annex I: Fig. A1).

We defined natural coverages (i.e., forests and pastures) using canonical correlation analysis (CCA) for (1) Landsat 8 for a 2 and 17 March 2017 mosaic and (2) Sentinel 2A; September 29, 2018. The synergy classification was based on 25 classes from Sentinel, defining 7 as the most relevant for this study and validated in six field trips during 2017 and 2018.

We also processed a digital elevation model (DEM) developed by the Remote Sensing Directorate and the Geodesy Directorate of the National Geographic Institute of Argentina. The DEM is based on aerial photogrammetric flights carried out between 2011 and 2016, having a spatial resolution of 5 m and a sub-metric vertical precision which allowed us to make a precise delimitation of the basin.

The model for waterlogging probability used (a) the rational formula for the runoff coefficient and calculated on the result of the CCA and the slope factor from the DEM; (b) the erodibility and (c) available water capacity were calculated with the USLE K Factor using the SPAW model with data from INTA (2014); (d) the infiltration rate was obtained directly from INTA (2014). Finally, we combined them into a unique map for understanding the spatial behavior of the mentioned variable.

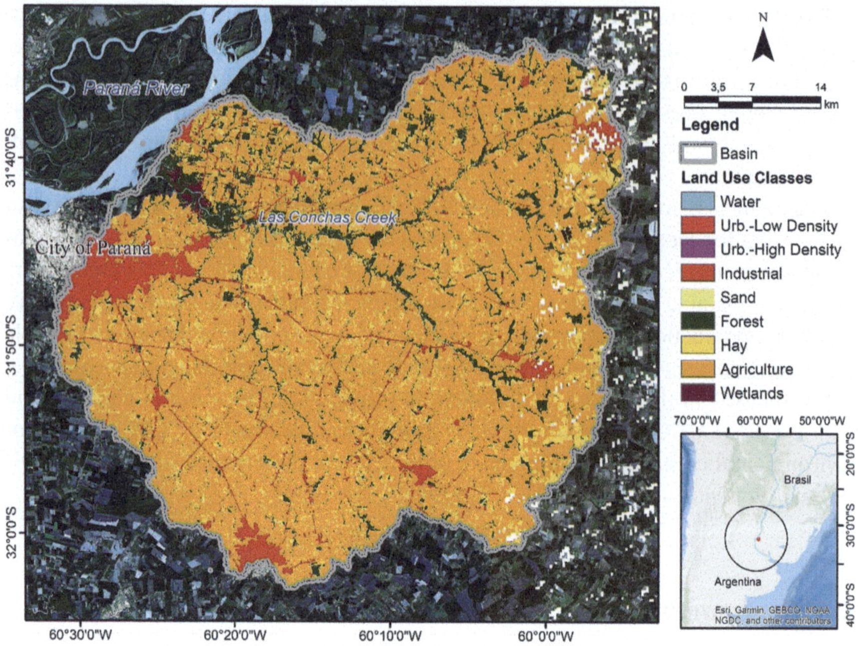

Fig. 1 LCC basin and land use cover (2017) using CCA on Landsat 8 imagery. Imagery from: *Land Processes Distributed Active Archive Center (LP DAAC), located at USGS/EROS, Sioux Falls, SD.* http://lpdaac.usgs.gov

We used this spatial model to depict an alternate state based on the findings by Farley et al. (2005). They state that afforestation reduces runoff being higher in drier regions. Using GIS modeling tools and based on the assumption that reforestation might be a mean for recovering environmental services such as waterlogging regulation, we generated a forest expansion taking a growth seed from the borders of the currently forested areas up to 100 m. Later, we transformed the runoff coefficient from agriculture to forests for those parcels which have their centroid inside the defined area. Finally, we compared the results from the (1) current waterlogging probability with the (2) waterlogging probability with a reforestation scenario, generating a map for the latter condition.

3 Results

Biophysical Determinants

Vegetation Elements

Félix de Azara (1742–1821), the military, naturalist, and cartographer, was commissioned to set boundaries between the Portuguese and Spanish territories of the Río de la Plata exploring the area of Las Conchas basin in 1784 and generating what might be the first specific reference of the physiognomy and vegetation of the basin (de Azara 1809). On his way from the Baxada (Paraná) to the Estancia de Vera Mujica (northern margin of LCC), he mentions that he crossed high and open algarrobales (*Prosopis* spp.) with aromos (*Vachellia aroma*) in the hills and that in the lower parts of the territory, it was observed willows (*Salix humboldtiana*) and ombuses (*Phytolacca dioica*), referring to the alluvial plain of "Las Conchillas" (Las Conchas), as low and marshy.

de Azara (1809) states the presence of 2 (or 3) inhabited places in the entire route of more than 20 km. This gives an idea that there was low pressure over natural resources limited to cattle and sheep. At that time, the Vera Mujica farm was the most important populated place north of the Las Conchas stream producing livestock. Crossing the north-west portion of the basin, heading to Asunción del Paraguay, the landscape is described as mixed, with algarrobales, open areas (grasslands) and areas with yatay palm trees (*Butia yatay*) (Aceñolaza 2003).

In 1827, the French naturalist Alcide D'Orbigny (1835–1947) disembarked near the mouth of the Las Conchas stream, describing the observed vegetation as covered by trees of acacia and aroma genera (*Prosopis* and *Vachellia*) and the presence of timbó (*Enterolobium contortisiliquum*). He also determined the presence of a Fan-shaped leave palm (*Trithrinax campestris*). Later, in the north-western portion of the basin, he mentions the presence of the yatay palm trees (*B. yatay*) and undulating fields with open forests without cattle, indicating that the area was practically uninhabited and with a single wheat crop (Aceñolaza 2003). This author mentions a drought and the presence of burnt forests, assuming fires were one of the most important variables of environmental impact.

From the second half of the nineteenth century, with the conformation of the Argentine State, land colonization was promoted with European immigrants. In 1853 with the foundation of Colonia Las Conchas town, many important changes took place in the area, associated with land demarcation, the concession of parcels to settlers, and their changes of low production to more intensive agricultural crops. The Paraná River with confluences to Las Conchas mouth (Villa Urquiza, Nuñez, Colonia Celina) was used for this purpose, taking the production to Buenos Aires urban markets.

Between 1912 and 1920 the railroad from Curuzú Cuatiá to Puerto de Diamante was built, crossing Las Conchas basin in its upper portion, generating environmental changes coming from the transport connectivity (Truffer 2010). With it, agricultural goods got a fast track for the output of regional production. Consequently, new towns were founded on the edge of the railway and meeting the increase in regional activities. The area reached its expansion peak between the 30s and 40s and implied the clear-cut of an important part of the original forests of the area (Truffer 2010). The arrival of technology in the countryside generated lasting droughts, the introduction of agricultural pests, increasing of fallows and cattle ranching (Leyes 2016).

In the late twentieth century, natural vegetation of the basin represented a boundary of different plant formations. Aceñolaza and Manghesi (1993) mention that three phytogeographic units were distinguished in the region: the association of the

Ñandubay District (within the Espinal Phytogeographic Province), and the Pampean grassland (Cabrera 1976). The third unit shows a limited spatial development within the basin, corresponding to cliff forests associated with the Paraná River.

In the basin portion dominated by the District of Ñandubay, patches of closed forest can be found, with different height and density, with the presence of *Prosopis affinis, Prosopis nigra, Vachellia caven, Aspidosperma quebracho-blanco*, and *Phytolacca dioica*. Towards the lower parts and closer to permanent (or temporary) watercourses, it can be usually found complementary *Sideroxylon obtusifolius, Fagara hyemalis, Myrcianthes cisplatensis, Sapium haematospermum, Trithrinax* campestris, and *Achatocarpus praecox*. It also can be found secondary successional situations dominated by *Vachellia caven*, shrubs of *Eupatorium* spp. or *Baccharis* spp. (Aceñolaza 2000; Bortoluzzi et al. 2008).

The other phytogeographic unit, Pampean grassland (Cabrera 1976), corresponds to a steppe of grasses with dominance of Estípeas, Poeas, and Eragrósteas tribes. The characteristic community of the open fields is composed by *Nassella neesiana, N. hyalina*, and *Jarava plumosa* among other species. The third and last unit corresponds to cliff forests associated with the Paraná River. These forests have a restricted distribution to the final portion of the Las Conchas stream, at its mouth to the Paraná River. It includes species like the Espinal (*P. nigra, P. dioica, A. quebracho-blanco*) and others associated with the watercourses of the Espinal, or the Paraná River corridor (*Coccoloba argentinensis, Myrsine laetevirens, Rupretchia laxiflora, Nectandra angustifolia, Erythrina crista-galli, Hexaclamys edulis*).

Economic Forcing

Current land use is associated with the fragmentation and reduction of continuous forest surfaces and patches with different degrees of degradation (Aceñolaza 2000; Maldonado et al. 2012, 2013; Muñoz et al. 2005). These forests are used as extensive grazing areas and selectively use for sawing (*Prosopis nigra*), posts (*Prosopis affinis*), or firewood (*P. affinis, P. nigra, V. caven*, etc.) (Roskopf et al. 2007). These processes have produced patterns of landscape conversion from forests to grasslands, especially in livestock production areas.

On grasslands, extensive cattle ranching also affects its composition and original structure since the use was based on the exploitation of natural fields with the sowing of mixes species of grasses and legumes for foraging or, totally replaced with pastures implanted. Finally, forested cliff areas show low and sporadic livestock use, generating degrees of alteration, due to the invasion of exotic tree species such as *Olea europaea, Morus alba*, and *Ligustrum lucidum*.

These species are distributed in the whole basin as a trend of environmental deteriorators inasmuch as from the second half of last century, *Melia azedarach* began colonizing roadsides and abandoned fields, accompanied by *Morus alba, Ligustrum lucidum, Gleditsia triacanthos, Broussonetia papyrifera*, and more recently *Ligustrum sinense* and *Maclura pomifera*. Each of these species has dispersal and proliferating characteristics allowing the invasion of many places in the basin. Thus, *M. azedarach* is often invading several secondary road edges in the northern center of the basin, with the appearance of *Bauhinia forficata, Robinia pseudoacacia*, or *Manihot flabellifolia* in the southern basin. *Morus alba* is a widely distributed by birds and invading most of the forested physiognomies. Much of the abandoned buildings in rural areas were invaded by *B. papyrifera, L. lucidum*, and/or *M. azedarach*, while there is an important nucleus of *Gleditsia triacanthos* associated with the lower portion of the LCC. This last species is invading lowlands, roadsides, and fences throughout the basin.

Currently, Las Conchas watershed has a percentage of natural coverage close to 7.8% (135 km^2). Natural areas are located mainly in lower areas of the basins and to a lesser extent associated with watercourses in higher regions. In this category, some patches of vegetation are untied at the upper stream of the Arroyo El Tala (in the vicinity of María Grande), in the middle stretch of LCC, corresponding to the protected area General San Martín Park, and the debouches into the Paraná River. The rest of the natural remnants correspond in general to small and scattered fragments, mainly associated with the margins of watercourses and headwaters of the basin (*see* Fig. 1). The invasion of fences by native species, such as *Schinus molle, P. nigra, Celtis tala*, and exotic as *M. azedarach*, has determined the conformation of important areas of refuge for the native flora and fauna.

During 2017–2018, agriculture covered more than 85% (1598 km^2) of the basin. Wheat is the main winter crop accounting up to 19% of the agricultural area and flaxseeds with 0.2% of the area, the rest of the extension kept on grasslands and fallow. Summer crops corresponded to soybean (61% of the agricultural area), corn (15%), sorghum (4%), and sunflower (0.2%) of the area. The three main crops (soybean, corn, and wheat) of the 2017/2018 productive season generated a total production of c. 334 ton (SIBER 2018), with a market value representing a total of c. USD 64 M.

Other agriculture activities are more related to livestock production and feedlots that emerged for increasing production and reducing economic losses. In 2013, pig farms reached almost half of the regional production (76 feedlots) of which 333 are indoor broilers. Regarding parceling, the area of the basin comprises approximately 8700 rural plots with an average surface area of 0.25 km^2. This pattern is typical of intensive use of the territory and a long history of productive use of the land.

Given the background, Sasal et al. (2011) applied the drastic model for the LCC basin, confirming the most vulnerable area to pollution is related to plains and the high basin of the creek. The lesser vulnerability was established in the area with deeper levels of the aquifer (from the surface) and low basin. Jergentz et al. (2005) demonstrated the effects of agrochemical residuals from soybean crops, especially by pesticides (Grondona et al. 2019) in those Pampean landscapes, including impacts on colonies of honeybees (Medici et al. 2019).

Political Reconfiguration Elements

In 2008, Bortoluzzi et al. (2008) determined that only 12.7% of the total area of Las Conchas stream basin corresponds to natural vegetation that coupled to our study; it shows a 5% of total loss in the last decade. The importance of forests at the basin level lies in the so-called water supply by forests (De Groot et al. 2002) which is described as "the filtration, retention and storage of water in estuaries, lakes and aquifers". The infiltration function is mainly related to the vegetation and biotic components of the soil. The ecosystem services associated with this function are linked to the supply and use of freshwater by households, agriculture, and industry (Oyarzún et al. 2005). On the other hand, forestry specialists estimate that a region with undulating land and numerous waterways (such as the studied basin) has to maintain 25% of its surface covered with forests to avoid land drags due to water erosion and the loss of the flow of their streams (Muñoz et al. 2005).

The forests of the LCC basin offer fundamental supply services that benefit the inhabitants of the region. The great diversity of plants, animals, and microorganisms offers a huge range of foods, energy sources, building materials, medicines, and these forests are used as extensive grazing areas and have been the subject of selective extraction of wood for sawing for poles or firewood (Roskopf et al. 2007).

Involved with those critical situations, the ecosystem or environmental services have become political elements when meeting the goals of sustainable development due to their definition as "the benefits that humanity obtains from ecosystems, conditions and processes in which ecosystems and the species that inhabit them meet the needs of people" (MEA 2005).

Many of these services do not have market prices, and therefore their economic value has not been incorporated into environmental policy decisions related to the management of natural resources. Accordingly, the economic valuation can be defined as an attempt to assign a quantitative and monetary value to goods and services provided by environmental resources or systems, whether there are market prices that can assist us, which results in Payments for Environmental Services (PES) (Lambert 2003).

Counting the PES schemes are still incipient in the country, they are being encouraged by non-governmental organizations, research institutes, companies, and the Argentinian State. The National Law 26,331 of Minimum Budgets for the Conservation of Native Forests, governed in 2009, provides the regulatory framework for the design and implementation of PES in all forestry provinces (Ministerio de Justicia y Derechos Humanos 2007). Entre Ríos, where the LCC basin is located, comes as one of the most important zones for this policy because of the 2.5 million hectares of native forests at the beginning of the twentieth century, with 1.4 million hectares remaining in 2005 (Muñoz et al. 2005).

Thus, PES strategies are focused on making changes in plant cover, which depend on a transformation of the aptitude of land use (from agricultural to conservation) or the reduction in the use of agrochemicals in the targeted areas (Jack et al. 2008; Castro-Díaz 2014).

Capitalization of Nature

An incipient mercantile valorization of natural resources is being organized in the area of the LCC basin. Both residents and government institutions proposed the region to be part of a regional tourism market, based on its natural and cultural resources. Entre Rios is the fourth province in infrastructure for tourism (8% of the touristic companies of the national total)[1] being more prominent around the thermal baths, sport fishing, beaches, carnivals, and religious tourism.

The LCC basin has few touristic itineraries promoted by the provincial government. One is the so-called Getaway to the North ("Escapada al Norte"), starting in Paraná city (west LCC) until the town of Cerrito (north-west LCC). The "Walk of the Colonies" ("Paseo de las Colonias"), which runs through villages and rural landscapes of the area, particularly on the National Route number 12. In the city of Villa Urquiza (west LCC) are the beaches and coastline on the Paraná River and La Balsa town with an ancient hand-operated raft crossing the LCC.

The San Martin Natural Park (6 km^2) is important in terms of ecological services, with natural riverside on the LCC with touristic attractions like native forest and beaches. This place is highly valuable by locals, proposing cultural values linked with the "local horse culture" and other folkloric elements. For this condition, it is required the recovery of the surrounding land to the LCC and the placement of infrastructure for recreational uses and intended to low impact tourism.

Natural Protected Areas

The San Martin Natural Park was established in 1950 as a recreational space for conserving a wild area representative of the Spinal Ecoregion (NT0801). The area holds 218 species of birds, 62% of the total of the Province of Entre Rios, and uses related to conservation such as a rural school, environmental production, and recreation is allowed. A land portion is managed by descendants of native people and intended to cultural and ceremony place.

This park is part of the proposal for the Ramsar Site Yjára (in Guaraní "Guardian of the Waters"), with an area of 64.5 km^2 and centered in the San Martin Natural Park. This Ramsar Site will be an area of conservation and rural sustainable production pursuing the preservation of the basin, promoting ecotourism and sustainable development.

Social Structuring Elements

Even with these administrative policies, the main impacts of human activities on the environmental services of the LCC basin keep on relating to agricultural activities. Watercourses are being affected by the agrochemical's residuals and discharges of

[1] Entre Ríos is mainly a destination of national tourism, given that more than 96% of the total hotel occupancy corresponds to Argentine tourists (Ministerio de Hacienda Argentina 2018).

wastewater of industrial, urban, and domestic origin, which have generated multiple socio-environmental conflicts in the region.

Inappropriate uses of pesticides cause pollution in the environment, affecting land, watercourses and crops, as well as causing damage to the health of farmers and rural as urban inhabitants (Díaz de Astarloa and Pengue Walter 2018). The effects on human health are related to long-term problems, such as cancer and chromosomal and reproductive damage (Arbuet et al. 2015).

The use of agrochemicals (e.g., glyphosate) in the region was the main topic of the first social mobilization in early 2000. Residents of small towns of the basin are in alert status because of the fumigations in nearby fields but their mobilizations prevail spontaneously, and only launched against specific cases, especially in situations of aerial spraying with agrochemicals.

The inhabitants of the area pointed out that since early 2010 the health problems have increased due to the use of agrochemicals and related to the expansion of soybean crops in the region. This increasing awareness is due to the dissemination of cases of diseases and health problems arising from the use of agrochemicals (Verzeñassi 2014) and by the presence in the area of diverse environmental institutions represented by the San Martin Natural Park, which protects the lower stream of LCC, and also the Provincial College of Agronomists focused on problematics from food production.

Since 2011, the San Martin Natural Park has become a cohesion place for mobilizations of neighbors concerned about the state of the environment. That year, several meetings were held, resulting in the formulation of claims that not only concern the environment and health but also to build a different lifestyle. Residents demanded the State to determine radius of non-application of agrochemicals around schools, inhabited places, and natural protected areas and requested to provide public information on the consequences of the use of agrochemicals. They also asked for the forestation of the area with native trees and to start the elimination of invasive flora. Finally, neighbors claimed for the improvement in health services and waste treatments.

In December 2014, the Courts of Concepción del Uruguay set a precedent by sentencing three defendants by the aerial fumigation over the rural School No. 44 of Colonia Santa Anita (Uruguay Department) (Uno Entre Ríos 2017a, b). This became a milestone in local struggles against the use of agrochemicals, considered by the inhabitants a legitimation for their mobilization.

A new chapter on the struggle over fumigations took place in 2018 when agro-industrial sectors claimed for a reform of the 1980 Provincial Law of Pesticides No. 6.599 (CO.P.A.E.R. 1996). This Law (and its regulatory and complementary regulations) regulates the use, sale, and transportation of products, establishing that the most toxic agrochemicals (Ia, Ib, and II) can be sprinkled out of a radius of 3000 m from the urban area and minors (III and IV) in 500 m. However, the reform demanded by agro-industrials specifies that each fumigation case requires a decision from the Provincial Department of Production, who could even be able to suppress a sanitary distance in areas with human populations. The same year, social movements and environmental organizations struggled to stop the reform of the law. Currently, there still are disputes on this topic, with the participation of LCC basin's neighbors.

In 2012, garbage dumping near La Picada town, close to the San Martin Natural Park, motivated neighbors to argue against the establishment of a sanitary landfill of organic waste in the flooded valley of LCC. The main risk of this project was visible during floods because of the dispersion of toxins in a wide area that includes households, schools, and rural infrastructure. Its installation was finally rejected due to the local mobilization, but the struggle is anchored in the memory of the inhabitants and mentioned as a background against new environmental problems. In turn, this specific conflict led to a discussion in the Municipality of Paraná about the kind of treatment given to the waste generated by each urban core.

The areas surrounding the Las Conchas and Sauce Grande Creeks and their tributaries act as a drainage space for urban areas and rural productive areas. Another problem mentioned by settlers of the coasts of these creeks, is the periodic floods on fields and affect houses. According to San Martin Natural Park officials, the occurrence of prolonged floods caused by the overflow of the creek has been aggravated by the invasion of "Black Acacia" (*Gleditsia triacanthos*), which makes it difficult to drain the water, flooding productive, recreational, and touristic areas.

Spatial Modeling of Waterlogging Probability

Being one of the main areas of regional colonization of Entre Rios (Argentina), the LCC faced land use changes from livestock to agriculture in the nineteenth century. Most of these transformations were originated from the settlement of mall colonies and the increasing use of agrotechnologies to meet demands within the internal market for food, wood, fibers, which although produced well-being for a large number of people, they were accompanied by the deterioration of natural resources and the loss of biodiversity at the local level.

With the decrease of natural land cover, as a result of this agrotechnology production and the use of agrochemicals, the first indicators of environmental stress were established. This situation was accelerated by the international valuation of the commodities at the beginning of the 1990s, producing a negative impact on hydrological environmental services mostly observed in the decrease of the regulation of water surface and the potential risks of soil erosion.

In this context, the current natural surface within the Las Conchas basin is scarce and endangered by the productive private exploitations. Some of these vegetation remnants are a key habitat for the conservation of biodiversity, demanding environmental protection actions to prevent their complete loss.

Some of these actions should be addressed given the recognition of the intrinsic value of the ecosystem services provided by the basin, taking into account the high economic losses from waterlogging. Indeed, the conservation and increase of natural areas will only increase the adaptive mechanisms of the LCC.

Those mechanisms were not established in the LCC, as observed in the Decree of Agricultural Emergency, issued by the "Poder Ejecutivo de Entre Ríos" (2018), the loss of economic production up to 70% (500 MM) of soybean and corn crops

(Mesquida 2018) and the frequent impacts caused by extreme precipitations during 2015 and 2016.

We modeled the waterlogging (WL) probability (WLP) through a multivariate analysis using the runoff coefficient, erodibility, water capacity, and infiltration rate as described in the Methods section. We calculated the runoff coefficient using LULC and slope for the LCC. Low values were caused by low and plain lands in the basin (Annex I: Fig. A2). Agriculture, as shown in Fig. 1, contributes with the most proportion of this variable. Low erodibility values for the upper basin were changing to the west area in the south of the biggest urban area (Annex I: Fig. A3). Several areas with high values are located close to LCC increasing potential sedimentation.

We also modeled the available water capacity using the SPAW model and finding a medium to medium-high values for the basin (Annex I: Fig. A4). And, finally, the infiltration rate based on INTA (2014) defines the waterlogging basis, which is the lowest in at least half of the basin and the remaining keeps on slow rates (Annex I: Fig. A5). This is determining the most probable areas for flooding and establishing the worst areas for agricultural production.

Figure 2 shows the waterlogging probability resulted from the socioeconomic change factor depicted as runoff coefficient and biophysical variables such as available water capacity, erodibility, and infiltration rate. The numbers on the map represent (1) flood plain and wetlands of the low LCC; (2) urban and suburban area of the city of Paraná and (3) the Ponciano Creek Complex, with no soil data included in INTA (2014).

We assumed the runoff variable as an outcome from economic production related to social structure for the basin because the type of crop destination is mostly based on reflections from international prices of commodities and the typology of stakeholders oriented to satisfy this demand (as seen in *Economic forcing elements*). This waterlogging probability (WLP) is fully related to the natural structure of the basin as mentioned in the historical records as "low and marshy" in eighteenth century. The northeast and east of the LCC basin are clearly the most probable area with WL due to a less capacity of infiltration. The use of these areas, as seen in Fig. 1, is agriculture which is plausible for suffering losses as seen during 2017–2018, leading us to determine that at least half of the basin does not have a vocation for the current uses. In the west side of the basin, the situation is more heterogeneous but also with the same conclusion. The worst condition is held by the erodibility factor which increases sedimentation in the lower basin.

As referred by Jobbágy (2011), there is a high risk of flooding related to the water table dynamic based on the assumption of crops variability. They state that there is a conditional feedback between plant roots and close-to-surface phreatic levels, e.g., the hydric requirements of the land cover in their different phenological stages and radicular depth of the plants are affecting the water table (Jobbágy et al. 2008; Nosetto et al. 2005).

Figure 3 resulted from the comparison between the (1) waterlogging probability in the current state and (2) the waterlogging probability in the reforestation scenario. It shows that the increase in the forested area would reduce the waterlogging

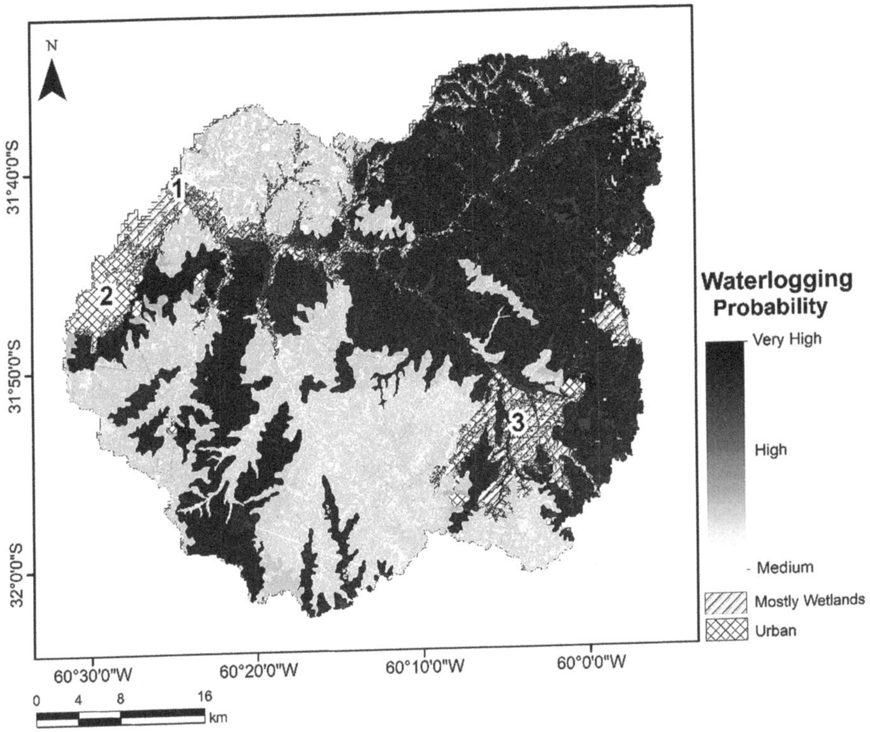

Fig. 2 Waterlogging probability in LCC based on multivariate analysis of runoff, erodibility, water capacity, and infiltration

probability as a result of changes in the runoff coefficient. Changes occur in areas with a very slow-to-slow infiltration rates, little slopes, medium-to-high erodibility, and medium water capacity (northeast of the basin), showing a medium-high reduction of the WP around the water drainages and the northeast of the basin. Figure A6 (Annex I) shows the frequency of the values in a 1 (lowest) to 10 (highest) scale and demonstrates that the forests displace the WLP values from 9 and distributed to 8, also a general reduction of the WLP in the basin.

The reduction of the waterlogging probability in the reforestation scenario implies a strategic resource for policymaking. With a reduction of the WP, it is possible to reduce the negative impacts caused by WL in agriculture production, even if it would allow only a better capacity of rain capture and runoff reduction. It is plausible the sedimentation reduction from the upper basin and an increase in the chances of conservation of ecosystem services such as biodiversity, hydrological reduction, or carbon sequestration.

In this context, the infiltration emerges as a key natural process in the loss of hydrological environmental services for regulating waterlogging because of its greater implications in water table dynamics and its relationship with the geomorphological configuration of the basin. Because of it, the permeability attributes and

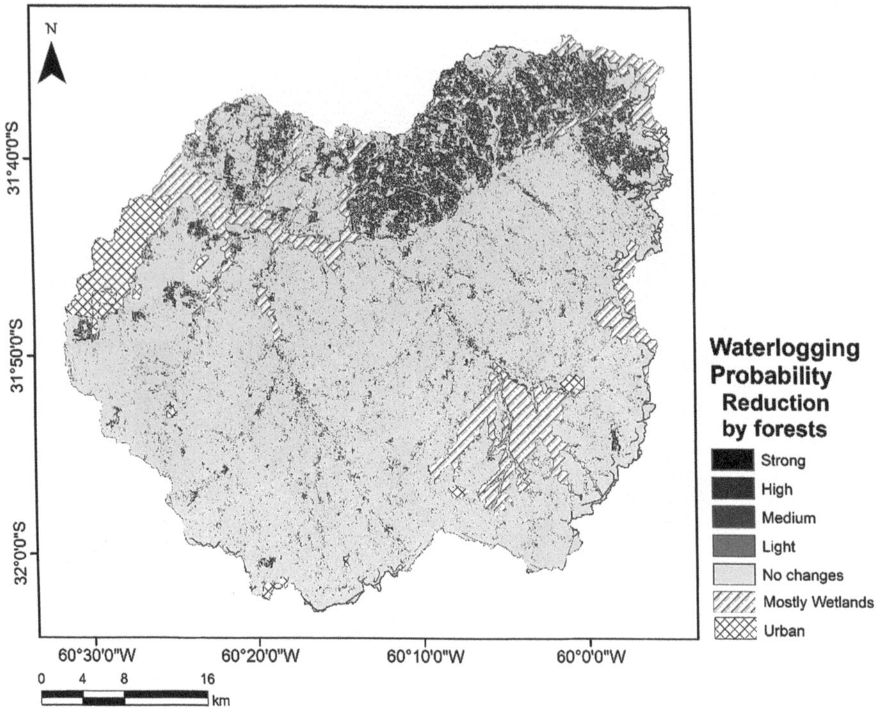

Fig. 3 Waterlogging probability changes in a reforestation scenario

soil saturation influences the degree of responses to the rain regime and meteorological extreme events (Maasland 1959; Simonson and Boersma 1972; Tanco and Kruse 2001). The loss of the natural hydrologic regulation has negative effects on agricultural activities because the water excess increasing the groundwater height, leads to crops flooding (Shimada et al. 1995). Not only very slow-to-slow infiltration rates configure a determinant for waterlogging. Indeed, when the meteorological conditions change to extremely dry, these areas become more vulnerable to drought, causing millionaire economical losses up to 50 MM as those occurred during 2017–2018 (PEER 2018).

The agriculture pressure is also changing the river landscape, considering the natural sedimentation dynamic of the basin. For example, the rise of a bowed delta in the estuary of this creek in the Paraná River is a consequence of erosive processes in the basin and sediments transport of principal drainage (Fig. 4).

This sedimentation is explained by the permanent, intensive, and constant removal of soil, agricultural machinery, accelerated elimination of vegetation cover during harvest, and abandonment of land increasing the erosive effect of the rain (Müller 1995; Power 2010; Wilkinson and McElroy 2007). In addition, the slope average values of the basin (2.8% ± 2.7) turns into an important factor in the soil degradation in the LCC and near basins. Gaitán et al. (2017) estimate the real hydric

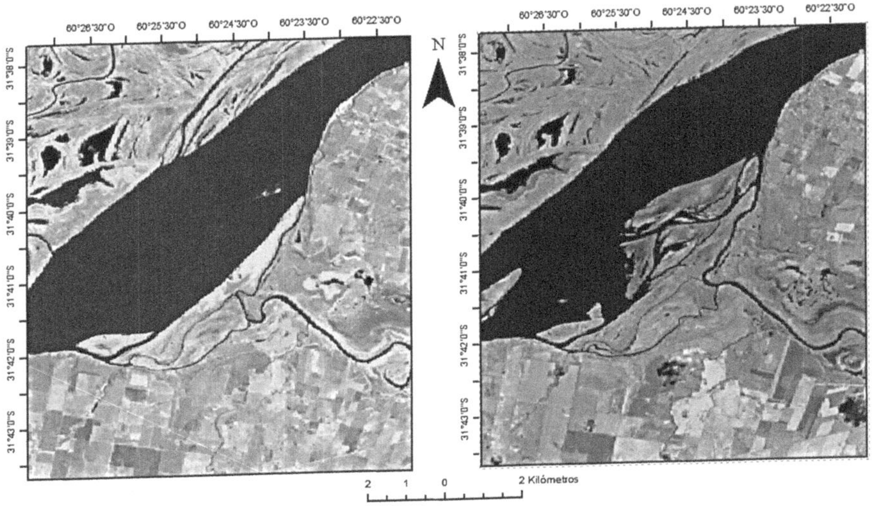

Fig. 4 Changes in the estuary of the LCC in Paraná River between 1987 and 2017 (Landsat 4 and 8 imagery). Source: *Land Processes Distributed Active Archive Center (LP DAAC), located at USGS/EROS, Sioux Falls, SD.* http://lpdaac.usgs.gov

erosion in 1000–3000 ton/km^2/year and the most probable causal factor for the rise of the bowed delta as seen in Fig. 4.

Nevertheless, this land degradation process is mostly related to specifications in agriculture. The monoculture results are key for understanding the loss of soil in Pampean basins like the LCC (Duval et al. 2016), and their impacts on microbial communities (Gil et al. 2011). Indeed, the monoculture dynamic based on soybean configures special conditions in Argentina, e.g., Galafassi (2005) uses the term “pampeanization” to explain the dynamics of the agricultural production in similar regions, referring to its configuration like a prevailing phenomenon pushed by the “soybeanfication.”

These current conditions, along with a negative climate change scenario, suggest the upsurge of the implicated factors increasing flooding areas with the reduction of the field capacity, sedimentation increase, and a clear probability of intensification of the waterlogging situations.

4 Conclusions

The spatial modeling of socio-ecological systems for hydrological environmental services analysis has become a challenging field of study. Current events of extreme climate impact in diverse ways the world population. Their economic activities are endangered, and the ecosystem services turn out to be one of the most promising frameworks to find sustainable solutions.

In Las Conchas Creek (LCC) basin, the complex dynamic of the interdependence between economical production and the sustainability of the hydrological regulation is in a critical situation. With millions of losses in crops during 2016–2017 caused by extreme events of precipitation and 2017–2018 of drought, and with a national difficult economic context, the questions were multiple, and answers required.

We used different methodologies for understanding the complex dynamics underlying the hydrological regulation as an environmental service. It depends on the agricultural production, urban and growing population, and their derived activities, and at the same time, its feedback results in natural control of flooding. Nevertheless, the thresholds were reached, and the ecosystem became unable to support such intensive exploitation.

Our method was based on the contextualization of socio-ecological system conditioners such as biophysical determinants, economic forcing, social structure, and political reconfiguration. And finally, we modeled the waterlogging probability for the LCC basin.

We found that the economic forces threaten the cycle of hydrological regulation, and the historical relationship from the local with their territory are energizers of this critical situation. However, the boomerang effect is causing havoc and more drivers are making the situation unsustainable for the local people. The need for more agrochemical is one of the largest struggles in the LCC and the agriculture producers are carrying the highest responsibility.

The LCC is naturally a basin exposed to very slow infiltration rates, with high erodibility, medium water capacity in the soils, and influenced by the land cover on the runoff process. Indeed, the current LULC reflects the results of decades of agricultural intensity and have implications with the waterlogging. We used multiple GIS and remote sensing techniques for classifying LULC, infiltration erodibility, available water capacity and runoff. We modeled the waterlogging probability based on their combination and generating the current state of the waterlogging probability (WLP).

We established a comparison between the WLP in two scenarios: (1) the current state and (2) a hypothetical reforestation scenario based on the replanting of 100 m surrounding the current forest areas. We confirmed that forest influence the WLP, reducing the vulnerable areas and their degree of vulnerability to WLP. The reduction was also observed in the displacement of very high values of WLP to high WLP.

These current conditions, along with a negative climate change scenario, suggest the upsurge of the implicated factors increasing flooding areas with the reduction of the field capacity, sedimentation increase, and a strong probability of intensification of the waterlogging situations. More alternate scenarios will help to find better ways to understand the complexity of the social-ecological systems in the ecosystems services conservation and their future.

Annex I

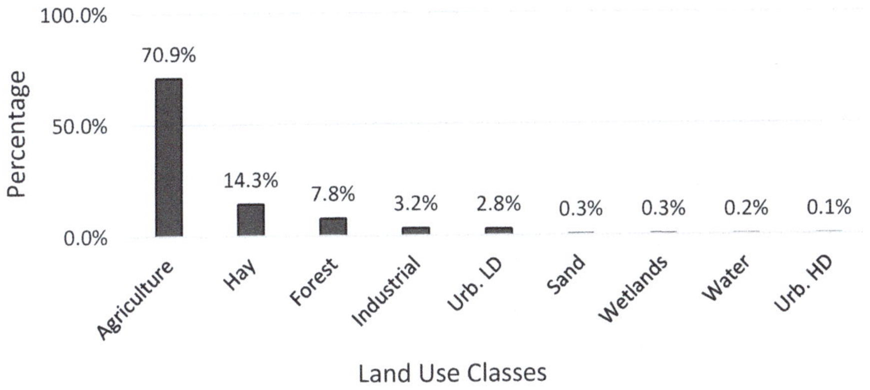

Fig. A1 Land use classes in percentage for the LCC basin from satellite imagery classification as shown in Fig. 1 of the text

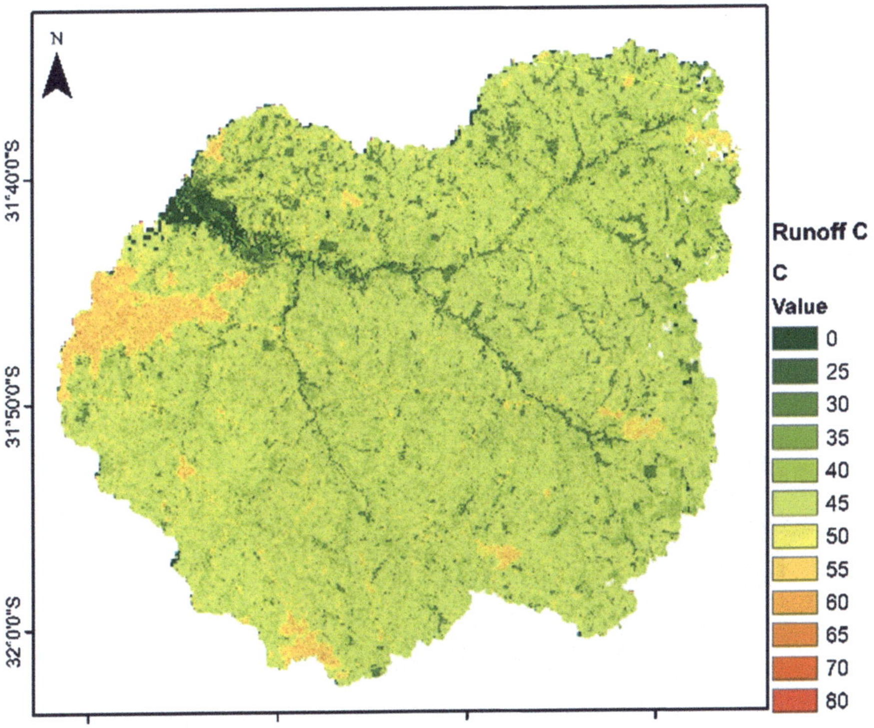

Fig. A2 Runoff coefficient (RM)

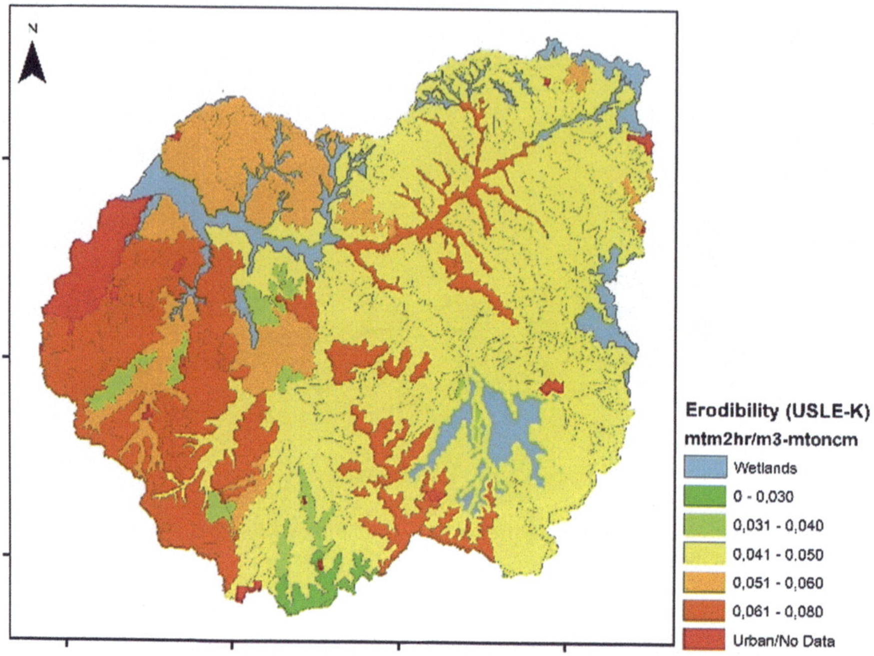

Fig. A3 Erodibility (USLE K Factor)

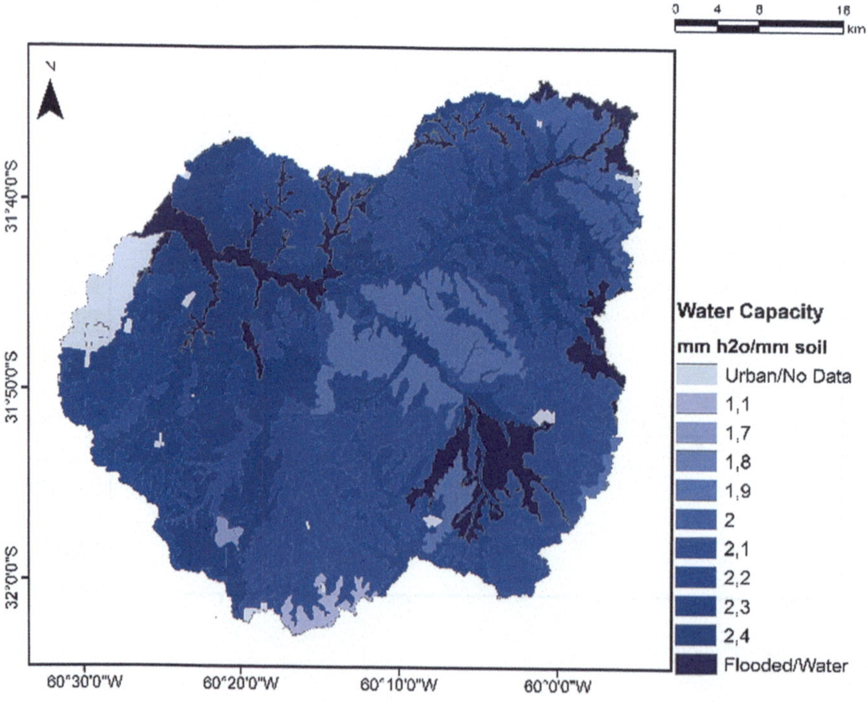

Fig. A4 Water capacity (SPAW Model)

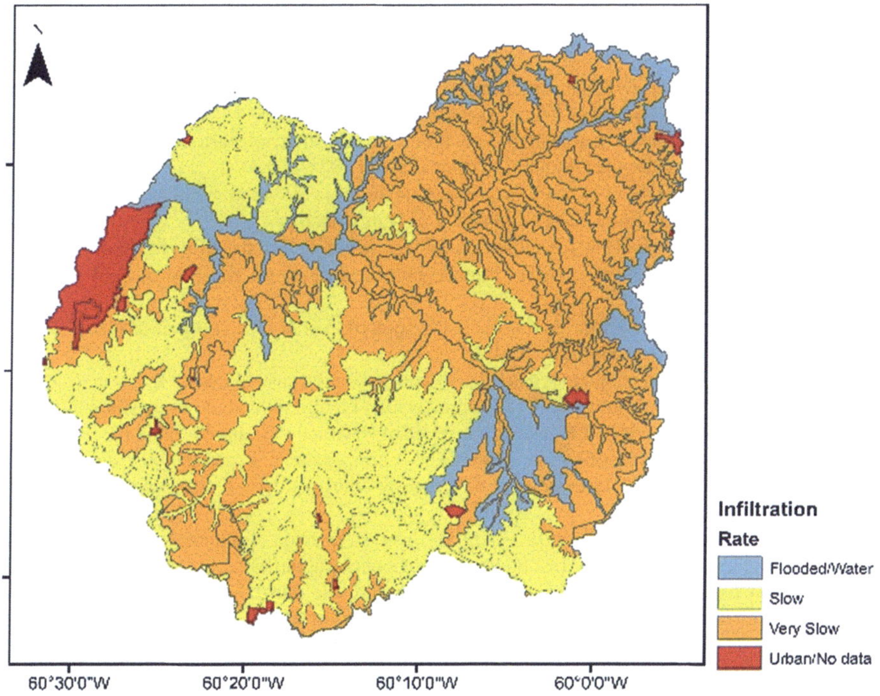

Fig. A5 Infiltration rate (raging criteria)

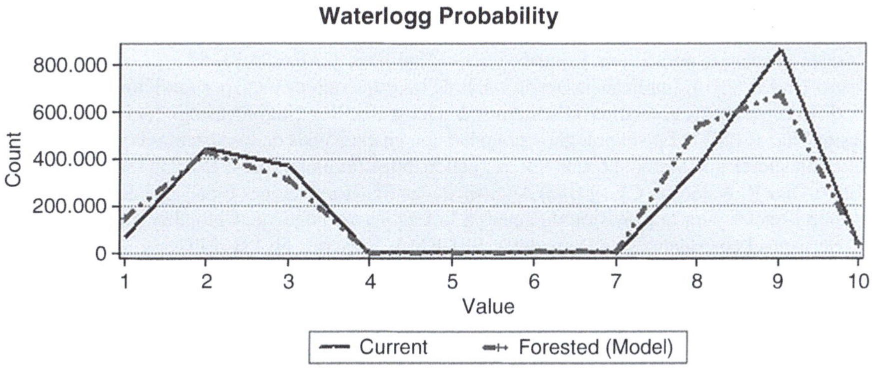

Fig. A6 Waterlogging probability changes frequencies comparing the current status vs. the reforestation scenario

References

Aceñolaza PG (2000) Variabilidad estructural de una comunidad forestal sobre suelos vérticos de la provincia de Entre Ríos. Rev Facult Agron 20(1):123–130

Aceñolaza FG (2003) Relatos de un Pago Viejo. Edición del Sesquicentenario Fundación "Colonia de la Conchas" Villa Urquiza. Ediciones Magna. ISBN 987-9390-48-2

Aceñolaza P, Manghesi E (1993) Flora arbórea de la zona de Villa Urquiza, Distrito Tala, Departamento Paraná, Entre Ríos. Ed. Fund. Col., Las Conchas

Aceñolaza FG, Sayago J (1980) Análisis preliminar sobre la estratigrafía, morfodinámica y morfogénesis de la región de Villa Urquiza, provincia de Entre Ríos. Acta Geol Lilloana 15(2):139–154

Alonso-Betanzos A, Sánchez-Maroño N, Fontenla-Romero O et al (2017) Agent-based modeling of sustainable behaviors. Springer, Berlín

Arbuet MA et al (2015) Herramientas estratégicas para la gestión sustentable de los recursos hídricos en una cuenca rural antropizada. Informe Tecnico Final. Proyecto de Extensión de Interés Social (PEIS). Secretaría de Extensión, Universidad Nacional del Litoral, Santa Fe

Benegas L, Faustino J (2008) Cogestión de cuencas hidrográficas: experiencias y desafíos. In: Seminario Internacional de Cogestión de Cuencas Hidrográficas: Experiencias y Desafíos. CATIE, Turrialba

Bortoluzzi A, Aceñolaza P, Aceñolaza FG (2008) Caracterización Ambiental de la Cuenca del LCC, provincia de Entre Ríos. Inst Super Correl Geol (Insugeo) Misc 17:219–230

Boumans R, Costanza R, Farley J et al (2002) Modeling the dynamics of the integrated earth system and the value of global ecosystem services using the GUMBO model. Ecol Econ 41(3):529–560

Brenner L (2012) Desarrollo local participativo y buena gobernanza ambiental: ¿una combinación viable? La experiencia de una reserva de biosfera Mexicana. Medio Ambiente y Urbanización IIED–AL 76:211–242

Bruckmeier K (2016) Social-ecological systems and ecosystem services. In: Bruckmeier K (ed) Social-ecological transformation. Springer, London, pp 183–234

Cabrera AL (1976) Regiones fitogeográficas Argentinas. Enc. Arg. De Agric. Y Jard. 2da. edn. Editorial ACME

Castro-Díaz R (2013) Implicancias de la resiliencia espacial en la prestación de servicios ambientales en cuencas norandinas. Contribuciones Científicas. GAEA 25:71–87

Castro-Díaz R (2014) Implicancias territoriales de los esquemas de pago por servicios ambientales (PSA) en cuencas norandinas. Cuadernos de Geografía. Rev Colomb Geogr 23(1):61–74

Castro-Díaz R (2017) Epistemología y pragmatismo en el análisis de los sistemas complejos. Rev Latinoamericana Metodol Cienc Soc 7(2):e026. https://doi.org/10.24215/18537863e026

Castro-Díaz R, Natenzon CE (2018a) Análisis de las transformaciones espaciales del suelo y sus implicancias para la provisión de agua en la Laguna de Fúquene, Colombia. In: Díaz I (ed) Servicios Ecosistémicos en Humedales. SEDEMA, Veracruz, pp 181–217

Castro-Díaz R, Natenzon CE (2018b) The social vulnerability and ecosystem services feedback: approaching social-ecological analysis in water supply for Andean communities. Paper presented at the world social science forum "security and equality for sustainable futures", Fukuoka

Castro-Díaz R, Zamboni LP, Sione W et al (2018) Elementos territoriales claves para la gestión del fuego en el Paraná medio. El ojo del Cóndor 9(1):28–30

Christensen AA, Brandt J, Svenningsen SR (2016) Landscape ecology. In: International encyclopedia of geography: people, the earth, environment and technology. pp 1–10

CO.P.A.E.R Ley de Plaguicidas No. 6.999. Decretos Reglamentarios No. 4.483/95 y 5.575/95 y Normas Complementarias (1996). https://www.entrerios.gov.ar/ambiente/userfiles/files/archivos/Normativas/Provinciales/Ley_6599_80_PLAGUICIDAS.pdf. Accessed 28 Dec 2018

Cumming GS (2011) Spatial resilience in social-ecological systems. Springer Science & Business Media, Berlin

Cumming GS, Collier J (2005) Change and identity in complex systems. Ecol Soc 10(1):29

D'Orbigny A (1835–1847) Voyage dans l'Amérique méridionale exécuté pendant les années 1826, 1827, 1828, 1829, 1830, 1831, 1832 et 1833. Ed. P. Bertrand, Paris

Dawson TP, Rounsevell MD, Kluvánková-Oravská T et al (2010) Dynamic properties of complex adaptive ecosystems: implications for the sustainability of service provision. Biodivers Conserv 19(10):2843–2853

de Azara F (1809) Voyages dans l'Amérique Méridionale par don Félix de Azara. Pub. Dentu, París

De Groot R, Wilson M, Boumans R (2002) A typology for the classification, descriptions and valuation of ecosystem functions, goods and services. Ecol Econ 41:393–408

Díaz de Astarloa D, Pengue Walter A (2018) Nutrients metabolism of agricultural production in Argentina: NPK input and output flows from 1961 to 2015. Ecol Econ 147:74–83

Dourojeanni A, Jouravlev A (1999) Gestión de Cuencas y Ríos Vinculados con Centros Urbanos. División de Recursos Naturales e Infraestructura, CEPAL

Duval ME, Galantini JA, Capurro JE et al (2016) Winter cover crops in soybean monoculture: effects on soil organic carbon and its fractions. Soil Tillage Res 161:95–105

Entre Ríos (2006) Ley Provincial N° 9757/06. Gobierno de la Provincia de Entre Ríos, Paraná

Farley KA, Jobbágy EG, Jackson RB (2005) Effects of afforestation on water yield: a global synthesis with implications for policy. Glob Chang Biol 11(10):1565–1576

Gaitán J, Navarro MF, Carfagno P, Tenti L (2017) Estimación de la pérdida de suelo por erosión hídrica en la República Argentina. INTA Investigación, Desarrollo e Innovación. diciones INTA, Buenos Aires

Galafassi GP (2005) La pampeanización del Delta: sociología e historia del proceso de transformación productiva, social y ambiental del Bajo Delta del Paraná. Extramuros Ediciones, Buenos Aires

Gencer E, Folorunsho R, Linkin M, Wang X, Natenzon CE, Wajih S, Mani N, Esquivel M, Ali Ibrahi S, Tsuneki H, Castro R, Leone M, Panjwani D, Romero-Lanako P, Solecki W (2018) Disasters and risk. In: Rosenzweig C, Solecki WD, Romero-Lankero P et al (eds) Climate change and cities: second assessment report of the urban climate change research network. Cambridge University Press, Cambridge, pp 61–98

Gil SV, Meriles J, Conforto C et al (2011) Response of soil microbial communities to different management practices in surface soils of a soybean agroecosystem in Argentina. Eur J Soil Biol 47(1):55–60

Grondona SI, Gonzalez M, Martínez DE, Massone HE, Miglioranza KSB (2019) Assessment of organochlorine pesticides in phreatic aquifer of Pampean region, Argentina. Bull Environ Contam Toxicol 102(4):544–549

Hardin G (1968) The tragedy of the commons. Science 162(3859):1243–1248

INTA (2014) Cartas de Suelos de Entre Ríos. Proyecto PFIP ER 07–08. Base de datos geoespaciales para la toma de decisiones estratégicas en las Cadenas de Valor sustentadas sobre Recursos Naturales. Map: Geointa

Jack BK, Kousky C, Sims KR (2008) Designing payments for ecosystem services: lessons from previous experience with incentive-based mechanisms. Proc Natl Acad Sci 105(28):9465–9470

Jergentz S, Mugni H, Bonetto C et al (2005) Assessment of insecticide contamination in runoff and stream water of small agricultural streams in the main soybean area of Argentina. Chemosphere 61(6):817–826

Jobbágy EG (2011) Servicios hídricos de los ecosistemas y su relación con el uso de la tierra en la llanura Chaco-Pampeana. Valoración de Servicios Ecosistémicos Conceptos, herramientas y aplicaciones para el ordenamiento territorial. Ediciones INTA, Buenos Aires, pp 163–183

Jobbágy EG, Nosetto MD, Santoni CS et al (2008) El desafío ecohidrológico de las transiciones entre sistemas leñosos y herbáceos en la llanura Chaco-Pampeana. Ecol Austral 18(3):305–322

Lambert A (2003) Valoración económica de los humedales: un componente importante de las estrategias de gestión de los humedales a nivel de las cuencas fluviales. Convención de Humedales, Ramsar

Laterra P, Castellarini F, Orúe M (2011) ECOSER: Un protocolo para la evaluación biofísica de servicios ecosistémicos y la integración con su valor social. Valoración de Servicios Ecosistémicos: conceptos, herramientas y aplicaciones para el ordenamiento territorial. Ediciones INTA, Buenos Aires

Leyes R (2016) Caravanas de hombres marchaban: el éxodo obrero en Entre Ríos. 1925-1945. Folia Histórica del Nordeste 27:9–38

Maasland M (1959) Water table fluctuations induced by intermittent recharge. J Geophys Res 64(5):549–559

Maldonado FD, Sione WF, Aceñolaza PG (2012) Mapeo de desmontes en áreas de bosque nativo de la Provincia de Entre Ríos. Ambiência 8:523–532

Maldonado FD, Rodriguez EE, Aceñolaza PG et al (2013) Monitoreo histórico de los desmontes en áreas de bosque nativo de la Provincia de Entre Ríos, Argentina. Anais do 16ª Simpósio Brasileiro de Sensoriamento Remoto. pp 1–8

Millennium Ecosystem Assessment (MEA) (2005) Ecosystem and their services. In: Millennium ecosystem assessment: strengthening capacity to manage ecosystems sustainably for human well-being. World Resources Institute

Medici SK, Blando M, Sarlo E et al (2019) Pesticide residues used for pest control in honeybee colonies located in agroindustrial areas of Argentina. Int J Pest Manag 1–10. https://doi.org/10.1080/09670874.2019.1597996. *In press.*

Mesquida F (2018) Soja: las enfermedades de fin de ciclo pueden afectar hasta un 30% del rinde. http://www.infocampo.com.ar/soja-las-enfermedades-de-fin-de-ciclo-pueden-afectar-hasta-un-30-del-rinde/. Accessed 30 Apr 2018

Ministerio de Hacienda Argentina (2018) Entre Rios, Informes Productivos Provinciales. Secretaría de Política Económica Subsecretaría de Programación Microeconómica. Buenos Aires

Ministerio de Justicia y Derechos Humanos (2007) Presidencia de la Nación. Ley 26.331 "Presupuestos mínimos de protección ambiental de los bosques nativos". http://servicios.infoleg.gob.ar/infolegInternet/anexos/135000-139999/136125/norma.htm. Accessed 5 May 2019

Molina Garate E (2008) El Consejo de Gestión de Aguas de la cuenca del Paute: experiencias y lecciones. In: Seminario Internacional de Cogestión de Cuencas Hidrográficas: Experiencias y Desafíos. CATIE, Turrialba

Müller N (1995) River dynamics and floodplain vegetation and their alterations due to human impact. Archiv für Hydrobiologie. In: Large rivers, vol 9. pp 477–512

Muñoz JDD, Milera S, Romero C, Brizuela A (2005) Bosques nativos y selvas ribereñas en la provincia de Entre Ríos. Temas de la biodiversidad del litoral fluvial argentino. INSUGEO. Miscelánea 14:169–182

Nahuelhual L, Laterra P, Villarino S et al (2015) Mapping of ecosystem services: missing links between purposes and procedures. Ecosyst Serv 13:162–172

Nelson E, Mendoza G, Regetz J et al (2009) Modeling multiple ecosystem services, biodiversity conservation, commodity production, and tradeoffs at landscape scales. Front Ecol Environ 7(1):4–11

Nosetto MD, Jobbagy EG, Paruelo JM (2005) Land-use change and water losses: the case of grassland afforestation across a soil textural gradient in central Argentina. Glob Chang Biol 11(7):1101–1117

Ostrom E (1990) Governing the commons: The evolution of institutions for collective action. UK: Cambridge university press. 298 pp

Ostrom E (2009) A general framework for analyzing sustainability of social-ecological systems. Cambridge University Press, Cambridge

Oyarzún CE, Nahuelhual L, Núñez D (2005) Los servicios ecosistémicos del bosque templado lluvioso: producción de agua y su valoración económica. Ambiente y Desarrollo 20-21(3). pp 88–97

Poder Ejecutivo de Entre Ríos (PEER) (2018) Decreto 373 373. Paraná

Pochat V (2005) Entidades de gestión del agua a nivel de cuencas: experiencia de Argentina. CEPAL, Santiago

Power AG (2010) Ecosystem services and agriculture: tradeoffs and synergies. Philos Trans R Soc B Biol Sci 365(1554):2959–2971
Ramsar's Manual (2010) Manejo de cuencas hidrográficas. Manuales Ramsar Para El Uso Racional de Los Humedales. Secretaría de la Convención de Ramsar, Gland
Roskopf R, Riegelhaupt E, Aceñolaza P et al (2007) Patrón local de uso para combustibles de madera en una localidad rural entrerriana. In: XXII Jornadas Forestales de Entre Rios
Ruiz De Galarreta VA, Noriega RBB, Barranquero RS et al (2010) Análisis integral del sistema hídrico, uso y gestión. Cuenca del arroyo Langueyú, Tandil, Argentina. Bol Geol Min 121(4):343–356
Sasal C, Wilson M, Santi M et al (2011) Aplicación del modelo DRASTIC en la cuenca del arroyo Las Conchas, Entre Ríos. Rev Invest Agropecuarias 37(2):150–158
Shimada S, Kokubun M, Matsui S (1995) Effects of water table on physiological traits and yield of soybean: I. Effects of water table and rainfall on leaf chlorophyll content, root growth and yield. Jpn J Crop Sci 64(2):294–303
SIBER (2018) Informe Final de Producción—Campaña 2017/8. http://www.bolsacer.org.ar/Fuentes/siberd.php?Id=1030. Accessed 13 Nov 2018
Simonson G, Boersma L (1972) Soil morphology and water table relations: II. Correlation between annual water table fluctuations and profile features. Soil Sci Soc Am J 36(4):649–653
Stoll-Kleemann S, Bender S, Berghöfer A et al (2006) Linking governance and management perspectives with conservation success in protected areas and biosphere reserves. Perspectives on Biodiversity Governance and Management. Humboldt-Universität zu Berlin, Berlin
Tanco R, Kruse E (2001) Prediction of seasonal water-table fluctuations in La Pampa and Buenos Aires, Argentina. Hydrogeol J 9(4):339–347
Truffer I (2010) Análisis de la dinámica territorial del territorio: Feliciano, Entre Ríos. INTA, Programa Nacional de apoyo al desarrollo de Territorios. Doc. N° 7:114–142
Turner MG (2010) Disturbance and landscape dynamics in a changing world 1. Ecology 91(10):2833–2849
Turner KG, Anderson S, Gonzales-Chang M et al (2016) A review of methods, data, and models to assess changes in the value of ecosystem services from land degradation and restoration. Ecol Model 319:190–207
Uno Entre Ríos (2017a) Este martes dictan el fallo en la causa por la escuela fumigada de Colonia Santa Anita. Uno Entre Ríos. https://www.unoentrerios.com.ar/la-provincia/este-martes-dictan-el-fallo-la-causa-la-escuela-fumigada-colonia-santa-anita-n1480421.html. Accessed 28 Dec 2018
Uno Entre Ríos (2017b) La muerte lenta en las escuelas fumigadas de San José. Uno Entre Ríos. https://www.unoentrerios.com.ar/la-provincia/la-muerte-lenta-las-escuelas-fumigadas-san-jose-n1573940.html. Accessed 28 Dec 2018
Verzeñassi D (2014) El modelo agrosojero y su impacto en nuestras vidas. In: Melón D (ed) La patria sojera: el modelo agrosojero en el Cono Sur. Editorial El Colectivo, Buenos Aires, pp 31–48
Wilkinson BH, McElroy BJ (2007) The impact of humans on continental erosion and sedimentation. Geol Soc Am Bull 119(1–2):140–156

Analyzing Social Vulnerability to Natural Disasters in Northeast Brazil: Catastrophic Flooding Cycles at Alagoas Littoral Zone

Neison Cabral Ferreira Freire and Claudia Eleonor Natenzon

Abstract The process of innovation and technological development have generated situations of risk to post-industrial societies of the twenty-first century, becoming what several authors have called a "risk society." However, risks do not spread homogeneously among societies. Some less developed regions never share the goods, but the risks, attributed to the intensification of technological and productive processes arising from the globalization of the markets. The objective of this chapter is to discuss and analyze the generation of risks to national catastrophes, such as flooding, in tropical areas with late capitalism. We developed our ideas using Alagoas littoral zone (Brazil) as a case study.

Keywords Social-ecological systems · Latin America · Complexity · Brazil · Social vulnerability · Natural disasters

1 Introduction

The process of innovation and technological development have generated situations of risk to post-industrial societies of the twenty-first century, becoming what several authors have called a "risk society" (Beck 1992, 1993; Giddens 1990; Luhmann 2006). However, risky decisions within our daily life promote situations of catastrophe. Although several authors manage the concept that our contemporaneous society does not share the goods anymore but the risks, we should keep in mind that in some less developed regions, especially in Latin America, goods were never shared,

N. C. F. Freire (✉)
Diretoria de Pesquisas Sociais—DIPES/CIEG, Fundação Joaquim Nabuco, Recife, Brazil
e-mail: neison.freire@fundaj.gov.br

C. E. Natenzon
Programa de Investigaciones en Recursos Naturales y Ambiente (Natural Resourses and Environmental Research Program), Instituto de Geografía, Facultad de Filosofía y Letras, Universidad de Buenos Aires, Buenos Aires, Argentina
e-mail: natenzon@filo.uba.ar

L. E. Delgado, V. H. Marín (eds.), *Social-ecological Systems of Latin America: Complexities and Challenges*, https://doi.org/10.1007/978-3-030-28452-7_12

only the risks. These risks were intensified by technological and productive processes inherent to the globalization of markets.

Brazil is not any different. New hierarchies of richness generation and appropriation show that capitalism has perfected its instruments, easier management of scales, and the utilization of the built environment. Yet, social inequalities follow their course including the uncertainty situations of the populations marginalized from the economic development process. Observed emigrations from old rural areas of the poorest states of the country (such as Alagoas and Piauí) between 1980 and 2001, toward small towns close to their rural work, occurred without planning or urban control. The result was an intensive exposition of the new homes to natural dangers such as floods close to rivers that go through towns and villages.

Thus, a general risk social picture was configured in the states of Alagoas and Pernambuco in three dimensions: dangerousness, vulnerability, and exposition. From this, uncertainty was derived as a fourth dimension in its political and perception aspects of all involved social groups with their values and interests at stake (Natenzon 2005).

We should add to this social picture the changes that have been occurring in climatic processes in the intertropical convergence zone in the North Atlantic. For example, in June 2010 those changes generated a 3-day powerful storm, a natural disaster situation that generated many material damages and several deaths in both states, especially in the watersheds of the Mundaú and Paraíba rivers (Freire et al. 2014).

When in that moment the Brazilian President Luis Ignacio Lula da Silva visited the flooded areas in Alagoas said that "there was no explanation for this tragedy…it was a fortuitous event of nature and the divine," was like that? In this chapter, we seek to clarify the circumstances in which the tragedy occurred and how an analysis based on the social theory of the risk can contribute to the risk management of natural disasters in tropical areas in a more effective and efficient way.

Some questions emerge from the observed facts; identification and understanding territorial interrelationships are fundamental in social research to contribute preventing new tragedies and supporting decision-making.

For example, it has been shown that the greatest rainfall really occurred in the State of Pernambuco. However, the greatest disaster, with the largest social-environmental consequence occurred in Alagoas. How did this happen? If Brazil has orbiting satellites and zero-cost climate forecast, why the civil defense of each state did not have the capacity to foresee what was coming to them in order to minimize the effects of the phenomenon? Why the responses of the public sector of both Alagoas and Pernambuco were so different? In this game of interests, who loses and who wins? Finally, and no less important, to what extent the changes resulting from the globalization and restructuring of the world's economy interfere in this process in less developed regions such as the mentioned states?

Although climatic events do not respect geopolitical boundaries, based on the current social, historical, and environmental conditions from the areas where the phenomenon occur, the generated natural disasters may propitiate rather different situations. Thus, damages are directly related to the organization of the public power, society's response capacity, and whether or not there are prevention processes. What are at stake are the dimensions of the uncertainty and risk in each society. It is the role of the State, at least, to monitor the fulfillment of the environmental legislation and to provide the necessary infrastructure to the Civil Defense so the civil society may have ways to react to early warnings.

This chapter has been organized into two parts in order to advance on the issues presented in this introduction. The first part is a review of the main theoretical concepts about the risk and its four dimensions: dangerousness, vulnerability, exposure, and uncertainty. The second part is a detailed case study of the State of Alagoas (Brazil), with the purpose of understanding its social and historical configuration, and the adequate connections with the already exposed theory. We also address the conditions that generated uncertainty in the flooded regions. Finally, we summarize the most significant conclusions.

2 Conceptual Frameworks

Human's vertiginous technical, scientific, and information development, generated since the end of the Second World War, has brought contemporaneous society to face new dangers in daily life. Although human progress and knowledge advances have generated a new world of possibilities with supposed individual liberties unimaginable before, it is also true that new threats created by this development have emerged (López Cerezo and Luján López 2000).

The notion of daily life risk is increasingly present and the more we know it, the better we appreciate our ignorance. And, paradoxically, the more we try to control it, the larger is the generated risk in other parts of the system. Currently, people have lost a little of their illusions and glare about the technology. It perceives that they have not yet created the instrument for their freedom but new jails (Ellul 1964). Thus, we could say that today we live in what Beck (1993) called a high-risk society, referring to the fact that today greater damages are generated affecting most humankind. However, regarding risks we are all not equal. Or, paraphrasing George Orwell (1945: 112), we are all equal, but some are more equal than others.

This notion about the urgency of the risk is part of the contemporaneous society and consequently many academic debates, especially within the social sciences, are taking place in political and governmental agendas. Risk is also an issue in many scientific kinds of research in several knowledge areas given its great potential to understand catastrophes. Yet, may be due to its magnitude and spatial and cultural diversity, experts have not yet arrived at a unanimous definition regarding a general theory of the risk.

The likelihood of risk and dangerous exposure in our current society would allow us to successfully accomplish a specific understanding of our research. Our objective is to understand the basic principles that may clarify the issues that occurred in 2010 in our research area, the watershed of the Mundaú River, partly located in one of the less developed areas of Brazil: The State of Alagoas.

In order to understand the events that occurred, we will analyze the distinctions and similarities of the general concept also including its particular European colonization history. That means, what can we observe as particular characteristics of the researched region? Beck (1993: 19–20, 153) states that we are moving toward a new modernity where the axis structuring our industrialized society is no longer the classic good's distribution, but the distribution of evils that means the distribution of risks.

However, the allocation of risks in Alagoas was never preceded by the allocation of goods or wealth. The total population of Alagoas, according to the 2000 Population Census, was of 2,822,621 people and in 2004 the Economically Active Population (EAP) was of 1,133,203 people (PNAD).[1] de Carvalho (2005), analyzing the concentration of income within Alagoas EAP, highlights the polarization between a small group of people (4%) with more than five minimum salaries and a larger group with no income (20%) or surviving with up to two minimum salaries (66%). On the contrary, risks emerging from the economic development generated by the cane agriculture business and the recently incorporated biofuel generated by ethanol were always shared but never was shared the richness historically generated since the sixteenth century, when the Portuguese colonization of northern Brazil started up to our days in the Atlantic zone of Mata *alagoana* (Furtado 2003; Freyre 2006).

In fact, globalization has imposed a role on emergent economies, such as Brazil and Argentina, in terms of guaranteeing the provision of commodities and energy to central countries, and currently other emergent economies like China, last years has generated several land use changes in rural areas as soy, sorghum, and sugarcane, now strategic exportation actives (Neiman 2017; Rebizo and Rodriguez 2011; Reboratti 2010; Wilkinson 2010—among other authors). Beyond losses in biodiversity and generalized land clearance, those changes have generated high income in multinational enterprises related to agriculture businesses. Now, the most important thing is the increase in the production and export of agricultural products and, for that, the incorporation of new innovation processes and new cultivated areas is essential. The population outside this process is left in marginalized areas with high risks of suffering natural catastrophes since they build their houses in exposed zones where they live together with dangerousness. The wealth generated by this process is still concentrated in the hands of the social elites that live away from those uncertainty and dangerousness, thanks to their low standards of vulnerability.

Most social scientists analyze a disaster situation from the point of view of the decision-makers. It starts with the presence of a natural phenomenon, then it moves to the emergency phase and it ends with the reconstruction when the authority determines that it is time to return to normality (Calderón Aragón 2011). But the risk is associated with the production of geographic space, in its multiple dimensions and scales. It is therefore under economic, cultural, and political determinism. It is the product of a given society in its time and space, and the decisions of a given society regarding the use, occupation, and organization of their space will create the

[1] Pesquisa Nacional de Amostragem por Domicílio/Brazilian Institute of Geography and Statistics.

situations of risk and disasters. It is not the "divine" or the "fortuitous condition of nature" that make natural phenomena becoming natural disasters, if not the social system itself that generates the dimensions of risk (Giddens 1990). But they do it selectively, directed to certain sectors of the population that are the most socially vulnerable.

Indeed, production social relationships (and not the natural or the techno-industrial phenomena) produce a situation of natural disaster. The phenomena only expose the vulnerability that it has a certain stratum of the population whose origins are hidden by the day-to-day life, imposed by social relationships (Calderón Aragón 2011).

We still have to make a distinction between a *natural phenomenon* and a *natural disaster*. Although natural phenomena such as earthquakes, floods, and cyclones are highly destructive, they do not necessarily generate disasters. Disasters occur when they affect a society directly and its activities in a given space and time.

According to Maskrey (1989: 14), "Natural disasters are generally considered as a coincidence between natural hazards (such as floods, cyclones, earthquakes and drought) and conditions of vulnerability." Then, he proposes the equation: Risk = danger × vulnerability, meaning that the risk is directly associated with the simultaneity of natural processes and social structures. In the words of Blaikie et al. (1996: 11), "The natural and the human are linked so tightly in almost all disaster situations, especially when looked from large time and space frameworks, that it is impossible to understand that disasters may frankly be natural."

Wilches-Chaux (1993) proposes to consider the disaster as the convergence of risk and vulnerability. He understands risk as any phenomena of natural or human origin that generates changes in the environment. The vulnerability would be determined by the incapacity of a given society to adapt to a particular change in its environment. Yet, the notion of disaster has multiple meanings and the same is true for risk, dangerousness, and vulnerability. Thus, it is convenient to clarify these concepts for a better understanding of the socioeconomic characteristics that generate risk spaces in a given society.

Funtowicz and Ravetz (1993) propose that "there is a risk when we can quantify it" or when we have a probability as to what will happen, even if it is an approximation, a statistic. López Cerezo and Luján López (2000) consider risk as all possible although uncertain events that may generate damage. Thus, the risk would be an attenuated modality of insecurity. This concept allows postulating that it is possible to face the danger by searching more information and knowledge, investing more money and/or time, promulgating new laws, using the wisdom developed by the communities, exercising the population so that they may know how to act in critical situations and, fundamentally, implementing development policies that decrease social vulnerability.

The risk is a feature of modernity and of the technological development processes of our society. Still, if there is no probability calculation, previous knowledge of where they are and what are the physical, political, and socioeconomic conditions of the people potentially affected by a natural disaster, then it is not possible to decide with certainty about what is to come. Thus, risk converts into uncertainty.

Risk, for Natenzon and González (2010), implies complexity, hence having multiple dimensions. In this way, the comprehension of a particular situation of the risk of a catastrophe involves four dimensions: dangerousness, exposure, vulnerability, and uncertainty. The required knowledge for each one of them is different, derived from the social and natural science fields including both applied and theoretical knowledge; their interrelationships allow characterizing the risk and anticipating, foreseeing, and mitigating the catastrophe. These dimensions acquire different configurations in the disaster cycle (before, during, and after). The risk is configured by the first three dimensions. The lack of knowledge over any one of these dimensions will configure uncertainty as the fourth dimension.

Thus, the dangerousness evaluates the potential that something may happen. However, to know it is required learning about the physical–natural aspects of the involved processes, while its exposure relates to the material impacts that those dangers may generate in the territory and, as a result, it requires knowledge about the geographic distributions of goods and people. In turn, vulnerability is located within the social structures, being necessary to know verifiable socioeconomic characteristics on the involved social groups (Herzer et al. 2002). Regarding uncertainty, it is the dimension generated at the expenses of ignorance about the other three dimensions. Then, social issues such as perception and decision-making, their political aspects, and multiple and legitimate, but partial, values and interests at stake (Funtowicz and Ravetz 1993) enter to play.

3 Case Study: Background and Context

The Historical Situation of Alagoas

The state of Alagoas, located in the northeast of Brazil (Fig. 1), is part of a regional context of profound social inequalities. Its agricultural space—locus of its productive activity since the colonization of Brazil—has always been scenario of sugarcane monoculture. Its European occupancy occurred during the second half of the sixteenth century in three fronts: to the north up to the Camaragibe River; to the south along the banks of the San Francisco River up to its mouth in the Atlantic Ocean and in the central coastal region, around the Mundaú–Manguaba estuarine-lagoon complex where is currently located the city of Maceió, the capital of the State.

The occupancy of these vast colonial lands was based upon a possession regime of large rural properties (latifundios) distributed by the captaincy donors within colonists. A determinant factor in the need of large estate was the fact that sugarcane monoculture was only profitable, from an economic perspective, in large culturing areas. The environment suffered the consequences of this vast occupation. From early times the *Mata Atlântica (Atlantic forest)* was completely dismantled to serve the plantation regime in these fertile areas of *massapê* (clay land). According to

Fig. 1 Location map of the study area

Dean (2004), the most recent estimations show that less than 10% of the forest has survived until today either in its native form or slightly changed. The Atlantic forest, according to the available scientific evidence, was the largest tropical jungle destroyed by human activities in historic and prehistoric times.

Few spaces of this forest remained in the sides and bottom of the valleys, given their inadequacy for sugarcane culturing. However, this "environmental protection" occurred due to the need to ensure low-cost sources of clean water, necessary for the industrial process of sugar production and, starting in the 1970s, local alcohol distilleries.

Consequently, the estate's structure in northeast Brazil was, since very early, marked by large rural properties, with few and powerful landowners, high-income concentration, and large social vulnerability. Indeed, northeast Brazil was recognized as a region only since half of the nineteenth century and mostly during the twentieth century. Accordingly, there were several versions of the northeast, differentiated by their economic activities and how high social classes appropriated the capitals: Bahia and Serpige as one "region," Pernambuco, Alagoas, and Río Grande do Norte as another, separated from Ceará and Maranhão (de Oliveira 1981).

The sugarcane monoculture region, distributed along the coast of Alagoas, has its own accumulation characteristics, but no expanded reproduction in the same region where the production occurs. The accumulation served and still does, to maintain the *status quo* benefiting the dominant social classes. The truth is that the *alagoan* economy still produces nothing more than sugar, alcohol, cassava, milk, tobacco, coconuts, and chemical elements derived from rock salt; and the products that *alagoan* consumers need are imported from other Brazilian states or from abroad. This means that few years into the twenty-first century, Alagoas still has not gone beyond the first phase of capitalist development. In other words, it has not been able to produce the consumer goods needed by its internal market (Freire 2012).

If we analyze the social indicators of the last 30 years, it can be shown that although some industrial installation owners are in excellent financial conditions, the general *alagoan* social picture remains extremely critical. The Human Development Report from 2003[2] prepared by the United Nations Development Program (UNDP), when describing income inequalities state that "Alagoas has transformed into the most unequal states of Brazil, and its Gini Index elevated from 0.63 to 0.69 moving *alagoans* ten positions higher in that ranking."

Other social indices also help to understand the precarious living conditions in Alagoas, especially if compared with other Brazilian states and regions. Maybe the most serious concern is the infant mortality rate. For example, in 2008, 50 children died before reaching 1 year, for every 1000 born, the worse rate in Brazil.

Another data reflecting the precarious way of life in Alagoas refers to its sewage system. Almost 50% of households (650,000 during the year 2000) had individual wells for wastewater. Only 15% of the houses in Alagoas were connected to the general sewage system of the city (2000 Census, PNAD).

The origin of this social vulnerability can be found in the historically determined economic matrix. Sugarcane economy is highly vertical and furthermore, plantations have not generated the emergence of significant activities that may benefit the productivity in the northeast of Brazil, which is based on large rural properties with few horizontal economic interdependence.

This large income concentration has prevented the development of a middle class strong enough to guarantee profitable scales and dynamics to the necessary commercial relationships to expand the local market, also reducing the salary of the non-specialist worker. Consequently, the low socioeconomic indicators of this state are directly related to these characteristics present in all regions.

What is left to the socially vulnerable population is the occupation of inadequate residential sectors, mostly near the banks of rivers where the vegetation has gradually been eliminated due to the expansion of sugarcane plantations, particularly in the 1980s when new chemical fertilizers and industrial techniques began to be used at large scale. These technological innovations increased the saccharose content of sugarcanes and, as a result, industrial installations generated larger revenues. The main consequences were a depletion of the sugarcane land in Alagoas and a decrease of rural work market due to the intensive mechanization, especially in *Zona da Mata*. These "new" informal residences of the formerly rural and now urban population became "spatial traps," where the individual perspective and exploitative

[2] http://hdr.undp.org/en/content/human-development-report-2003

relationships are superimposed on the collective interest, highlighting the absence of the State in almost the entire region.

On top of this picture, we should add the political struggles of social movements for access to rural land, intensified after the end of the military dictatorship in Brazil during 1984. Fear of losing their lands moved landowners to gradually disintegrate the small rural nuclei inside their estates, forcing old inhabitants to move to the periphery of cities. However, these rural workers kept their field activities under the supervision of the same landlords showing a new and important social dimension of the risk within the region.

This condition may contribute to explaining the population migrations between the 1991 and 2000 censuses when most municipalities within the region became from being eminently rural to be dominated by an urban population. The nearly 600,000 urban inhabitants in Alagoas cities in 1980 became two million in 2006. Since most non-qualified workers could not find jobs within the incipient economies of the cities of *Zona da Mata* in the North and the Mundaú River watershed in Alagoas, rural workers' camps without land on the sides of the roads and within the river flooded areas multiplied. Thus, cities saw a decrease in their quality of life generated by the increase in population size because migration from the countryside to the city, generating strong pressures when looking for urban services and infrastructure that the public local power could not provide.

Structuring Problems: Risk and Uncertainty in Alagoas

The sugarcane monoculture, despite several crises in the last decades, is still dominant in the agrarian space of Alagoas, concentrating its activities in *Zona da Mata* located in the littoral zone of the State. It is economically hegemonic, representing 87% of the state exports. Social indicators are basically homogeneous in this territory, characterized by low economic dynamism and a low-income population due to the historically inherent production model (de Carvalho 2005; Freire 2012; Prado 1982; Diegues 1964).

The model early generated different structuring problems in the *alagoan* society. The population, in general, has low educational levels, political awareness of their rights, and a generalized poverty condition. This generates a situation of high uncertainty over the possibilities that people would have when facing natural catastrophes. Furthermore, risks are unknown for the migrant populations. Without reacting capabilities to the exposure or the capacity to technically recognize the dangerousness of the disorganized occupation of the new urban flooded areas, the poor population is at mercy of the derived uncertainties of their precarious living conditions.

It is important to point out that given the implications in the strategies to be implemented, floods in tropical regions such as the study area have high destructive power. It is not a mild increase in water levels as in template prairies. In this case, given the high kinetic energy of the floods, every building is destroyed almost immediately. Thus, it is not reconstructing but constructing as if it was the first time.

Considering this historical and social configuration, an extreme climatic event that occurred between the 18th and 20th of June 2010 produced a natural-origin disaster with serious social-environmental repercussions mainly in the Mundaú River watershed, between Alagoas and Pernambuco, also affecting the watersheds of the Una, Sirinhaém, Piranji, and Canhoto rivers. In only 3 days of strong rain, 26 municipalities from Alagoas were declared in a state of public calamity and other 34 in a state of emergency (Fig. 2). Fifty-five people died and nearly one hundred and fifty thousand were left homeless. The consequences of this disaster are still far from being solved. For example, 6 months after the event the building of only 9000 houses, out of the 50,000 to build, had started.

Only 11 of the 26 municipalities affected by the floods have an operating Municipal Civil Defense coordinator. The existence of a civil defense system is essential not only to assist directly the people affected by tragedies but above all for prevention. That is, educating the population to face disasters, since, what institutional response can be expected to face these events? In addition, municipalities are legally entitled to receive financial aid from the nation's emergency funds.

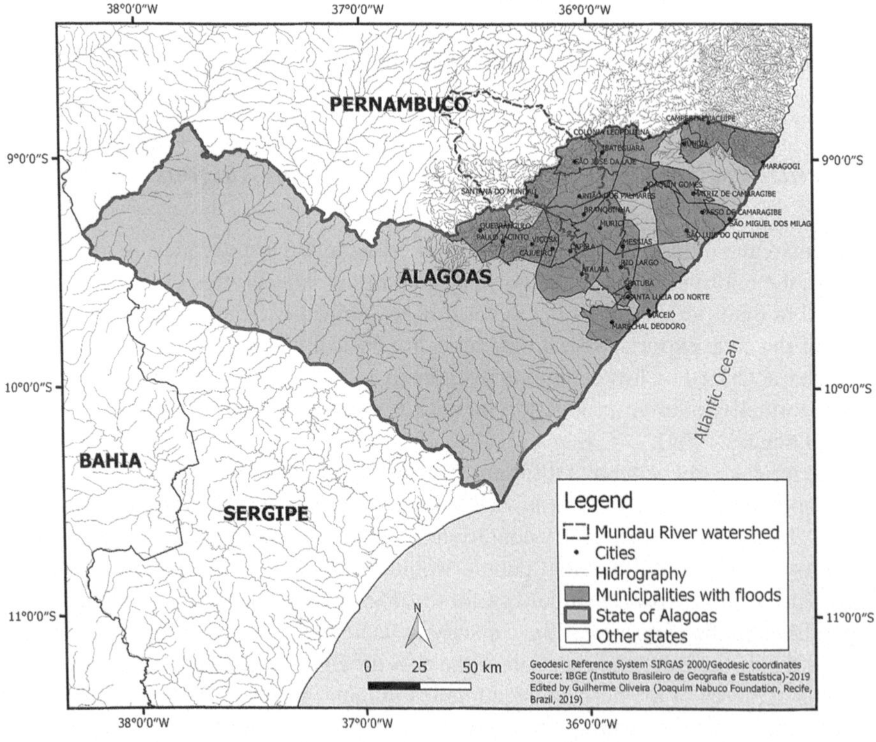

Fig. 2 Map of Alagoas (Brazil), the Mundaú River watershed, and the municipalities flooded during 2010

Fig. 3 Broken dam resulting from the 2010 floods in the Río Largo city. (Source: http://veja.abril.com.br/noticia/brasil/a-origem-do-tsunami-que-varreu-o-nordeste)

The devastating floods in Alagoas and Pernambuco in 2010 were also related to the precarious conditions of the dam maintenance in the watersheds of the Mundaú and Paraíba rivers (Fig. 3). The dams filled rather fast due to the strong storms, breaking one after the other, generating a cascade effect.

But flooding is nothing new for these two Brazilian states. According to Valmir Pedrosa, professor at the Federal University of Alagoas, every 10 years there are catastrophic floods in the watersheds of these rivers.[3] For example, nearly 1000 people died in the 1969 flood, and strong floods occurred in the Mundaú River in 1987, 1988, and 2000.

The indifference of the government is such that, according to the Integrated System of Financial Management of the Federal State (SIAFI) in 8 years, the Brazilian president spent only 0.74% of the US$ 250 million available in the 2010 funds "to prevent and prepare for disasters." Furthermore, according to SIAFI, only one million dollars were invested in disaster prevention actions in northeast Brazil. The June 2010 floods generated a loss of R$ 1.27 thousands of million (nearly USD$ 305 million). According to vice governor, Tomás Nonô, the Nation sent US$ 540 million and an additional US$ 727 million for house reconstruction; but still, after 2 years, only 10% had been rebuilt.

[3] http://www.anovademocracia.com.br/no-67/2909-alagoas-e-pernambuco-apos-as-terriveis-enchentes. 18/06/2012.

The Mundaú River Watershed

The watershed of the Mundaú River has a surface area of 4126 km^2 of which 52.2% correspond to the state of Pernambuco and 47.8% to the state of Alagoas. Nearly 50% of the watershed municipalities are located in Alagoas. The Mundaú River is the most used water system of the watershed, where all environmental activities have been exhausted. Furthermore, there are four sugarcane industrial compounds located along the river, yet it is still used for irrigation and to provide water for sugarcane and alcohol industrial compounds. It also generates energy, provides sand for constructions, is an area for relaxation and fishing, and provides water to all cities that cross.

The watershed is a zone of high primary productivity but also of fragile environments. It hosts a variety of activities such as wetland-related ecological activities, sugarcane culturing, cultural and historic patrimony preservation activities, especially in the city of Marechal Deodoro.

Although knowing the ecological limits of the watershed is important, it is having into consideration socioeconomic, territorial, and political aspects of the population in areas with the potential of suffering catastrophic floods. This knowledge is an important challenge for urban planning and risk management in underdeveloped regions. Contemporaneous geo-technologies and abundant available spatial data would allow having success with this challenge. But, is it technology the greatest challenge in this configuration? Or the policies to manage the uncertainties?

4 Final Discussion

> "Del dicho al hecho hay mucho trecho"[4]
> Popular expression.

The risk is in our daily life. Our postmodern life brings us exposures to different dangerousness. The scientific and technological development of our modern society changed the way in which we manage our activities, developing great confidence in the expert's wisdom (Giddens 1990). However, although the risk is present for all, the social production relationships make different social groups react differently toward dangerousness. The social equation of the losses generated by natural disasters is inversely proportional to recovery and adaptation. Meaning, the lower the income and wealth, the higher will be the produced damages and the difficulty of returning to the "normality" in our daily life. Although currently there may be more wealth than in past times, there are still social inequalities, poverty, and the consumption of environmental activities, especially in the less developed regions of our planet.

[4] There is a long way from the saying to the fact.

In our current world, the economic and technological dependence context of less developed countries can clearly be seen. These dependencies not only submit those countries to the commercial interests of large transnational corporations but also determine its insertion in the globalization process. This is a process causing a profound transformation in the means of production and a division of work, with a systematic incorporation of new forms to design, create, and organize space. The old production structures have been affected, and also new ones are being created, changing the use of space and, probably, generating a renovation in the old power structures. The new hierarchies in the generation and appropriation of wealth suggest that the capitalist model has perfected its instruments, most agile handling of the scales and the usability of the constructed environment.

Some governments, in the name of socioeconomic development and the consequent increase in the gross national income, prioritized the agriculture-export businesses, especially in the emergent Latin American nations. This model, in the last years of the twentieth century and the first two decades of the twenty-first century, has generated much wealth to the system, still not solving the historical capital concentration. But, supposedly unexpected effects have occurred in other parts of the system. We observe that the consumption of environmental activities, on top of other changes in global climate, has decreased cities' resilience to climatic phenomena. That is the capacity of the urban system to return to its original state after being affected by the action of external perturbations.

Changes in the hydrological systems associated with human activities are interfering in the water cycle of hydrographic watersheds. The building of dams for protection against drought, energy generation, irrigation and human consumption and their associated rules causes changes in the spatial and temporal distribution of river fluxes. This, in turn, affects the evaporation and infiltration in areas close to river courses and surrounding biota (Christofoletti 1999).

Catastrophic floods have generated significant human and material losses around the world. But the different anticipation, response, and reconstruction capacities of each society when facing a natural phenomenon generate the magnitude of the disaster. It is the knowledge of the dimensions of the risk that allow making decisions of a given social group about the current and future life. The uncertainty is a negative factor for the life of socially marginalized peoples, not allowing them to exercise their rights and to have access to the knowledge about their own risk conditions in which they live.

Knowledge is the key to modern civilization. We cannot live without it. So, social sciences, in particular, are interesting to explore how new technologies (including those that capture, analyze, and show spatial solutions to social problems) can help larger groups of society into decision-making. This means that for old problems we need new solutions based not only on technology but, fundamentally, on politics. It is the later we have to change. This is clear in our case study, where we see that during each flooding in Alagoas the same responses are repeated: the militarization of the help to the homeless, liberation of emergency funds, and promises of reconstruction of houses, schools, and public buildings. Authorities declare war to the environment. "The river is guilty!" hiding the fact that it is the social system working within its own normal condition that determines the disaster. That means, what is unexpected is expected from such social-spatial configuration.

Thus, the disaster was installed due to multiple causes: the precarious maintenance conditions of the dams in the Mundaú and Paraíba watersheds and cutting down the vegetation on slope especially for the sugarcane monocultures. The result was a loss in the capacity to retain water, soil erosion, and sedimentation of rivers.

Another serious problem is the lack of adequate urban planning in cities along wetlands occupied by the rivers during periodical episodes of intense rain. These factors, along with extreme rains related to global climate change, mean that the probability of new disasters in the area is now larger than it was before. Thus, spatial traps still exist for the inhabitants of these cities.

However, it is still possible to prevent and mitigate the catastrophe, taking measures that allow avoiding negative impacts over the population, goods, services, and the environment including those destined to attenuate and reduce negative impacts (Natenzon and González 2010). In this case, the idea is a participative risk management to act over the social vulnerability during normality and decreasing the uncertainty in precautionary decision-making.

That is where geo-technologies may mean an advance in terms of the necessary knowledge for the elaboration of public participative policies that may allow considering the plurality of interests and perspectives of the different actors and social groups at stake (Funtowicz and Ravetz 1993). Digital cartography has expanded its accessibility in the last decades with the objective of satisfying the specific interests of traditionally marginalized social groups.

Nevertheless, as with other sciences, cartography is not neutral. As in a tension game, it can serve both to show and to hide objects, to enforce or to hide rights, and to empower or subdue social groups. Thus, truly, the limit of technology is politics.

It is beyond arguments that social vulnerability is involved with the political conditions that allow or not people's rights to decide about their life with the greatest and best possible knowledge about present and future risks. Right to live in secure areas within cities, to have access to social security, work, health, education, and environmental protection. Prevention is possible, it is only a matter of fact to have the decision and carry it out.

References

Beck U (1992) La sociedad del riesgo. Paidós, Barcelona.

Beck U (1993) De la sociedad industrial a la sociedad del riesgo. Cuestiones de supervivencia, estructural social e ilustración ecológica. Revista de Occidente, N° 150 México

Blaikie P, Cannon T, David I et al (1996) Vulnerabilidad: El entorno social, politico y económico de los desastres. La Red, Lima

Calderón Aragón G (2011) Construcción y reconstrucción del desastre. Plaza y Valdés, México DF

Christofoletti A (1999) Modelagem de Sistemas Ambientais. 1a. edn. São Paulo, Edgard Blücher, SP

de Carvalho CP (2005) Economia popular: uma via de modernização para Alagoas. EDUFAL, Maceió
de Oliveira F (1981) Elegia para uma Re(li)gião. Editora Paz e Terra, Rio de Janeiro
Dean W (2004) A ferro e fogo. A História e a devastação da Mata Atlântica brasileira. Companhia das Letras, São Paulo
Diegues JM (1964) Propriedade e uso da terra na "Plantation" brasileira. In: Sistemas de Plantaciones en El Nuevo Mundo. União Panamericana (Estúdios y Monografias, 17), Washington
Ellul J (1964) The technological society. Vintage Books, New York
Freire NO (2012) Geonegócio: as geotecnologias e a emergência de novos mercados em regiões periféricas. Editorial Académica Española, Madrid
Freire N, Bonfim C, Natenzon C (2014) Vulnerabilidade socioambiental, inundações e repercussões na Saúde em regiões periféricas: o caso de Alagoas, Brasil. Rev Ciênc Saúde Coletiva 19(9):3755–3762
Freyre G (2006) Casa Grande & Senzala. 51ª. ed. Global Editora, São Paulo
Funtowicz S, Ravetz J (1993) La ciencia posnormal. Ciencia con la gente. Grupo Editor de América Latina, Buenos Aires
Furtado C (2003) Formação Econômica do Brasil. 32ª. ed. Companhia Editora Nacional, São Paulo
Giddens A (1990) Consecuencias de la modernidad. Alianza Editorial, Madrid
Herzer H, Rodríguez C, Celis A, Bartolomé M, Caputo MG (2002) Convivir con el riesgo o la gestión del riesgo. https://www.researchgate.net/publication/237638971_CONVIVIR_CON_EL_RIESGO_O_LA_GESTION_DEL_RIESGO1
López Cerezo J, Luján López J (2000) Ciencia y política del riesgo. Alianza Editorial, Madrid
Luhmann N (2006) Sociología del Riesgo. Universidad Iberoamericana, Universidad de Guadalajara, México
Maskrey A (1989) Disaster mitigation: a community based approach, development guidelines, 3rd edn. Oxfam, Oxford
Natenzon C (2005) Inundaciones catastróficas, vulnerabilidad social y adaptaciones en un caso argentino actual. Cambio climático, elevación del nivel medio del mar y sus implicancias. In: Climate change impacts and integrated assessment EMF workshop IX. Colorado
Natenzon C, González S (2010) Riesgo, vulnerabilidad social y construcción de indicadores. Aplicaciones para Argentina. In: Argentina e Brasil: Possibilidades e obstáculos no proceso de integração territorial. Org. Arroyo, Mónica. Zusman, Perla. São Paulo: Humanitas; Buenos Aires: Facultad de Filosofía y Letras
Neiman G (2017) Agrarian restructuring and changes in the demand for labour in Argentina. Rev Agrar Stud 7(1). http://ras.org.in/feb533d7d42a64e93ff6f61b340ff74a
Orwell G (1945) Animal farm: a fairy story. Secker and Warburg, London
Prado JC (1982) História econômica do Brasil. 27a. edição. Editora Brasiliense AS, São Paulo
Rebizo MM, Rodriguez AT (2011) Balance de inserción internacional de las cadenas agroindustriales argentinas. CEPAL, Santiago de Chile
Reboratti C (2010) Un mar de soja: la nueva agricultura en Argentina y sus consecuencias. Rev Geogr Norte Gd 45:63–76
Wilches-Chaux G (1993) La vulnerabilidad global. In: Maskrey A (ed) Los desastres no son naturales. La Red, Colombia, pp 11–44
Wilkinson J (2010) Transformações e perspectivas dos agronegócios brasileiros. Rev Bras Zootec 39:26–34

San Rafael Reserve, Paraguay: Key Social Stakeholders and Sustainability Scenarios Through Environmental Governance Approaches

Stella Mary Amarilla R., Alberto Yanosky, and Janet Villalba

Abstract San Rafael Conservation Area is one of the greatest conservation challenges in Paraguay, as it is an area of historical socio-environmental conflicts, dating back to the Rio Summit in 1992 when the area was proposed. It has a relict fragment of Atlantic Forest, representative of the Alto Paraná ecoregion combined with natural pastures. These ecosystems are considered as areas of high biodiversity and numerous ecosystem services (the Guarani aquifer, a world reserve of freshwater). From a social point of view, the area includes large private estates, small communities or colonies, and ancestral-indigenous communities (The Mbyá Guarani). They use the ecosystems and their services, including biodiversity, for various purposes: productive uses on a large scale (agriculture, livestock), conservation, subsistence (hunting and gathering), illegal extraction (wood and firewood), and also constituting part of the identity of ancestral-indigenous groups. In this chapter, we analyze the importance of the role of local and national social actors, their interests, use of ecosystem services and biodiversity, as well as their characteristics (influence-importance). The central idea is to conceive and accept this area as a socio-ecosystem of multiple uses and to propose a new approach to conservation that considers social, economic, and ecological aspects.

Keywords Social-ecological systems · Latin America · Complexity · Paraguay · Nature reserve · Environmental governance · Stakeholders

---

The original version of this chapter was revised. The correction to this chapter is available at https://doi.org/10.1007/978-3-030-28452-7_24

S. Mary Amarilla R. (✉) · J. Villalba
Facultad de Ciencias Agrarias, UNA, San Lorenzo, Paraguay
e-mail: stella.amarilla@agr.una.py; janet.villalba@agr.una.py

A. Yanosky
Guyra Paraguay CONACYT-Pronii, Asunción, Paraguay
e-mail: yanosky@guyra.org.py

L. E. Delgado, V. H. Marín (eds.), *Social-ecological Systems of Latin America: Complexities and Challenges*, https://doi.org/10.1007/978-3-030-28452-7_13

1 Introduction

San Rafael Reserve (SRR) is a result of the proposal of Paraguay to the 1992 Rio Summit. It was originally created with the management category of "National Park," changing later to "Reserve of Managed Resources," and then taking it back to "National Park." This area represents a high priority for conservation, as one of the most important remaining fragments of the Alto Paraná Atlantic Forest ecoregion, located in the southern part of the Eastern Region of Paraguay. It has the potential as part of a biological corridor, linking other protected areas; and, in addition, it hosts a sample of the Southern Grasslands or Pampas (Yanosky 2002).

The protected area is composed of private properties which, due to the lack of rules for access and use of natural resources, in several cases have suffered unsustainable uses and a historical degradation process (Cartes and Yanosky 2005, 2008). In addition, socioeconomic conditions (poverty levels in some stakeholders, lack of economic alternatives, and lack of coordination among key stakeholders) increase the dependence on resources (mainly from forests) and generate the main conflicts related to their use (Amarilla 2004).

Several types of social actors are responsible for the degradation of the reserve: landowners, indigenous communities, and farmers; some public and private institutions and non-governmental organizations with interests in the conservation of the area. However, they could also contribute to its sustainability (Amarilla 2004). Thus, identifying the reserve's key stakeholders and analyzing their decisions regarding resource use is a way to reconcile conservation and sustainable use of resources.

The relevance of the area, in terms of biodiversity conservation within the National System of Protected Areas of Paraguay (SINASIP), is highlighted in terms of species composition, ecosystems, and area. The SRR and its significant fragment of forest (currently about 50%) and its natural grassland systems conserve headwaters of important watersheds in the region, in the catchment areas of Paraguay and Paraná's rivers (Fig. 1), which also contribute to the watershed of the Guaraní Aquifer, one of the main freshwater reservoirs on the planet. This is one of the most important ecosystem services. The high biodiversity in the different ecosystems of the San Rafael area, in addition to the forest, such as the natural pastures and the wetlands, represents important habitats for diverse species of fauna and flora. Because of this, the growing complexity has made this area a real challenge for conservation at a country level (Cartes and Yanosky 2005). For example, a lack of definition regarding land tenure versus property and customary rights, the State's figure as the entity responsible for the definition of criteria and clear rules for sustainable use, its monitoring and control, and the diversity of social actors with different objectives and resource needs.

SRR is located in the Alto Paraná Atlantic Forest (APAF) ecoregion, southeast of Eastern Paraguay, which represents one of the most threatened tropical forests in the world, with 9% remaining of its original forest cover, in a highly fragmented and

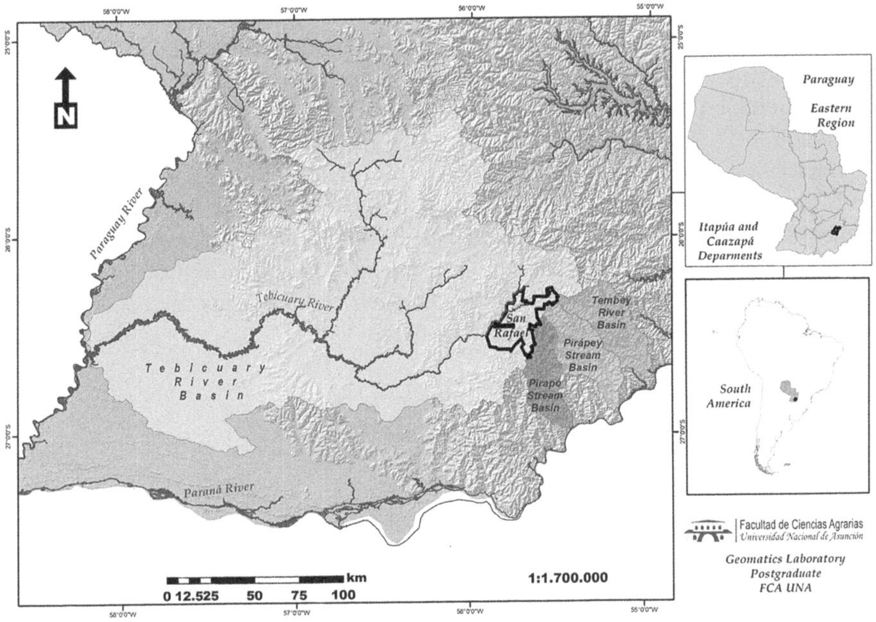

Fig. 1 Study area and river basins' location map

degraded condition (WWF 2016; Da Ponte et al. 2017). It is located between 26°10′ and 26°40′ south and between 55°30′ and 55°55′ west (Fig. 1).

Analyzing the history of SRR management in socioeconomic, ecological, environmental, and legal terms is not an easy task. There is a great deal of technical and scientific information in various thematic areas, carried out by institutions or sectors interested in the subject and by the very nature implied by the context of protected wild areas.

However, in order to propose robust sustainability strategies, a historical management view is necessary because the conservation status of ecosystems depends and continues depending on responses, actions, and interests that their social actors have given towards ecosystems and their resources. This set of social actors and fields has changed over time, in accordance with relevant management processes such as the declaration as a protected area with strict protection and conservation objectives, since the 1992 Rio Summit, without prior consultation with local landowners, indigenous peoples, and other organizations. Then, the change to a less strict and confusing category in its implementation, understood by some social actors with less strict and more permissive goals towards the use of resources ("Managed Resources Reserve") in 2002, to finally and currently back to the category of "National Park" since 2005. This change of category could express the recognition by the enforcement authority and the Paraguayan State of their inability to implement a reservation area for a national park as expressed after 27 years of inaction.

The change of management category could already mean a series of irreversible impacts to ecosystems and the provision of their services. For example, the loss of some endemic or threatened species (or their populations) at site level, the general degradation of the various ecosystems presents in the areas declared as "areas reserved for..."[1] (FCA 2012).

International cooperation agencies have historically invested in SRR, starting with the government's commitment to applying for a World Bank loan (PARN Project), at least two GEF grants through the United Nations Development Program (UNDP) and other bilateral grants and investments. The results of these investments have been unsustainable, and the fulfillment of the commitments and conditions set to get the loan and/or grant hasn't been assured.

2 Main Conflicts in the Area

The restriction of resource use on properties is one of the main conflicts, followed by land invasions, especially in forest areas where wood is the main attraction. Farmers take over properties of indigenous communities and some other non-indigenous landowners. The situation of poverty in which the farmers find themselves makes these populations look for alternative ways to survive using natural resources of the area. Policies used to manage the problems with the farmers are, in fact, threats. There is a social degradation of the indigenous people as a consequence of the dispossession of their natural resources and the growing fragmentation of their communities. These and other aspects have been addressed by Amarilla (2004) and by De Egea Juvinel and Balbuena (2011).

These conflicts happen easily due to a situation of documented "free access" in the protected area, increasing problems between land ownership (e.g., of private landowners) and property rights (of ancestral communities). The same situation of free access in the territory makes possible the presence of other "private social actors" or better called "opportunists," who are responsible for the presence of illegal activities such as illicit crops, timber, firewood, and charcoal (Amarilla 2004; López 2017) and, possibly, wildlife poaching, as well as cattle raising. In recent years, there has been an increase in activities on clandestine landing strips and even the fall of an aircraft that can be seen as evidence of the free action occurring in the reserve, with no control or surveillance, or even the involvement of local authorities in those illegal activities.

Therefore, the area is characterized by conflicts related to property rights, insecurity in land ownership, strict control of property, and diverging interests in the land and its natural resources. Overlapping of rights is another feature, as indigenous

[1] Reserve Area: Article 10, Law 352/94: of Protected Wild Areas: private property that has been declared as such by the respective decree and that will remain under that denomination until the process of conversion into a Protected Wild Area under public domain is finalized.

communities' own ancestral rights on extensive forest areas which are in private ownership in the SRR area.

3 The Key Players in the San Rafael Reserve

Characterization of Key Stakeholders and Their Importance

Some studies (Amarilla 2004; Cabrera 2008; FCA 2012; López 2017) carried out in SRR propose the following social groups as main social actors (Fig. 2):

- Private landowners (small, medium, and large-sized, in favor of conservation, in favor of production) prioritized by the direct impact of their activities on the resources and positive or negative effect towards the supply of ecosystem services, in addition to their property rights and proximity to the resources.
- The indigenous communities in the area (with and without property titles) prioritized by their dependence and proximity to the forest, their property rights because they represent ancestral communities in the area and claim for the entire area. These stakeholders use their properties and they also move in the territory considered by them as the *Tekoha Guasu* (ancestrally claimed).

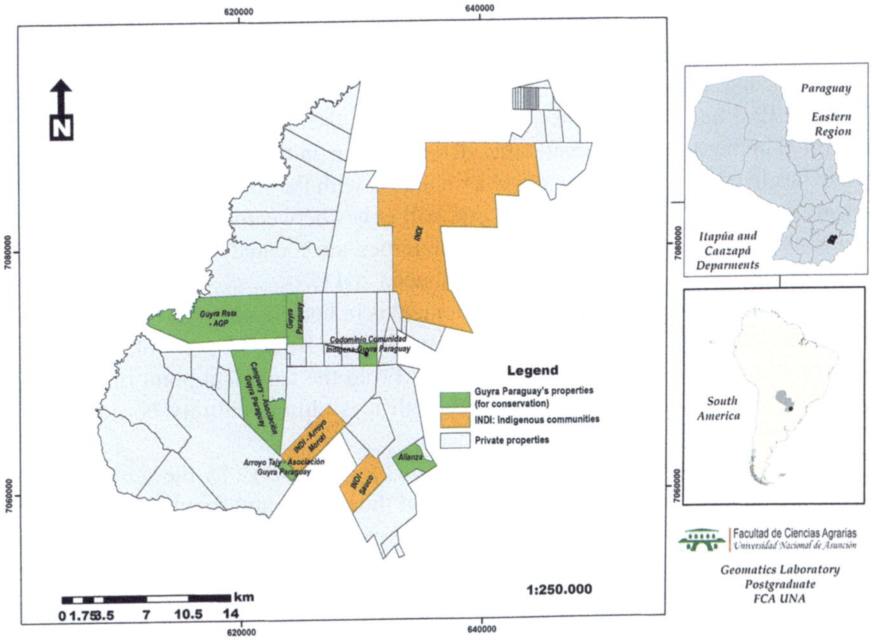

Fig. 2 Location and classification map of San Rafael Reserve's properties

- Farmers settled within the reserve's boundaries and other groups with interests in land occupation (whether it is derived from private owners or indigenous communities).
- Civil society organizations locally present.
- Public institutions that are involved in environmental issues, including the academy.
- The Government.

Key stakeholders for SRR are those social groups with different interests, objectives, and circumstances regarding the use or non-use of natural resources in the protected area, with positions for or against both conservation and sustainable use of natural resources.

Regarding indigenous communities: approximately 16 communities of the *Mbyá Guarani* ethnic group (a Guaraní linguistic family), where only four communities maintain relations with other non-indigenous groups. Some communities hold property titles, either in the name of the community or from the National Indigenous Institute (INDI). The other communities are located within private properties.

Numerous indigenous communities of the Mbyá ethnic group, also known as *Kayngua* or Monteses, find daily sustenance in their forests and also the source of their cultural identity, as part of what constitute their ancestral territory (a relevant aspect as a cultural ecosystem service) giving bases to the area of reserve claimed by the Mbyá people as their "*Tekoha Guasu*" (De Egea Juvinel and Balbuena 2011) since its creation in 1992.

The *Tekoha Guasu* identified within the boundaries of the Reserve San Rafael and beyond involves 13,000 ha (titled by the Paraguayan Indigenous Institute (INDI) and recognized by its leaders) of traditional management by the Mbyá Guaraní people according to their customary rights.

In the IInd Indigenous Census (DGEEC 2004), 27 indigenous settlements were identified in the study area, inside the SRR as well as in the buffer zone. This ethnic group has lived for centuries in close association with the forests of the APAF. This census reveals the close relationship that still exists between the indigenous settlements and the forest. In fact, many communities are located in the heart of large forest remnants (De Egea Juvinel and Balbuena 2011).

In the case of farmers, "Amistad" settlement is indicated exclusively. The reasons people recognize this group as a key stakeholder are: from the point of view of the reserve's integrity, the farmers are part of it. From the technical point of view, the settlement constitutes a buffer zone although it is almost entirely bordered with private properties (Amarilla 2004).

The presence, activity, or impacts generated to the reserve by the settlement are not conditioned by the administrative limits of the reserve and should be considered fundamental in its management. There are approximately 110 families in the settlement, distributed in land plots of 8 ha each. However, it is important to consider that this settlement didn't always exist and resulted from a disaffection carried out in 1997 by the Paraguayan State to give land to "landless farmers" settled in the micro-center of the country's capital. This clearly shows the fragility of the system and the promotion of settlements with people who were not at the time linked to a land. From this group, 33 families (30%) benefit from the Paraguayan Forest Conservation Project which gets funding from the voluntary carbon market to mitigate climate change and produce additional social and environmental benefits. While there are recommendations for this project, it has demonstrated that preserving forests is an economic alternative for farmer communities (González and Achinelli Báez 2017; Goh and Yanosky 2015).

There are roughly 38 private owners. A stratum of small- and medium-sized owners and another of large-sized owners have been differentiated by the area of the properties. Areas larger than 1000 ha were considered as large properties. There are absent owners, in some cases, the property is in charge of contracted managers or administrators. In a study carried out by Guyra Paraguay (2005), small owners were differentiated from large owners using a threshold of 500 ha. At that time, 18 properties were considered, excluding the organization's properties. The distribution of properties in the SRR can be seen in Fig. 2.

It is important to point out that private landowners, farmers, and indigenous people all have poor infrastructure within the protected area, without the governmental presence and that most of it is maintained by the private initiative. More recently, a cobblestone road is being built connecting the south of San Rafael and the town called *Caronay* with Amistad, and a bridge over the *Tebicuary Mi* River is currently being rehabilitated, which would re-establish the lost road connectivity between Amistad and Lima.

Although not considered as key stakeholders per se, a group of "linked" social actors in the San Rafael area, known as "opportunists," cannot be overlooked. Also referred in the literature as "*free-rider*" agents (Amarilla 2004), they are responsible for the generation of negative externalities such as the illegal extraction of resources, illicit crops, animals' theft, burning of pastures, among others. The presence and increase of the actions from this group of opportunists in San Rafael must be analyzed and regulated.

Among the civil society organizations, Guyra Paraguay Association stands out for its presence in the area, which has made efforts to create conservation core zones in San Rafael, with a total of 11 properties that add up to 6760 ha. Pro Cordillera de San Rafael (PROCOSARA) is an important association which has carried out some monitoring and patrolling activities and coordination between some private landowners. Several studies emphasize the leading role for conservation actions in San Rafael to these two institutions, which also initiated a process to create the San Rafael Alliance, which finally couldn't prosper; it sought to bring other institutions to seek an alternative to destruction.

The list of public and private institutions could be more detailed. However, the degree of their involvement, or the type of effective involvement with the conservation cause of SRR, differs widely among them (Table 1).

Table 1 Key stakeholders in SRR

Local institutional key stakeholders
– MADES: Ministry of the Environment and Sustainable Development, former Secretariat of the Environment (SEAM)
– Public Prosecutor's Office: Itapúa's Environmental Prosecutor's Office
– Itapúa's Environmental Secretariat
– INDERT Institute for Rural and Land Development
– INDI: Paraguayan Indigenous Institute
– INFONA: National Forest Institute
– Technical Secretariat for Planning (known as STP)
– Ministry of Agriculture and Livestock (MAG)
– Districts of Alto Vera and, *Itapúa Poty*
– Yacyretá Binational Entity *(Hydroelectric binational company)*
Key international institutional stakeholders
– UNDP Paraguay
– WWF Paraguay
Other relevant organizations
– National University of Asunción
– Paraguayan Network of Conservation in Private Lands
– Paraguayan Tropical Forest Conservation Fund
– ACIDI (Association of indigenous communities of Itapúa)
– IDEA (Institute of Environmental Law and Economics)

Interests and Circumstances of Key Stakeholders

In more than 27 years of conservation management history in SRR, the typology of stakeholders has changed under a dynamic process if the whole management period is considered. Perhaps few social actors are at the beginning of the management's history, multiplicity, and complexity of social stakeholders in the last 10 years, taking into account the temporary functionality of a San Rafael Area Management Committee, which came to operate actively at least for 2 years. However, it is important to consider that any process of participation, involvement, and exchange of information have been valid to maintain, over time, the clear idea that the territory called San Rafael area or rather the fragment of forest called San Rafael must be preserved in the long term.

The synthesis of interests and circumstances among the main key stakeholders is summarized in Table 2. A synthesis of the types of relationships analyzed between the different groups of key stakeholders is shown in Tables 3 and 4.

The institutions most directly related to the SRR are: San Rafael Reserve Management Committee, PROCOSARA, Itapúa's Government, Guyra Paraguay, Institute of Environmental Law and Economics (IDEA), CECTEC, MADES, School of Agrarian Sciences/UNA, School of Exact and Natural Sciences/UNA, other universities and research centers (FCA 2012; López 2017).

These institutions have been linked with a direct relationship to San Rafael based on their institutional goals, conservation goals in the reserve, commitment to the administration and management of the protected area, and commitment to the local population. Among these institutions, only two operate locally and currently

Table 2 Key stakeholders' interests and circumstances in the Reserve San Rafael area

Types of stakeholders	Interests and circumstances	Level of organization	Land usage
Indigenous communities	Objectives (a) Ensure the recognition of their ancestral rights to their lands (b) Maintain and improve their subsistence production (c) Obtain small incomes for medical treatment and/or mobilization outside their communities through the sale of wood, posts, honey, and handicrafts. The forest is their livelihood	They are organized with the leadership of a "Cacique," who has the power to make decisions on productive activities, and in the social, political, and cultural aspects They live in the forest in small and rustic wooden houses They are distributed in small cores called "*Tapýi*" that are made up of three to four houses each The level of organization has improved in the last 10 years, nucleated to claim their lands "*Tekoha guazú*"	Intervened secondary forests and forests without intervention. A small part is used for crops. The average area under cultivation is 2 ha with one or two products (cotton, soybeans, corn, beans) Some communities are dedicated to handcrafts in wood or palms although this activity doesn't have an established market. They usually commercialize ornamental forest species such as orchids and epiphytes
Farmers	Objectives: to improve and expand its income from farming. They devote enough time and attention to the most profitable crops such as soybeans, cotton, cassava, and corn The crop to be cultivated is the one that has more profitability in the year	Regarding the "Amistad" settlement, they are organized through a Neighborhood Commission to prioritize needs and procedures. Although the level of organization is better than in the other groups, their community management is still weak and they need the support of institutions. These farmers' way of life is based on the search for land suitable for agriculture and rooting	On average, between 40 and 50% of the farm is agriculture, which represents between 3 and 4 ha. They dedicate up to 2 ha to income crops (cotton, soybean, cassava) and between 0.5 and 1 ha to subsistence crops (corn, beans, and others)
Owners	Different situations, there are owners interested in selling their property. Others rely on maintaining the property, securing and improving agriculture, and even expanding if possible. They are interested in generating the best and highest income for the property by engaging in the activity that is currently known to be the most profitable. In certain cases, the forest represents a "problem" for them because they cannot use it, care for it, or know what to do with it	The organization between owners is a worrying void, they don't have any association that coordinates their actions or serves as a channel of communication between them. Each owner works and decides individually no matter the conditions or requirements in the neighboring properties, which leads to conflicts of interest and negative externalities (clogged streams, erosion, burning fields, etc.)	Large landowners have a main agricultural activity and forest activity is secondary; the small- and medium-sized landowners have multi-activity (livestock, crops, and forestry) Soybeans and rice generate the highest expectations. Several hectares have been implanted for soybean growth; this was given in natural grasslands and are considered experiments to prove their profitability
ONGs with local presence	Conservation of as much of the San Rafael area as possible	High and stable, with concrete results to date, but with a need of greater support from other stakeholders and institutions linked to the San Rafael cause	Monitoring and control of conservation areas Long-term protection and conservation

Adapted from Amarilla (2004), López (2017)

Table 3 Synthesis of the relationship between key stakeholders

	Indigenous people		Farmers		Landowners	
Stakeholders	+	−	+	−	+	−
Indigenous communities	Communication, agriculture information exchange, shared management	Conflicts over administrative boundaries	Natives recognize farmer's needs	Natives blame farmers over resources extraction, hunting, illegal crops, wood extraction	Hiring of indigenous workers by landowners	There isn't minimal communication between them
Farmers	Farmers recognize indigenous rights	Farmers blame indigenous people for the extraction of natural resources and hunting	Clear rules among farmers Projects unification or plans to be managed[c]	Differences between groups due to organizational issues	Farmers use roads in properties as entries and/or exits to the settlement	Landowners blame farmers[b] for timber extraction, cattle theft, and hunting of wild animals
Landowners	Indigenous communities in private lands	Landowners blame natives for timber extraction and hunting of wild animals	Landowners allow farmers to use roads located inside their properties	Landowners blame farmers[b] for forest extraction (timber or wood)	Exchange of information on agriculture and/or man power Loan of agricultural implements	There is no landowner's organization Externalities because of the use of agro-toxics, dams of water courses
Opportunist stakeholders[a]	Unconnected	Natives rent their lands to opportunists or stowaways, which are used for illegal crops	Unconnected	They rent their lands to opportunists for illegal crops, convinced by stowaways or in cooperation with them	Unconnected	Opportunists stakeholders or stowaways occupy properties to use them for illegal activities (illegal crops plantation)

[a]Opportunistic stakeholders, stowaways, or "free-riders" mainly responsible for negative externalities towards the use of resources, whose presence is given by free access or weakness in the control or definition of physical boundaries between properties within the territory
[b]In this case, "farmers" refers to any farmer either inside or outside the reserve and not exclusively to those of the Amistad Settlement
[c]Exclusively among those of Amistad Settlement

Table 4 Synthesis of the relationship between stakeholders and institutions

	Institutions	
Landowners	+	Changeable dialogues, but with possibilities of getting to an agreement Contracts for patrolling and surveillance between some owners and Procosara (until 2007) Proposals for conservation usufructs and negotiated land purchase between landowners and NGOs In reality, there are no solid and efficient relationships until this day (formal agreements)
	–	Some projects created mistrust to the owners due to the lack of continuity. Institutions fail to inform owners sufficiently about their actions or projects. The owners lack representation before institutions, which creates a void in the management
Indigenous people	+	Indigenous people recognize conservation interests common to them and the institutions. Few institutions exchange information with indigenous people or provide minimal transportation services
	–	Some institutions pressure the indigenous people to not use the resources, especially the forest It creates pressure which annoys most indigenous people
Farmers	+	Some farmers initiate demands before institutions searching for improvements in infrastructure. Some conservation organizations have initiated actions to involve farmers in their projects through dialogue and exchange of information
	–	Farmers reject projects or institutions with occasional or temporary relationships. Some institutions pressure farmers for conservation and reserve regulations

(PROCOSARA and Guyra Paraguay). Guyra Paraguay carries out land purchase actions, declaring these secured lands for conservation in perpetuity.

Given the inaction of the Government and the justification of physical presence because of lack of public lands in San Rafael, Guyra Paraguay donated 500 ha to the Paraguayan State for the development of a core conservation area. For the first time, with the support of Itaipú Binational, a control post related to Colonia Amistad and the *Tebicuary Mi* River has been installed there.

4 Possible Sustainability Scenarios

Sustainability Based on Collective Action Approaches and Institutional Arrangements

Our analysis has its theoretical and intellectual basis in Ostrom (1990, 1994) regarding the use of common goods, collective action, and socio-ecological systems, as a framework for the design of institutions related to the management of common property resources. Successful self-administration in community management is based on shared agreements and rules' definition, as has happened in communities or organized stakeholders' groups, who have been able to establish their own "local institutional arrangements" and achieve a successful and coordinated action regarding the management of these resources (Ostrom 2010), called "sustainability scenarios."

These guidelines for institutional arrangements have also allowed it to transcend towards newer governance of natural resource concepts and the rules to access and use them (see also chapter "A New Environmental Governance"), strategically combining those governmental mechanisms (formal laws) with local institutional (generally non-formal) designs, accepted and shared by the majority of stakeholders participating in a common goal (Amarilla 2004; Solano 2018).

The case studies pointed out by Ostrom have shown that not all individuals (or communities) who use a resource achieve its overexploitation. In cases, they are able to create their own management institutions (internal arrangements) to allow them to avoid overexploitation and manage resources sustainably. These regimes or institutional arrangements regulate one or more of the following aspects (Ostrom 1997):

- people that are authorized to appropriate, use, or exploit the resource
- time, quantity, location, and technology needed for the use or exploitation of resources
- people forced to contribute to maintain this system
- modalities for supervision and monitoring of these activities
- the way conflicts are handled and resolved
- the way in which "rules" change over time according to the functioning of the management system and the participants' strategies

As for rules of the game and codes of conduct, the institutions are defined under the approach applied in this study to make viable the interactions and transactions between individuals and groups and to acquire effects considered desirable for society at some point in its development. An institution is a set of rules used to establish: who is eligible to make decisions within a group of individuals, what actions are allowed or restricted, and what are the sanctions for those who break the rules (Madrigal 2003).

The main functions of social institutions focus on regulating social relations (individuals, local groups, organizations), establishing the rules of coexistence necessary to relate, adapt, or even solve the problems or conflicts that are generated in the social, economic, and environmental setting (Madrigal 2003).

The same author also points out that the social functions of institutions aim to cut down uncertainty in social traffic, make possible interactions and transactions in the social and economic area (without game rules, players cannot play), and channel possible social conflicts.

In the San Rafael case, it is key to design and define "rules of the game" that can provide information about the actions that a stakeholder and institution have to carry out (an obligation), have to avoid (prohibition), or cannot carry out (permissions). It can be defined as "operational norms" in practice, also as flexible complements to fill a system of laws (formal legal framework). The rules must include rights and obligations that effectively regulate stakeholders' behavior and promote certain activities desirable to maintain resource productivity or conservation interests (Amarilla 2004).

The definition of these operational rules is the responsibility of the stakeholders, those users of the resources since they are the ones to really coordinate actions through the definition of rules entirely acceptable to them. This is one possible vision to coordinate the actions around SRR. That is, the possibility for its key stakeholders to define their own rules of appropriation with consensual goals towards either sustainable production and/or long-term conservation. In other words, the change in individual behavior allows assuming that the few institutional arrangements that people or stakeholders use to govern and manage resources for the collective use and public goods contain a series of incentives and learning opportunities (Ostrom 2010), feasible for the construction of common goals.

These institutional arrangements happen today, but without being recognized as such. For example, recognizing the conservation efforts of some NGOs, with concrete results in terms of conservation cores established in San Rafael, allows cooperative relationships with some private owners, support and management shared with indigenous communities, and a stable relationship and exchange of information with other public and private institutions interested in the SRR.

Sustainability Based on Complex Socio-Ecological Systems Approach

Collective action and institutional arrangements, in addition to the ecosystem approach, the understanding of situations presented such as the San Rafael case as a clear scenario of a complex ecological system (diversity of ecosystems) as well as a complex socio-ecological reflexive system (Delgado and Marín 2005) is a source of debate at both the academic and scientific levels.

From this point of view and starting from the concept of ecosystem and/or natural areas, the appreciation of these concepts varies widely according to the benefits that each type of stakeholder gains; as also applied for benefits obtained from ecosystem services. Delgado et al. (2007) and Bachmann (2006) mention that this complexity is based on cultural diversity, education and socioeconomic level, location, levels of communication between members of the local society, governmental policies, economic interests, and others. For the case of San Rafael, other aspects such as property rights, access to information, and levels of participation from stakeholders, which are not considered, or their considerations are changing and sporadic need to be added.

Therefore, while studying and managing ecological-environmental problems, human societies and their interaction with natural ecosystems cannot be left out. These should be incorporated as another component for research, through its diverse forms of relations with nature and its interventions in the conservation and use of resources (Delgado et al. 2007).

A scenario for sustainable strategies for SRR, from the ecological systems approach (see chapter "Simplifying the Complexity of Social-ecological Systems with Conceptual Models") is based on the idea of a social, economic, and environmental valuation of how key stakeholders use their resources. It may be ancestral uses and customs, basic subsistence requirements, interests in production at different scales, cultural interests, specific interests on biodiversity protection and conservation, political and institutional interests, and on research and knowledge generation, among others.

On the other hand, Ostrom (2009) further argued that the framework of analysis of socio-ecological systems would make possible a greater accumulation of scientific knowledge, through the systematic comparison of data collected from multiple studies on sustainability in various countries systems; and their development would be decisive to diagnose such systems, as well as for the design of sustainability strategies (Solano 2018). In this way, SRR has a reliable scientific information base and a documentary part accumulated in the last 25 years of management (FCA 2012, and all the information generated by Guyra Paraguay), which might be a key input in the documentation process towards the in-depth study of socio-ecological systems in practice and their advantages. This clearly indicates that new lines of research should be applied in the area.

Sustainability Based on Environmental Governance Approaches

The mandatory question in this section would be why to refer to environmental governance approaches for the SRR as a possible strategy to achieve sustainability. The focus on the degree of participation and the interaction of the different key stakeholders around the usage of natural resources obtained from the different ecosystems appears to be evident.

According to Piñeiro (2004), environmental governance is the process of decision-making and exercise of authority in the field of public goods, in which government services intervene at different levels or in decision-making instances. This is similar to other interested parties that belong to civil society or the business world, related to the fixation of regulatory frameworks and the establishment of boundaries and restrictions over the use of natural resources and ecosystems.

As pointed out by Delgado et al. (2007), in terms of San Rafael, environmental governance strategies could contribute to the effectiveness of sustained local development or minimum sustainability scenarios. This is done through the involvement of all social actors in and integrated ecosystem management, as has been developed in other countries, in the form of co-management or co-administration. In these cases, participation has been crucial for the success of the process, with a bottom-up development and carried out in an adaptive manner (according to specifics political and social conditions).

The concept of governance conveys the idea that management (e.g., conservation in protected areas and their ecosystem services) is no longer an exclusive Government monopoly; it is also a monopoly of other stakeholders. Through environmental governance, decision-making processes on access to and use of natural resources have tended to be decentralized for more effective use and management plans (Delgado et al. 2007).

In fact, SRR has shown these forms of management and has produced different results; conservation management has been fulfilling, thanks to the effort put on by the most direct and involved sectors with the need to preserve the area.

Some landowners have always conserved or used their available resources in a sustainable way. This is also the case for indigenous communities that have also respected their habitat with traditional uses or with clear positions regarding their interests and needs. There are also the civil society organizations that, in San Rafael's history, either collectively or individually, have managed to position San Rafael's importance as a relevant conservation site at a country level in the first place. This includes tangible outputs such as the purchase of land and conservation sites (relatively safe and with high conservation value) to continue studying the biodiversity of the area, among other related goals.

To this typology and characterization of key stakeholders, other social actors linked or interested in the San Rafael cause, such as academics and international organizations, are to be added. These have cooperated since the beginning of the management with various studies, scientific work, and project funding for the San Rafael area, stakeholder thus giving more support, visibility, and raising the profile of this important site for biodiversity and its people. This synergy and degree of collaboration is a relevant precedent for environmental governance approaches that increase the potential for sustainability in the reserve.

Potential positive pillars of governance (social actors with characteristics of cooperation and participation) must be added to the other social actors and their different roles. For example, the institutional role of the Public Institutions, which created the protected area in 1992 but failed to accompany the management process to the extent of its institutional responsibility and also failed to follow the dynamics of conflicts.

San Rafael's main key stakeholders hope to know what is the governmental vision towards the SRR today? Is it a governmental priority to preserve SRR? Taking into account answers to these kinds of questions, a whole dynamic of interactions must be analyzed again, in order to prioritize as much as possible one common sustainable goal for the greatest number possible of relevant stakeholders in San Rafael; without losing sight of those long-term sustainable development objectives, maintaining their clear principles of intergenerational and intragenerational equity.

5 Conclusions

The roles of key stakeholders in the management of natural resources are relevant, especially in complex territories such as protected areas with different land situations and property rights under discussion. Guidelines of rules that shou established by the stakeholders are suggested. The search for sustainability scenarios is not a simple task given the stakeholders' diversity in the area, the differences in their perceptions, the different future visions for the reserve, and the way they decide the use of resources. This requires the consideration of different mechanisms to make possible some regulations for their actions and/or economic incentives which in turn motivate sustainable uses. These are pointed out from the approaches of socio-ecological studies and environmental governance schemes that are under development and consolidation at the site level.

The SRR can be sustainable based on the possible existence of agreements and instances of solid and stable participation over time. Political will and sectoral support are required to ensure the principles of shared governance, especially with governmental participation and leadership in processes related to management; also greater support from other social actors and people interested in the potential San Rafael, not only for its conservation, but also for its potential for sustainable use of resources and the valuing of its cultural and ancestral wealth.

Stakeholders who use resources in San Rafael will establish, accept, and follow rules only if they are duly represented, there is cooperation, and also there is trust-building among all the users. Economic alternatives are to identify, replace, or to compensate for the use of resources, there are institutions in charge of monitoring and controlling the area and the presence of "opportunists" is also regulated.

The strengthening of governance must be based on a learning process that motivates participation, as well as on an agreement for its practice in local government, under the premise that enough and consistent governance strengthens governability, improves social trust, increases the environmental management effectiveness, and makes possible the sustainability of the environment in the orbit of territorial development (Gutiérrez and Morales 2017).

All the points presented above indicate that the scope of an institutional arrangement in San Rafael is a process that should not take long if the area is to be conserved and the sustainability of the area is to be consolidated. With more than 25 years of experience in the management process of the area, with known and already experienced ups and downs and setbacks, successive and participatory improvements must be sought out to contribute to creating an atmosphere of trust between the different stakeholders. Short-, medium-, and long-term action strategies should be established taking into account the prevailing reality. The reserve is moving towards cooperative work searching for social welfare and sustainability in general, based on its potential as a multiple-use socio-ecosystem.

From the viewpoint of environmental governance, promoting its potential in this type of scenario is necessary and to combine participation and market strategies that favor reciprocal relations among the different social actors, favoring the consolidation of the social capital in the area. Thus, strengthening capacities for leadership in conservation management and developing it jointly with broader and more democratic economic, social, and environmental benefits are key to environmental governance.

References

Amarilla S (2004) Acceso y uso del recurso forestal y escenarios de sostenibilidad en la reserva de recursos manejado San Rafael, Paraguay. Tesis M.Sc. CATIE, Turrialba

Bachmann P (2006) La participación ciudadana y el manejo integrado: un análisis eco-social de la cuenca del río Aysén (XI región, Chile). Seminario de título Carrera de Biología Ambiental, Facultad de Ciencias, Universidad de Chile.

Cabrera E (2008) Campaña por el Orgullo Área de Reserva para Parque Nacional San Rafael. Paraguay. Informe Final. Asociación Guyra Paraguay. Campaña de Educación para la Conservación de los Recursos Naturales. http://guyra.org.py. Accessed 11 June 2019

Cartes J, Yanosky A (2005) La Cordillera San Rafael: un capítulo particular en la conservación del BAAPA en Paraguay. In: Cartes JL (ed) El Bosque Atlántico en Paraguay: biodiversidad, amenazas y perspectivas. Asociación Guyra Paraguay, Asunción, pp 177–201

Cartes J, Yanosky A (2008) Situación general de las IBAs en Paraguay: estrategias de conservación efectivas. In: Guyra Paraguay. Áreas de Importancia para la Conservación de las Aves en Paraguay, 1° Edición, Guyra Paraguay/BirdLife Internacional, Asunción, pp 315–334

Da Ponte E, Mack B, Wohlfart C et al (2017) Assessing forest cover dynamics and forest perception in the Atlantic Forest of Paraguay, combining remote sensing and household level data. Forests 8(10):1–21

De Egea Juvinel J, Balbuena C (2011) Adopción de los Criterios De Altos Valores de Conservación de Recursos Naturales en la Reserva San Rafael. WWF, Guyra Paraguay, Asunción, PY. https://www.researchgate.net/publication/289372448_Adopcion_de_los_Criterios_de_Altos_Valores_de_Conservacion_de_Recursos_Naturales_en_la_Reserva_San_Rafael_Itapua_y_Caazapa_Paraguay

Delgado LE, Marín VH (2005) FES-sistema: un concepto para la incorporación de las sociedades humanas en el análisis medioambiental en Chile. Rev Ambiente Desarrollo 21(3):18–22

Delgado LE, Bachmann PL, Oñate B (2007) Gobernanza ambiental: una estrategia orientada al desarrollo sustentable local a través de la participación ciudadana. Rev Ambiente Desarrollo 23(3):68–73

DGEEC (2004) Anuario estadístico del Paraguay. https://www.dgeec.gov.py/Publicaciones/Biblioteca/Anuario2004/Indice.pdf Accessed 2 aug 2019

FCA (2012) Proyecto Sistematización de la Información sobre San Rafael. Informe Final para el Fondo de Conservación de Bosques Tropicales. http://fondodeconservaciondebosques.org.py. Accessed 11 June 2019

Guyra Paraguay (2005) Identificación y valoración de inmuebles incluidos en el área de reserva para el parque Nacional San Rafael. PAR98/G33 "Iniciativa para la Protección de Areas Silvestres Protegidas del Paraguay". Programa de las Naciones Unidas para el desarrollo - PNUD, Asunción.

Goh TY, Yanosky AA (2015) Payment for Ecosystem Services works, but not exactly in the way it was designed. Glob Ecol Conserv 5:71–87

González PD, Achinelli Báez MF (2017) Mercado Voluntario de Carbono y su impacto socioeconómico en la Colonia La Amistad, Departamento de Itapúa, Paraguay. Periodo: 2010 – 2015. Población Desarrollo 23(44):45–51

Gutiérrez A, Morales T (2017) Evaluación de la gobernanza ambiental local en Risaralda. Rev Luna Azul 45:309–328

López C (2017) La gobernanza ambiental. Problemática de la consolidación de la Reserva para Parque Nacional San Rafael. Memoria fin de master. Universidad de Rennes 1. Facultad de Ciencias Económicas

Madrigal (2003) Efectos de los incentivos económicos y la acción colectiva sobre el uso del agua de irrigación en el Distrito de Riego Arenal Tempisque, Costa Rica. Tesis Mag. Sc. Turrialba, Costa Rica, CATIE

Ostrom E (1990) Governing the commons: the evolution of institutions for collective action. Cambridge University Press, Cambridge

Ostrom E (1994) Diseño de instituciones para sistemas de riego auto-gestionarios, traducción de Adriano Miguel Tejada. Institute for Contemporary Studies, San Francisco

Ostrom E (1997) Esquemas institucionales para el manejo exitoso de recursos comunes. Conferencia sobre instituciones locales y manejo de los bosques: cómo puede la investigación hacer la diferencia. Bogor

Ostrom E (2009) A general framework for analyzing sustainability of social-ecological systems. Science 325:419–422

Ostrom E (2010) Analyzing collective action. Agric Econ 41:155–166

Piñeiro D (2004) Movimientos sociales, gobernanza ambiental y desarrollo territorial rural. Departamento de Sociología, Facultad de Ciencias Sociales, Universidad de la República. Uruguay, RIMIS

Solano (2018) Sistemas socio-ecológicos, bienes comunes y acción colectiva. Análisis crítico para su aplicación en la Zona de Reserva Campesina del Valle de Río Cimitarra. Universidad Nacional de Colombia Instituto de Estudios Ambientales—IDEA—Bogotá, Colombia. Tesis Magister Medio Ambiente y Desarrollo

WWF (Word Wide Fund for Nature) (2016) Bosque Atlántico. https://goo.gl/qCT57g. Accessed 11 2019

Yanosky AA (2002) Conservation in Paraguay. The Cordillera de San Rafael: Paraguay's Interior Atlantic Forest. Sylvanet 15(1):13–15

Water Supply Valuation of High Andes Wetlands, Chinchiná River High Watershed, Colombia

Gloria Yaneth Flórez Yepes, Jhon Fredy Betancur Pérez, Jorge Andrés Ospina, and Angela María Alzate

Abstract The valuation of the ecosystem services in the high area of the Chinchiná River watershed started with those related to the wetlands, with important community participation who considered that the water supply has the greatest value in relation to other prioritized services, being used for economic valuation. We first calculated the collection and the recovery values, considering variables such as forest areas and collected water volume, the forest importance value, the productive system opportunity costs, and the ecosystem restoration cost. Finally, we used the total value for a comparison between the obtained values and those corresponding to the costs, per cubic meter, that the water service company charges to the community. Results show wetland's generated water would have a value of 4.28 USD per cubic meter if they were valued at public service rates.

Keywords Social-ecological systems · Latin America · Complexity · Colombia · Water valuation · Community participation

1 Introduction

Water recharge areas, especially wetlands, are affected by multiple anthropogenic conditions. Excessive natural resource consumption has caused deterioration, and by extension, a global environmental crisis. Water resources reduction figures worldwide have become ever more worrisome. The lack of both water quality and

G. Y. F. Yepes (✉) · J. A. Ospina · A. M. Alzate
Facultad de Ingeniería y Arquitectura, Universidad Católica de Manizales, Manizales, Colombia
e-mail: gyflorez@ucm.edu.co; jaospina@ucm.edu.co; aalzate@ucm.edu.co

J. F. B. Pérez
Facultad de Economía y ciencias contables, Universidad de Manizales, Manizales, Colombia
e-mail: jbetancur@umanizales.edu.co

L. E. Delgado, V. H. Marín (eds.), *Social-ecological Systems of Latin America: Complexities and Challenges*, https://doi.org/10.1007/978-3-030-28452-7_14

quantity is on the rise; consequently, it is necessary to perform studies of the water ecosystem services and their evaluation for ecosystem planning processes. Furthermore, investing in water is necessary in order to allow for economic growth, employment, and inequality reduction (UNESCO 2016).

Assessment is important whenever it leads to the generation of usage or non-usage values, contributing to the recognition of ecosystem services by decision-makers, motivating communities to recognize their environment, the importance of their assets, and to understand their relationships with nature. Thus, assessments allow for the generation of conservation incentives, which are fundamental for public participation in the protection of natural resources. Similarly, those who decide to transform their productive agricultural and livestock systems into conservation activities require substantive plans, which may be based on the said incentives. Still, our research was not to provide a sale value for water resources ecosystem services, but rather to generate a value which may express the economic benefits generated by the water supply in the studied area.

Many authors have questioned whether or not environmental services should be economically assessed. Authors, including Nasi et al. (2002), believe it to be improper making these types of assessments, while authors such as Hoyos and Mariel (2010) mention that assessment is a useful decision-making tool.

One of the methods that have been used most frequently for assessment is contingent valuation. Roldán (2016) found that the first theoretical approaches to assessment were initiated in the 1940s, and the first application in the 1950s when an American company was hired to consult about a community's ability to pay the entrance to a park. When market price assessments are implemented, the value assignment is more direct and may be easier to achieve. Contingent valuation, on the other hand, requires the inclusion of many variables for the adjustment of its viability, as well as to guarantee that methodological assumptions correspond to social, economic, and environmental realities. In this chapter, we describe a direct economic valuation of the water supplied by wetlands as a result of the prioritization of ecosystem services conducted by the community.

2 Methods

Community Participation in the Identification and Prioritization of Ecosystem Services

A community participation exercise was carried out to assess the water supply in the studied area (Chinchiná River basin). The main purposes were the identification and hierarchization of wetland ecosystem services. According to Flórez et al. (2014: 276), "community participation is important for proposal creation processes, that may help to decide actions for the sustainable use of natural resources, based on their understanding and recognition as parts of their environment." It should be emphasized that the study area is the primary water recharge region for the aqueduct in Manizales, Caldas, Colombia. Environmental assessment may be performed to either establish ecosystem protection costs or for the benefits obtained from ecosystems. The study area belongs to the moorland ecosystem, highly relevant in terms of ecological conservation. In 2016, the Constitutional Court of Colombia issued a decree (C-305), in which the third section discussed the regulation of extractive activities in moorland ecosystems, demanding that higher conservation standards be employed therein. In the same year, Law 126 which protects Colombian moorland came into effect.

The starting point was the National Policy for the Integrated Management of Biodiversity and its Ecosystem Services (PNGIBSE) of the year 2012 (MADS 2012). This policy establishes the characterization of ecosystem services and the guidelines of wetland valuations (Convención Ramsar 2007). We classified ecosystem services according to the following criteria:

- Provisioning services: food, water, fibers, biochemical products, etc.
- Regulating services: air regulation, climate regulation, pollution control, protection against erosion, water supply
- Cultural and recreational services: cultural identity, recreation, and tourism opportunities.

Social participation is of great importance since it is the inhabitants who develop daily dynamics that are favored by these functions or "services" provided by the forests, moorlands, or wetlands (Flórez-Yepes and Betancourt-Perez 2019).

Spatial Distribution and Hierarchization of Ecosystem Services

The participative identification of ecosystem services provided by wetlands in the studied area included their generation of maps containing both wetlands location and the services they provide using the cognitive mapping methodology proposed by Vélez et al. (2012). Next, the community assigned importance values and frequency of use to the services according to the following choices:

Importance	Frequency
1. Low importance	1. Infrequent
2. Medium importance	2. Sometimes
3. High importance	3. Very frequent

We assessed the water supply and recovery costs following the methodology described by Barrentes and Castro (1999), Silva (2007), and Encalada (2006). This methodology was selected since it comes closest to the determination of a value per cubic meter, per year, generated in the studied area. The Aspar–Laguna Negra, Letras—El Nueve, and La Favorita—Rio Blanco—Martinica sectors were afforded special attention because they correspond to zones showing deterioration processes in wetland ecosystems.

Determining water catchment values The catchment value is a value of the water supply which is related to the forest coverage as the determinant factor of the hydrological regulation. This value corresponds to the cubic meters caught in a given area during a year. However, increasing the catchment value requires generating vegetation coverage, which implies giving up the income generated by the productive systems of the studied area. Thus, determining its value entails finding the opportunity cost for the area. We determined this cost for potatoes and milk production, which are the main productive systems of the study area. We used primary data from the community and the inquiries of the Agriculture Secretariat of the Caldas department which coincided on the type of collected information.

We reclassified the vegetation coverage into natural and non-natural areas, using it to calculate fragmentation. We further considered the collected volume of water for the catchment value using the method of Silva (2007). We estimated runoff using geographic information systems, adjusting its value by a factor obtained from natural forest area divided by the total studied area.

We used a weighting factor for the catchment value obtained from the community of the study area asking the importance they give to conservation of forests in relation to the water resource by means of a ranking survey. The number of respondents corresponded to those participating in the meetings (@ 24). Then, catchment value was calculated using the formula from Barrentes and Castro (1999) and ratified by Silva (2007):

$$\mathrm{VC} = \sum_{i=1}^{n} \frac{\alpha_i B_i \mathrm{Ab}_i}{\mathrm{Oc}_i}$$

where

VC Forest hydrological catchment (¢/m^3) (quantity + quality)

B_i Opportunity cost of any economic activity competing for land use with the forest at watershed i (¢/ha/year)

Ab_i Forest area at watershed i (ha)

Oc_i Volume of water caught watershed i (m^3/year)

α_i Importance of the forest at watershed i as a function of the quality and quantity of the hydrologic resource

The recovery value corresponds to the costs which must be considered for reforestation and maintenance activities in the area. We considered the area 3000 m above sea level, corresponding to clean grasses that form part of moorlands, which, in accordance with Law 126 of 2016, must be preserved and conserved.

We then calculated the recovery value (per cubic meter per year) according to the following formula:

$$VR = \sum_{t=1}^{n} \frac{\alpha_i C_{ij} \mathrm{Ar}_i}{\mathrm{Oc}_i}$$

where

VR: Recovery value for hydrological watersheds
α_i: Importance of forest in the watershed (%)
C_{ij}: Costs of activity j destined to watershed recovery
Ar_i: Area to be recuperated
Oc_i: Water catchment value for the watershed

3 Results

Prioritization of Ecosystem Services with Community Participation

Community participation generated the following insights:

1. The community is conscious of the importance of future ecosystem cleanliness and the importance of waste prevention around water mirrors.
2. They are also aware that channels contain a wide variety of aquatic species.
3. Also, that representative flora and fauna of different areas must be cared for, and they desire a healthy environment.
4. Additionally, they recognize that they must stop the destruction of animal habitat, prevent further deforestation, so existing forests may continue to provide oxygen for the planet, that the mountain should always be maintained because of its vital importance in providing water to the population.

Creation of a Hierarchy for the Ecosystem Services of the Upper Chinchiná River Basin

The case of the El Nueve sector In this sector, there were four groups which reached consensus over the assessment of each matrix. Regulation services were expressed to have the greatest level of importance, followed by provisioning services, including fish production, provision of irrigation water for irrigation, human consumption, lumber and firewood production, and obtention of medicinal products. The services with lowest ratings were cultural services, including the use of wetlands as patrimony and part of the region's cultural identity, recreational activities, landscape appreciation, and educational training.

The importance of regulation services was quite high. The community recognizes the importance of this ecosystem service because it permits ecosystem sustainability over time. That is, water regulation ensures the amount of water necessary to sustain other ecosystem services. Provisioning services were nearly as important as regulation services in this sector. Cultural services were assigned a value nearer to the middle, as it is service of the least value to this community.

Río Blanco, Martinica, and Casa Larga sectors Three groups participated and achieved consensus about the degree of importance and frequency of use of ecosystem services. Services which were assigned the highest level of importance were provisioning, followed by cultural and regulation. Notably, in these areas, a greater number of services were identified, related to water provision for basic food crops, other crops such as potatoes, and for reforestation processes. It is important to emphasize that part of livestock farming in the area is dedicated to beef cattle, and part is dedicated to dairy farming.

Ecosystem services, in terms of the frequency of wetland use, are most often used for crops and recreation, followed by livestock farming and procurement of medicinal plants, of which, the majority in these areas are: wolf's bane and paramo rosemary.

In these sectors, there is greater recognition of provisioning services, with an importance value rating above two-third, which is the maximum rating. Regulation services rated below two, the medium assessment value, rated the lowest importance value. Similarly, cultural services had a value below two, but above regulation services rating.

The Laguna Negra sector In this sector, three groups participated, and reached consensus on the rating of each service. This area is of great importance, owing to its proximity to the Natural National Park Los Nevados and the presence of wetlands such as La Laguna Negra.

Regulation services, in this sector, were rated with the highest importance value, followed by provisioning and cultural services. Still, cultural services rating was above the median.

In summary (Fig. 1), the sectors with the highest water supply ratings were the Laguna Negra and the El Nueve sectors, from which the two principal branches of the Chinchiná River emerge. The fact that the neighboring community recognizes that water regulation is the ecosystem service of greatest value may facilitate awareness processes for ecosystem conservation.

Cultural ecosystem services showed the highest value in sites such as la Laguna Negra, where wetlands have a lagoon-like water mirror, as this typology provides an improved scenario for landscape appreciation, recreation, rituals, etc. We economically valuated water supply since it was the highest service valued by the communities. We show the results in the following section.

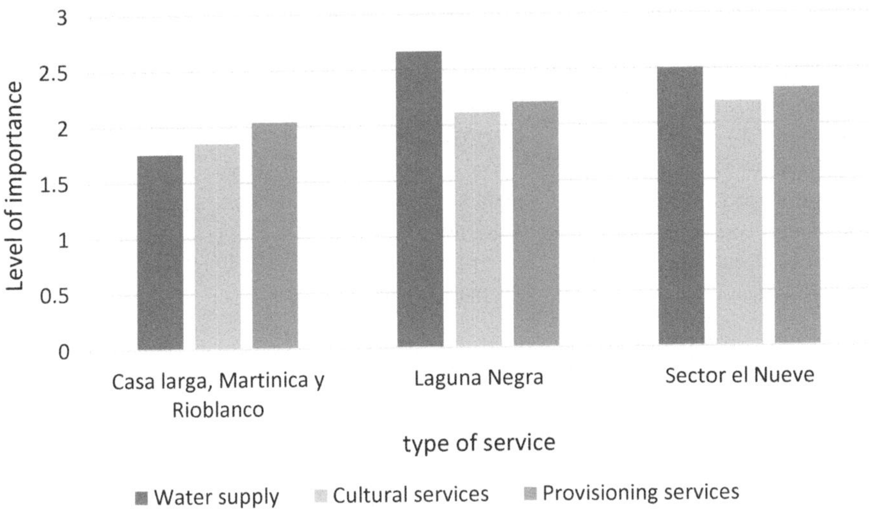

Fig. 1 Community assigned importance values of the ecosystem services provided by wetlands

Valuation of the Water Supply in the Study Area

Opportunity costs Given the results obtained for the collection value and opportunity costs, as mentioned previously, the following was determined for potato and livestock farming:

Opportunity costs for livestock farming

Utility: 40%
Number of livestock per hectare: 1
Liters per day: 10
Vr per liter: 0.26 USD
Daily utility: 1.025 USD
Days per year: 360
Annual utility: 369.23 USD ha^{-1}

Opportunity costs for potato farming:

Crops per hectare per year: 1
Total income: 480.76 USD ha^{-1}
Utility per hectare: 20%
Utility: 96.15 USD ha^{-1}

The results displayed in Tables 1 and 2 are a product of community discussions as well as the departmental Secretariat of Agriculture's investigation.

Table 1 Collection value in the study area studied with opportunity costs for livestock farming agriculture

Variable	Livestock farming	Agriculture
Total area (ha)	20,545.83	20,545.83
Forest area (ab)	8320.64	8320.64
Importance of water resource (*a*)	70%	70%
Volume of water collected (Oc_i)	973,847.69	973,847.69
Opportunity cost in USD (*b*)	369.10	96.11
Value of collection per year	2.21 USD m^{-3} $year^{-1}$	0.57 USD m^{-3} $year^{-1}$
Total collection value	2.78 USD m^{-3} $year^{-1}$	

Table 2 Comparison of the current fee system vs. what should be paid, by strata, including the monthly per cubic meter collection and recuperation values for the first year

Concept	Collection value/ month USD	Recuperation value/ month USD	Current fee/ month USD	Total value that should be charged USD
Strata 1	0.23	0.56	0.29	1.08
Strata 2	0.23	0.56	0.35	1.14
Strata 3	0.23	0.56	0.50	1.29
Strata 4	0.23	0.56	0.58	1.37
Strata 5	0.23	0.56	0.87	1.66
Strata 6	0.23	0.56	0.93	1.72
Industrial	0.23	0.56	0.76	1.55
Commercial	0.23	0.56	0.87	1.66
Official	0.23	0.56	0.58	1.37

Collection and recuperation values, per month, were obtained by dividing annual values by the 12 months of the year

Determination of Natural and Productive System Areas

We performed our valuation with 2010 areas (20,545.83 m^2 total area and 8320.64 m^2 for natural area).

Volume of intercepted water We used a runoff of 2.4×10^6 m^3 $year^{-1}$ which multiply by the weighing factor of forest areas divided by total area (0.405) generated a total value of 0.97×10^6 m^3 $year^{-1}$ (Silva 2007).

Community-generated weighing factor The definition of the forest weighing factor for the conservation of water not only considered sampling points, but also comments from the community. The result, considering 0% as unimportant and 100% as very important, was of 70%.

The total collection value obtained was 2782 USD m^{-3} $year^{-1}$ (Table 1). Barrentes (2010) and Barrantes and Vega (2002) determined a collection value of 1.67 USD m^{-3} for the Tempisque River basin in Costa Rica. Barzev (2000), in Achuapa,

Nicaragua, with the same methodology, determined a collection value of 3.12 dollars USD m^{-3}. In the case of Silva (2007), there is smaller difference than those obtained in the above-mentioned studies. This may be due to the variation between collection volume and low opportunity costs calculated for livestock farming and agriculture.

Reforestation costs The recovery value was determined both for the time of establishment and for the four following years of maintenance thereof. This period of time is considered, as recommended by the environmental authority for moorland zones. Restoration costs considered 555 trees $ha^{-1}$of native species, manpower, supplies, and each one of the activities required for restoration. Total cost for the first year was 1273.66 USD and 969.08 USD starting the second year. Calculations considered the area of 3000 m above sea level since it is the zone where moorland ecosystems are located and where livestock farming and agriculture is most important.

We next show the vegetation coverage areas classified as non-natural that served to define the restoration areas. We adopted the criteria of using clean grass areas only since the conservation index showed that cattle areas had the lowest indices, even lower than potato culturing areas.

Type of coverage above 3000 m:

Potatoes: 10.76 ha.
Fragmented forest with grass and crops: 17.23 ha.
Grass and crops mosaic: 31.29 ha.
Pasture: 748.18 ha.
Grass mosaic with natural space: 1533.55 ha.
Clean grass: 4221.05 ha.

If we apply the recovery value formula, we obtain the following result:

Recovery values

Year 1

Importance of the hydrological resource: 70%
Restoration costs per hectare: 1273.67 USD
Maintenance costs per hectare: 969.21 USD
Area to be restored: 4221 ha.
Volume of water caught: 973,847.69 $m^{3.}$
Total recovery value for year 1: 6.36 USD

Years 2–4:

Importance of the hydrological resource: 70%
Maintenance costs per hectare: 484.60 USD
Area to be restored: 4221 ha.
Volume of water caught: 973,847.69 $m^{3.}$
Total recovery value years 2–4: 1.47 USD

Total Valuation of the Water Supply in the Study Area

Results showed that 1 cubic meter of raw water for the first year (excluding driving and infrastructure costs) for the study area would cost 6.38 USD m^{-3} $year^{-1}$ for the first year. As of the second year, the value would be 4.25 USD m^{-3} $year^{-1}$. This data is interpreted as the economic-ecological benefit provided by the water supply, where a decrease of 46.5% of the value of the resource is observed.

Furthermore, it should be noted that in accordance with that contemplated in the Manizales aqueduct fee system, a comparison was made between the value that inhabitants of Manizales pay per cubic meter of water and what they should pay, including the value of the water supply. Table 2 compares the difference between that paid by each inhabitant, by socioeconomic strata, per cubic meter per month, and what they should pay, considering collection and recuperation values. The fee for the first year is considered, and subsequently, so too is that for later years.

In order to determine the recuperation and collection values per month, the total cubic meter value was divided by the 12 months of the year, which revealed the amount by which the fee system should be increased for ecosystem conservation purposes.

Total valuation of the cubic meter per year:

Catchment value year 1 1: 2.78 USD
Recovery value year 1: 6.81 USD
Total value year 1: 9.59 USD
Catchment value years 2–4: 2.78 USD
Recovery value years 2–4: 1.47 USD
Total value year 2 and beyond: 4.25 USD

We then compared the values paid in the Manizales fee system, per cubic meter of water, with what they should be paying including the water supply values, which includes how much should the fee system be incremented to include the conservation of the ecosystem (Table 3).

The increase observed in the column for the total value that should be charged, in the two tables above, should be invested in incentives for conservation in the area studied.

Wetland ecosystem services have been affected by productive systems, especially those located in the moorland zone (livestock and potato farming, Flórez Yepes 2015). For this reason, the economic valuation of these services has taken new strength worldwide using the economy as its main pillar. This science has different perspectives, according to Ávila-López and Pinkus-Rendón (2018): environmental economy, natural resources economy, green economy, and ecological economics. These first two approaches are based on the interactions between the economy and the environment, while the latter two in the interactions between economic, social, and environmental dimensions, the main pillars for sustainable development.

Table 3 Comparison of the current fee system vs. what should be charged, by strata, including the monthly per cubic meter collection and recuperation values for subsequent years

Concept	Collection value/ month USD	Recuperation value/ month USD	Current fee	Total value that should be charged
Strata 1	0.23	0.12	0.29	0.64
Strata 2	0.23	0.12	0.35	0.70
Strata 3	0.23	0.12	0.50	0.85
Strata 4	0.23	0.12	0.58	0.93
Strata 5	0.23	0.12	0.87	1.22
Strata 6	0.23	0.12	0.93	1.28
Industrial	0.23	0.12	0.76	1.11
Commercial	0.23	0.12	0.87	1.22
Official	0.23	0.12	0.58	0.93

The economic valuation has also been addressed from the perspective of the economic benefit that a community may have when decreasing pollution levels in a given place. However, Costanza et al. (2017) state that the traditional economy has valued and still does, ecosystems where products have been harvested and sold in supermarkets. In our case, we valued a product without a market, i.e., water in its natural state. Consequently, our valuation challenge required extra effort and research.

Barrentes and Castro (1999), in the case of Heredia (Costa Rica), estimated the total value of the water supply at USD 4.67 per m^3. In the case of Barzev (2000) study, for the Achuapa micro-basin in Nicaragua, it summed to USD 9.55/m^3 annually. In accordance with the water supply values obtained in the present study, a number of small differences were found, which could be conditioned by the volume of water collected, opportunity costs for each production system, or restoration costs, depending on the area.

Gutierrez (2002) developed a study to evaluate economically the hydrological environmental service of the Molino Norte and San Francisco sub-watersheds, with a proposal to incorporate them in the water rate from Matagalpa (Nicaragua). Using contingent valuation, he calculated a willingness to pay (WTP) of 2.46 USD per family per month. The proposal included a mechanism for the payment of environmental services related to WTP and a follow-up system for the program in the sub-watersheds. One advantage of the contingent valuation is that it provides information on a visible and realistic market, proved with simulated markets (Loomis 1989 in Gutierrez 2002). The main disadvantage is that the procedure is open to biases during the direct (Romero 1997 in Gutierrez 2002).

Other authors have used methods different than contingent valuation. Retamal (2006) valued economically the offer of the hydrological ecosystem service for human consumption at Copán Ruinas municipality (Honduras) through a scheme of payment for hydrological ecosystem services (PHES) for three 5-year scenarios. PHES are usually chosen for watersheds since they support financial strategies promoting integrated equitable management of hydrological resources (Jiménez and Faustino 2005 in Retamal 2006).

Lowest socioeconomic strata are those which would be most affected by the fee adjustment that the increase would generate. It should be noted that, with this assessment of the water supply in the area studied, the goal was to estimate how much fees should increase, in order to generate the income necessary to craft conservation incentive schemes. The funds collected from the increase should be used for purposes of ecosystem conservation and recuperation.

In accordance with the studies analyzed by Barrentes and Castro (1999), Silva (2007), and Barzev (2000), following the determination of the value to be paid, considering the fee adjustment per cubic meter of water, a resource management system was created for the generation of the payments for environmental services to the farmers charged with conservation.

Although the topic of ecosystem assessment has just begun gaining momentum in the past few years, both on national and international levels, there is still not clearly established criteria for the best methodology to employ for said purpose. The most commonly used method at present is contingent valuation. In Roldán (2016) doctoral thesis, as in many others, the contingent assessment is posed as a viable assessment alternative. However, in the present investigation, it is posited that there should be both a social perception process promoting the importance of wetland ecosystem services and the determination of a direct water supply value. In this case, the method of the applied value clearly demonstrates the way in which an adjustment might be made to a fee system for the aqueduct in Manizales. Similarly, geographic information systems are incorporated as an important tool for the determination of the final value.

4 Conclusions

The need to conduct a valuation of the hydrological resource lies in the existence of a marked trend toward the decreasing of wetlands' water mirrors in the study area. There is coincidence between this phenomenon and the decrease in vegetation coverage, as shown in our mapping of natural and non-natural areas for the years 1998 and 2010.

The participation of the community in the identification of ecosystem services was fundamental to understand the recognition of their environment, and their priorities served to decide the valuation method. When the environment is appropriated, processes may be more sustainable in time. Regulation services, water supply, and provisioning are the most important services for the community. Cultural services are of less importance.

Contingent valuations are difficult to perform in the area since the communities interested in water conservation are mostly those that live in lower or medium altitude zones, being those mostly benefited by having a recharge zone at higher altitudes. In other words, willingness to pay would not relate to the direct influence area of this study but with the communities at Manizales city.

Our results about what people actually pay as compared with the value they should pay (Tables 2 and 3) show that the increment would be above 70%, a situation that could generate disagreements within the community. The surplus should be oriented to protection actions and we propose that it should be given to the peasant communities from moorlands as a compensation for the substitution of productive agricultural systems into conservation areas.

Although we may agree with Martínez-Alier et al. (1998) about the analysis of the flows of energy and materials when conducting evaluation studies, such measurements require higher investments that were not available in this research. We may also disagree with the willingness to pay methodology due the large quantity of subjective variables affecting the results. However, the participation of the community is vital when prioritizing ecosystem services.

Finally, water has been a strategic resource for humankind ever since, historically, it has been fundamental to develop all its activities. Indeed, in some past scenarios, water was conceived as a great allied in many aspects. However, water as a natural patrimony is currently perceived in a different way because if the vital liquid which is deteriorating. In the last years, different valuation methods, structures, and planning have emerged allowing the recovery and conservation of ecosystems affected by the decrease in water supply and its quality, both for human consumption as for the proper functioning of productive systems. Therefore, a big social responsibility should exist in order to manage water through governance, transparency, equity, and well-defined medium and long-term conservation criteria.

Society is called to know its natural environment, its function within, identifying the ecosystemic value of the natural patrimony, especially water, with the hope that starting from there, conservation decisions may be generated since it is not possible to conserve something that is not known. For this to happen, it is necessary to consider in an integrated way society and nature, or in this case society and water given their tight relationships. The problems and solutions for the humans–nature relationships must be addressed with a holistic perspective so as to guaranty life quality for society and the preservation of our natural patrimony.

References

Ávila-López C, Pinkus-Rendón M (2018) Teorías económico-ambientales y su vínculo con la dimensión social de la sustentabilidad en Áreas Naturales Protegidas. Ciencia UAT 13(1):108–122

Barrantes G, Vega M (2002) Evaluación del servicio ambiental hídrico en la cuenca del río Tempisque y su aplicación al ajuste de tarifas. Preparado para ASOTEMPISQUE. Financiado por el Fondo Canje Deuda Costa Rica Canadá, el Programa de Pequeñas Donaciones (PPD) de PNUD y la Fundación CR-USA

Barrentes G (2010) Evaluación del servicio ambiental hídrico en la cuenca del río tempisque (costa rica) y su aplicación al ajuste de tarifas. Rev Lebret 2(132):160

Barrentes G, Castro E (1999) Implementación de un Esquema de Cobro y Pago por servicio ambiental hídrico: El Caso de la empresa de servicios públicos de Heredia S.A. ESPH, Costa Rica

Barzev R (2000) Estudio de Valoración Económica de la Oferta y Demanda Hídrica del Bosque en que nace la Fuente del Río Chiquito (Finca El Cacao, Achuapa) para determinar la factibilidad de mantener el bosque en vista de garantizar la calidad y cantidad del recurso hídrico. Programa para la Agricultura Sostenible en Laderas de América Central (PASOLAC), Achuapa

Convención Ramsar (2007) Valoración. Lineamientos para valorar los beneficios derivados de los servicios de los ecosistemas de humedales. https://www.ramsar.org/sites/default/files/documents/pdf/lib/lib_rtr03_s.pdf. Accessed 11 June 2019

Costanza R, de Groot R, Braat L et al (2017) Twenty years of ecosystem services: how far have we come and how far do we still need to go? Ecosyst Serv 28:1–16

Encalada G (2006) Pago por servicios ambientales del Recurso Hídrico como una alternativa de conservación. Tesis de Maestría en economía, Facultad Latinoamericana de Ciencias Sociales Flasco, Ecuador

Flórez Yepes G (2015) Servicios ecosistémicos y variables socioambientales determinantes en ecosistemas de humedales altoandinos. Sector el ocho y paramo de letras Manizales Colombia. Rev Mexicana Cienc Agrícol 1:173–179

Flórez G, Alzate Á, Rincón A (2014) Participación comunitaria para la construcción de lineamientos de uso y conservación de humedales altoandinos. Experiencia piloto en el sector El Ocho y Páramo de Letras. Rev Luna Azul 38:274–296

Flórez-Yepes G, Betancourt-Perez J (2019) State of conservation index for high Andean wetlands. J Nat Conserv 49:45–53

Gutierrez J (2002) Valoración económica del servicio ambiental hídrico en las subcuencas Molino Norte y San Francisco, y propuesta para su incorporación en la tarifa hídrica, Matagalpa, Nicaragua. Tesis M.Sc. Turrialba, CR, CATIE, San José

Hoyos D, Mariel P (2010) Contingent valuation: past, present and future. Prage Econ Papers 19(4):329–343

MADS (Ministerio de Ambiente y Desarrollo Sostenible) (2012) Política Nacional para la Gestión Integral de Biodiversidad y sus Servicios Ecosistémicos-PNGIBSE. http://www.humboldt.org.co/images/documentos/pdf/documentos/pngibse-espaol-web.pdf. Accessed 11 June 2019

Martínez-Alier J, Roca J, Sanchez J (1998) Curso de economía ecológica. Programa de las Naciones Unidas para el Medio Ambiente Oficina Regional para América Latina y el Caribe. Serie Textos Básicos para la Formación Ambiental N° 1. Mexico D.F.

Nasi R, Wonder S, Campos JJ (2002) Servicios ecosistémicos forestales, Podrían ellos pagar para detener la deforestación? Serie técnica, informe técnico No 331. Catie. Turrialbal, Costa Rica

Retamal MR (2006) Valoración económica de la oferta del servicio ecosistémico hídrico para consumo humano en el municipio de Copán Ruinas, Honduras. http://repositorio.bibliotecaorton.catie.ac.cr/handle/11554/4362. Accessed 11 June 2019

Roldán DF (2016) Valoración económica de recursos hídricos para el suministro de agua potable. Caso de estudio Parque Nacional Cajas La cuenca del Rio Tomebamba Ecuador. Universidad de Alicante. Tesis doctoral programa de economía aplicada. Alicante, España. https://rua.ua.es/dspace/bitstream/10045/54425/1/tesis_roldan_monsalve.pdf. Accessed 11 June 2019

Silva R (2007) Valoración económica de los servicios ambientales hidrológicos el Egido, La Victoria, Pueblo viejo, DGO. Instituto Politécnico Nacional, Tesis de Maestría en Ciencias en Gestión ambiental, Durango México

UNESCO (2016) Informe de las Naciones Unidas Sobre el Desarrollo de los Recursos Hídricos en el Mundo 2016. Agua y Empleo. https://unesdoc.unesco.org/ark:/48223/pf0000244103. Accessed 11 June 2019

Vélez I, Rátiva S, Varela D (2012) Cartografía social como metodología participativa y colaborativa de investigación en el territorio afrodescendiente de la cuenca alta del río Cauca. Cuadernos de Geografía. Rev Colombiana Geogr 21:59–73

Participative Management and Local Institutional Strengthening: The Successful Case of Mangrove Social-ecological Systems in Ecuador

Fernando Félix and Mario Hurtado

Abstract In 1999, the government of Ecuador established a conservation strategy for ancestral and traditional communities settled in mangrove areas to promote the conservation and management of this productive ecosystem, also giving users an opportunity to improve their quality of life. Local fishing associations can request mangrove areas for protection and utilization under a "Sustainable Use and Mangrove Custody Agreement." This legal instrument guarantees beneficiaries exclusive access to fishing resources in these areas.

The beneficiaries are committed to custody mangrove areas, keep the legal status up to date, submit semi-annual reports to the environmental authority, implement a management plan, and observe the legal provisions of the Agreement. Up to October 2018, 79 areas were given in custody of which 53 are operative (67%). The total surface under custody amounts to 6836.3 km^2, which is approximately 44% of the mangrove area in the country. The initiative has demonstrated to be a useful management alternative for rural areas usually lacking governmental control. In most cases, the performance of fishers associations has been good and concrete benefits have been achieved. This mechanism has also helped to stop mangrove degradation, and in some cases its surface area has increased. In cases of poor performance, more assistance by governmental authorities is recommended. In this chapter, we describe and discuss this participative management scheme.

Keywords Social-ecological systems · Latin America · Complexity · Ecuador · Mangroves · Participative management

F. Félix (✉)
Southeast Pacific Action Plan, Comisión Permanente del Pacífico Sur, Guayaquil, Ecuador
e-mail: ffelix@cpps-int.org

M. Hurtado
Hurtado & Hurtado Asociados, Guayaquil, Ecuador

L. E. Delgado, V. H. Marín (eds.), *Social-ecological Systems of Latin America: Complexities and Challenges*, https://doi.org/10.1007/978-3-030-28452-7_15

1 Introduction

Mangroves are highly productive ecosystems, located at the land–sea interface on the tropics and temperate areas around the world. It is an ecological community in which organisms are adapted to extreme conditions including exposure to desiccation, changes in water salinity, and anaerobic soils (Kathiresan and Bingham 2001). The term "mangrove" describes both the ecosystem and the families of plants that have evolved adaptations to living in such an extreme environment (Tomlinson 1986).

The mangrove ecosystem characterizes by a complex network of ecological interactions. Mangrove swamps constitute the breeding and feeding grounds for many invertebrate and vertebrate species of fish, birds, reptiles, and mammals (Kathiresan and Bingham 2001). They are also vital for the food security and livelihood of human populations at providing a variety of ecosystem services derived from biological resources such as fish, forest products (timber and non-timber), as well as recreational activities and tourism (López-Angarita et al. 2016; Barbier et al. 2011). Mangroves are among the ecosystems with the highest capability to store carbon, and their importance for mitigating climate change has been highlighted (Donato et al. 2011). Additionally, they are important to maintain the integrity of coastal zones by preventing and mitigating the impacts of coastal erosion and other natural phenomena, including storms and tsunamis (Barbier et al. 2011; Spalding et al. 2014).

Despite their ecological, cultural, social, and economic importance for the well-being and development of coastal communities, mangroves have experienced in recent decades an accelerated decrease in their cover as a result of different human activities, including urban development, aquaculture, dredging, filling, pollution, etc., which threatens the long-term provision of ecosystem services (López-Angarita et al. 2016). It is estimated that in the last two decades of the twentieth century, 35% of the total mangrove area in the world was lost, particularly in Asia (Valiela et al. 2001). In some Latin American countries, more than 25% of the mangrove cover was lost by the end of the twentieth century (Yáñez-Arancibia and Lara-Domínguez 1999). The conversion of swamps into agricultural or aquaculture areas has been the main threat to this ecosystem. Nearly 38% of the total mangrove loss is attributed to the construction of ponds for shrimp farming, especially in South Asia and Latin America (Valiela et al. 2001). Pollution by domestic and industrial discharges has increased the pressure for its deterioration. The loss of services from mangroves and other coastal and marine ecosystems is of concern since the economic and environmental costs have not yet been sufficiently well evaluated (Duke et al. 2007; Barbier et al. 2011; Huxham et al. 2015; López-Angarita et al. 2016).

The mangrove ecosystem is a good example of the complexity and dynamics of a socio-ecological system, where users can lead to a rapid degradation if there are no governance mechanisms or agreements to guarantee the provision of ecosystem services and preserve the way of life of coastal communities. Understanding the complexity of human interactions with their environment, involving biophysical and social aspects in all their dimensions, economic, political, and cultural, through interdisciplinary approaches is essential to ensure sustainability (Lerner and Charli-Joseph 2018). Problems in mangrove systems seem to be common in different parts of the world. Factors include changes in tenure and land use to favor lucrative activities such as aquaculture, lack of technical and economic capacity at the local level, population growth, low educational level and poverty, among others, which increase the vulnerability of ancestral communities (Orchand et al. 2015; Quinn et al. 2017). It is, therefore, necessary to implement management schemes that allow reverting or at least halting the deterioration of mangroves in favor of communities highly dependent on their resources, for which it has been suggested, among others, to develop productive alternatives and promote more equitable distributions of its benefits (Orchand et al. 2015). It has been demonstrated that the local population is key for the management of forest resources and to maintain the capacity to preserve habitats and environmental services. Therefore, they must be a primary part of the strategy to be implemented (e.g., Glaser and da Silva 2004; Durán et al. 2018).

This chapter describes a community management experience for the exploitation of benthic resources in the mangroves of Ecuador. For this purpose, the Ecuadorian government has developed a management strategy called "Sustainable Use and Mangrove Custody Agreement," which grants exclusive rights to access aquatic resources to traditional communities, with the purpose that beneficiaries improve their quality of life and conserve this important ecosystem. Mangrove woods are public property in Ecuador, where no real rights on land tenure can be exercised not even by ancestral communities. The evaluations carried out on this initiative show mixed opinions on its application; notwithstanding, results in general terms have been positive, particularly because users see this local management alternative as an opportunity for their development. However, the high dynamics of socio-ecological interactions at different scales present new scenarios for social conflicts, which represent new challenges for the governance of marine and coastal for environmental, fishing, and maritime authorities. Although the initiative was initially implemented almost two decades ago in Ecuador, curiously, there is no published information describing the process and results of internal evaluations. A problem we want to help to solve.

2 Methods

The Mangroves on the Coast of Ecuador

Ecuador, with a predominantly subtropical climate, is located in the northwestern part of South America. It is a small country with a total surface of 270,670 km^2 and a coastline that stretches along 650 km (CPPS 2014). The country borders on the north with Colombia, on the south and east with Peru, and the Pacific Ocean on the west. Despite its relatively small coastline, there is a variety of biomes and microclimates due to the influence of the cold and productive Humboldt Current from the south and warm tropical waters of the Panama Bight from the north. Both currents form the Equatorial Front, which moves from north to south along the coast of Ecuador depending on the strength of the Southeast Pacific anticyclone winds (Cucalón 1996). The Gulf of Guayaquil, the largest estuary on the west coast of South America with about 12,000 km^2 (Stevenson 1981), is located in the southern part of Ecuador. Rivers, islands covered by mangroves and channels characterize most of the gulf. The central part of Ecuador is characterized by extensive beaches and low cliffs and deciduous forests on the coast, typical of a semi-arid climate, with a short rainy season. In contrast, the northern part of the country is characterized by an increased rain regimen, as typical of the Choco region, and extensive beaches surrounded by rain forests.

Ecuador possesses around 15,600 km^2 of mangrove distributed in four large blocks: Cayapas-Mataje estuary, Muisne-Cojimíes estuary (Esmeraldas province), Chone River estuary (Manabí province), and the Gulf of Guayaquil (provinces of Guayas and El Oro) (Fig. 1). The Gulf of Guayaquil is by far the most important mangrove area in Ecuador, with 10,500 km^2 amounting to 70% of the country's total mangrove surface (Urquizo et al. 2011). However, its ecological characteristics are slightly different than those mangroves located in the center-north of the country (i.e., Esmeraldas), where trees grow higher and the ecotone gives way directly to rainy forests as in the Chocó and Darién regions. Instead, in the Gulf of Guayaquil, the ecotone gives way to saline areas and then to tropical dry forests. Mangrove deforestation has been particularly high in Ecuador with a net loss of 5,300 km^2 between 1969 and 2006 (MAE 2017).

Socio-economic benefits provided by mangroves to local communities include fishing resources (both industrial and artisanal), aquaculture, charcoal, timber, tannins, and tourism. Currently, the public policy privileges non-extractive and traditional community uses. As in other countries, mangroves in Ecuador supply food for both local and national consumption. Most important artisanal fisheries in Ecuadorian mangroves include the red crab (*Ucides occidentalis*), black clams (*Anadara tuberculosa*, *A. similis*, and *A. grandis*) and dozens of fish species (Fig. 2). The red crab fishery supports 2,215 fishers in the Gulf of Guayaquil (13% of the families in the area) organized in at least 41 associations and cooperatives. The gross contribution in dollars of the entire value chain surpasses USD 40 million per year (USAID 2012). The main problem faced by artisanal fishers, whether associated or working independently, is the inability to be linked into the value chain in a sustainable way to achieve their economic and financial insertion, while at the same time allowing to preserve the mangrove ecosystem and fishing resources of which their economy depends on (BID 2017).

Industrial and artisanal fisheries coexist in the mangroves of Ecuador. White shrimp (*Litopenaeus vannamei*) aquaculture is the most important industry within mangrove areas, particularly in the productive waters of the inner estuary of the Gulf of Guayaquil. Their exportation in 2017 grew 17%, reaching 426,000 tons and 2,800 million dollars (Cámara Nacional de Acuacultura 2018). Recreational

Fig. 1 Location of mangrove areas on the coast of Ecuador

Fig. 2 Red crab (left) and black clams (right), two of the most important fishing resources extracted by artisanal fishers in the mangroves of Ecuador

activities and ecotourism are also raising in protected areas within mangroves throughout the country in sites such as Puerto Hondo, the Ecological Reserve "Manglares-Churute" and the Wildlife Refuges "Isla Corazon y Fragatas" and "Manglares el Morro," which receive thousands of visitors interested in bird and dolphin watching, sport fishing, and other activities associated to mangroves and their resources.

Sources of Information for the Case Study

Due to the lack of published information, the authors reviewed available gray literature and requested information directly from the Ministry of the Environment of Ecuador through the Under Secretariat of Marine and Coastal Management, the entity in charge of the implementation of this initiative. The analyzed information included databases and technical reports generated by the Ministry of the Environment, technical consultancies, and project reports, and included topics such as management effectiveness, competitiveness, and financial sustainability.

In addition, we included a case study of a small fishing community within the Gulf of Guayaquil named "Puerto El Morro," located approximately 70 km southwest of Guayaquil, the largest city on the coast of Ecuador. Puerto el Morro was characterized in the 1990s, as a small village dedicated almost exclusively to artisanal fishing on red crabs, and complementarily as a shipyard of the shrimp wooden fishing fleet (PMRC 1993). In the national census of 2010, Puerto el Morro had 5,019 inhabitants with an Economically Active Population (EAP) of 1,825 persons, of which about half of them (845 persons, 46%), belonged to primary sectors of the economy such as agriculture, livestock, forestry, and fishing. Only 19% of them were women. The census also showed a high rate of Unsatisfied Basic Needs (NBI)[1] (84.3%), which was 24.2% above the national average. This situation would have improved significantly in recent years due to the intervention of the national and municipal governments and the diversification of employment through ecotourism and associated services, as a result of the creation of the coastal–marine protected area "El Morro Mangroves Wildlife Refuge" in 2007. In fact, the Plan of Territorial Ordering and Development (PDYOT) (2015–2019),[2] currently in force, describes tourism as one of the main sources of income in Puerto el Morro.

The case study is based on the re-analysis of data from a survey carried out by the consulting company BIOTICA for Conservation International Ecuador and the

[1] Multidimensional indicator that according the Institute of Statistics of Ecuador INEC considers the economic income, access to basic education, access to home, access to basic services, and overcrowding.

[2] JJ&KM Consulting Associated International Cia. Ltda. (s/f). http://app.sni.gob.ec/sni-link/sni/PORTAL_SNI/data_sigad_plus/sigadplusdocumentofinal/0968538740001_PLAN%20DE%20DESARROLLO%20Y%20ORDENAMIENTO%20TERRITORIAL%20RURAL%20DEL%20MORRO%2028%2010%202015%20(1)_30-10-2015_16-38-46.pdf.

Ministry of Environment of Ecuador over a sample of 42 artisan fishers constituted by 179 family members (BIÓTICA 2017a, b, c). Results are interpreted based on public information available. Over three-quarters of interviewed fishers (79%) identified themselves with the red crab fishery as their main activity, followed by black clam fishers (10%), white fish fishers (5%), and other secondary activities (6%).

Legal Aspects of Mangrove Management in Ecuador

The existing regulations and institutional documents on mangroves in the country, including those giving legal support to the Sustainable Use and Mangrove Custody Agreement, are reviewed below.

The Forestry and Conservation of Natural Areas and Wildlife Act of 1981, reformed in 1990,[3] established that mangroves are part of the nation's Forest Patrimony and considered them as State assets that can only be exploited by concessions regulated in the framework of this law.

The exploitation of mangroves was banned in 1995 and all remaining declared Protected Forest the following year. The regulation on the management, conservation, and use of the mangrove was issued in 1995, which allowed granting concessions for use of mangrove outside the protected areas as well as the construction of channels, docks, and roads for aquaculture facilities. In addition, the regulation expressly stated that local communities will be subject to agreements for the use of mangroves and that for this purpose they will count with financial support from the government. The regulation also defined the following unique activities to be conducted in mangrove areas: (1) tourism and recreation, (2) traditional non-destructive activities, and (3) non-traditional, artisan, non-destructive activities (BIÓTICA 2017a). Such public policy was the basis of the subsequent regulation referred to as the Sustainable Use and Mangrove Custody Agreement adopted in 1999.

The Ecuadorian Constitution (2008) defines mangroves as fragile and threatened ecosystems to be regulated by the government to promote their conservation, management, and sustainable use and recovery (Article 406). Additionally, it recognizes rights to both nature and communities and nationalities to use and benefit from the environment and its resources within the framework of the citizen's right to live in a healthy and ecologically balanced environment through effective mechanisms for the sustainable management of natural resources (Article 397.2). In fact, the Ecuadorian constitution is a pioneer at global level in granting rights to nature and in harmonizing the collective rights of communities and nationalities with international treaties (BIÓTICA 2017a, b). In 2011, the Ministry of the Environment established USD 89,273.01 per hectare[4] as the cost of environmental services loss and restoration due to cut, harvesting, alteration, transformation, or destruction mangrove forests.

---

[3] Law 91. Official Register 495, 7 August 1990.

[4] Ministerial Resolution N° 56, 28 January 2011. Official Register N° 496, 21 July 2011.

The new Organic Environmental Code (COA) (2018) establishes that no rights can be acquired on the mangrove ecosystem (referring to the natural formation of mangrove forest and associated vegetation), while fishing resources inhabiting and depending on the mangrove (fish, mollusks, and crustaceans, among other species) can be used and be protected by ancestral communities. From a conceptual perspective, this law recognizes the right of use by ancestral communities over mangrove and biological resources inhabiting the intertidal and subtidal zone and the surrounding bodies of water, which is clear from a biological perspective. However, it may still be subject to future legal analyzes on institutional competence since the COA, regarding sectoral competences, indicates that the fishing authority may grant concessions for the sustainable use of the mangrove forest, while the environmental authority may regulate the use of aquatic resources dependent on the mangrove, which includes water bodies (BIÓTICA 2017a).

Sustainable Use and Mangrove Custody Agreements as a Management Model

The Management Plan for the Churute Mangroves Ecological Reserve (INEFAN/NATURA/ECOLAP 1996) was the first public instrument recognizing exclusive access rights to ancestral users (red crab fishers) on traditional fishing areas. This was a pioneer initiative through which the government of Ecuador accepted the commitment to identify and deliver zones for fishers defined through a participative process. The need to incorporate local actors in planning, control, and decision-making had been identified in parallel through the Coastal Zone Management process (Bodero and Robadue 1995). This was also fostered by two primary issues: (1) the Churute Mangroves Ecological Reserve was part of the National System of Protected Areas (SNAP) within a category that allowed multiple uses and management flexibility, and (2) the IUCN, at the same time (1994), had redefined the management categories of protected areas at the global level, adopting Category VI: "habitat species management areas, aiming at promoting sound management practices for sustainable production."

In 1999, the Ecuadorian government decided to launch a new management model for mangroves, scaling up the initiative implemented in Churute, granting ancestral communities, from the entire country, the exclusive use and exploitation of fishing resources in mangrove areas through the mechanism named Sustainable Use and Mangrove Custody Agreement,[5] hereinafter the Custody Agreement. Such a mechanism made justice to the communities settled in the mangroves whose rights over the territories in which they had lived for generations were affected by the concessions granted during the past two decades to shrimp farmers. In many cases, shrimp ponds were extended to mangrove areas, restricting access to traditional users and putting in risk people's food security.

[5] Executive Decree N° 1102. Official Register N° 243, 28 July 1999.

In August 2010, the Ministry of the Environment of Ecuador issued the Ministerial Regulation No. 129,[6] detailing the procedures for the approval and granting Custody Agreements in favor of ancestral communities and traditional users.[7] This instrument redefined the scope, identifies the beneficiaries, the conditions, and obligations of those who aspire to sign a Custody Agreement, under a transparent scheme. Agreements are granted to communities and user groups that, having legal status and organized in legally recognized associations or unions, request such a privilege. The following seven activities are allowed in areas under Custody Agreements:

1. Extracting fish and invertebrates
2. Breeding or rising fish, invertebrates, mammals, reptiles, or other species of native fauna through practices that do not affect the mangrove forest or dynamics of water bodies
3. Controlled forest management
4. Mangrove reforestation
5. Ecological tourism and non-destructive mangrove recreation activities
6. Conservation and protection
7. Education and scientific research

In addition, applicant communities must have a Management Plan approved by the regional environmental authority (Undersecretary of Coastal Marine Management), describing the baselines of the requested area and plans for exploitation, control, monitoring, and evaluation, as well as documents related to the organizational structure, members, and internal regulations. Obligations are also established. Lack of compliance may terminate the Custody Agreement. Responsibilities assumed by beneficiaries include: (1) guarding the mangrove against any aggression and reporting to the environmental authority, (2) implement the Management Plan fully, (3) comply with sectoral legal provisions, (4) prepare semi-annual reports, (5) an external and independent evaluation on the first year of the agreement, (6) a second independent evaluation that includes the first 9 years of the agreement, and (7) maintain legal status.

Thus, the initiative promotes associativity, self-control, and accountability by regularly delivered progress reports to the environmental authority. Custody Agreements are granted for 10 years and their renewal will depend on the performance evaluation made by the environmental authority. The management of mangroves in Ecuador is aligned with a regional initiative promoted by the Permanent Commission of the South Pacific (CPPS)—a regional intergovernmental maritime body—through a regional initiative named Action Plan for the Conservation of Mangroves in the Southeast Pacific or PAR-Manglares (CPPS/UNESCO/CI/Hivos 2016). This management tool aims to help participant governments to strengthen policies and programs for the protection, recovery, and sustainable use of mangroves in the region.

[6] Official Register N° 2833, 21 September 2010.

[7] Modified on 9 August 2011, Ministerial Agreement N° 144.

3 Results

According to the information provided by the Ministry of Environment of Ecuador, 79 Custody Agreements were issued between 1999 and October 2018 of which 53 are currently operational (67%), 16 timed out, and 10 are extinct. Additionally, there are 13 new applications in review. The total assigned surface through this instrument is nowadays 683.6 km^2, which is approximately 44% of Ecuador's mangrove. Likewise, there are six Custody Agreements that were terminated by the Ministry of Environment when beneficiaries did not accomplish commitments in any form. Most Custody Agreements issued (86.5%) are concentrated in two provinces of the Ecuadorian coast, Guayas, and El Oro, both in the Gulf of Guayaquil, and the rest are in two other provinces in the center and north of the country, Manabí and Esmeraldas (Table 1). The total area in custody is also concentrated in the two provinces within the Gulf of Guayaquil with 98.7% of the surface and 88.6% of the direct beneficiaries.

The size of the areas delivered in custody and the number of beneficiaries is highly variable. On average, areas have about 131 km^2 (SD = 189.1, range 1.1–1,080 km^2). The average number of beneficiaries per association is 66 (SD = 68.9, range 15–400).

The mangrove area under Custody Agreements increased threefold since 2010 (Fig. 3). This sudden increase suggests that over time the confidence and interest of users in implementing this management model has increased.

One of the requirements for granting a Custody Agreement is that users must be associated; thus, fishing associations have been strengthened during this process. For example, associations have a directive to meet in general assemblies once or twice a month and offenses are punishable by fines. Likewise, most organizations generate income through membership fees ranging between USD 4 and 56 a month. Funds are allocated generally to control and surveillance, but also for institutional strengthening (UTPL 2017).

Table 1 Number of Custody Agreements, surface, and number of direct beneficiaries in four provinces of the coast of Ecuador

Province	Agreements	%	Surface (km^2)	%	Beneficiaries	%
Esmeraldas	5	9.6	8.3	1.2	140	4.5
El Oro	22	42.3	143.1	20.9	881	28.4
Guayas	23	44.2	531.6	77.8	1867	60.2
Manabí	2	3.8	0.6	0.1	212	6.8
Total	52	100.0	683.6	100.0	3100	100.0

Source: database of the Under Secretariat of Coastal and Marine Management of the Ministry of Environment of Ecuador

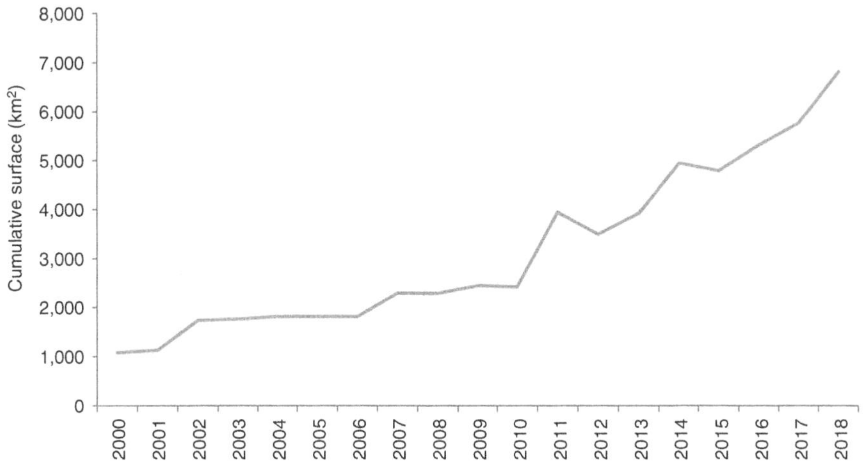

Fig. 3 Evolution of mangrove area under Custody Agreements between 2000 and 2018. Source: Database of the Under Secretariat of Marine Coastal Management of the Ministry of Environment of Ecuador

Management Effectiveness of the Custody Agreements of Use

Available information shows different levels of effectiveness regarding Custody Agreements. In general terms, there are achievements and limitations in the application of this innovative mechanism in which local communities have the most important role; still, sustainability needs to be consolidated (Hurtado et al. 2010).

The Ministry of Environment as the institution responsible for granting Custody Agreements has carried out several studies to evaluate their effectiveness. The first known evaluation was made in 2005 on the Management Plan of the Cayapas—Mataje Ecological Reserve (MAE 2008). During this process, there were discrepancies between stakeholders and evaluators regarding the lack of participation of real representatives of local communities during the evaluation process. However, positive and negative aspects of the Custody Agreement implementation were recognized.

A second evaluation was carried out in 2008 to assess the level of implementation in the first 10 years of this initiative (Coello et al. 2008). This study evaluated the performance of 26 areas under custody and its main conclusions were the following:

1. All stakeholders agreed that Custody Agreements is a positive tool for the conservation of mangroves as well as a legal instrument that guaranteed ancestral users exclusive access to these areas.
2. Most of the beneficiaries complied with commitments and managed to conserve their areas. In several areas, social and economic benefits were generated and living conditions improved.

3. In other areas, no benefits were evident, which does not imply that the tool in these areas was deficient, but rather the application process did not receive the same attention and support by the government.
4. In at least nine areas (35%), a rapid intervention by the environmental authority is required to strengthen the process.
5. Twenty-five management plans showed progress and one Custody Agreement did not work (3.8%).
6. Some areas under custody lost more than 60% of mangrove cover due to logging and natural erosion processes. In some cases, because organizations were weak.
7. Most concessions showed difficulties in managing the area.
8. In some cases, particularly in the province of Esmeraldas, the access to other fishers was not restrained.
9. A good understanding of the implications and responsibilities associated with mangrove custody is lacking.

According to this exercise, Coello et al. (2008) concluded that there are six key elements to guarantee the success of a Custody Agreement: (1) a beneficiary user organization; (2) technical assistance; (3) realistic management plan; (4) control and surveillance system in place; (5) a fishing resources management system in place; and (6) support from authorities.

A third exercise using a similar methodology to assess 20 Custody Agreements was carried out by the Technical University of Loja in 2017 in the province of El Oro (UTPL 2017). The analysis evaluated the implementation of the activities as defined in the management plans including programs on fishing, control, monitoring, and evaluation, among others. The analysis showed that four associations (20%) had a level of compliance higher than 75%. In seven associations (35%), the level of compliance was between 50 and 75%. In eight associations (40%), the level of compliance was between 25 and 50% and only one association (5%) had less than 25% of compliance. It was recognized that some of the organizations had less than 2 years of implementation, and this may affect results since they did not have enough time to implement their programs.

Regarding the semi-annual reports to the Ministry of the Environment, only eight associations (40%) complied satisfactorily. Additionally, only seven associations (35%) had valid technical assistance arrangements, an important requirement to obtain a mangrove area under custody.

The evaluation of the UTPL (2017) revealed, among other things, the following:

1. Some management plans were not realistic for the local organizational level either because they were extensive, complex, and/or with budgets beyond associations capabilities.
2. During the period 2010–2017, the mangrove cover recovered at an average annual rate of 4.13%.
3. All organizations reported compliance with environmental regulations in relation to catching sizes, closure seasons, and fishing techniques.
4. Between 60 and 80% of the budget was destined to control and surveillance.

5. There is a direct relationship between the level of compliance and organizational commitment.
6. Deficiencies are observed in relation to technical assistance which results in lower management plan accomplishing.

Case Study: The Association of Artisanal Fishers of Puerto el Morro

According to data from the Ministry of Environment analyzed by BIÓTICA (2017a, b, c), only 75 artisanal fishers from Puerto El Morro (39%) belong to associations, so the majority (61%) are independent fishers, in some cases accredited in any form by fishing, maritime, and/or environmental authorities. This created tension among fishers of Puerto El Morro with fishers from neighboring villages that used the same fishing areas without restrictions, except within marine protected areas. Interviewed independent fishers indicated they do not find benefits to belong to a social organization, nor in the regularization of their activities, as this may involve costs and waste of time in bureaucratic procedures.

Facing this scenario of potential social conflicts, the three associations of artisanal fishers at Puerto El Morro, (Marine, Shapers of the Future and the Cooperative of Artisanal Fishers Porteño) joined and created the Commonwealth of fishing associations of Puerto El Morro for the conservation of mangroves and their fishing resources (Table 2). This commonwealth requested the custody of 38.6 km^2 of mangrove where traditionally have deployed their fishing activities around Puerto El Morro and in the northwest of Puná Island (Fig. 4). The objective of this process was to implement a management model that would encourage the responsible use of fishing resources and the long-term maintenance of mangrove ecosystem services, basis of the local economy.

The Management Plan for Puerto El Morro Commonwealth was built based on three principles: (1) a participative planning process, (2) a zoning system based on available resources and uses, and (3) a management framework according to the socio-cultural characteristics of future beneficiaries.

A first step during the planning stage was to define the extent of the zone to be requested. The zoning was based on the identification and compatibility of current and potential uses, also considering the presence of the protected area El Morro Manglares Refuge (REVISEM). Thus, three major uses zones were defined

Table 2 Artisanal fishing associations participating in the Commonwealth of Puerto El Morro that applied for a joint Custody Agreement

Description	Marine	Future	Porteño	Total
Fishers	24	16	35	75
Areas requested in custody (km^2)	11.4	18.4	8.8	38.6

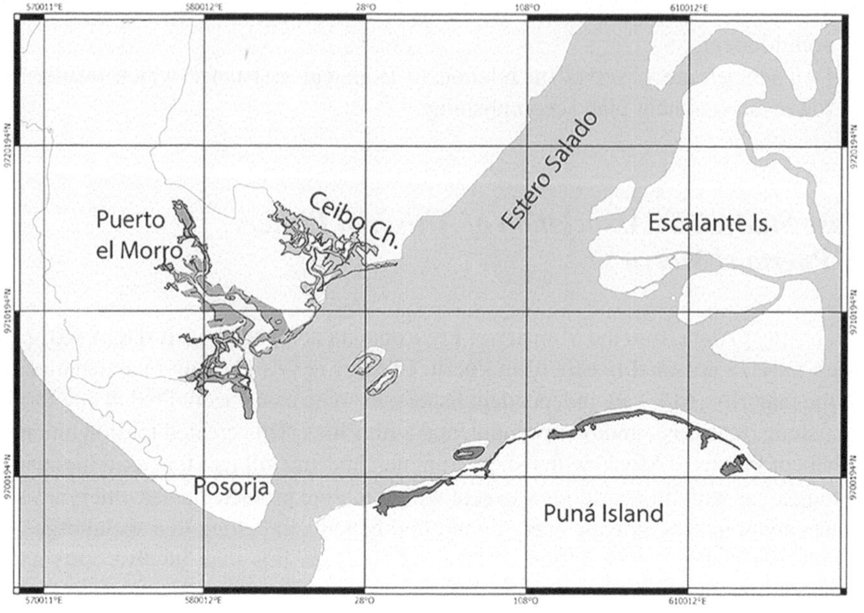

Fig. 4 Location of the areas granted with a Custody Agreement to the Commonwealth of fishing associations of Puerto El Morro, province of Guayas, Ecuador. Source: Adapted from BIÓTICA (2017a, b, c)

according to the four management objectives accorded between users and authorities (BIÓTICA 2017b):

1. *Extraction Zone*. In this area, the extraction of regulated fishing resources such as crabs, clams, and white fish is allowed. Uses are based on compliance with legislation and self-imposed rules which are in fact stricter than those established by law. Compliance refers to: (1) minimum size of clams (4.5 cm) and male crabs (7.5 cm cephalothorax width), (2) stop fishing crab in the breeding and molting seasons (1–31 March and August 15–September 15, respectively), and (3) the use of mesh traps for crab is forbidden, as well as the capture of gravid females. The self-imposed norm establishes a maximum daily catch quota per person based on the current catch average (84 crabs and 300 clams).
2. *Rotation Zone*. Zone established to allow the recovery of fishing resources through spatial-temporal extractive closures, regulated by the environmental authority or self-imposed by fishers in overexploited areas. Activities oriented to diversify production with added value (aquaculture, recreational fishing) are allowed.
3. *Reference and recovery Zone*. No extraction is allowed since the purpose of these zones is to maintain pristine areas for research and mangrove restoration.

In general terms, designed zones vary according to the characteristics of each area requested in custody, but the extractive area should not exceed two-thirds of the total area requested in custody. The assigned area to the Commonwealth of fishing associations of El Morro is divided as follows: extraction zones (59% of the area), rotation zones (39% of the area), and reference zones (2% of the area). Affecting mangrove forest is explicitly forbidden for beneficiaries, except for fishing purposes after authorized by the environment authority.

4 Discussion

General Features

Custody Agreements have been referred in several reviewed reports as "valid and effective" mechanisms of local governance to manage benthic resources associated with the mangroves in Ecuador in a more efficient way; resources that have been exploited for generations by fishermen (López-Angarita et al. 2016). Although the results of the evaluating their effectiveness are not satisfactory in all areas for different reasons (e.g., Coello et al. 2008; UTPL 2017), the mechanism undoubtedly implies important advances in local participation and management of artisanal fisheries in a critical and vulnerable ecosystem, which has traditionally been freely available and scarcely regulated by competent authorities.

Improving governance of coastal–marine areas is still a pending challenge in Ecuadorian public policy. Although there is no question about the competence of the Ministry of the Environment in the protection of the mangrove forest, the jurisdiction in the different strata of the water column, including intertidal and subtidal resources are not well defined, remaining intersectoral and even intra-institutional discrepancies.

The complexity of mangroves and associated resources has been evidenced through several initiatives promoted by the government, academia, and NGO to develop more effective management mechanisms for these areas (e.g., CPPS/UNESCO/CI/Hivos 2016; Maldonado et al. 2014). As a social-ecological system, in which human beings are directly affecting the ecosystem through extractive activities, the future of this initiative deals with understanding its dynamics in space and time in order to incorporate an integral and adaptive approach to decision-making (see Petrosillo et al. 2015; Quinn et al. 2017). Mangrove forest management is currently a cross-cutting issue within the country's coastal and marine policies (SETEMAR 2014). That is the reason why the Ministry of Environment has given institutional and political support to this initiative. Such support is necessary for model's sustainability because when processes do not have political support, they tend to become short-term efforts without real impact and/or benefit for local communities (Barragán 2014).

Technical assistance and financing are key issues for the success of this management scheme (UTPL 2017; BIÓTICA 2017a). Currently, they depend more on external cooperation than on long-term institutionalized mechanisms. Communities that are better organized have demonstrated a greater capacity to obtain technical assistance and government funding through initiatives such as the Mangrove Partner Program (UTPL 2017). It is possible that for other communities, elements such as distance, the difficulty of access (particularly in distant islands), higher levels of unsatisfied basic needs, poverty and lower level of schooling, also have influenced model's implementation. In this regard, the current model includes common agendas for all areas such as control, monitoring, and evaluation programs, which should be complemented with other initiatives to address specific previously identified needs. Today, fishers continue depending entirely on intermediaries located in a low position of the productive chain. Thus, a priority should be strengthening internal capacities to improve their administrative skills to reach organizational maturity.

Despite the fact that Custody Agreements give users exclusive access to marine resources, increasing productivity and improving market systems in those communities where the model has been better implemented, it is necessary to continue making efforts to generate new or alternative productive activities in order to improve their competitiveness (Coello et al. 2008; BID 2017). It is estimated that only 15% of the total income generated by the commercialization of fishing products remains in the communities; the rest is distributed along the commercial intermediaries chain (USAID 2012). Some efforts are being made to generate collective ventures with a gender focus, aimed at markets of greater scale and value, such as the processing crab meat to sell directly to consumers and reducing the chain of intermediaries, as well as the certification for good practices (BID 2017). Financing productive initiatives will be crucial for communities to obtain tangible benefits from this initiative.

Custody Agreements have helped to organize extractive activities and reduce conflicts between users. Nevertheless, conflicts of different types remain in most areas including problems between associated fishers that do not meet their quotas or disregard minimum sizes of capture, invasion of areas under custody by users from other sectors, use of forbidden fishing gear, insecurity, conflicts with industrial fishing and with shrimp trawlers, among others (Coello et al. 2008; UTPL 2017). In some cases, these problems end in complaints to the authorities and in other cases they are resolved within the association with sanctions. The mediation of the Ministry of the Environment has helped to reduce or minimize conflicts between neighboring associations, but it is more difficult when it occurs with users who are not associated or belong to other community.

Sustainability of the Model

An essential aspect for the sustainability of Custody Agreements is that the area in custody generates tangible benefits (economic, social, or environmental), significantly greater than the costs associated with surveillance and management. In this regard, in 2014 the Ministry of Environment developed the Mangrove Partner

Program (Programa Socio Manglar),[8] as part of a sustainability strategy to complement and consolidate the results achieved. This mechanism transfers money, conditioned to comply with management plans. It consists of a fixed amount between USD 7,000 per year for areas between 1 and 5 km^2, USD 10,000 per year for areas between 5.01 and 10 km^2 and USD 15,000 per year for areas larger than 10 km^2 USD. Authorities have estimated that this monetary contribution corresponds to approximately 50% of the operational costs, and the other 50% corresponds to co-financing by beneficiaries.

Such monetary benefit, granted to the associations by the government through the Mangrove Partner Program, can be spent only in the following activities:

1. Control and surveillance
2. Administration
3. Technical assistance for the implementation, monitoring, and/or evaluation of management plans
4. Organizational strengthening
5. Total or partial financing of productive or social initiatives

The Mangrove Partner Program has helped many associations in the country to finance activities that in another form would not be possible to accomplish. In the case of the associations of El Oro province, ten of the twenty evaluated organizations received this incentive, and their destination was mainly control and surveillance and institutional strengthening (UTPL 2017). According to the Ministry of Environment, 23 associations and 2,674 fishers were benefited with USD 210,000 between 2014 and 2015 at a national level with this program.

The Mangrove Partner Program is a pioneering scheme that has helped to scale up the Custody Agreements initiative, and it has been widely recognized by the communities. It is not clear, however, if the Ecuadorian government will maintain and eventually extend this benefit to all Custody Agreements currently issued and those to be issued in the future. In this sense, the Ministry of the Environment must conduct a detailed economic analysis on this and other initiatives that could eventually provide additional funds in a sustained manner. Initiatives associated with mitigating climate change including the blue carbon initiative[9] promoted by Conservation International and the United Nations Collaborative Program on Reducing Emissions from Deforestation and Forest Degradation in Developing Countries REDD +,[10] could be interesting to explore. Within the framework of the regional initiative PAR-Manglares, the need to create an experimental account for mangroves within the system of environmental and economic accounts of the country was evaluated (Orellana et al. 2018). Despite a few studies in the country on valuation mangrove environmental services, the current legal framework offers an opportunity to internalize this economic concept.

[8] Ministerial Agreement N° 198, 9 July 2014.

[9] http://thebluecarboninitiative.org/

[10] http://www.un-redd.org/

Any new mechanism that may be implemented to strengthen the initiative has to come from the government or from institutions that currently provide technical support to beneficiaries. Some mechanisms can be complex to implement or require extra funds for evaluations or baseline information. Thus, ensuring financial sustainability for this initiative should be a priority for the Ministry of Environment in order to respond to expectations of beneficiaries, create resilience and take the process to another level.

Benefits for the Environment

Official data on mangrove coverage in Ecuador indicate that in less than three decades (1980s–2000s), more than a quarter (27.1%) of country's mangrove forest was lost due to a change in the land use, mainly for shrimp farming and urbanization. This situation led social and non-governmental organizations to request from environment authorities the adoption of public policies for mangrove protection, including the implementation of the Custody Agreement as management mechanism and the inclusion of mangrove forest as fragile and threatened ecosystem protected by the State in the Constitution 10 years later (2008). These and other policies and regulations finally would manage to revert the trend in 2010 (SENPLADES 2017).

Thus, Custody Agreements have not only been successful in generating socioeconomic benefits for coastal communities but also producing environmental benefits. The recovery of mangrove areas, among others, is producing the following positive impacts: (1) improving habitat quality for a diversity of marine species in different trophic levels, including many commercial species of fish, mollusks, crustaceans, and other taxa, that spend part of their life in the mangrove; (2) improvement of water quality due to the increased filtration capacity of sediments, nutrients, and contaminants; and (3) maintaining ecological processes that sustain marine biodiversity including iconic fauna such as birds and mammals, offering new opportunities for diversification of productive activities through nature-based tourism. Unfortunately, the quantitative information available is limited and represents a challenge to be addressed in the future.

References

Barbier E, Hacker S, Kennedy C (2011) The value of estuarine and coastal ecosystem services. Ecol Monogr 81(2):169–193

Barragán JM (2014) Política, gestión y litoral. Una nueva visión de la gestión integrada de áreas litorales. Tébar Flores, Madrid, p 685

Banco Interamericano de Desarrollo (BID) (2017) Competitividad de las comunidades concesionarias de manglar en el golfo de Guayaquil. Banco Interamericano de Desarrollo, Fondo Multilateral de Inversiones. Consultancy report

BIÓTICA (2017a) Plan de Manejo para el uso y custodia de 1,138.49 hectáreas de manglar solicitada por la Asociación de Pescadores Artesanales Marine en el sector de Puerto el Morro. Ministerio de Ambiente de Ecuador, Conservación Internacional Ecuador, Instituto Humanista para la Cooperación con los Países en Desarrollo, organización de las naciones Unidas para la Alimentación y la Agricultura y Fondo para el Medio Ambiente Mundial. Guayaquil, Ecuador. Consultancy report

BIÓTICA (2017b) Plan de Manejo para el uso y custodia de 882,16 hectáreas de manglar solicitadas por la cooperativa de pescadores artesanales manglares porteño en el sector de Puná. Ministerio del Ambiente de Ecuador, Conservación Internacional Ecuador, Instituto Humanista para la Cooperación con los Países en Desarrollo, Organización de las Naciones Unidas para la Alimentación y la Agricultura y Fondo para el Medio Ambiente Mundial. Guayaquil, Ecuador. Consultancy report

BIÓTICA (2017C) Plan de Manejo para el uso y custodia de 1839.6 hectáreas de manglar solicitadas por la cooperativa de pescadores artesanales Forjadoes del Futuro. Ministerio de Ambiente de Ecuador, Conservación Internacional Ecuador, Instituto Hunabista para la Cooperación con los Países en Desarrollo, Organización de las Naciones Unidas para la Alimentación y la Agricultura y Fondo para el Medio Ambiente Mundial. Guayaquil, Ecuador. Consultancy report

Bodero A, Robadue D (1995) Estrategia para el manejo del ecosistema de manglar en el Ecuador. In: Ochoa M (ed) Manejo Costero Integrado en Ecuador. Fundación Pedro Vicente Maldonado. Guayaquil, Ecuador. Programa de Manejo de Recursos Costeros

Cámara Nacional de Acuacultura (2018) Camarón, reporte de exportaciones ecuatorianas totales. http://www.cna-ecuador.com/estadisticas/. Accessed 11 June 2019

Coello S, Vinueza D, Alemán R (2008) Evaluación de desempeño de los acuerdos de uso sustentable y custodia de manglar en la zona costera de Ecuador. Ministerio de Ambiente de Ecuador/Conservación Internacional/Unión Mundial para la Naturaleza (UICN)/Comisión Mundial de Áreas protegidas de IUCN/programa de apoyo a la gestión descentralizada de los recursos naturales en las tres provincias del norte de Ecuador (PRODERENA)/Ecobiotec. Consultancy report

Comisión Permanente del Pacífico Sur (CPPS) (2014) Estado del medio Ambiente Marino y Costero del Pacífico Sudeste. Guayaquil, Ecuador. Serie Estudios Regionales 4

CPPS/UNESCO/CI/Hivos (2016) Plan de acción regional para la conservación de los manglares en el Pacífico Sudeste (PAR-Manglares). Guayaquil, Ecuador

Cucalón E (1996) Oceanografía y Sistemas Físicos. In: Sistemas Biofísicos en el Golfo de Guayaquil. Comisión Asesora Ambiental CAMM. Ecuador

Donato DC, Kauffman JB, Murdiyarso D et al (2011) Mangroves among the most carbon-rich forest in the tropics. Nat Geosci 4:293–297

Duke NC, Meynecke JO, Dittmann S et al (2007) A world without mangroves? Science 317:41–42

Durán E, Gumeta-Gómez F, Olguín L (2018) Manejo comunitario en paisajes forestales. In: Ávila Foucat VS, Perevochtchikova M (eds) Sistemas socio-ecológicos: marcos analíticos y estudios de caso en Oazaca. Universidad nacional Autónoma de México, Instituto de Investigaciones Económicas, México, pp 245–268

Glaser M, da Silva R (2004) Prospects for the co-management of mangrove ecosystems on the north Brazilian coast: whose rights, whose duties and whose priorities? Nat Res Forum 28:224–233.

Hurtado M, Hurtado-Domínguez MA, Hurtado-Domínguez LM et al (2010) Áreas Costeras y Marinas Protegidas del Ecuador. Ministerio del Ambiente—Fundación Natura

Huxham M, Emerton L, Kairo J et al (2015) Applying climate compatible development and economic valuation to coastal management: a case study of Kenya's mangrove forest. J Environ Manag 157:168–181

INEFAN/NATURA/ECOLAP (1996) Plan de Manejo Reserva Ecológica Manglares Cayapas Mataje – REMACAM. San Lorenzo, Esmeraldas

Kathiresan K, Bingham BL (2001) Biology of mangroves and mangrove ecosystems. Adv Mar Biol 40:81–251

Lerner AM, Charli-Joseph L (2018) Marcos teóricos para el estudio de los sistemas socio-ecolçogicos y las ciencias de la sostenibilidad. In: Ávila Foucat VS, Perevochtchikova M (eds) Sistemas socio-ecológicos: marcos analíticos y estudios de caso en Oazaca. Universidad nacional Autónoma de México, Instituto de Investigaciones Económicas, México, pp 83–100

López-Angarita J, Roberts CM, Tilley A et al (2016) Mangroves and people: lessons from a history of use and abuse in four Latin American countries. For Ecol Manag 368:151–162

Ministerio del Ambiente del Ecuador (MAE) (2017) Guía de derechos y deberes de las organizaciones custodias del manglar. MAE, CI, HIVOS, FAO, GEF. Guayaquil, Ecuador

Ministerio de Ambiente de Ecuador (2008) Plan de Manejo de la Reserva Ecoógica Manglares Churute. Cuatro tomos. Quito, Instituto Ecuatoriano Forestal y de Áreas Naturales y Vida Silvestre (INEFAN), Fundación Natura (NATURA) e Instituto de Ecología Aplicada (ECOLAP)

Maldonado M, Torres J, Carvajal R (2014) Plan de Fortalecimiento para el Desarrollo Sustentable del Sitio Ramsar Manglares del Estuario Interior del Golfo de Guayaquil "Don Goyo". Conservación Internacional, Gobierno Autónomo Descentralizado Provincial Guayas, Ministerio del Ambiente, Junta de Manejo Participativo Comunitario "Puerto Libertad—Cerrito de los Morreños—Santa Rosa". Guayaquil, Ecuador. Consultancy report

Orchand SE, Stringer LC, Quinn CL (2015) Mangrove system dynamics in Southeast Asia: linking livelihoods and ecosystem services in Vietnam. Reg Environ Chang. https://doi.org/10.1007/s10113-015-0802-5

Orellana Y, Viteri C, Albán M (2018) Evaluación rápida de las condiciones para la implementación de las cuentas ecosistémicas experimentales de manglar en Ecuador. Conservación Internacional. Unpublished report. http://www.par-manglares.net/images/docs/informes/Informe_cuentas_ambientales_Ecuador_OCT2018.pdf. Accessed 11 June 2019

Petrosillo I, Aretano R, Zurlini G (2015) Socioecological systems, reference module in earth systems and environmental science. pp 1–7. https://doi.org/10.1016/B978-0-12-409548-9.09518-X

Programa de Manejo de Recursos Costero (PMRC) (1993) Plan de Manejo de la ZEM Playas—Posorja—Puerto el Morro. Publicación preparada por la Fundación Pedro Vicente Maldonado. Guayaquil

Quinn CH, Stringer LC, Berman RJ et al (2017) Unpacking changes in mangrove social-ecological systems: lessons from brazil, Zanzibar, and Vietnam. Resources 6:14. https://doi.org/10.3390/resources6010014

Secretaría Nacional de Planificación y Desarrollo (SENPLADES) (2017) Plan de Ordenamiento del Espacio Marino Costero-POEMC. http://www.planificacion.gob.ec/wp-content/uploads/downloads/2018/07/Plan-de-Ordenamiento-del-Espacio-Marino-Costero.pdf. Accessed 11 June 2019

Secretaría Técnica del Mar (SETEMAR) (2014) Políticas públicas oceánicas y costeras: Diagnóstico y propuesta de implementación. BIÓTICA Cía. Ltda. El Telégrafo, Guayaquil, Ecuador

Spalding M, McIvor A, Tonneijck FH et al (2014) Mangroves for coastal defense. Guidelines for coastal managers & policy makers. Published by Wetlands International and The Nature Conservancy

Stevenson MR (1981) Seasonal variations in the Gulf of Guayaquil, a tropical estuary. Boletín Científico y Técnico 4(1):1–133

Tomlinson PB (1986) The botany of mangroves. Cambridge University Press, Cambridge

Urquizo R, Viejó L, Carvajal R et al (eds) (2011) Biodiversidad del Guayas: conociendo nuestra verdadera riqueza. Gobierno Provincial del Guayas-Dirección de Medio Ambiente. Guayaquil, Ecuador

USAID-Costas y Bosques Sostenibles (2012) Cadena de valor del cangrejo rojo en el Golfo de Guayaquil. Quito, Ecuador

Universidad Técnica Particular de Loja (UTPL) (2017) Evaluación de efectividad de manejo de los acuerdos de uso sustentable y custodia del manglar en la provincia de el Oro. Informe de consultoría de proyecto GEF para el Ministerio de Ambiente, Conservación Internacional, HIVOS & FAO. Consultancy report

Valiela I, Bowen JL, York JK (2001) Mangrove forests: one of the world's threatened major tropical environments. BioScience 51:807–815

Yáñez-Arancibia A, Lara-Domínguez AL (1999) Los manglares de América Latina en la encrucijada. In: Yáñez–Arancibia A, Lara–Domínguez AL (eds) Ecosistemas de Manglar en América Tropical. Instituto de Ecología A.C. México, UICN/ORMA, Costa Rica, NOAA/NM FS Silver Spring, pp 9–16

Integrated Evaluation of the Effects of the Payment for Hydrological Environmental Services Program in Ajusco, Mexico City

María Perevochtchikova, Iskra Alejandra Rojo Negrete, Estebania Teyeliz Martínez Jiménez, Zenia María Saavedra Díaz, and Sandra Edda Martínez

Abstract This chapter examines, in an integrated way, the effects of the implementation of the federal Payment for Hydrological Environmental Services (PHES) Program in Mexico, managed by the National Forestry Commission (CONAFOR) since 2003. The program involves economic compensations for carrying out conservation activities in areas of hydrological importance (and where the corresponding forestland is mostly collectively owned). The case study included the community of San Miguel and Santo Tomas Ajusco, located in the southwestern periphery of Mexico City. The analysis was based on the Community Capital Framework (CCF) and included five stages: (1) compilation and systematization of the available information, (2) social, (3) economic, (4) and ecologic effects analysis, and (5) result's integration. Socially, divergences and convergences in actors' perceptions were identified, with a general tendency to perceive positive effects. In economic terms, the low effectiveness of the program was evident, and in the environmental results, a great potential to provide hydrological services was observed. There is a pressing need to re-examine the design of the public policy instruments applied at local scale, while developing participatory monitoring schemes, to determine long-term effects.

M. Perevochtchikova (✉) · I. A. Rojo Negrete
Centro de Estudios Demográficos, Urbanos y Ambientales, El Colegio de México A.C. (CEDUA-COLMEX), Ciudad de México, Mexico
e-mail: mperevochtchikova@colmex.mx

E. T. Martínez Jiménez
Posgrado en Geografía, Universidad Nacional Autónoma de México, Ciudad de México, Mexico

Z. M. Saavedra Díaz
Produraduría Ambiental y de Ordenamiento Territorial de la CDMX, Ciudad de México, Mexico

S. E. Martínez
Autoridad de Cuenca Matanza Riachuelo, Buenos Aires, Argentina

L. E. Delgado, V. H. Marín (eds.), *Social-ecological Systems of Latin America: Complexities and Challenges*, https://doi.org/10.1007/978-3-030-28452-7_16

Keywords Social-ecological systems · Latin America · Complexity · Mexico · Hydrological services · Payment

1 Introduction

Since the Kyoto Protocol in 1997, setting economic compensation mechanisms for environmental conservation has been promoted worldwide, where a marked difference can be observed in the terms used, such as environmental services and ecosystem services (ES) (Sanders et al. 2013). It is important to point out that the former is frequently used in a political stance, that assumes a broader meaning in relation to fields such as economics, environmental management, and political ecology; whereas ES are derived from the ecological approach resulting from the concept of "ecosystem" (Perevochtchikova 2014, 2016). The economic compensation (or Payment) mechanism for Ecosystem Services (PES) was one of the public environmental policy tools developed to mitigate environmental degradation while seeking to improve human well-being by carrying out activities that ensure the long-term ecosystems' quality in exchange for economic compensation (Fregoso 2006; DDS-OEA 2008). In this way, this scheme is based on an ecological stance (i.e., a good condition of ecosystems is central to the conservation of its services) and an economic one; namely, markets and payments assume the internalization of externalities (Martínez-Alier and Roca Jusment 2001; McElwee 2012), which opens up possibilities for alternative management of natural resources (Cordero 2008; Kosoy et al. 2008).

PES schemes per se are defined as financial compensation instruments, which ensure the ES conservation at a local and global scale (Fischer et al. 2009), by granting them an economic value (NRC 2005). This scheme consists of a voluntary monetary transaction, in cash or in-kind (Brüschweiler et al. 2004: 23), for the ecosystem service, which becomes a "purchase" by at least one user of these services to the provider (Wunder 2005). It is important to note that there are a number of actors directly involved in the functioning of PES schemes: (1) the service providers (usually the landowners who own the natural resources and who benefit from the payment, giving up other economically more attractive potential land uses) and (2) the users of these services (which may be a population, an industrial sector or the government). In federal programs, the government directs and funds this type of mechanism, adapting the role of users (especially when services are not well defined). But, in the case of local-level initiatives, there may be other intermediaries, such as non-governmental organizations (NGOs) or state and municipal governments, with support coming from mixed funds (Perevochtchikova 2016).

Several compensation forms have been created in forest conservation schemes (which may include biodiversity, hydrological and carbon capture). For example, federal funding, co-investment of funds, including contributions from the governmental and private sector, such as those in Costa Rica, Ecuador, Nicaragua, etc. (Martínez Harms and Balvanera 2012), and conservation programs of global importance, in addition to forest certification and other forms of monetary compensation

(Engel et al. 2008). The Mexican Payment for Hydrological Environmental Services program (PHES) carried out by the National Forestry Commission (CONAFOR) from 2003 to the present, grants an economic compensation to landowners in the watersheds' upper areas that decide to preserve their forests, in order to provide hydrological ES to the lower located population, ensuring water consumption, water cycle regulation, water capture and infiltration, sediment reduction, etc. (Cordero 2008; Perevochtchikova et al. 2015, 2016).

The increasing implementation of these instruments in public policies has stimulated the interest of academia, with the literature addressing the topic from different theoretical and empirical approaches, such as ecology, ecological economics, neo-constitutionalism, power networks, among others (Perevochtchikova 2016). Hydrological ES, particularly, have been approached from disciplines like geology, hydrogeology (based on the theory of groundwater flow), and surface-water hydrology (with water balance calculations); biodiversity ES from biology; carbon capture ES from biology (molecular microbiology); and sociocultural ES from social theories, such as common-pool resources and collective action, among others (Perevochtchikova and Oggioni 2014). In Mexico, PES programs have been studied by academics and assessed from a public sector perspective as public policy instruments in order to analyze their effectiveness, efficiency, eligible areas, and limitations (Cortina and Saldaña 2014; Rojo Negrete 2018).

The development of this chapter was based on the theoretical-conceptual framework of "Community Capital" proposed by Brandon and Lombardi (2011). This framework refers to an integration of capitals in a pyramidal form, where the fundamental basis is occupied by natural capital, on which social and human capitals are founded, and finally, it is dominated by built and financial capital. In fact, this points to the creation and supply of social and economic benefits to society extracted from the use and exploitation of ecosystems and their multiple assets and services. Therefore, the key concepts in this chapter are hydrological ecosystem services (HES) and common use resource (CUR), within the binomial called "water-producing forest" (incorporated in the human actions performed to guarantee the supply of ES). In this way, and according to the framework, this work presents a case study of one Mexican forest community, developed in several stages according to the analyzed capitals (social, economic, and natural) and their integration for the evaluation of the PES effects.

2 Methods

Case Study

The community of San Miguel and Santo Tomas Ajusco, with a long history since pre-Hispanic times (Perevochtchikova 2016), is located in the southwest of Mexico City (CDMX) (Fig. 1); within the Conservation Land (CL) which has an extension

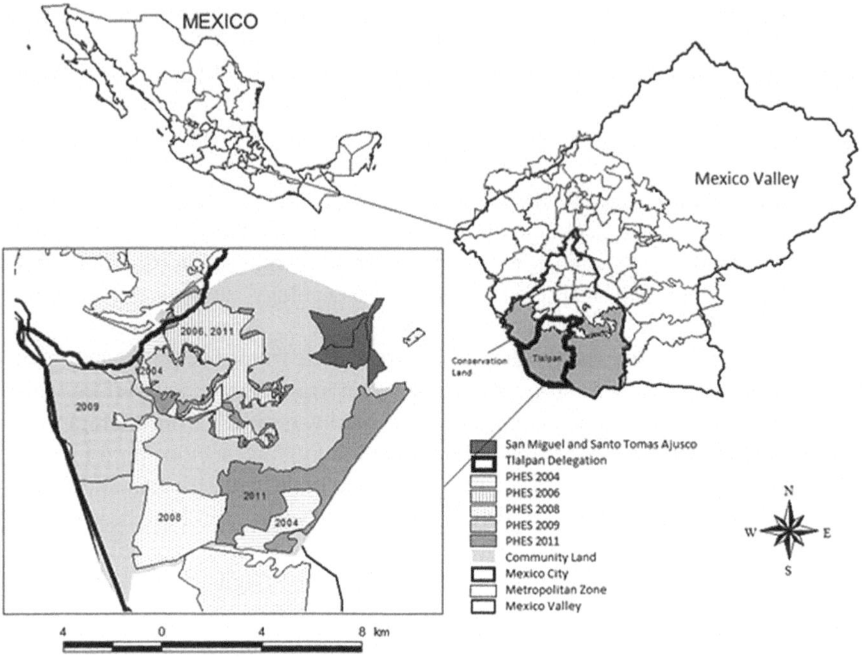

Fig. 1 Location of the community of San Miguel and Santo Tomas Ajusco (Source: CONAFOR 2015)

of 884 Km^2 (PAOT-SEDEMA 2012) and accounting for 12% of the national biodiversity (PAOT 2005).

According to the Department of Agrarian Affairs and Colonization (in Spanish: Departamento de Asuntos Agrarios y Colonización) (DOF 1975), the community divided into the San Miguel and Santo Tomás Ajusco villages, has 76.2 Km^2 and 604 landowners, called "comuneros." The community, in terms of administrative boundaries, belongs to the Tlalpan mayoralty, which in the last five decades has been strongly affected by urban expansion (Almaraz Vázquez 2014; Pérez-Campuzano et al. 2011, 2016). By their biological, cultural, political, social, and economic peculiarities, the community territory is an example of a complex social-ecological system, with stressors (with applied public policy instruments), internal and external processes, human–environment interactions (from ES flows and conservation activities), and outputs (as effects); with dynamics over time and space, at different scales.

The implementation of the federal PES program in the hydrological modality (PHES) started in the CL in 2003 with four agrarian nuclei (*ejidos* and communities[1]). Other territories, with important forest cover, have been added, incorporating

[1] Ejidos and communities refer to collective property lands, which have legal status and their own patrimony. They are the owners of the lands that have been endowed or of those that they have

Table 1 The PHES program in the community

Period	Incorporated area (Km^2)	Amounts granted for 5 years (in USD[a])
2004–2009	4.3	34,056
2006–2011	8.9	74,843
2008–2013	9.5	95,877
2009–2014	11.6	108,792
2011–2016	26.1	275,317

Source: Prepared with data from Sandoval and Gutiérrez (2012) and interviews with CONAFOR staff, 2016
[a]Current change of USD from Mexican peso by Bank of Mexico, 30th April 2019 (http://www.banxico.org.mx/)

by 2010 a total area of 390 Km^2, with an investment of about 2.4 million USD (Sandoval and Gutiérrez 2012). This shows a growing interest of the landowners for accessing to the PHES program. The San Miguel and Santo Tomas Ajusco (from here Ajusco) community joined the PHES program since the beginning, with the approval of the General Assembly (the maximum internal authority) and incorporating over the years more than 50 Km^2 as accumulative surface (Perevochtchikova and Rojo Negrete 2015) (Table 1, Fig. 1).

Furthermore, in 2012, the community participated with 2.2 Km^2 for a three-year period at the Concurrent Funds signed between CONAFOR and Associated Civil Engineers (ACE), with the highest economic transfer in the country (8447 USD per Km^2 per year), which was the first case in Mexico City. This scheme involved a mixed contribution from CONAFOR (that funded up to 50%) and another institution (in this case ACE) to increase the payment for ES conservation and to be more attractive to forest owners. It also allowed environmental monitoring (e.g., mushroom's observation proposed by the National Autonomous University of Mexico) without any other social or environmental commitment (CONAFOR 2012). It should be noted that this program only received two support installments and ended when the ICA declared bankruptcy (Perevochtchikova 2016).

On the other hand, due to the noncompliance of the technical requirements of another program granted by CONAFOR, the community was sanctioned, without being able to participate in the CONAFOR programs since 2013. This situation has only been solved by 2017, allowing the community to rejoin PHES mechanisms in 2018 (observations conducted in 2018). This entire process has demonstrated the importance of the PHES program, not only for the community, but also at the city level (and in general in the country); and, the need to evaluate the generated effects with the goal of improving its implementation.

acquired by any other title (Article 9, Agrarian Law). The difference between ejidos and communities is that the communities have a history of collective land possession since pre-Hispanic times; while the ejidos have been formed with agrarian distribution of the Mexican Revolution (https://archivos.juridicas.unam.mx/www/bjv/libros/2/913/11.pdf).

Methodology Proposal

In this chapter, we analyzed the effects of the PHES program in the Ajusco community in relation to the natural, social, and financial capital, using the framework of Community Capital (Brandon and Lombardi 2011). In this framework, the interactions between the three capitals are presented on multilevel and multimodal scales, with a proposal of several aspects that can be considered for spatial planning and socioeconomic and environmental studies. We started with the social capital as a central level of interaction, followed by the financial capital (its highest construction point), ending with a broad study of the natural capital, understood as the basis for other capital's interactions (Table 2). Each capital was studied with different techniques applied and diverse variables used (Table 3).

First, we performed an exhaustive literature review on ES and PES at the global and national levels (Perevochtchikova and Oggioni 2014), then we focused on the local scale analysis, as suggested by Poteete et al. (2012). This is because a case study allows visualizing the details of the causes and consequences of a complex social-ecological problem, also enabling the integration of multidimensional results. Consequently, the study was carried out by an interdisciplinary, interinstitutional, and international team.

Thus, methodologically, the work consisted of five stages of development, with preparative and integrative phases and capital analysis (Table 3):

1. Compilation and systematization of the available information, obtained from official sources, such as databases, technical reports, geographic information systems, as well as academic ones, like books, papers, and theses developed on the study area.
2. Social analysis based on the framework of Collective Action and interviews and surveys to community members and CONAFOR staff in central and regional offices.

Table 2 Capitals and their levels of interaction

Capital (first level)	Characteristic (second level)	Multimodal	Feature to consider
Natural capital	Quality	Physical	Physical environment (quality of the environment)
		Biological	Ecological protection and biodiversity
Social capital		Sensitive	Perception of people about the environment
	Development	Social	Social ambiance, social relationships, social cohesion
Financial capital		Economic	Economic efficiency
	Governance	Juridical	Rights and obligations (legal framework)

Source: Prepared by the authors based on Brandon and Lombardi (2011): 144

Table 3 Methodology used for the study of capitals

Capital and second level variables	Specific objectives	Applied theory	Method and techniques	References
Social capital (quality and development)	Understand how community organization works Determine the social and economic effects Contribute to the training of community technicians	Collective action	Interviews and surveys applied to key actors, *comuneros* (Assembly members and PES workers), Review of existing literature; Seminars	Poteete et al. (2012) Almaraz Vázquez (2014) Perevochtchikova and Rojo Negrete (2015)
Financial capital (development)	Economic-environmental value of the forest	Opportunity cost	Application of economic valuation techniques	De Groot et al. (2002) Martínez Jiménez (2015)
Natural capital (quality)	Understand surface and groundwater flows Measure the quantity and quality of water Generate a participatory monitoring scheme Deforestation forecasting	Theory of groundwater flows Integration of physical variables (climate, soil, topography, geology, water)	Field work: Obtaining of meteorological data, Water samples from springs and streams Tools for analysis of hydrochemical and isotopic data Application of Geographic Information Systems techniques (GIS)	Toth (2000) IG-UNAM (2005) Saavedra et al. (2011) Saavedra and Perevochtchikova (2017) Perevochtchikova et al. (2015, 2016) Zabala et al. (2017)

3. Economic analysis through an economic-environmental valuation based on the opportunity cost method.
4. Ecologic analysis based on measures of field data, analysis of data using different tools such as GIS, among others.
5. Integration of studied variables (results by capital).

We have established collaborative relationships with the community of San Miguel and Santo Tomas Ajusco (Ajusco) since 2012, formalized by signing an agreement, and finally consolidated over time through constant community support and exchange of information. These relationships have played an important role in the success of our research.

Methodology Operationalization by Type of Capital

Social analysis Social perceptions were analyzed comparing the vision of the main groups of actors involved in the implementation process of the PHES program in Mexico and community. The analysis was based on the application of surveys and interviews, which were designed considering the work of Ávila-Foucat et al. (2009), Pérez-Campuzano et al. (2011) and Perevochtchikova and Vázquez (2012), and Perevochtchikova and Rojo Negrete (2015). Four groups of criteria were determined: design and implementation, social effects, environmental effects, and economic effects. Actors were divided into two groups: (1) users of the SE, in this case, the federal government represented by CONAFOR; and (2) the beneficiaries of the payment, in this case, the community of Ajusco (Perevochtchikova 2016). Each group was then subdivided in subgroups according to their role in decision-making. The group of CONAFOR was divided in: (i) representatives of the highest strategic decision-making group, from the central office, and (ii) the State Management of the Federal District (in Spanish Gerencia Estatal del Distrito Federal now CDMX), as operational level. Ten interviews were applied to this group in 2013, and another ten in 2016.

In the group of *comuneros*, there were: (1) the General Assembly as the highest level of internal authority (with 131 surveys in 2012), and (2) the group of workers who directly participated in conservation activities established by the PHES program (with 108 surveys in 2013). In addition, community representatives were interviewed in the community members group (with 12 interviews between 2015 and 2016). With this data, a comparative analysis was conducted to identify coincidence and divergence points between the actors' views, in order to determine limitations and opportunities for PHES improvement (Perevochtchikova 2016).

Economic analysis In the last decades, the economic valuation has been a tool used to determine the value of different ES provided by multiple ecosystems, such as forests, wetlands (Sanjurjo 2001), coral reefs, and marine environments, among others (Maldonado and Cuervo Sánchez 2016); nevertheless, studies for peri-urban spaces are scarce (Martínez Jiménez 2015; Sylla et al. 2019). In existing literature, there are still few studies for Mexico City, for example Martínez-Jiménez et al. (2017) presented an economic valuation for the Conservation Land using the hedonic pricing method; Almeida-Leñero et al. (2007) referred to the importance of the Magdalena River Basin located west of the city; Arreguín-Sámano and Torres-Pérez (2012), provided a model of economic valuation of hydrological ES (HES) of the Magdalena Contreras district to analyze the willingness to pay for the service. For the economic valuation of the hydrological ES, Núñez et al. (2006) made an assessment of temperate forests in a region of Chile using the change in productivity method and Barrantes (2000) used the opportunity cost method for the Tempisque river basin in Costa Rica, among others (Martínez Jiménez 2015).

After reviewing different economic valuation methods for ES, we decided to apply the opportunity cost method. The latter included the sum of the values of water catchment (WC) and land restoration of areas (LR) with other land uses, based on the study of Barrantes and Castro (2002). According to this model, in order to assess the value of water catchment (as a determinant factor of forest water productivity), we need to know: the annual water volume captured and stored by the forest and the land opportunity cost of those areas, to assign a relative weight or value of the water production importance of the forest (Martínez Jiménez 2015).

First, the forest area within the community of Ajusco was calculated, using the land use and vegetation cover (LUV) information from the Atlas prepared by the Environmental and Territorial Ordinance Procurator's Office (In Spanish Procuraduría Ambiental y del Ordenamiento Territorial) (PAOT-SEDEMA 2012). Using the ArcGIS Software, the community polygon was established and the number of square kilometer for each land use category was calculated, namely: urban, agriculture, primary forest (native, without significant human interaction), disturbed forest (modified), induced forest, secondary vegetation (that grows after the modification of natural habitat), pasture, scree formation, and areas without vegetation. Within the category "primary forest," four forest types were considered: oak with 0.35 Km^2, sacred fir with 9.48 Km^2, pine with 12.1 Km^2, and 10.6 Km^2 of mixed forest. The total of primary, disturbed, and induced forest was 55.5 Km^2. On the other hand, pasture covers an area of just over 20% and agriculture by about 11% (Martínez Jiménez 2015).

Ecologic analysis This study combined the review of existing academic publications, official documents and databases about the community territory (CONAGUA 2011; PAOT-SEDEMA 2012; INEGI 2015; CONABIO 2015; CONAFOR 2015), with geographical information system (GIS) analysis (Saavedra and Perevochtchikova 2017) and field work for hydrological data compilation (Perevochtchikova 2016). It was subdivided into sections: general ecological characterization, water monitoring, and deforestation projection.

In general characterization, we reviewed the ecological conditions that determine the status of forest-hydrological ES (on which the PHES program is based), such as topography, climate and geological conditions. Therefore, these characteristics could be used as indicators for the evaluation of the environmental effects of the PHES program (Perevochtchikova et al. 2015). Furthermore, it is important to consider that land use change (LUC) reflects directly in the quality and quantity of water captured by the territory (Manson 2007; Jujnovsky et al. 2010). Because of this, it is necessary to project the LUC into the future, especially deforestation, given that the PHES program is applied only on the territories that have at least 50% forest cover.

For the water monitoring, we developed a field work during 2012–2016 period, measuring water quality and flow in springs and streams. The physical-chemical water characteristics were determined in situ considering basic parameters, such as temperature, hydrogen potential, alkalinity, hardness, turbidity, dissolved oxygen (from 2015 with La Motte technic and Global Water Watch certification). In the laboratory, we analyzed major and minor components, and isotopic analysis (Perevochtchikova et al. 2015, 2016). Also, we studied the groundwater flows dynamics and the recharge in the community's territory at the local and regional scales (Zabala et al. 2017).

For deforestation analysis, we considered the work of Saavedra and Perevochtchikova (2017), where the deforestation forecasting index was built with information from two previous studies carried out by PAOT (2010, 2014), commissioned by the government agency for environmental and territorial issues of Mexico City (In Spanish Procuraduría Ambiental y del Ordenamiento Territorial). The first study estimated the forest cover for the period 1986–2010 (PAOT 2010), using Landsat images[2] for the years 1986–2002 and SPOT images[3] for 2006–2010. It also included a forecast of the deforestation trend for 2020 and 2030, based on the deforestation rate for the historical period. The second study used supervised classification of images as the main technique, but only for the period 2006 to 2014 (PAOT 2014). After this, we crossed the deforestation information with the PHES recipient areas to determine the role of PHES in deforestation.

3 Results

Social Analysis

The actors involved in the implementation of the PHES feel that the program has generally had positive effects in the study area. Above all, the opinion of CONAFOR mentions many positive changes over the past 15 years concerning the operation of the program, including the establishment of consultation mechanisms, as well as an increase of the area incorporated in the program, the amount paid, and the number of sites supported.

Furthermore, since 2008, forest technicians have been involved in the development of the Management Practices Plan, and there have been workshops, training, etc. by technicians and CONAFOR staff. In addition, the CONAFOR has diversified PES schemes by creating Concurrent Founds and Local Mechanisms, among others.

The opinion within the *comuneros* group was divided between those directly involved in forest conservation initiatives and, consequently, more familiar with the program and its benefits, and the representatives of the Community General Assembly, who know little about the instrument and detect no tangible profit for them. The community also agreed to identify many positive effects, related to better organization and internal social cohesion for carrying out conservation activities, better environmental awareness, income benefit (albeit low and temporary) and improved forest conservation (the cleaning and maintenance of runoff and water

[2] The Spot satellites ("Satellites Pour l'Observation de la Terre") are a series of civil satellites for remote sensing (https://spot.cnes.fr/fr/SPOT/Fr/index.htm).

[3] Landsat series of satellites were put into orbit by the United States for the observation in high resolution of the Earth's surface (http://landsat7.usgs.gov//Landsat_Stories.php).

quality from springs and streams); even in the absence of links with other community sectors, such as young people, the elderly, or the so-called productive groups (Perevochtchikova 2016).

Nevertheless, the PES program has also negative effects on the community, such as the generation of internal and external conflicts due to the nontransparent distribution of economic resources, the fact of not controlling the development of extractive practices in the forest; marked preferences in the division of responsibilities and activities and the absence of unity for the development of forest activities between the population and beyond territorial boundaries with neighboring communities. One of the main internal threats has been the consequences of the change in community administration (which takes place every 3 years), which has led to a shift in priorities and in the lack of measures to control the extraction (wood, soil, water, and stone, etc.). This affects the execution of forest conservation activities, thereby contributing to its constant deterioration, and even leading to sanctions and enforcement by government agencies.

However, it is important to mention that, since the last change in community representation in 2016, the community's interest in seeking funding to support environmental conservation work has reappeared. This has even led to the creation of the "Community Round" Brigade in 2017, one of whose main tasks is participatory environmental monitoring based on pre-established criteria from the work of Rojo Negrete (2018).

Economic Analysis

The aim of this section was the economic-environmental valuation of the community forest based on the opportunity cost method and the calculation of the total forest value, as a sum of the water catchment (WC) and land restoration (LR) values. For this, we first determined the water catchment (WC) (Martínez Jiménez 2015). We used water infiltration data from PAOT-SEDEMA (2012) for the area, which gave an annual water infiltration value of 0.971 m^3 per year. Considering the territory with only the highest water infiltration rates (46.6 Km^2), the infiltration volume in the study area was estimated to 45.26×10^6 m^3. Furthermore, we estimated the economic value of the forest, considering water catchment as one of the most important ES.

The WC value includes the main economic activity competing with the forest (i.e., with its capacity to capture water), which for the case study was agriculture (maize cultivation). The 2012 survey showed that corn and oats are cultivated in the community; thus, we considered the average price per Km^2 per year (in rural areas), equal to 17,873 USD (SIAP—SAGARPA 2014). In this sense, the average yield of maize was 320 tons/Km^2. Thus, multiplying it by the price per ton of maize, the total amount of 9673 USD per year/Km^2 was obtained, which was then used to calculate the cost of forest catchment as 0.54 USD/m^3.

For the land restoration (LR) value we used the "Manual of practices for the protection, restoration and conservation of forest land" by the CONAFOR (DOF 2011; Martínez Jiménez 2015). We obtained, using the available information, an amount of USD 91,344 per Km^2 to carry out conservation activities in a temperate forest ecosystem. It was adjusted for an annual inflation rate of 4.05% (Chávez-Cortés and Mancilla-Hernández 2014). On the other hand, for the area to be recovered, we considered the Km^2 of agriculture, grassland, disturbed and induced forests, secondary vegetation, and no vegetation giving a total amount of 54.9 Km^2. In this way, the LR value needs for the first year an investment of 0.033 USD/m^3 and from the second to the tenth year an amount of 0.006 USD/m^3.

Therefore, the total forest value (TFV) for the first year was calculated as 0.061 USD/m^3 and 0.034 USD/m^3 for 2–10 years. Thus, the restoration cost in the first year is 77,362 USD per Km^2, and from the second year to the tenth, the maintenance costs correspond to 13,981 USD per Km^2. It is important to mention that the results turned out to be 26 times higher than the amount of 1900 USD per Km^2 (1478 USD in 2017) offered by CONAFOR for the PHES in this area.

In tune with these findings, a recent study by Rojo Negrete (2018) developed an analysis of the economic contribution of the PES program to the community and households. The author found the economic contribution index which seeks to determine the competitiveness of the annual income obtained from the PHES program to the community's families, showed a maximum 20% of total income. Other important economic activities were agriculture, livestock, trading, tourism and recreation activities, and, also, illegal activities (such as illegal logging, land and stone extraction). The index's result manifests: "...how the PHES income is non-competitive for the community, and therefore, unattractive and unlikely to generate a substantial economic effect" (Rojo Negrete 2018: 171).

Thus, summarizing the economic analysis, we observed that the opportunity cost of community land is very high given its location on the outskirts of the country's capital, as well as its natural characteristics. Therefore, the effects of implementing PHES in the context of urban pressure are limited, described as being low and temporary, related to a very low income for the activities developed under the program. This situation may have contributed to making land use changes unstoppable, losing forest cover, as we show in the next section.

Ecologic Analysis

General characterization The biophysical conditions of the community territory correspond to semi-cold and sub-humid climates Cb'(w2)(w), with a cool summer, an average annual rainfall of 1400–1600 mm, and a runoff coefficient higher than 55 mm, with rainy season from May to September and air temperature of 9–10 °C (PAOT-SEDEMA 2012). These characteristics can be explained by the geographical location of the site in the central part of the country and the topography of steep slopes on mountainous hillside higher than 3000 m above sea level.

In hydrological terms, the community is located in the XIII Hydrological Administrative Region of Mexico Valley Waters (In Spanish Aguas del Valle de México) and belongs to Hydrological Regions 18 Balsas River and 26 Panuco River (CONAGUA 2011). The hydrographic network consists of intermittent streams that use short paths due to high soil permeability. There are only two rivers (San Buenaventura and San Juan de Dios), with a plentiful flowing a few decades ago; and the Eslava River running through the Tlalpan and Magdalena Contreras districts (Perevochtchikova et al. 2015).

Estimated infiltration is between 10 and 15 mm/day and a runoff between 300 and 400 mm/year in half of the territory and 400–500 mm/year in the other (PAOT-SEDEMA 2012). Water erosion is 1200–5000 annual tons per Km^2 in 50% of the territory (considered "mild erosion") and less than 1200 annual tons per Km^2 in the rest, where the lowest erosion is observed. Wind erosion was estimated at less than 1200 ton/Km^2/year. This is linked to the restrictions of land use in the CL of Mexico City (PAOT-SEDEMA 2012), and the location of Protected Natural Areas, such as the Cumbres de Ajusco National Park and the Ecological Community Reserve (DOF 1936, 2011).

The low runoff coefficient (PAOT-SEDEMA 2012; INEGI 2015) evidences the relation to the high degree of water infiltration, associated with the pine-oak and fir forest cover, with low evapotranspiration values (CONABIO 2015). In this way, the area has a great potential for hydrological SE, mainly because of its contribution to groundwater recharge, given the topography, altitude, high percentage of forest cover and high values of precipitation and water infiltration (Perevochtchikova et al. 2015).

Water monitoring The physical-chemical water characteristics measured during field work in springs and rivers of the community were: pH 6.52 on average, water temperature between 9.7 and 18.5 °C, total dissolved solids between 20 and 80 mg/L, and the electrical conductivity between 40 and 260 μS. Analysis of physical and chemical characteristics indicates that groundwater streams are from local flow (Toth 2000) and comply with Mexican standards for drinking water (Zabala et al. 2017).

It is important to note that the location of the sampled sites does not correspond to the local dynamics that arise in response to conservation activities carried out in different places of the community in accordance with the requirements of the PHES Management Practices Plan, and necessary for participation in different governmental compensation programs, as PES (Perevochtchikova et al. 2015). This challenges the analysis of the management practices' influence required in this plan on the water conditions of each site. But for this, it is necessary to have long-term monitoring systems for the quantity and quality of water (which are absent throughout the country, except for participatory initiatives).

In this sense, during 2015–2017, participatory monitoring of water quality was carried out in the Ajusco community, under the training of Global Water Watch (Perevochtchikova et al. 2016). Results showed good water quality except for *Escherichia coli* bacteria that rise during the rainy season, which is directly related to the increasing flow rate and the organic matter dragging from the streams to water sources (Perevochtchikova 2016). Historical data on water quality was consistent with field obtained data. The evaluation of groundwater flows dynamics has further shown that the community's territory is an important water recharge area that has influence at the local and regional (CDMX) scales (Zabala et al. 2017).

Deforestation forecasting The available information from studies by Saavedra and Perevochtchikova (2017), PAOT (2010, 2014) and Perevochtchikova (2016) point out to a worrying trend of deforestation. Indeed, the projection from 1986 to 2030 presents a potential loss of 24.6 Km^2 of the forest, which represents one-third of the total area of the community (Fig. 2).

The PHES program, as shown in Fig. 2, has not been able to solve the problem of deforestation in the study area. Land use changes (LUC) toward urban and agri-

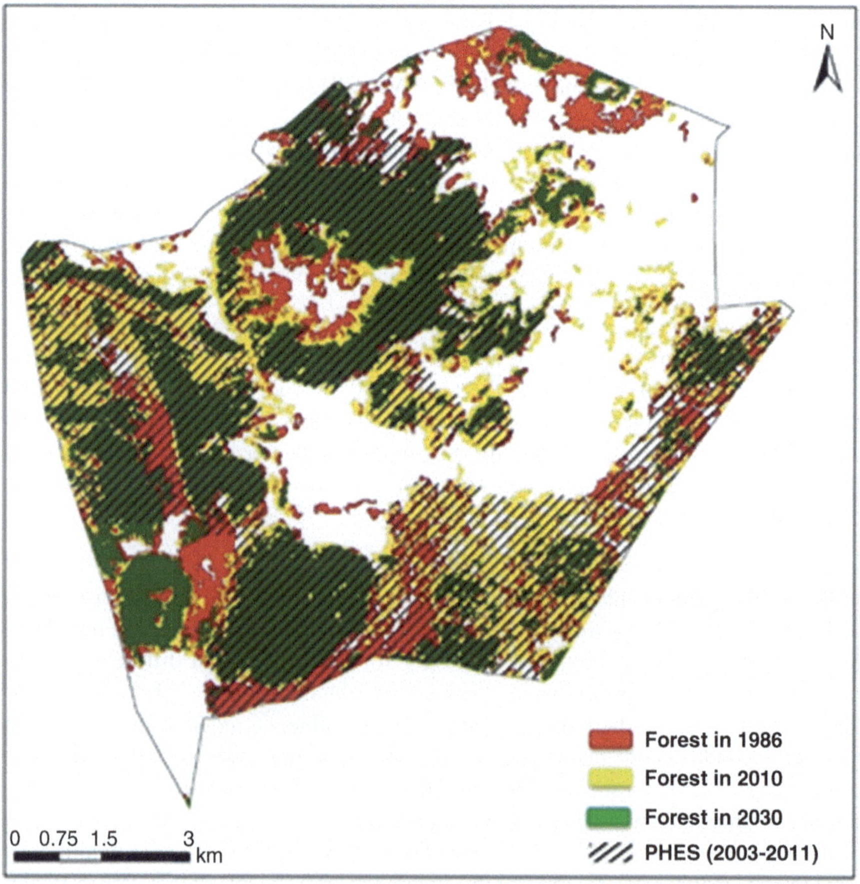

Fig. 2 Deforestation in the study area, with 1986–2030 projection and PHES program areas (Source: Prepared with data from PAOT 2010, 2014)

cultural uses continues, which is partly associated with the low income and temporary work generated by the program, and consequently with the search for better-paid jobs by the local population. In fact, deforestation projections show more LUC in areas receiving the PHES program than others, particularly in 2009 and 2011 (Saavedra and Perevochtchikova 2017), and in the southwest area of the community that has not received any type of forest conservation assistance and is located within the Community Ecological Reserve.

Rojo Negrete (2018), complementing this analysis, proposed other interesting ecological assessment indicators, which are related to the quality of habitat, soil, and functionality of the "soil-vegetation" complex. The first consists in the measurement of the population's ecological parameters, such as density, distribution, and excreta for three key species of the study area: the volcano rabbit also known as teporingo or zacatuche (*Romerolagus diazi*), wild cat or lynx (*Lynx rufus*), and mountain sparrow (*Xenospiza baileyi*). Data, recorded during the 2017–2018 fieldwork, showed that the zacatuche population has increased, the lynx is stable, but the sparrow has decreased. In terms of soil quality and the functionality of the "soil-vegetation" complex, the results of the soil quality indicator were negative, exceeding established optimal ranges, and the functionality indicator was positive (Rojo Negrete 2018).

Therefore, the Ajusco territory has a high potential for the provision of multiple ES, in addition to hydrological ES, at local and regional scales. Thus, the study area does not only present potentialities, but it is an entire ecosystem (the forest), which offers ES in a "package" at multiple levels. Consequently, the PES mechanism should be rethought in a sectoral (hydrological) modality. It is important to consider the dynamics of deforestation, supporting forest conservation initiatives, participatory environmental monitoring, and bottom-up community initiatives.

Result's Integration

Social Capital was referred from Community Capital framework in relation to the component of Quality with Sensitive variables, and Development with Social variables. In general terms, it was observed that perception of the key actors considered the different effects of the program on the community (social and economic) and on the environment (forest). Therefore, the perception of the environmental effects of the PHES implementation is generally positive. However, several local issues have been detected that emphasize the importance of the current community administration and its priorities for developing or not conservation actions.

As social effects, the perception analysis detected the improvement of cohesion in working groups, but with internal and external conflicts generated for several reasons (such as the distribution of resources and transparency). More positive effects have been identified among those directly involved in the implementation of PHES activities; where social cohesion is strengthened, and multiples benefits are detected (including temporal and low economic income).

Financial Capital was referred from the conceptual framework to the component of Development with Economic variables. The exercise of the economic-environmental valuation of the community's forest (based on opportunity cost analysis) estimated for the first year an investment required of 77,362 USD per Km^2, and for forest maintenance in the following years of 13,981 USD per Km^2. It is 26 times higher than the amount of actual PHES federal program (1.478 USD per Km^2 for 2017), representing a maximum of 20% of the total annual household incomes. As a result, the effects of PHES implementation in the context of urban pressure (and strong competition of land uses) are limited and described as being low and temporary, without meeting the basic needs of community members.

Natural Capital was referred from Community Capital to the components of Quality with Physical and Biological variables. On the physical level, at the regional and local scale, we detected, through the hydro-climatological characterization and the measurement of the water quality and quantity, that the community's territory offered an important potential for the provision of hydrological ecosystem services (supply and regulation). But it is necessary to establish constant monitoring of water quality and quantity, as well as other environmental characteristics (soil, flora, and fauna).

With regard to biological variables, serious problems of deforestation with projection on the future have been observed; with a dangerous dynamic of higher deforestation in the PHES zones. Thus, despite the fact that the area provides multiple and important multiscale SE, no impact of PHES has been detected.

4 Discussion and Conclusions

The results obtained on the effects of the implementation of the federal PHES program in a Mexican forest community show that, at the social level, the positive effects are mainly related to the social cohesion of the participants, the improvement of environmental awareness and the perception of income, albeit low and temporary. Similar results have been found in other studies (Kosoy et al. 2008; Rodríguez-Robayo et al. 2016) and other countries. For example, in Colombia, Moros et al. (2019) analyzed the effects of economic incentives on participants' behavior and motivations and showed that all types of PES, except for the crop-price premium payment, increased conservation behavior. As Hayes and Murtinho (2018) pointed out, it is also important to emphasize that to avoid internal conflicts, the role of communal governance mechanisms in promoting participatory and transparent decision-making processes is crucial.

Regarding the economic valuation, the payment of the program represents only 3% of the total value of the forest, based on the opportunity cost calculation. Furthermore, the direct income of the participants accounts for only a small part of the household's income. In Cambodia, two higher-paying market-linked PES programs had significant positive impacts, whereas a lower paying program that targeted biological diversity protection had no detectable effect on livelihoods

(Clements and Milner-Gulland 2015). Likewise, not all types of payments affected motivations equally (Moros et al. 2019), hence the importance of making programs attractive. Another problem is that the participation of young people or women is not encouraged, also pointed out by authors such as Corbera et al. (2007), Pascual et al. (2014) and Andeltová et al. (2019).

The ecological study shows great potential for the provisioning and regulating hydrological SE at the local and regional level, but also other ES. On the other hand, deforestation forecasting indicates that this worrying trend has not slowed down with the PHES program in the context of urban pressure. The problem of addressing how context influences an environment (driver changes) has been discussed in the publications of Krutilla et al. (1995) and Braud et al. (2013). Narain and Singh (2019) describe the impacts of urbanization on the water sources in a mountain context in India. In this sense, our findings are relevant to ecosystem services' studies, as it contributes to understanding a peri-urban case with its local implications.

For this reason, the PHES program should not be considered a panacea, but only as one of the public policy tools that should be accompanied by, and aligned with, other strategic actions, such as ecological and spatial planning, covering multiple administrative levels and geographic scales. Above all, given the growth of community commitment in these tasks, it could be used to make public policy instruments consistent with local initiatives for natural resources conservation and environmental control in this territory (Perevochtchikova 2016).

Finally, we encourage further studies on the processes and impacts generated by the PHES implementations, by developing participatory monitoring schemes to determine long-term integrated effects (Perevochtchikova et al. 2016). Specifically, a framework of indicators for a full assessment could be established, as pointed out by Rojo Negrete (2018). Different climate change scenarios, for example as a natural hazard for different types of ES (Muenzel and Martino 2018), and adaptation measures should also be considered (Perevochtchikova and Vázquez 2012; Huang et al. 2019). An example is a model provided by Scheiter et al. (2019) to analyze the interactions between vegetation, climate change, and economic aspects of land use in Savanna rangelands.

Other new areas of enquiry are: the analysis of collective action scenarios under various external stressors (Delgado et al. 2018), the influence of context on the success of conservation programs (Rodríguez-Robayo and Merino-Pérez 2017), the study of the process and satisfaction of the support by certified forest technicians and their performance in the communities (Martin-Ortega and Waylen 2018), in addition to training and actions developed within the power networks (Pérez-Campuzano et al. 2011), among others. In summary, there is still a lot to do.

Acknowledgements We want to express our acknowledgement to the Mexican National Council for Science and Technology (CONACyT), for the funding to the National Problems Project N° 246947 and the 290832 "TRASSE", ANR-CONACyT. We also thank to the community of San Miguel and Santo Tomas Ajusco for their constant support, to the CONAFOR for the interviews granted and to the entire team working on the projects and have participated in this process.

References

Almaraz Vázquez MN (2014) Servicios ambientales forestales y prácticas de aprovechamiento de Recursos de Uso Común en el Suelo de Conservación del Distrito Federal: Caso de estudio los B.C. San Miguel y Santo Tomás Ajusco. Tesis de licenciatura en Geografía, FFL-UNAM, México

Almeida-Leñero L, Nava M, Ramos A et al (2007) Servicios ecosistémicos en la cuenca del río Magdalena, Distrito Federal, México. Gaceta ecológica 84–85(número especial):53–64

Andeltová L, Catacutan DC, Wünscher T et al (2019) Gender aspects in action-and outcome-based payments for ecosystem services—a tree planting field trial in Kenya. Ecosyst Serv 35:13–22

Arreguín-Sámano M, Torres-Pérez J (2012) Modelo de valoración económica del servicio ecosistémico hídrico, delegación La Magdalena Contreras, Distrito Federal. Revista Internacional de Ciencias Sociales y Humanidades SOCIOTAM 12(2):9–23

Ávila-Foucat VS, Ramírez FR, Ortiz AM (2009) Indicadores para herramientas de conservación in situ. In: Pisanty I (ed) México; capacidades para la conservación y el uso sustentable de la biodiversidad. CONABIO, PNUD, México, pp 95–116

Barrantes G (2000) Metodología para la Valoración Económica del Recurso Hídrico como un Servicio Ambiental: caso de aplicación Cuenca del Río Tempisque. Instituto de Políticas para la Sostenibilidad (IPS), Costa Rica

Barrantes G, Castro E (2002) Experiencias replicables de valoración económica de bienes y servicios ambientales y establecimiento de sistemas de pago por servicios ambientales. Implementación de un esquema de cobro y pago por servicio ambiental hídrico: el caso de la empresa de servicios públicos de Heredia S.A. In: Barsev R (ed) Guía metodológica para la valoración de bienes, servicios e impactos ambientales. Un aporte para la gestión de ecosistemas y recursos naturales en el CBM. Serie Técnica 04, Corredor Biológico Mesoamericano-CCAD-PNUM/GED, Nicaragua, pp 108–112

Brandon PS, Lombardi P (2011) Evaluating sustainable development in the built environment. Wiley-Blackwell, Hoboken

Braud I, Breil P, Thollet F et al (2013) Evidence of the impact of urbanization on the hydrological regime of a medium-sized periurban catchment in France. J Hydrol 485:5–23

Brüschweiler S, Höggel U, Kläy A (2004) Los bosques y el agua: interrelaciones y su manejo. https://boris.unibe.ch/72600/3/bosques_y_agua.pdf. Accessed 11 June 2019

Chávez-Cortés MM, Mancilla-Hernández KE (2014) Esquema de cobro del servicio hidrológico que provee la cuenca alta del Pixquiac. Tecnología y Ciencias del Agua 5(5):161–177

Clements T, Milner-Gulland EJ (2015) Impact of payments for environmental services and protected areas on local livelihoods and forest conservation in northern Cambodia. Conserv Biol 29(1):78–87

CONABIO (Comisión Nacional para el Conocimiento y Uso de la Biodiversidad) (2015) Portal de geo-información. Sistema Nacional de Información sobre biodiversidad. http://www.conabio.gob.mx/informacion/gis. Accessed 11 June 2019

CONAFOR (2012) Se suman comunidades del Ajusco al esquema de Pago por Servicios Ambientales. CONAFOR, México. http://www.conafor.gob.mx:8080/documentos/docs/7/3238Iniciativa%20privada%20participa%20en%20%20PSA%20.pdf. Accessed 11 June 2019

CONAFOR (2015) Base de datos de los sitios receptores de Pago por Servicios Ambientales Hidrológicos en México, 2003-2015. CONAFOR, México

CONAGUA (Comisión Nacional del Agua) (2011) Estadísticas del agua en México, edición. CONAGUA, México

Corbera E, Brown K, Adger WN (2007) The equity and legitimacy of markets for ecosystem services. Dev Chang 38(4):587–613

Cordero D (2008) Esquemas de pagos por servicios ambientales para la conservación de cuencas hidrográficas en el Ecuador. Investigación Agraria: Sistemas y Recursos Forestales 17(1):54–66

Cortina SS, Saldaña EA (2014) Retos de la evaluación del Programa de Pago por Servicios Ambientales de la Comisión Nacional Forestal. In: Perevochtchikova M (ed) Pago por Servicios Ambientales en México. Un acercamiento para su estudio. COLMEX, México, pp 133–154

De Groot RS, Wilson MA, Boumans RMJ (2002) A typology for the classification, description and valuation of ecosystem functions, goods and services. Ecol Econ 41:393–408

Delgado LE, Tironi-Silva A, Marín VH (2018) Sistemas Socioecológicos y Servicios Ecosistémicos: Modelos Conceptuales para el Humedal del Río Cruces (Valdivia, Chile). In: Díaz Carrión I, Sedas Larrios EEJ et al (eds) Servicios Ecosistémicos en humedales. SEDEMA, VerAmbiente, México, pp 177–205

Departamento de Desarrollo Sostenible, Organización de Estados Americanos (DDS-OEA) (2008) Guía Conceptual y Metodológica para el Diseño de Esquemas de Pagos por Servicios Ambientales en Latino-América y el Caribe. DDS-OEA, Washington, DC

DOF (Diario Oficial de la Federación) (1936) Decreto que declara Parque Nacional "Cumbres del Ajusco", la porción de esa serranía que el mismo delimita. DOF 23/09/1936, México

DOF (1975) Resolución sobre reconocimiento y titulación de bienes comunales del poblado denominado San Miguel Ajusco, Delegación de Tlalpan, DF. GDF, DOF 03/01/1975, México

DOF (2011) Acuerdo por el cual se emiten los costos de referencia para reforestación o restauración y su mantenimiento para compensación ambiental por cambio de uso de suelo en terrenos forestales y la metodología de su estimación. GDF, DOF, 25/02/2011, México

Engel S, Pagiola S, Wunder S (2008) Designing payments for environmental services in theory and practice: an overview of the issues. Ecol Econ 65:663–674

Fischer B, Tuerner K, Morling P (2009) Defining and classifying ecosystem services for decision making. Ecol Econ 68:643–653

Fregoso A (2006) La oferta y el pago de los servicios ambientales hídricos: una comparación de diversos estudios. Gaceta Ecológica 78:29–46

Hayes T, Murtinho F (2018) Communal governance, equity and payment for ecosystem services. Land Use Policy 79:123–136

Huang L, Liao FH, Lohse KA et al (2019) Land conservation can mitigate freshwater ecosystem services degradation due to climate change in a semiarid catchment: the case of the Portneuf River catchment, Idaho, USA. Sci Total Environ 651:1796–1809

IG-UNAM (Instituto de Geografía, Universidad Nacional Autónoma de México) (2005) Definición de indicadores de impacto al recurso hídrico en zonas receptoras de pago por servicios ambientales hidrológicos 2003-2004. Instituto de Geografía, UNAM, México

INEGI (2015) Mapa Digital de México. Instituto Nacional de Estadística y Geografía. http://gaia.inegi.org.mx/mdm6/. Accessed 11 June 2019

Jujnovsky J, Almeida-Leñero L, Bojorge-García M et al (2010) Hydrologic ecosystem services: water quality and quantity in the Magdalena River, Mexico City. Hidrobiologica 20(2):113–126

Kosoy N, Corbera E, Brown K (2008) Participation in payments for ecosystem services: case studies from the Lacandon rainforest, Mexico. Geoforum 39:2073–2083

Krutilla K, Hyde WF, Barnes D (1995) Periurban deforestation in developing countries. For Ecol Manag 74(1–3):181–195

Maldonado JH, Cuervo Sánchez R (2016) Valoración económica del Parque Nacional Natural Corales de Profundidad. Investig Mar 45(1):99–121

Manson RH (2007) Efectos del uso del suelo sobre la provisión de servicios ambientales hidrológicos: monitoreo del impacto del PSAH. Informe final. INECOL, México

Martínez Harms MJ, Balvanera P (2012) Methods for mapping ecosystem service supply: a review. Int J Biodivers Sci Ecosyst Serv Manag 1:1–9

Martínez Jiménez ET (2015) Valoración económica del Suelo de Conservación en un contexto periurbano. Caso de estudio de la Delegación Tlalpan. Tesis de Maestría en Estudios Urbanos, CEDUA-COLMEX, México

Martínez-Alier J, Roca Jusment J (2001) Economía ecológica y política ambiental, 2da edición. Fondo de cultura económica de España, España

Martínez-Jiménez ET, Pérez-Campuzano E, Aguilar Ibarra A (2017) Hedonic pricing model for the economic valuation of Conservation Land in Mexico City. WIT Trans Ecol Environ 223:101–111
Martin-Ortega J, Waylen KA (2018) PES what a mess? An analysis of the position of environmental professionals in the conceptual debate on payments for ecosystem services. Ecol Econ 154:218–237
McElwee PD (2012) Payments for environmental services as neoliberal market-based forest conservation in Vietnam: panacea or problem? Geoforum 43:412–426
Moros L, Vélez MA, Corbera E (2019) Payments for ecosystem services and motivational crowding in Colombia's Amazon Piedmont. Ecol Econ 156:468–488
Muenzel D, Martino S (2018) Assessing the feasibility of carbon payments and Payments for Ecosystem Services to reduce livestock grazing pressure on saltmarshes. J Environ Manag 225:46–61
Narain V, Singh AK (2019) Replacement or displacement? Periurbanisation and changing water access in the Kumaon Himalaya, India. Land Use Policy 82:130–137
NRC (National Research Council) (2005) Valuing ecosystem services. Toward better environmental decision-making. NRC, National Academies, The National Academy Press, Washington DC
Núñez D, Nahuelhual L, Oyarzún C (2006) Forests and water: the value of native temperate forests in supplying water for human consumption. Ecol Econ 58:606–616
PAOT (Procuraduría Ambiental y del Ordenamiento Territorial) (2005) Elementos para una gestión adecuada del Suelo de Conservación del Distrito Federal. Documento de trabajo, PAOT, México. http://www.paot.org.mx/centro/paot/suelodeconservacion06.pdf. Accessed 11 June 2019
PAOT (2010) Modelo de análisis tendencial sobre la pérdida de cubierta forestal en el Suelo de Conservación del Distrito Federal. Primer Informe, México. http://centro.paot.org.mx/index.php/publicaciones-paot/16-estudios-tecnicos/37-modelo-de-analisis. Accessed 11 June 2019
PAOT (2014) Ampliación, consolidación e integración de los estudios, aplicaciones, iniciativas y recomendaciones realizadas por CentroGeo a PAOT. Informe final, México http://centro.paot.org.mx/index.php/publicaciones-paot/16-estudios-tecnicos/2271-recomendaciones_centrogeo-paot. Accessed 11 June 2019
PAOT-SEDEMA (2012) Atlas geográfico del Suelo de Conservación del Distrito Federal. PAOT, SEDEMA, México
Pascual U, Phelps J, Garmendia E et al (2014) Social equity matters in payments for ecosystem services. Bioscience 64(11):1027–1036
Perevochtchikova M (ed) (2014) Pago por Servicios Ambientales en México. Un acercamiento para su estudio. México, Ed. El Colegio de México, México
Perevochtchikova M (2016) Estudio de los efectos del Programa de Pago por Servicios Ambientales. Experiencia en Ajusco, México. Ed. El Colegio de México, México
Perevochtchikova M, Oggioni J (2014) Global and Mexican analytical review of the state of art on Ecosystem and Environmental services: a geographical approach. Investigaciones Geográficas 85:47–65
Perevochtchikova M, Rojo Negrete IA (2015) The perceptions about payment schemes for ecosystem services: study case of the San Miguel and Santo Tomás Ajusco community, Mexico. Ecosyst Serv 14:27–36
Perevochtchikova M, Vázquez A (2012) The federal program of Payment for Hydrological Environmental Services as an alternative instrument for Integrated Water Resources Management in Mexico City. Open Geogr J 5:35–46
Perevochtchikova M, Rojo Negrete IA, Martínez S et al (2015) Análisis hidro-climatológico para la evaluación de los efectos del programa de Pago por Servicios Ambientales Hidrológicos. Caso de estudio de la comunidad de San Miguel y Santo Tomás Ajusco, México. Revista Latinoamericana de Recursos Naturales 11(1):46–55

Perevochtchikova M, Aponte Hernández NO, Zamudio Santos V, Sandoval-Romero GE (2016) Monitoreo comunitario participativo de la calidad del agua: caso Ajusco, México. Tecnología y Ciencias del Agua 7(6):5–22

Pérez-Campuzano E, Perevochtchikova M, Ávila Foucat VS (eds) (2011) Suelo de Conservación del Distrito Federal: ¿hacia una gestión y manejo sustentable? Serie de Estudios Urbanos. Instituto Politécnico Nacional, Miguel Ángel Porrúa, México

Pérez-Campuzano E, Avila Foucat VS, Perevochtchikova M (2016) Environmental policies in the peri-urban area of Mexico City: the perceived effects of three environmental programs. Cities 50:129–136

Poteete AR, Janssen MA, Ostrom E (2012) Trabajar juntos. Acción colectiva, Bienes comunes y múltiples métodos en la práctica. UNAM, México

Rodríguez-Robayo K, Merino-Pérez L (2017) Contextualizing context in the analysis of payment for environmental services. Ecosyst Serv 23:259–267

Rodríguez-Robayo KJ, Perevochtchikova M, Ávila-Foucat S (2016) Payment for hydrological services programme effects in rural and peri-urban communities: comparison of two experiences in Mexico. Int Rev Soc Sci 4(5):146–160

Rojo Negrete IA (2018) Evaluación integral de efectos del Programa de Pago por Servicios Ambientales en el Suelo de Conservación del DF. Tesis de doctorado en Geografía, UNAM, México

Saavedra DZM, Perevochtchikova M (2017) Evaluación ambiental integrada de áreas inscritas en el programa federal de Pago por Servicios Ambientales Hidrológicos: caso de estudio Ajusco, México. Investigaciones Geográficas 93:76–94

Saavedra Z, Ojeda Revah L, López Barrera F (2011) Identification of threatened areas of environmental value in the Conservation Area of Mexico City, and setting priorities for their protection. Investigaciones Geográficas 74:19–34

Sanders J, Dendoncker N, Keune H (2013) Ecosystem services—global issues. Local practice. Elsevier, New York

Sandoval E, Gutiérrez J (2012) Servicios Ambientales, experiencia federal en el Distrito Federal. In: Pérez Campuzano E, Perevochtchikova M, Ávila-Foucat VS (eds) Hacia un manejo sustentable del Suelo de Conservación del Distrito Federal. IPN, MA Porrúa, México, pp 74–79

Sanjurjo E (2001) Valoración económica de servicios ambientales prestados por ecosistemas: Humedales en México. INE-SEMARNAT, México

Scheiter S, Schulte J, Pfeiffer M et al (2019) How does climate change influence the economic value of ecosystem services in savanna rangelands? Ecol Econ 157:342–356

Servicio de Información Agroalimentaria y Pesquera (SIAP-SAGARPA) (2014) Producción de maíz en México. Disponible en web: http://www.financierarural.gob.mx/informacionsectorrural/Panoramas/Panorama%20Ma%C3%ADz%20(may%202014).pdf. Accessed 8 Apr 2015.

Sylla M, Lasota T, Szewrański S (2019) Valuing environmental amenities in peri-urban areas: evidence from Poland. Sustainability 11(3):570

Toth J (2000) Las aguas subterráneas como agente geológico: causas, procesos y manifestaciones. Bol Geol Min 111(4):9–26

Wunder S (2005) Pagos por servicios ambientales: Principios básicos esenciales. Centro Internacional de Investigación Forestal (CIFOR). Occasional Paper 42(s):2

Zabala ME, Martínez S, Perevochtchikova M et al (2017) Hydrochemical assessment of Hydrological Environmental Services in the recharge area in the Southwest of Mexico City. Environ Earth Sci 76:113

Influence of the Rural/Urban Context in the Implementation of Forest Conservation Programs in Mexico: Two Case Studies from Oaxaca and Mexico City

Karla Juliana Rodríguez-Robayo, Maria Perevochtchikova, Veronique Sophie Ávila-Foucat, and Gabriela De la Mora-De la Mora

Abstract Payment for ecosystem services (PES) is one of the most commonly used economic instruments developed for biological conservation. Evidence has shown the importance of the local context in PES design and implementation, and the complexity of defining and specifying this context. We developed a proposal to narrow the local context, using the social-ecological system's framework, through two case studies: San Antonio, an indigenous community in Oaxaca, Mexico, and El Ajusco, a community in the periphery of Mexico City. This chapter discusses ten variables drawn from a series of interviews to approach the local context. Four of these variables stand out because of their incidence on the way PES outcomes are perceived, and on the local context definition: opportunity costs, confidence and cooperation, internal organization, and the presence and experience of NGOs.

Keywords Social-ecological systems · Latin America · Complexity · Mexico · Forest conservation · Local context

---

K. J. Rodríguez-Robayo (✉)
Consejo Nacional de Ciencia y Tecnología, Centro de Investigación en Ciencias de Información Geoespacial, Merida, Mexico
e-mail: krodriguez@centrogeo.edu.mx

M. Perevochtchikova
Centro de Estudios Demográficos, Urbanos y Ambientales, El Colegio de México A.C., Mexico, Mexico

V. S. Ávila-Foucat
Instituto de Investigaciones Económicas, Universidad Nacional Autónoma de México, Mexico, Mexico

G. De la Mora-De la Mora
Centro Regional de Investigaciones Multidisciplinarias, Universidad Nacional Autónoma de México, Cuernavaca, Mexico

L. E. Delgado, V. H. Marín (eds.), *Social-ecological Systems of Latin America: Complexities and Challenges*, https://doi.org/10.1007/978-3-030-28452-7_17

1 Introduction

Ecosystems, as providers of environmental services and the problems associated to their degradation, are issues globally acknowledged (Daily 1997; Metzger et al. 2006; Sterner 2008). These services can be defined, integrating the definitions of ecosystem (or environmental) services from MEA (2005), Costanza et al. (1997) and Daily (1997), as the processes from ecosystems, and the species that constitute them, that generate benefits to society, expressed as satisfactions of human needs.

According to MEA (2005), there are four categories of environmental services: supporting, regulating, provisioning, and cultural services. Recently IPBES has advanced a new analysis of ecosystem services, which acknowledges that many services may fit into more than one of the four categories (Pascual et al. 2017).

Current services' conditions show an unequal distribution of the benefits and costs of provisioning services. Usually, conservation imposes land use and natural resources restrictions to the entity holding the property rights; thus, conservation implies costs (mainly associated with the reduction of production in traditional activities such as agriculture and livestock raising) that must be assumed by the owner. On the other hand, benefits are enjoyed not only by the owner but by society as a whole. Still, there are different conservation initiatives to preserve and provide ecosystem services. Yet, correcting these inequalities, where environmental public policies play an important role in their design and implementation. The basic classification of conservation instruments divides them in two groups: direct approaches (command and control instruments) and indirect approaches (economic or market instruments), (CEPAL et al. 1998; Dietz and Vollebergh 2002; Hanley et al. 2007; Pérez et al. 2010).

Payment for ecosystem services (PES) is one of the most commonly used economic instruments of environmental public policy (Fisher et al. 2009; Balvanera et al. 2012; McElwee 2012). Sticking to a contract, with specific rules, PES is a payment that direct or indirect users make to ecosystem services providers (Wunder 2015). Scholars have noticed PES is often made to collectivities or communities rather than individual landowners (Kaczan et al. 2017; Alix-Garcia et al. 2018). The most common public policies related to PES are subsidies or payments based on the performance of the group's management practices (Matthew and Segerson 2019). Payments are also frequently used as an incentive to reduce deforestation by communities (Pagiola et al. 2002).

The benefits of PES in accomplishing conservation have been widely addressed. PES is recognized as a potentially efficient mechanism to attain ecosystems protection (Alix-García et al. 2004; Sierra and Russman 2006; Engel et al. 2008; Wendland et al. 2010). However, evidence also shows that PES has been designed and developed by academics or elites with little consultation with local communities (Mañez 2011), responding to the worldview of urban societies (Pascual and Corbera 2011; Gómez-Baggethun et al. 2010).

PES design and implementation must be rooted into a deep understanding of the local context of community beneficiaries (Frost and Bond 2008; Cranford and Mourato 2011; Lapeyre et al. 2015; Van Hecken et al. 2015; Rodríguez-Robayo et al. 2016). Nevertheless, there is hardly any clarity on how local context conceptualization could be addressed. Environmental, socioeconomic, and political contexts change over time, as well as incentive-based mechanisms. Furthermore, considering changes in context for PES design is a key issue to achieve effective performance of a given policy. They determine whether it is possible to maintain a high degree of cost-effectiveness, environmental effectiveness, and equity over time, keeping in mind the accumulated knowledge about these environmental public policy instruments (Jack et al. 2008).

Mexico has a 15-year experience in implementing PES schemes. The federal PES program started in 2003, having as its main objective to preserve forest cover in areas threatened by deforestation and changes in land use (Perevochtchikova and Ochoa 2012). Landowners (mostly collective entities, but also single individuals), having agreed to federal government payments for a 5-year period, contract the obligation to maintain forest lands and engage in land management actions (e.g. controlling pests, fighting fires or patrolling to avoid illegal activities). This federal program has two main types of payments: one for hydrological services conservation and another for ecosystem services associated with biodiversity conservation.

Initially, the program was located in priority regions to guarantee water recharge; but later it also aimed to address the complex problem of deforestation and poverty in these territories (Perevochtchikova and Ochoa 2012). Since 2008, the Mexican government has developed another type of PES scheme called "Matching funds" (or concurrent). The government, through the National Forestry Commission of Mexico (CONAFOR), may contribute up to 50% of the required funds, while the remaining 50% is covered by the consumers of the ecosystem services, local governments, or nongovernmental organizations.

Financial incentives for conservation may appear as a highly controversial issue. Still, Alix-Garcia et al. (2018), examining the impact that the Mexican PES federal program has had on land cover management and communal social capital, have shown that this program fostered an increase in the levels of community social capital, but no changes in household trust or participation. Thus, they conclude that it is relevant to further explore the social impacts of all types of participatory development programs, examining different contexts.

Currently, there are different frameworks for analyzing social and/or ecological systems. Binder et al. (2013) compared ten types (driver-pressure-state-impact-response, earth system analysis, ecosystem services, human–environment systems framework, material and energy flow analysis, management and transition framework, social-ecological system framework, sustainable livelihood approach, the natural step, and the vulnerability framework) and found that the social-ecological system (SES) framework is the most general, and that data collected within its structure can potentially be used in any of the other frameworks analyzed. They argue that SES provides a framework for selecting the variables necessary to describe the dynamics of social and ecological systems.

In this sense, McGinnis and Ostrom (2014), described SES in terms of eight categories (i.e., groups of variables). The first five are related to the context: the resource system, the resource units, governance, actors, and exogenous conditions and related ecosystems. Furthermore, the authors defined three additional categories: stressors, interactions, and outcomes. The presence of "stressors," which can include the implementation of PES schemes, modify the "interactions" between community members and their natural resources through (or because of) PES "outcomes," which can be divided into environmental, social, and economic outputs.

According to Rodríguez-Robayo and Merino-Pérez (2017), forest cover and hydrological services present in a community (the system and its resource units) determine the eligibility of an area for a PES program. The governance system and the performance of the involved actors are critical for the implementation and results of the PES. The way in which a PES scheme is implemented may affect the SES, generating new interactions among categories and new outcomes. For example, systems of resources and resource units can change, probably as a result of PES implementation, due to an increase of the forest mass or an improvement of water quantity or quality. Governance systems can change also because PES can have an influence on strengthening local organizations or the confidence between actors. Moreover, actors' attributes can be modified since a PES can foster changes in socioeconomic conditions.

Rodríguez-Robayo and Merino-Pérez (2017), defined "context variables" as the characteristics describing the local conditions in which a PES is implemented, with the purpose of unifying the diverse context approximations in PES analysis. The authors proposed nine "focal variables" as the most usual local context characteristics analyzed in PES drawn from (1) a PES literature review of 46 scientific articles containing, in the discussion and conclusion sections, the words: context, conditions, local factors or local characteristics, and payment for environmental services or payment for ecosystem services, (2) 30 online surveys of nongovernmental organizations (NGOs) with certified experience in the implementation of PES programs in Mexico, and (3) employees from The National Forestry Commission (CONAFOR) directly engaged in the implementation of the program. The focal variables defined by the authors were: (1) forest cover, (2) opportunity cost, (3) livelihood and productive diversification, (4) motivation and attitudes toward conservation, (5) confidence and cooperation, (6) traditional management practices, (7) internal organization, (8) land tenure and (9) rules for the management and use of natural resources.

The objective of this chapter was to analyze these nine context variables in two case studies in Mexico, corresponding to two forestry communities that have been implementing PES programs since 2004, reaching different results (Perevochtchikova 2018; Rodríguez-Robayo et al. 2019). We propose that the validation of context variables may contribute to the implementation and evaluation processes (and possibly to subsequent reformulations) of environmental public policy instruments.

2 Methods

Study Area: San Antonio and El Ajusco as Forest SES

This study was performed in two communities, San Antonio del Barrio (hereafter San Antonio), located in the municipality of San Felipe Usila (Fig. 1a); and San Miguel y Santo Tomás Ajusco (hereafter El Ajusco; Fig. 1b).

There have been previous analyzes of PES programs in the selected regions. Perevochtchikova (2016) and Saavedra and Perevochtchikova (2017) provide information about the effects of the PES program in El Ajusco. Denham (2017), Uscanga (2018) and Rodríguez-Robayo et al. (2019) analyze PES program outcomes in San Antonio and nearby communities in Oaxaca. Differences in PES results suggest the relevance of the local context in the analysis.

San Antonio and El Ajusco are indigenous and agrarian communities, and both are considered pre-Hispanic, with defined social and natural characteristics in its territories. Unlike El Ajusco, San Antonio has a low population density (7.2 inhabitants per square mile) and is located in a region of difficult access, isolating it from other communities.

The PES experience shows that both communities started at the same time the implementation of the "Federal PES" and "Matching funds" programs. However, El Ajusco has had difficulties with actors involved "Matching funds," so they went back to the "federal PES program." Furthermore, the way PES income is invested differs in both communities. In San Antonio, PES activities are carried out by community members as "tequio[1]" and PES income is distributed to households' heads (male and female). On the other hand, the money in El Ajusco is spent on wages and salaries (Table 1).

Comparative Analysis

There are different conceptual frameworks for PES analysis (Balvanera et al. 2012); one of them focuses on the perceptions of the actors involved regarding PES implementation and results. Previous studies have explored qualitative methods based on the analysis of social perceptions (Kosoy et al. 2008; Corbera et al. 2009; Perevochtchikova and Rojo Negrete 2015). In this chapter, we compare the local context and the outcomes of PES by means of interviews to key internal actors (i.e., community members with an incidence in PES participation) and external actors (PES intermediaries), who had closely followed conservation experiences of the communities.

The interview structure consisted of four sections: (a) general data, (b) history and community organization, (c) productive activities and sources of income, and (d) conservation of natural resources. We conducted 26 interviews (13 in each com-

[1] Collective work practice, it is a mandatory work without remuneration.

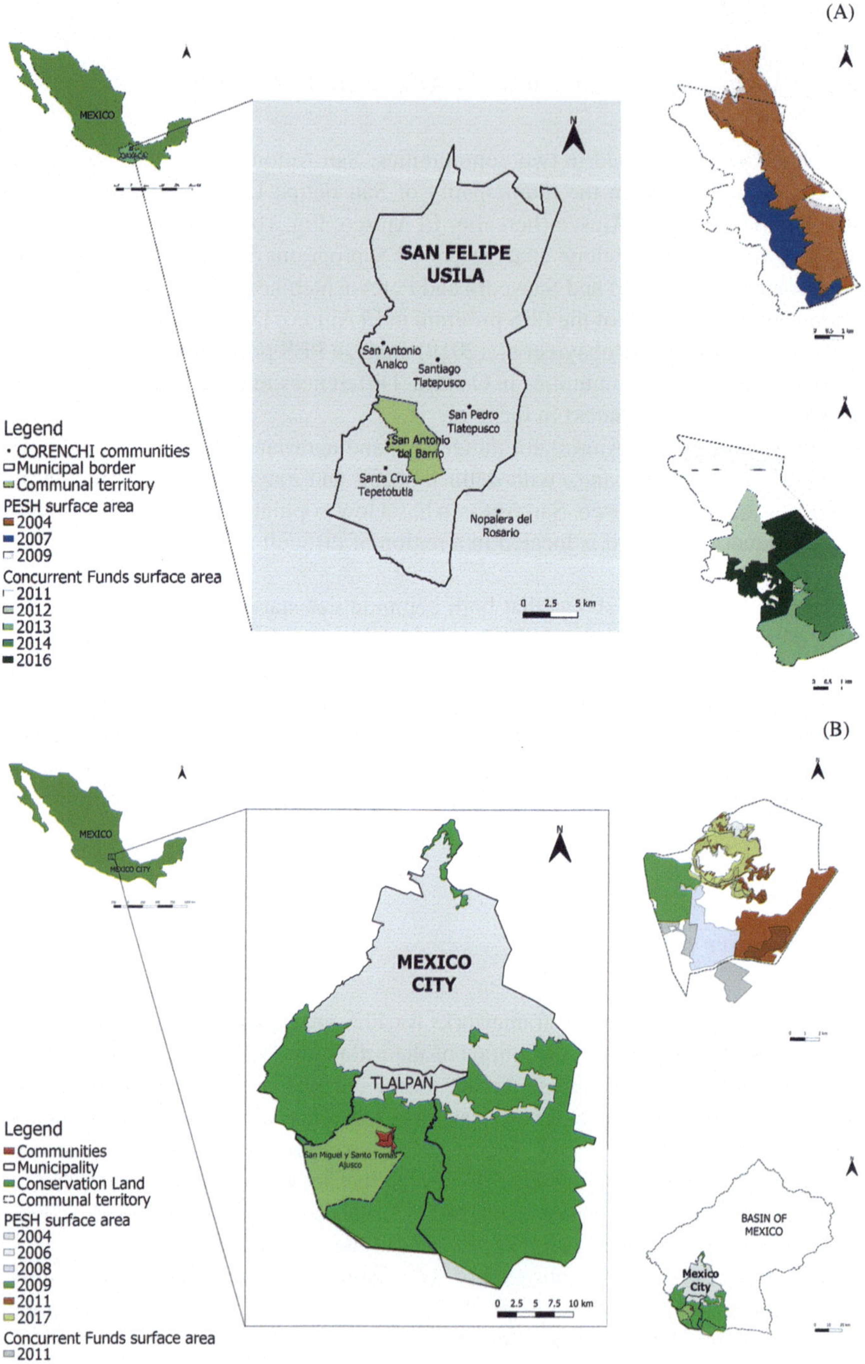

Fig. 1 (**a**) San Antonio Community location and PESH program zones, (**b**) El Ajusco Community location and PESH program zones

Table 1 SES characteristics by locality

SES part		San Antonio	El Ajusco
Exogenous conditions	Year of foundation	1817	1531
	Area (km^2)	25.7	76.2
System and units of resources	Forests	Perennial tropical forests, cloud forests and pine-oak forests	Pine-oak forests, oyamel forests
	Forest area (%)	90	59
Actors	Number of households	46	4476
	Inhabitants speaking indigenous language (%)	100	2
	Livelihoods	Multi-cultivation system for subsistence (maize, beans, and cassava), shade coffee for commercial exchange	Dominates the tertiary sector (61%)
	Average monthly household income (USD)	39.2	159.3
	Determinant historical events	The fall in coffee prices in the late 1980s	Forestry closed season in 1947
Governance	Govern system	"Uses and customs"[a]	"Uses and customs"[a]
	Land tenure	Communal land	Communal land
	Protected areas	More than half of this territory has been declared protected area by community agreement	There are two protected areas: A community ecological reserve and the National Park "Cumbres del Ajusco"
	Civil society organization in conservation	NGO. Geoconservacion	None
	Regional community organizations	Regional Committee of Natural Resources of Chinantla Alta—CORENCHI (six communities)	None
PES schemes (federal PES and matching funds) as stressors	Starting year	2004	2004
	Ending year	2016	In progress
	Actors involved	CONAFOR Geoconservación	CONAFOR
	Investment of PES income	– Distribution among household heads – Common fund to solve community expenses	– Payment of wages and salaries – Common fund to cover community expenses

Sources: San Antonio: Molina-González (2011), Bray et al. (2012), CONANP (2016), and data from fieldwork in 2016

El Ajusco: DOF 1975, Toscana (1998), SIDESO 2000, Chávez (2011), Sandoval and Gutiérrez (2012), Almaraz (2014), Martínez (2015), Perevochtchikova and Rojo Negrete (2015), Perevochtchikova (2016), and data from fieldwork in 2012, 2016

[a]Traditional system of governance, in which the general assembly is the highest political-administrative authority and in which "tequios" and the system of "cargos" (social choice positions with responsibility to the community) define the identity of the community

munity). Of the 15 internal actors interviewed, 11 were local authorities, three were local organizations' members, and one was a regional environmental organization member. As for the 11 external actors, eight were governmental actors, two were nongovernmental organizations' members, and one was an academy sector member. Interviews were conducted during August 2016 and January 2017. All interviews were recorded and later transcribed for systematization and analysis.

We created a matrix of answers with the information collected for the qualitative exploration of the following three topics: (a) description of the nine focal variables of the local context, (b) local perceptions of PES outcomes, and (c) relationships between the focal variables of the local context and the perceived PES outcomes.

The Nine Focal Variables of the Local Context

This chapter adopts the nine focal variables suggested by Rodríguez-Robayo and Merino-Pérez (2017) and described by Rodríguez-Robayo and Merino-Pérez (2018) to characterize the local context in the analysis of PES schemes.

According to their definitions, we establish a way to approximate to each variable:

1. Forest cover: The actor's perception of natural resource availability in the community and the perception of forest conservation state.
2. Opportunity cost: The forest use if PES schemes do not exist and household conservation costs.
3. Livelihood: Changes in economic activities throughout the history of the community and the main productive activities of families.
4. Pro-conservation attitudes: The interest in conservation activities and the history of conservation projects in the community.
5. Confidence and cooperation: The relationship between community members, local authorities, and external actors; the description of activities developed collectively, such as tequio[5].
6. Traditional management practices: Changes in the use of natural resources, and the description of the current use of natural resources.
7. Internal organization: Periodicity, attendance, and accountability in assemblies and the description of the system of cargos[6].
8. Land tenure: History of land tenure in the community and number of landowners in the community.
9. Rules for the management and use of natural resources: Description of the development of Community Statute and Land Use Planning and the most important rules and agreements in the management and use of natural resources.

Perception of PES Outcomes

We defined a group of variables to analyze PES outcomes in terms of the perception of SES providers regarding environmental, and social and economic PES outcomes (Table 2).

Table 2 Approach to PES outcomes

PES outcome variables	Approach
Environmental outcomes	– Perceptions of program contributions to forest protection
Economic outcomes	– Perception of program contributions to family income, community income – Perception of PES income distribution
Social outcomes	– Perception of program outcomes in social terms (strengthening of public and common goods, confidence, cooperation) – Perception of problems generated by the program in the community

Relationship Between Local Context Variables and PES Outcomes

The relationship of causality between the local context and the perceived outcomes of PES schemes was evaluated analyzing the causes of the perceived effects and the motivations of the community to participate for more than 10 years in the implementation of PES schemes.

3 Results

The Nine Focal Variables of the Local Context

The comparison of the nine focal variables showed that both communities recognize the ecosystem services provided by forests in their community. A high opportunity cost is perceived in both communities in relation to the time invested in conservation activities. Land tenure is collective and has been affected by various territorial conflicts with neighboring communities. Main differences between communities are summarized in Table 3.

Perceptions of PES Outcomes

Results showed similarities in the perception of the overall results of PES schemes in both communities, such as: (1) strengthening of environmental awareness in the community (social effect), and (2) general perception that the benefits of conservation are high for the society versus the low amounts of income afforded by PES schemes (economic effect). Table 4 summarizes the differences between San Antonio and El Ajusco.

Table 3 Focal variables differences between San Antonio and El Ajusco

Focal variable	San Antonio	El Ajusco
1. Forest cover	Well-conserved forest	Intervened forest, trend to deforestation
2. Opportunity cost of land use	Low. Traditionally, the forest has been preserved	High. There are strong pressures from illegal logging and urbanization
3. Livelihood	Subsistence agriculture, coffee and textile crafts	Tertiary sector with activities such as trade, food service, ecotourism, formal and informal employment
4. Pro-conservation attitudes	Conservation tradition, intrinsic value of natural resources	Conservation of natural resources as an alternative to generate income
5. Confidence and cooperation	Assembly, system of "cargos" and tequio very consolidated	Mistrust among community members. Existence of privileged groups
6. Traditional management practices	Strict compliance of established agreements	Practices depend on the commissariat in turn. Prohibited practices persist
7. Internal organization	Strength of the assembly and system of "cargos" The commissariat is elected in assemblies based on the merits of the different comuneros (holders of rights in an agrarian community)	Assemly and system of "cargos", Divisions among community members affect their participation in assemblies The election of the commissariat obeys an electoral process, in which each candidate presents his own proposals
8. Land tenure	Land tenure has been maintained as collective	It is collective, but parceling exert constant pressure on the community
9. Rules for the management and use of natural resources	Rules of strict compliance, defined in a communal statute and in a territorial ordinance communal	Statute it is not fulfilled. Absence of sanctions for not complying with established agreements

Table 4 PES outcomes differences between San Antonio and El Ajusco

PES outcomes	San Antonio	El Ajusco
Environmental effects	Conservation tradition is a key element in the environmental results obtained by the program	PES activities have favored forest conservation, but deforestation persist
Social effects	Strengthening of public and common goods There is no consensus on how to invest PES resources Restrictions on the use of natural resources (land area available for cultivation and types of crops)	Perceptions of unequal distribution of benefits, Social cohesion in the working group Perception of affectations to autonomy in the management and use of natural resources
Economic effects	Economic income for the whole community partially cushions coffee cultivation problems	Perception of inequitable distribution of benefits, low and temporal incomes

Relationship Between Local Context Variables and PES Schemes Outcomes

The answers to the question: "What do you consider are the causes of the results obtained by the PES schemes in the community?" were organized according to SES variables:

1. Opportunity cost variable showed low opportunity costs of conservation due to traditions in forest preservation, and high financial costs of deforestation in San Antonio; contrary to El Ajusco where people stated the high prices of the land and the high financial costs of environmental damage.
2. Livelihood variable evidenced the general perception to the need to generate more family income in San Antonio; thus, conservation activities such as PES programs could be a livelihood. Similar to El Ajusco, where people perceived that conservation activities generate income for their households.
3. Pro-conservation attitudes variable showed that from past generations San Antonio preserves its resources, and in El Ajusco the community recognizes that conservation is important and at the same time generate income.
4. Confidence and cooperation variables exposed the strong confidence, and strong collective work practices in San Antonio; contrary to the internal divisions that prevent the fulfillment of agreements in El Ajusco.
5. Internal organization, the interviews in San Antonio showed the relevance of the decision-making process where collective agreements are obeyed by all community members, and the strength of the community organization through the assembly. Contrary to El Ajusco, actors interviewed mention besides social cohesion weak internal organization and internal and external conflicts.
6. Land tenure, actors from San Antonio perceive that characteristics such as the history of land tenure in the community and the number of landowners in the community have favored the PES outcomes. In El Ajusco, no one mentioned characteristics related to this variable.
7. Rules for the management and use of natural resources. Strict sanctions described in Community Statute are argued as a cause of PES outcomes in San Antonio. In contrast to the variance in the interests and visions on conservation and production by local authorities in El Ajusco.

Finally, a new variable was stated as a cause of PES outcomes in San Antonio, the presence and experience with external actors, they emphasized the permanent accompaniment of the NGO GeoConservacion in community activities, not only PES.

Synthesis

The variables analyzed are organized in Fig. 2 according to their incidence in PES outcomes and the differences in the local context in both communities.

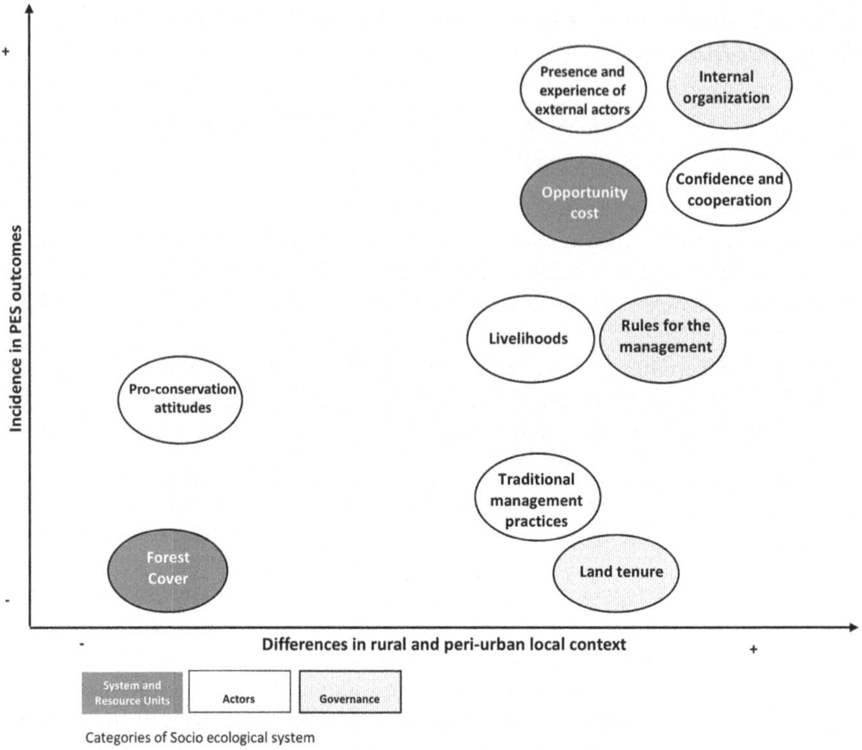

Fig. 2 Local context and causes for the perceived PES outcomes in rural and peri-urban areas

We evidenced three groups of variables. First, left lower quadrant, a group of two variables with a low incidence in PES outcomes and similar in both contexts: the forest cover and pro-conservation attitudes.

A second group, right lower quadrant, with four variables with a low incidence in PES outcomes and different in both contexts: livelihoods, traditional management practices, rules for the management and land tenure.

The third, and most important group, right lower quadrant, a set of four variables with most incidences in the PES perceived outcomes and very different in both local contexts: opportunity cost, confidence and cooperation, internal organization and the presence and experience of the actors. The discussion section emphasizes this group of variables.

4 Discussion

The main differences between both communities can be summarized in three elements, based on the characteristics of peri-urban areas (Ruiz and Delgado 2008; Pérez-Campuzano et al. 2016): (1) Opportunity costs. El Ajusco is affected by the pressure of urban expansion and land use change, which reflects a strong presence of inhabitants, plots and illegal land sales. (2) Productive economic activities highlight the relevance of the services sector (i.e., commerce and tourism) in El Ajusco. (3) Proximity to the metropolis has reconfigured traditional systems of governance in El Ajusco; although the community follows "uses and customs" system, disorganization, conflicts, and lack of strict compliance to rules are widespread.

According to Ortega (2010), El Ajusco settlements are characterized by being historical communities, with differentiated cultural identities. However, the proximity to the city has reconfigured its governance systems. In tune with studies carried out in various peri-urban areas (Abelairas-Etxebarria and Astorkiza 2012; Bicudo da Silva et al. 2017), we observed some evidences in El Ajusco of the presence of strong market distortions promoting land use change to residential and commercial uses, and changes in rural production systems and population dynamics.

On the other hand, San Antonio has the characteristics of the indigenous rural communities from the Mixe, Chinanteca, and Zapoteca ethnic groups. According to some studies (Toledo 1999; Sastre 2008; Carrasco and Barkin 2011), these communities have a successful management of the forest due to seven characteristics: (1) The presence of ecosystems and natural resources of regional and national importance, due to high biological diversity, endemism, water resources supply, (2) autonomy and defense of natural resources based on collective action management practices with a solid social organization based on the system of "uses and customs," (3) the assembly as a body that regulates access and use of natural resources through clear rules and sanctions, (4) organized community work, without monetary compensation, (5) definition of functions within the community structure, (6) the community as the owner of the land, and (7) efficient and transparent internal and external control.

There is a general perception, in both communities, that monetary PES benefits are insufficient and there are opinions highlighting negative social effects associated with divisions due to the distribution and investment of program resources. However, San Antonio reflects a better state of forests, and greater recognition of the strengthening of its public and common assets, as well as family and community incomes.

These evidences can be related to the fact that San Antonio shares characteristics of its local context that have been recognized as favorable in the successful implementation of PES schemes, such as (1) strong local organization (George et al. 2009; Bosselmann and Lund 2013; Huber-Stearns et al. 2013), (2) the presence of intrinsic motivations to conserve woodland (Lapeyre et al. 2015; Leimona et al. 2015; Page and Belloti 2015), (3) history in conservation (Bray et al. 2012; Denham 2017), and (4) management practices and use of natural resources that denote a close relationship with their natural resources (Hejnowicz et al. 2014; Kumar et al. 2014).

Finally, the comparison of the rural and peri-urban local context with the causes of the perceived results of the PES suggests that there is a relationship between these two sets of elements. This comparative study, following Jack et al. (2008), also show the relevance of considering specific contexts in policy design and implementation in order to achieve policy goals, because "no single policy is right for every scenario." Our evidence shows that a group of four variables of the local context has the most incidence on the PES perceived outcomes (Figs. 1 and 2).

The variable "opportunity costs" reflects the difference between rural and peri-urban and at the same time has a causal role on the perceived results of the program. Strong traditions in conservation and the high costs of deforestation have played a part in preserving forest cover in San Antonio, generating the perception that the schemes contribute to such conservation, whereas in El Ajusco the strong pressure to change land use triggers the constant perception that the results of the program are minimal. Variables "Trust and cooperation" and "internal organization" are evidenced as shared causes of the results obtained from the PES, showing high levels of confidence and organization in San Antonio, and disorganization and distrust among the actors in El Ajusco.

Finally, the new variable "presence and experience of the actors" was decisive in the results with key actors constantly mentioning it. In San Antonio, the constant presence (not only in PES) of a Civil Society Organization is relevant, as opposed to in El Ajusco, where this kind of organization has failed to accompany firmly, throughout the process, their participation in PES schemes and where the importance of the capacities and interests of the Commissariat is highlighted.

These findings allow us to conclude on the relevance of local context in the design, implementation and outcomes' perception of public policy instruments for environmental conservation. Although local context has an extensive conceptualization, it is possible to identify key elements for the analysis of these instruments through an initially small group of ten context variables, more particularly emphasizing four of these variables: opportunity costa, internal organization, presence and experience of external actors, and confidence and cooperation. These variables, we found have a strong influence on the radial difference effects generated by public policy programs in rural and peri-urban contexts.

Acknowledgments We wish to thank the interviewees (internal and external actors) for sharing their PES experience with us. We also wish to express our acknowledgments to the National Problems Project N° 246947 of the Mexican National Council for Science and Technology (CONACyT), and the 290832 ANR-CONACyT projects. Finally, we are grateful to Arturo Ramos for the elaboration of Fig. 1a, b.

References

Abelairas-Etxebarria P, Astorkiza I (2012) Farmland prices and land-use changes in periurban protected natural areas. Land Use Policy 29:674–683

Alix-García J, de Janvry A, Sadoulet E (2004) Payments for environmental services: to whom, where, and how much? In: Annual meeting, American agricultural economics association, Denver, 1–4 Aug

Alix-Garcia J, Sims K, Orozco-Olvera V et al (2018) Payments for environmental services supported social capital while increasing land management. PNAS 115(27):7016–7021

Almaraz M (2014) Servicios ambientales forestales y prácticas de aprovechamiento de recursos de uso común en el suelo de conservación del Distrito Federal: Caso de estudio los B.C. San Miguel y Santo Tomás Ajusco. Tesis de Licenciatura en Geografía, FFL-UNAM

Balvanera P, Uriarte M, Almeida-Leñero L et al (2012) Ecosystem services research in Latin America: the state of the art. Ecosyst Serv 2:56–70
Bicudo da Silva R, Delgado M, Aparecida S et al (2017) Perspectives for environmental conservation and ecosystem services on coupled rural-urban systems. Perspect Ecol Conserv 15(2):74–81
Binder C, Hinkel J, Bots P et al (2013) Comparison of frameworks for analyzing social-ecological systems. Ecol Soc 18(4):26
Bosselmann A, Lund J (2013) Do intermediary institutions promote inclusiveness in PES programs? The case of Costa Rica. Geoforum 49:50–60
Bray D, Duran E, Molina O (2012) Beyond harvests in the commons: multi-scale governance and turbulence in indigenous community conserved areas in Oaxaca, Mexico. Int J Commons 6(2):151–178
Carrasco M, Barkin D (2011) Concesiones forestales, exclusión y sustentabilidad. Lecciones desde las comunidades de la Sierra norte de Oaxaca. Desacatos 37:93–110
CEPAL, PNUMA, SEMARNAP (1998) Instrumentos económicos para la gestión ambiental en América Latina y el Caribe. SEMARNAP, México
Chávez C (2011) Identidad y luchas por las tierras en San Miguel y Santo Tomás Ajusco. Tesis de licenciatura en Arqueología, Escuela Nacional de Antropología e Historia. Instituto Nacional de Antropología e Historia
CONANP (2016) Áreas destinadas voluntariamente a la conservación. https://www.gob.mx/conanp/acciones-y-programas/areas-destinadas-voluntariamente-a-la-conservacion. Accessed 21 June 2018
Corbera E, González C, Brown K (2009) Institutional dimensions of payments for ecosystem services: an analysis of Mexico's carbon forestry programme. Ecol Econ 68:443–761
Costanza R, D'Arge R, De Groot R et al (1997) The value of the world's ecosystem services and natural capital. Nature 387:253–260
Cranford M, Mourato S (2011) Community conservation and a two-stage approach to payments for ecosystem services. Ecol Econ 71:89–98
Daily G (1997) Nature's services. Societal dependence on natural ecosystems. Island Press, Washington
Denham D (2017) Community forest owners evaluate a decade of payment for ecosystem services in the Mexican cloud forest: the importance of attention to indigenous sovereignty in conservation. Soc Nat Resour 30:1064–1079. https://doi.org/10.1080/08941920.2017.1295495
Dietz F, Vollebergh H (2002) Explaining instrument choice in environmental policies. In: Van den Bergh J (ed) Handbook of environmental and resource economics. Edward Elgar Publishing, Cheltenham
DOF (1975) Resolución sobre reconocimiento y titulación de bienes comunales del poblado denominado San Miguel Ajusco, Delegación de Tlalpan, DF, 03/01/1975. México
Engel S, Pagiola S, Wunder S (2008) Designing payments for environmental services in theory and practice: an overview of the issues. Ecol Econ 65:663–674
Fisher B, Turner R, Morling P (2009) Defining and classifying ecosystem services for decision making. Ecol Econ 68:643–653
Frost P, Bond I (2008) The CAMPFIRE programme in Zimbabwe: payments for wildlife services. Ecol Econ 65:776–787
George A, Pierret A, Boonsaner C et al (2009) Potential and limitations of payments for environmental services (PES) as a means to manage watershed services in mainland Southeast Asia. Int J Commons 3(1):16–40
Gómez-Baggethun E, de Groot R, Lomas P et al (2010) The history of ecosystem services in economic theory and practice: from early notions to markets and payment schemes. Ecol Econ 69:1209–1218
Hanley N, Shogren J, White B (2007) Environmental economics in theory and practice. Palgrave Macmillan, New York
Hejnowicz A, Raffaelli D, Rudd M et al (2014) Evaluating the outcomes of payments for ecosystem services programmes using a capital asset framework. Ecosyst Serv 9:83–97

Huber-Stearns H, Goldstein J, Duke E (2013) Intermediary roles and payments for ecosystem services: a typology and program feasibility application in Panama. Ecosyst Serv 6:104–116

Jack B, Kousky C, Sims R (2008) Designing payments for ecosystem services: lessons from previous experience with incentive-based mechanisms. PNAS 105(28):9465–9470. https://doi.org/10.1073/pnas.0705503104

Kaczan D, Pfaff A, Rodriguez L et al (2017) Increasing the impact of collective incentives in payments for ecosystem services. J Environ Econ Policy 86:48–67

Kosoy N, Corbera E, Broen K (2008) Participation in payments for ecosystem services: case studies from the Lacandon rainforest, Mexico. Geoforum 39:2073–2083

Kumar P, Kumar M, Garrett L (2014) Behavioral foundation of response policies for ecosystem management: what can we learn from payments for ecosystem services (PES). Ecosyst Serv 10:128–136

Lapeyre R, Pirard R, Leimona B (2015) Payments for environmental services in Indonesia: what if economic signals were lost in translation? Land Use Policy 46:283–291

Leimona B, van Noordwijk M, de Groot R et al (2015) Fairly efficient, efficiently fair: lessons from designing and testing payment schemes for ecosystem services in Asia. Ecosyst Serv 12:15–28

Mañez M (2011) A participatory framework for conservation payments. Land Use Policy 28:423–433

Martínez T (2015) Valoración económica del Suelo de Conservación en un contexto periurbano. Caso de estudio de la Delegación Tlalpan. Tesis de Maestría en Estudios Urbanos, CEDUA-COLMEX, México

Matthew J, Segerson K (2019) On the use of group performance and rights for environmental protection and resource management. PNAS 116(12):5285–5292. https://doi.org/10.1073/pnas.1802881115

McElwee P (2012) Payments for environmental services as neoliberal market-based forest conservation in Vietnam: panacea or problem? Geoforum 43:412–426

McGinnis M, Ostrom E (2014) Social–ecological system framework: initial changes and continuing challenges. Ecol Soc 19(2):30

MEA (2005) Ecosystem and human well-being: a framework for assessment. Island Press, Washington

Metzger M, Rounsevell M, Acosta-Michlik L et al (2006) The vulnerability of ecosystem services to land use change. Agric Ecosyst Environ 114(1):69–85

Molina-González O (2011) Análisis de las instituciones de gobernanza multi-escala para la acción colectiva de conservación de la naturaleza en la Chinantla, Oaxaca, México. Tesis de Maestría en Ciencias. Centro Interdisciplinario de Investigación para el Desarrollo Integral Regional Unidad Oaxaca. Instituto Politécnico Nacional, México

Ortega M (2010) Pueblos originarios, autoridades locales y autonomía al sur del Distrito Federal. Nueva Antropología 23(73):87–117

Page G, Belloti B (2015) Farmers value on-farm ecosystem services as important, but what are the impediments to participation in PES schemes? Sci Total Environ 515–516:12–19

Pagiola S, Bishop J, Landell-Mills N (2002) Selling forest environmental services: market-based mechanisms for conservation and development. Earthscan, London

Pascual U, Corbera E (2011) Pagos por servicios ambientales: perspectivas y experiencias innovadoras para la conservación de la naturaleza y el desarrollo rural. Revista Española de Estudios Agrosociales y Pesqueros 228:11–29

Pascual U, Balvanera P, Díaz S et al (2017) Valuing nature's contributions to people: the IPBES approach. Curr Opin Environ Sustain 26–27:7–16

Perevochtchikova M (2016) Estudio de los efectos del programa de Pago por Servicios Ambientales. Experiencia en Ajusco, México. El Colegio de México, México

Perevochtchikova M (coord) (2018) Pago por servicios ambientales desde el enfoque de sistemas socio-ecológicos: casos de estudio en Oaxaca y Ciudad de México. El Colegio de México, México (in edit process)

Perevochtchikova M, Ochoa M (2012) Avances y limitantes del programa de pago por servicios ambientales hidrológicos en México, 2003–2009. Revista Mexicana de Ciencias Forestales 3(10):89–112

Perevochtchikova M, Rojo Negrete I (2015) The actors' perception about payment schemes for ecosystem services. Study case of San Miguel and Santo Tomás Ajusco community, Mexico. Ecosyst Serv 14:27–36

Pérez R, Ávila S, Aguilar A (2010) Introducción a las economías de la naturaleza. UNAM, Instituto de investigaciones económicas, México

Pérez-Campuzano E, Avila-Foucat S, Perevochtchikova M (2016) Environmental policies in the peri-urban area of Mexico City: the perceived effects of three environmental programs. Cities 50:129–136

Rodríguez-Robayo K, Merino-Pérez L (2017) Contextualizing context in the analysis of payment for environmental services. Ecosyst Serv 23:259–267

Rodríguez-Robayo K, Merino-Pérez L (2018) Preserve and produce: experience in implementing payments for environmental services in two indigenous communities in the northern and southern ranges of Oaxaca, Mexico. J Sustain For 37(5):504–524

Rodríguez-Robayo K, Ávila-Foucat V, Maldonado J (2016) Indigenous communities' perception regarding payments for environmental services programme in Oaxaca Mexico. Ecosyst Serv 17:163–171

Rodríguez-Robayo K, Perevochtchikova M, Ávila-Foucat S et al (2019) Influence of local context variables on the outcomes of payments for ecosystem services. Evidence from San Antonio del barrio, Oaxaca, Mexico. Environ Dev Sustain. https://doi.org/10.1007/s10668-019-00321-8

Ruiz N, Delgado J (2008) Territorio y nuevas ruralidades: un recorrido teórico sobre las transformaciones de la relación campo–ciudad. Revista. EURE XXXIV(102):77–95

Saavedra Z, Perevochtchikova M (2017) Evaluación ambiental integrada de áreas inscritas en el programa federal de Pago por Servicios Ambientales Hidrológicos: caso de estudio Ajusco, México. Investigaciones Geográficas 93:76–94

Sandoval E, Gutiérrez J (2012) Servicios Ambientales, experiencia federal en el Distrito Federal. In: Pérez-Campuzano E, Perevochtchikova M, Ávila-Foucat S (coord) Hacia un manejo sustentable del suelo de conservación del Distrito Federal, IPN. M.A. Porrúa, México, pp 74–79

Sastre S (2008) Análisis de la gestión forestal comunitaria y sus implicaciones sociales en Ixtlán de Juarez. Tesis para optar el título de Ingeniero de montes en la Escuela técnica superior de ingenieros de montes, Universidad Politécnica de Madrid

SIDESO (2000) Siste de Información de Desarrollo Social. http://www.sideso.df.gob.mx/index.php?id=68. Accessed 12 June 2019

Sierra R, Russman E (2006) On the efficiency of environmental service payments: a forest conservation assessment in the Osa Peninsula, Costa Rica. Ecol Econ 59:131–141

Sterner T (2008) Instrumentos de política económica para el manejo del ambiente y los recursos naturales. CATIE, Turrialba

Toledo V (1999) El otro zapatismo: Luchas indígenas de inspiración ecológica en México. Ecología Política 18:11–22

Toscana A (1998) Análisis geomorfológico detallado del Volcán Ajusco y zonas adyacentes. Tesis de Geografía, Facultad de Filosofía y Letras, UNAM, México

Uscanga L (2018) De pago por servicios ambientales hidrológicos a fondos concurrentes: el análisis de la percepción social en la comunidad de San Antonio del Barrio, Oaxaca. Tesis de Maestría en Sostenibilidad, UNAM, México

Van Hecken G, Bastiaensen J, Windey C (2015) Towards a power-sensitive and socially-informed analysis of payments for ecosystem services (PES): addressing the gaps in the current debate. Ecol Econ 120:117–125

Wendland K, Honsák M, Portela R et al (2010) Targeting and implementing payments for ecosystem services: opportunities for bundling biodiversity conservation with carbon and water services in Madagascar. Ecol Econ 69:2093–2107

Wunder S (2015) Revising the concept of payments for environmental services. Ecol Econ 117:234–243

Exploring Traps in Forest and Marine Socio-Ecological Systems of Southern and Austral Chile

Laura Nahuelhual, Gonzalo Saavedra, Cristobal Jullian, María Amalia Mellado, and Felipe Benra

Abstract Traps in social-ecological systems depict situations where human actors and institutions interact with ecological dynamics and unintentionally steer development into vulnerable paths difficult to reverse. We use the social-ecological trap (SET) metaphor and path-dependence analysis to describe the emerge of trap situations in two contrasting cases: (1) Panguipulli municipality, representative of the significant land inequalities that dominate the rural landscape of southern Chile, and (2) southern king crab artisan fishery (*Lithodes santolla*) of the Magellan region, a semi-open access fishery of high economic value, where illegal extractions are a pressing problem. In Panguipulli, the system is caught in a "trilogy of inequalities" (land, forest, and ecosystem services) that together conform an inequality trap. Government policies surrounding land and forest tenure since the imposition of colonial rule and the modern State have interacted with other factors to concentrate economic power in large landowners, marginalize small peasants, and weaken customary management institutions. In the Magellan case, the trap could be erroneously confounded since there are no apparent human losers. As 3 years of interviews and participant observations reveal, the apparent absence of a trap rests on the confidence that "there are still resources for all" and that illegal fishing is not pressing the size of the stock.

L. Nahuelhual (✉)
Centro de Investigación en Dinámica de Ecosistemas Marinos de Altas Latitudes (IDEAL), Universidad Austral de Chile, Valdivia, Chile

Facultad de Ciencias Económicas y Administrativas, Instituto de Economía, Universidad Austral de Chile, Valdivia, Chile
e-mail: lauranahuel@uach.cl

G. Saavedra
Instituto de Estudios Antropológicos, Universidad Austral de Chile, Valdivia, Chile
e-mail: gonzalo.saavedra@uach.cl

C. Jullian · M. A. Mellado
Centro de Investigación en Dinámica de Ecosistemas Marinos de Altas Latitudes (IDEAL), Universidad Austral de Chile, Valdivia, Chile

F. Benra
UFZ—Helmholtz Centre for Environmental Research, Leipzig, Germany

L. E. Delgado, V. H. Marín (eds.), *Social-ecological Systems of Latin America: Complexities and Challenges*, https://doi.org/10.1007/978-3-030-28452-7_18

Keywords Social-ecological systems · Latin America · Complexity · Chile · Social-ecological traps · Forests · King crabs

1 Introduction

Over the last three decades, sustainability sciences have witnessed an ontological change in understanding human–nature relationships (Ingold and Palsson 2013). The social-ecological systems (SESs) approach, as part of these new conceptions, has challenged four prevalent features of past scientific research: (1) the notion of nature as the mere context in which social interactions take place, (2) the idea of human actions simply as external disturbances affecting ecosystems, (3) the narrow and reductionist view of the linkages between humans and nature, and (4) the traditional equilibrium-based models of disciplines such as economics and ecology (Schoon and Van der Leeuw 2015; Virapongse et al. 2016).

In this new setting, scholars have widely recognized that situations may arise in SESs where human actors and institutions interact with ecological dynamics and unknowingly steer development into vulnerable pathways (Tidball et al. 2016). In other cases, interactions reinforce the resilience of an already undesirable social-ecological state. These situations have been broadly conceptualized as "management syndromes."[1] Many of the world's most pressing problems represent management syndromes including overfishing, overgrazing of common property, overpopulation, and deforestation. Given their scope, implications, and interdisciplinary nature, management syndromes have motivated a vast development of literature in several disciplines, among which two schools of thought stand out. The first and oldest one focuses on "social dilemmas" or "social traps" (e.g., Platt 1973; Barry and Bateman 1996), which are described as situations in which a noncooperative course of action is attractive for each individual since it yields superior (often short-term) personal outcomes. Yet, if all actors pursue this noncooperative attitude, all end up (often in the longer-term) worse off than if they had cooperated (Capraro 2013; Van Lange et al. 2013).

The second school uses a system's approach to identify and solve or avoid what they call SESs traps (e.g., Lebel et al. 2011; Enfors 2013). SESs traps describe persistent, adverse situations that result from interactions among human actors, institutions, and ecological dynamics (Cinner 2011; Tidball et al. 2016), where feedbacks between social and ecological systems lead to undesirable states difficult to reverse (Steneck et al. 2011).

Comparing group with individual outcomes (positive or negative), five fundamentally different kinds of management syndromes arise, that are relevant to natural resources conservation (Cumming 2018): (1) give-some dilemmas, (2) take-some dilemmas, (3) SESs traps (lose-lose situations), (4) win-win situations (which can

[1] Management syndromes are collections of co-occurring actor and system behaviors that negatively impact natural resources and/or the communities that depend on them (Cumming 2018).

nonetheless lead to uncooperative or maladaptive behavior in the future), and (5) give or take some dilemmas, which share elements of (1) and (2) (McCarter et al. 2011). Give-some problems involve a higher cost or loss to the individual than to the group, while take-some problems involve a higher cost or loss to the group (see Cumming 2018 for a thorough revision of examples). SESs traps do not necessarily involve true dilemmas, in the form of conflicts between an individual or near-term outcome and collective or long-term consequences.

In practice, social dilemmas and SESs traps are difficult to disentangle. Furthermore, since most environmental problems are indeed social problems, some social traps (e.g., poverty trap, inequality trap) are an integral part of SESs traps. In this chapter, we explore dilemmas and traps, their drivers, reinforcing mechanisms and human responses, focusing on two contrasting SESs:

1. Panguipulli municipality, in Los Ríos Region; representative of the peasant dominated territories of southern Chile and of what Coomes et al. (2016) call "landscapes of social inequalities." In Panguipulli, small farmers have coexisted in the past with large operations dedicated to native forest logging and, more recently, large private protected areas and non-native tree plantation companies, in an institutional setting that has historically reinforced inequality and poverty (Benra and Nahuelhual 2019).
2. Chilean king crab (*Lithodes santolla*) artisan fishery, where formal laws have created access barriers to the entrance of new fishers, which added to increasing international demand and low law enforcement capacity, can at least partially explain the persistence of illegal fishing practices (Nahuelhual et al. 2018a).

A deeper understanding of the pathways in and out social dilemmas and traps is critical to inform the development, application, and adaptation of natural resource governance arrangements (Baker et al. 2018), particularly in developing countries where such syndromes may have the most detrimental consequences.

2 Methods

Case Studies

The municipality of Panguipulli (39°5′–40°5′S; 72°52′, and 71°59′W) is located in the Andes Range of Los Ríos Region (Fig. 1). It covers an area of 3292 km^2 of which less than 0.5% is classified as urban land. Total population reaches 33,273 people, of which 55.8% is rural. Forest degradation[2] and non-native tree plantation expansion are the main land use changes in the last two decades (Corporación Nacional Forestal, CONAF 1998, 2013).

[2] Forest degradation can be defined as the reduction of a forest's capacity to produce goods and services (ITTO 2002). Capacity includes maintenance of ecosystem structure and functions (ITTO 2005).

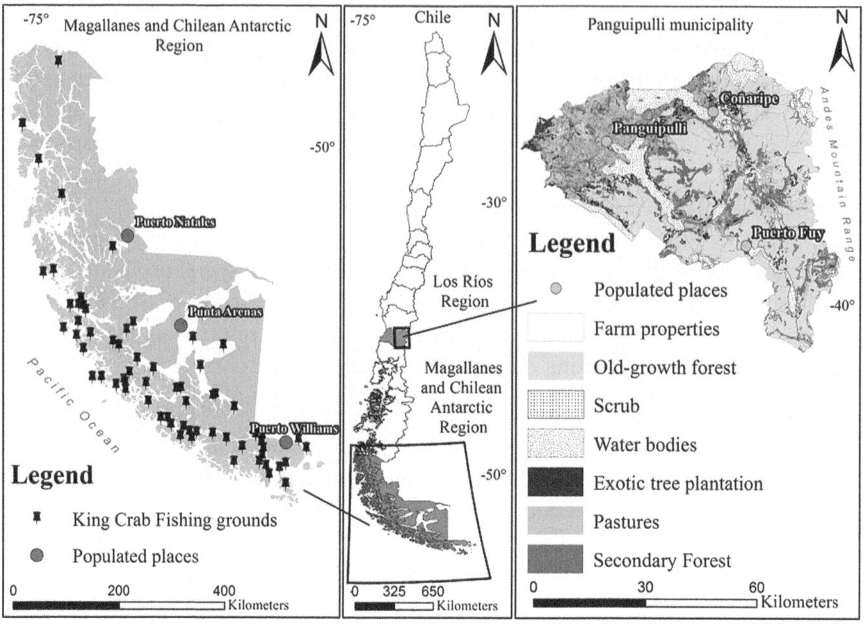

Fig. 1 Location of the study cases in the Magellan region (left) and Panguipulli municipality (right)

The Magallanes and Chilean Antarctic Region (54°5′–56° 9′S; 73°13′ and 66°2′W) (Magellan region hereafter) is the southernmost region of Chile and comprises a territory with more than 60,000 km of coastline along gulfs, channels, estuaries, and fjords (Fig. 1). The regional population comprises 165,593 residents (Instituto Nacional de Estadísticas, INE 2017).

The Chilean king crab fishery (centolla hereafter) is the second most important in terms of regional landings. According to Servicio Nacional de Pesca y Acuicultura (SERNAPESCA) (2018), 591 active small and midsize extractive and carrier vessels were authorized to extract centolla in 2018. We will focus on the largest operative centolla fleet which belongs to the Punta Arenas municipality (Fig. 1).

Dilemmas and SESs Trap Analysis

Social dilemmas and SESs traps have been described through four key features (see Haider et al. 2018 for details), which we used here:

Path Dependency

Path dependency refers to the observation that a system's dynamics depends strongly on its starting point. Path dependency can be characterized by a sequence of several distinct phases (Mahoney 2001): (1) antecedent conditions, which refer to factors in the past that determine the options available in later situations, (2) critical juncture, which is a circumstance in time where a particular option is chosen and where the system could have avoided falling into the trap or dilemma in years to come, (3) self-reinforcing mechanisms (e.g., institutionalization or habituation) triggered by the choice of an option, which causes (4) the structural persistence of the situation that in turn activates (5) reactive sequences, that consist of a series of reactions and counter-reactions of the actors to the structurally persistent situation.

Drivers

Dilemmas and traps are often triggered or maintained not only by endogenous processes (e.g., knowledge), but also by external drivers, including historical legacies, such as colonialism (Rudel et al. 2013).

Cross-Scale Interactions

Interactions can occur across actors or temporal and spatial scales (Costanza 1987; Barrett and Swallow 2006) and depend critically on the size of the actors' group and the structure of interactions (Cumming 2018). Larger and scattered groups might, for example, be more vulnerable to entrapment as long as their social capital is reduced by distance and segregation.

Diversity of Social-ecological Linkages

Diversity plays a role in adaptation and transformation out of undesirable states. Boonstra et al. (2016) propose a diversity of human responses that can explain the persistence or disappearance of the dilemma or trap, which include the following: (1) thick conformity, where actors have neither the ability nor the desire to change the situation, (2) thin conformity, where actors have the ability to change adverse situations but lack the desire to do so, (3) resignation, where actors have a desire to change the situation but lack the ability to do so, and accept it as such, (4) innovation, where actors have a desire to change the situation and the ability to do so and is the type of action that can lead to the dissolution of dilemmas and traps, and (5) rebellion, where actors have a desire to change the situation but lack the ability to do so, and do not accept that this is so.

Methodological Strategy and Data Sources

We combined methodological approaches to cover the past and present dynamics of the SESs under study, basically process tracing and ethnography. When path-dependent processes are present, an adequate explanation requires identifying key historical courses that set SESs on particular trajectories of development, even if these processes lie on the distant past (Mahoney 2001). Process tracing is a type of social science approach that historically links an outcome with the key events or processes that have produced it and searches for the observable implications of hypothetical causal processes within the framework of a single case study (Bennett 2008). The purpose is to establish a causal chain between an independent and dependent variable (George and Bennett 2005); that is, to highlight causal linkages between events in particular historical sequences.

In the context of a qualitative empirical research strategy, we also deployed an ethnographic record based on participant's observations, field notes, in-depth interviews (two from each site) and open interviews (one from each site), combined with the revision of secondary sources. A non-probabilistic sample criterion was used, based on a map of relevant actors and institutions for both cases (see Nahuelhual et al. 2018a, b and Benra and Nahuelhual 2019 for methodological details).

3 Results

A Trilogy of Inequalities in Panguipulli Municipality: Land Ownership, Forest Cover, and Ecosystem Services Distribution

An inequality trap portrays a circular causal relation between wealth, income, sociocultural capital, power, the institutions which contributes to the persistence of initial adverse conditions, and self-reinforcing components (Bourguignon et al. 2007). This concept underlines the lack of opportunities, which is characteristic of poverty traps (Azariadis and Stachurski 2004), but it is also accompanied by stagnation in society's income distribution structure.

The causal loop diagram in Fig. 2 shows how a diversity of endogenous and exogenous drivers and a set of SESs variables (e.g., forest cover, land tenure rights) have interacted to drive the system's path over the last 100 years. To facilitate the reading of the analysis, each interaction described in the text refers to one of the letters marked in Fig. 2.

Colonization policies in Chile advanced during the late 1800s and beginning of the 1900s by opening new grounds for settlers ("colonos"[3] in Spanish) and confining indigenous populations to reservations ("reducción indígena"[4] in Spanish).

[3]Colonos: European immigrants who settled in the rural colonies of Chile between 1883 and 1890 (Zavala 2008).

[4]Reducción indígena: indigenous reductions or reservations are lands assigned to the heads of

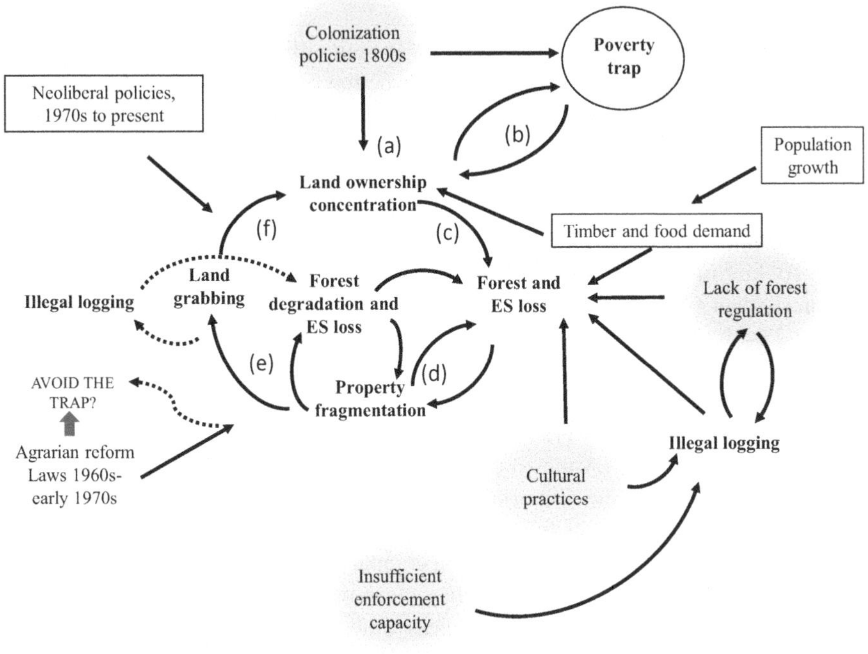

Fig. 2 Inequality trap in the forest-dominated rural landscape of Panguipulli, southern Chile. Rectangles correspond to external drivers and light grey circles to internal drivers. Remaining elements are outcome variables. *ES* ecosystem services

Thereafter, the process of land ownership concentration (acreage and value) set in motion, along with the concentration of native forest cover and economic and political power (a) (Clapp 1998; Almonacid 2009; Holmes 2015).

The concentration of land ownership and forest cover, preceded by poverty and vulnerability conditions of most marginalized landowners and rural dwellers (contained in the poverty trap) (b), in combination with an increasing demand for agricultural grounds for cattle raising and native timber (1930s on) and the lack of forest policies, led to deforestation, forest degradation, and increasing extractive pressure on the remaining native forest (c) (Barrena et al. 2016). The loss of assets and inheritance dynamics, in turn, led to property fragmentation (d). In later years, the system could have avoided the deepening of the trap through a land reform; instead, new forms of tenure (non-native tree plantations and large private conservation areas) (e) led to further concentration of land ownership, forest area, and ecosystem services (f), consolidating a trilogy of inequalities (Benra and Nahuelhual 2019).

Details of the entrapment process are provided in the following sections.

---

indigenous family groups in common and hereditary tenure through a document called "título de merced" (Schkolnik 1994).

Rural Poverty, Inequality, and Loss of Native Forests

Although there is no specific information for Panguipulli, it is well documented that by the end of the nineteenth century the destruction of the country's native forests and forest scarcity had reached wide recognition and was regarded as a public alarm (Camus 2006). Reactions from the private and public sectors were materialized in the first forest law (Law Decree N°656, the year 1925) aimed at regulating the indiscriminate destruction of native forests and promoting non-native timber plantations as a solution. It is estimated that by the beginning of the twentieth century, 130×10^3 km^2 of native forests had been destructed; that is, nearly half of the original forest cover (Albert 1911).

Along with deforestation, forest high-grading[5] became an important driver of native forest degradation and loss, affecting mainly those forests located in the central valley and accessible areas of the Andes and Coastal ranges of the southern regions of the country. High-grading was carried out without any management strategy or legal restrictions (Armesto et al. 1994, 2010). Forest cutters coming from neighboring areas became important players in forest degradation during this period, as well as the colonists and the illegal occupants of State forest reserves (Susaeta 1989; Miller 2006). Colonization companies, in the southern regions, concentrated large tracks of land where they cleared forests to expand pastures for livestock and installed sawmills to process timber (Miller 2006; Armesto et al. 2010). The following testimony reflects the role that these companies had on native forest loss and degradation in Panguipulli: *The entrepreneurs leased the forests, behind "closed doors," to exploit what they wanted. Between 1950 and 1955, they exploited all the raulí and coigue,*[6] *and now there is nothing left. They rented the forest, but they did not know how to control. The contracts were made with disadvantaged illiterate and ignorant indigenous people* (Member of Comunidad Manuel Curilef, Paillahuinte, 2015).

Speculators did the same, taking advantage of colonization laws and limited monitoring by State officials in the southern regions. A 1911-congressional report stated that as the railroad advanced to the provinces of the south, "abusive deforcement of public property and dispossession of indigenous property" took place. For large landowners on the frontier region, forest logging was a cheap and rapid way to make profits out of the land that required time to clear and cultivate (Miller 2006). The state railway was brought to Panguipulli in 1954, accelerating transportation of lumber and timber out of the municipality, to the rest of the country and the world, especially, Europe and the United States (Bize 2017).

[5] High-grading: Also called cutting limit diameter, occurs when all trees larger than a certain diameter are harvested, leaving only the lower quality individuals (Stewart and Dawson 2013).

[6] Raulí (*Nothofagus alpine* (Poepp. and Endl.) Oerst.) and coigue (*Nothofagus dombeyi* (Mirb.) Oerst.) are species of the Temperate Forest of Southern South America and the most heavily logged native species during the first half of the twentieth century in the study area.

Table 1 Properties across size range for the last five agricultural censuses (1955–2007[a])

Agricultural census (INE)[b]	Farms counted in Chile	Farm properties in Panguipulli	0–60 ha[c] (%)	60–1000 ha[c] (%)	>1000 ha[c] (%)
1955	151,000	757	374 (49.4)	340 (44.9)	43 (5.6)
1965	258,657	1900	1567 (82.5)	329 (17.3)	23 (1.2)
1976[d]	311,324	2207			
1997	312,302	3092	2747 (88.8)	313 (10.1)	32 (1)
2007	278,660	2796	2536 (90.7)	229 (8.1)	31 (1.2)

[a]Chile does not have a unifying farm size classification. For agrarian governmental agencies, small farms are those with less than 12 ha of basic irrigation (equivalent to a hectare of the most productive lands of the country). In turn, forestry governmental agencies consider small properties as those with a size up to 200 ha, which contain mostly forest cover (Law 20, 283 of 2008), whereas there are no clear size limits for medium and large farms

[b]INE 1955, 1965, 1976, 1997, and 2007

[c]Farm size categories were extracted from Benra and Nahuelhual (2019)

[d]Numbers correspond to the province level and there is no information at the municipality level

One can see in these dynamics, the late expression of internal colonialism (González 2003), where certain areas, prolific in natural resources and inhabited by traditional communities, particularly indigenous, function as enclaves of extractivism and exploitation of the local labor force to meet the demand of national and foreign markets. In Chile, this has been particularly evident in areas of mining and forestry aptitudes.

By the 1950s, the municipality's territory (and the country) was still divided in "latifundios"[7] administered by settlers and in "títulos de merced" managed by indigenous communities (Schkolnik 1994). The agricultural census of 1955, which covered 151,000 farms nationwide, reported that by that year Panguipulli comprised 757 properties (Table 1). Ten years later, the number of farms in the country increased by over 100 thousand, whereas in the municipality the amount more than doubled, along with a significant rise in small properties.

Official census data does not allow examining the evolution of forest cover since forests are aggregated with shrubland. As shrubland is the result of forest degradation, the real magnitude of forest loss during these years is impossible to estimate. Furthermore, there is no official record of forest loss and extraction prior to 1974, the year in which forest management plans[8] were officially implemented as a national policy. However, in Panguipulli, management plans started to be required for timber extraction only at the beginning of the 1990s. Thus, for almost 100 years, indiscriminate forest extraction proceeded without formal accountability.

[7] Latifundio: "latifundio" is described as the semi-subservient and low productivity work, which survives in the context of a "Developing State," as a support for the project of industrialization of the national economy and as a source of supply for the urban masses (Rosenblitt et al. 2001).

[8] Forest Management Plan: it is the instrument that, meeting the requirements established in the current regulations, guides the management of the ecological heritage or the sustainable use of the forest resources of a specific land, protecting the quality of the waters and avoiding the deterioration of the soils (Art. 2°, Law N° 20.283, 2008).

A Juncture Point: The Agrarian Reform

In the second half of the twentieth century, the problems derived from resources concentration led to a crisis, not only in Chile but throughout Latin America. In response, countries began to test methods and policies aimed at jointly improving land distribution and productivity (Garrido et al. 1988).

The Agrarian Reform in Chile (1967–1973) is probably the single most relevant event signaling a juncture point, where the system could have avoided the trap that persists until these days. The Agrarian Reform put an end to almost 400 years of productive and social structure that had hegemonized a significant portion of the Chilean territory (Bengoa 1990). In Panguipulli, the creation of the Panguipulli Logging Complex (Complejo Maderero Panguipulli) in 1971 is an iconic expression of these land reforms. The Complex managed an area of more than 4000 km^2 through a comanagement scheme in which workers and the State decided together on the production, commercialization, and use of forest resources (Corporación de Promoción y Defensa de los Derechos del Pueblo, CODEPU 1991; Rivas 2006).

However, the Agrarian Reform was abruptly interrupted after the military coup in September of 1973, which prompted a counter-reform process. This process, shaped by the dogmas of the neoliberal doctrine (Harvey 2005), was marked by new nuances regarding the ways in which farm improvement would be carried out. From then on, the focus of the agrarian and rural development would no longer be the redistribution of the land, but the modernization of local production systems, as well as the intensification of resources and capital use (Gómez and Echeñique 1991).

Mechanisms Reinforcing the Trap and New Pulses of Concentration

Several mechanisms reinforcing inequality were implemented during the Dictatorship (1973–1989) regime, framed in an economic-political design radically open to the market and to private investment. Two of the most emblematic were: (1) the consolidation of a non-native tree plantation model, supported by a series of economic incentives and franchises, being the Decree Law 701 of 1974, the most significant of all. This decree promoted the establishment of non-native tree plantations, which predominantly benefited large forestry companies (Lara and Veblen 1993) and prompted another pulse of large-scale deforestation and land concentration. While non-native tree plantation forestry has become the second most important extractive activity in the country, after mining, recent studies have shown that the geographic expansion of tree monocultures has contributed significantly to socio-environmental problems, questioning the sustainable forestry management that proponents of the industry claim it develops. Particularly, soil erosion and water scarcity in densely forested areas inhabited by rural Mapuche and non-Mapuche peasant communities (Bengoa 1999; Montalba and Carrasco 2005; Huber et al. 2010; Klubock 2014; Torres et al. 2015). (2) the establishment of a water market, which granted most water rights to large national and international agents (hydroelectric, sanitary, mining, and forestry) at the expense of rural inhabitants and indigenous communities (Nahuelhual et al. 2018b). Testimonies, such as the following, reflect the views of indigenous people regarding the privatization and concentration of water in Panguipulli: *And why are they going to give us the water? If we, Mapuche, were the first to arrive here, the Chileans and Spaniards arrived last. They took all the wood; now they want to take the water and leave the Mapuche in misery (Member Comunidad Inalafquen, Lago Neltume, 2016).*

During the last decades, deforestation decreased in Panguipulli, but degradation through firewood and sleepers' extraction without management plan became the main driver of forest loss (CONAF 1998, 2013), particularly in small and medium properties. Between 1998 and 2013, the area of adult native forest that was degraded in Panguipulli reached 198.8 km^2 (13% of the 1998 total forest cover), occurring at an annual rate of 1.5%. Furthermore, it is estimated that in Panguipulli near 67% of landowners, extract timber without a management plan, which is considered an illegal practice. Of this total, 48% declare to extract firewood that is commercialized within the municipality and in neighboring areas (Reyes et al. 2016). Yet, illegal timber logging remains unmeasured and ignored.

The high inequality in land ownership and forest cover distribution has unequivocally led to the concentration of biodiversity and ecosystem services in larger farms, which use resources less intensively than smaller farms (Table 2). In the study area, a single privately-owned protected area comprises 12.5% of the land, 15.9% of forests, 19.7% of timber stock, 12.4% of recreation opportunities, and 17.6% of water regulation. This property is located in the highland of the Andes range and preserves the majority of the remaining old growth forest of the municipality and region. Within its limits, it contains unique landscape attributes such as waterfalls, lakes, and a volcano (Benra and Nahuelhual 2019).

Table 2 Land ownership, forest cover, and ecosystem services concentration across farm size categories

Variable	Small ≤60 ha	Medium 61–1000 ha	Large >1000 ha
Number of farms in sample (80% of total farms)	2512 (88.7%)	289 (10.2%)	30 (1.1%)
Average farm size (ha)	12	185	1163
Accumulated land area (ha)	30,032 (9.2%)	53,318 (16.3%)	243,793 (74.5%)
Accumulated forest area (ha)	10,125 (5.7%)	24,484 (13.8%)	142,950 (80.5%)
Proportion of old growth forests	21.8%	44.1%	80.6%
Proportion of pastures	71.6%	52.7%	9.4%
Forage provision (tons of dry matter)	38.8%	50.1%	11.2%
Water regulation capacity (m^3)	9.2%	5.3%	75.5%
Recreation opportunities (number of persons)	7.6%	24.1%	68.3%

Source: Adapted from Benra and Nahuelhual (2019). Farm typology and ecosystem services indicators are explained in Nahuelhual et al. (2018b) and Benra and Nahuelhual (2019)

In this context of land and forest unequal distribution across farms, new conservation policies such as Payments for Ecosystem Services (PES) and REDD+ could endorse further inequalities. Recent studies show that in developing countries PES and REDD+ incentives are progressively being allocated to larger owners who can guarantee ecosystem services supply at lower costs (McDermott et al. 2013; Lansing 2014). This focalization in larger owners can lead to "green grabs" (Jiao et al. 2015; Tura 2018), involving the appropriation of land and the exclusion of local people from natural resources on the basis of "green" credentials (Fairhead et al. 2012) or the "ecosystem service curse" described by Kronenberg and Hubacek (2013) as the undesirable socioeconomic consequences of PES due to the unequal bargaining power of large landowners as compared to small tenants.

Human Responses and Policy Formulation

The actors in the inequality trap have always been the most disempowered and vulnerable of society: poor settlers, forest workers, small landowners, and indigenous communities. The powerful and advantaged actors have varied little over time, from latifundios to large forest companies and, more recently, large protected areas and real estate speculators. In the beginning of the entrapment processes, actors had probably the desire to change the situation but lacked the ability to do so and accept their fate. Instead, the 1960s with the Agrarian Reform and more recently the 1990s with peasant and indigenous movements marked eras of defiant responses, where actors manifest their desire to change their situation, and although they may lack the ability to do so, they do not accept their fate. The emergence of customary organizations such as the Coz Coz Mapuche Indigenous Parliament is a reflection of such responses (Nahuelhual et al. 2018b). They incarnate a fundamentally different ontology of nature, the "good living" (buen vivir en español) and forward a perspective of environmental governance that places the stewardship of the land and natural resources at the center of the discussion.

Whereas policies and market forces have been decisive in creating the entrapment, the cultural condition of local communities (peasant and indigenous) cannot be ignored. The extractive appropriation of nature precedes or does not derive entirely from the colonial and postcolonial capitalist expansion (Angosto-Fernández 2018). As it happens in other latitudes, extractivism is not only a matter of large corporations or Western agents; endogenous variables (e.g., habit and customs) may also promote local practices that are not sustainable in the medium and long term. Then, apart from explicit statements about the ecological virtues of local worldviews (Escobar 2008), and especially that of the Mapuche (Skewes et al. 2012), several investigations have demonstrated that in contexts of deep and systematic economic liberalization, the market logic end up permeating local ways of life, reformulating their contents and giving rise to multiple strategies for linking to capitalism (García Canclini 1990; Comaroff and Comaroff 2009).

Illegal Fishing of Chilean King Crab in the Magellan Region: A Win-Win Situation or a Trap in the Making?

Paradoxically, despite the large number of fisheries' regulations, Chile stands out for the high incidence of illegal fishing in several species of high commercial value (Servicio Nacional de Pesca y Acuicultura (SERNAPESCA) 2018). Centolla (king crab) is one of such species. The magnitude of stocks is unknown, and most observers believe that they are probably not sufficient to support greatly expanded commercial operations. At present, the fishery is under a semi-open access regime and it is managed through sex, size, and season measures, fishing gear regulations (only iron traps), and one access restriction, which is the Artisanal Fishing Registry (RPA hereafter) that grants vessels and fishers the authorization to extract centolla.

The artisan's fishing fleet is composed of: (1) ship-owners, in 2017 they reached 516 and handled 591 operating vessels; (2) transporters; in 2017 there were 151 records assigned to transporters (as part of the 591 operating vessels), and (3) fishers of the crew, who reached 3576 in 2017; most of them do not exclusively fish for king crab and, in a high proportion come from other regions of the country (Servicio Nacional de Pesca y Acuicultura (SERNAPESCA) 2018). At present, illegal fishing involves three main sets of practices that infringe all the established regulations (Nahuelhual et al. 2018a).

The Development of the Fishery: The Initial Role of Poverty and Migration

The development and consolidation of the fishing activity in the Magellan region were preceded by a series of natural, social, and economic underlying and interacting forces (Fig. 3). On one side, an already impoverished rural class caught in a poverty trap, the harsh impacts of the potato blight (*Phytophthora infestans*) in the 1950s (Urbina 1998) that collapsed Chiloé Island peasant agriculture, and the 1960s earthquake (the largest in Chilean and world history) and tsunami that devastated coastal areas in the central-south (Martinic 1999), prompted the migration of unemployed and impoverished people.

By 1970, 17,592 people had migrated from Chiloé, representing 19.67% of the Magellan region at that time (Muñoz and Zamora 1975; Martinic 1999). On the other hand, trade liberalization policies of the 1970s, together with an unparalleled boom in global fish demand, had significant effects on Chilean fisheries in the northern and central regions (Peña-Torres 1997; Jarvis and Wilen 2014). By the end of this decade, most pelagic fisheries were showing signs of collapse, and the interest started to focus on the southernmost fisheries. The following testimony reflects these collapses: *In 1975, in Calbuco (Los Lagos, X Region of Chile), the sierra disappeared, everything, everything disappeared, also the hake* (Migrant fisher, Punta Arenas, 2016).[9]

---

[9] Both the sierra (*Thyrsites atun (*Euphrasén, 1791)) and the Chilean hake (*Merlucius gayi gayi*

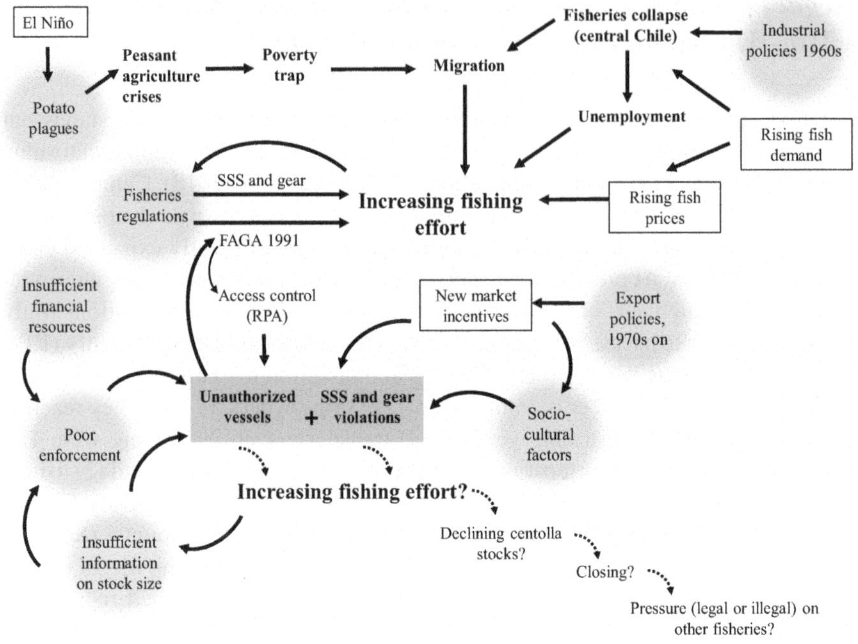

Fig. 3 Development of centolla fishery and the rise of illegal fishing. Rectangles indicate external drivers, light gray circles indicate internal drivers, and the remaining are outcome variables. The question marks (bottom) highlight the fact that these are potential rather than observed outcomes

Although the commercial centolla fishery began in 1928, the industry developed slowly until the 1950s. Most of the catch was initially canned and marketed domestically. During the 1960s, freezing plants were opened and the increasing demand and high prices of the Alaskan king crab (*Paralithodes camtschaticus* (Tilesius 1815)) served as a stimulus for its development (Instituto de Fomento Pesquero, IFOP 1966).

All previous factors, added to the open access regime of the fishery, led to increasing levels of fishing effort (Fig. 3). In 1961, a landing of 202 tons was recorded. By 1979, the landing had increased ten-fold, reaching 2254 tons (Fig. 4). The pick of the period was reached in 2012 with 5193 tons, after which the catches have systematically contracted. Although data on the number of operating vessels is incomplete and contradictory (Subsecretaría de Pesca y Acuicultura, SUBPESCA 2004), available information indicates that in 1965 there were 75 operating vessels.

(Guichenot, 1848)) are historical fisheries in Chile (with records of fishing prior to the 1950 s). While sierra has maintained its captures, destined to local consumption, the hake fishery grew to become one of the main exported species (the United States and Germany are main destinations). The hake fishery recorded its historical highs in 1968 with 127,800 tons; currently, the landings do not surpass 20,000 tons (75% decrease). In Calbuco city (the cove that the fisher mentions) in 1975, 99 tons of hake were caught, while in 2016 there were no landings of this species.

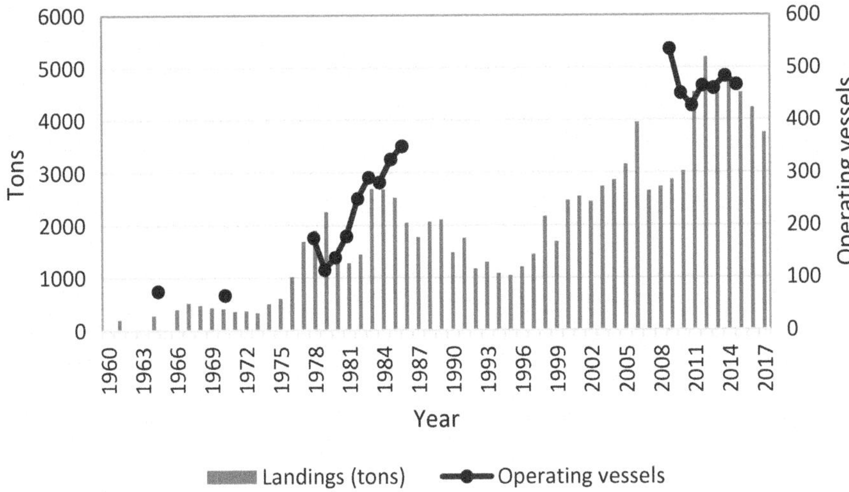

Fig. 4 Evolution of centolla landings and operating vessels since 1960, based on data from Servicio Nacional de Pesca y Acuicultura (SERNAPESCA 2018)

Between 1979 and 1987, vessels increased from 177 to 351. After more than 20 years without data, the number was recorded again at 536 in 2011, fluctuating between 450 and 470 vessels after that (Fig. 4).

At present centolla is destined almost entirely to foreign markets in Asia, whereas local supply is suspected of coming from illegal captures in its majority.

The Institutional Limits of the First Fisheries and Aquaculture General Act (FAGA)

During the latter half of the Chilean fishing boom, at the end of the 1980s, the country's fisheries began to show signs of overexploitation as the Tragedy of the Commons would predict (Bitrán 1989). The Chilean government, like others around the world, implemented regulatory measures for the industry and small-scale fisheries. Initially, these efforts were applied mainly through the use of top-down, command-and-control regulatory instruments (e.g., minimum landings sizes, closed seasons, and gear restrictions; Peña-Torres 1997). But the new Fisheries and Aquaculture General Act (henceforth FAGA[10]) ended open access to fisheries identified as "emerging," "fully exploited" or "recuperating." Among others, FAGA established that fishing authorizations should be made through the RPA for vessels with an assigned owner, thus creating access restrictions to all artisanal fisheries. The RPA forced that each artisanal fisher could only develop his/her activities in the

[10] Ley N°18.892, Ley General de Pesca y Acuicultura (Fishing and Aquaculture General Act, FAGA) texto refundido, coordinado, y sistematizado por el D. S 430 del 28 de septiembre de 1991 del Ministerio de Economía, Fomento y Reconstrucción.

region where he/she was registered. This slowed the traditional migration between regions to capture the resources that were better paid and more abundant (Ceballos and Ther 2011).

The RPA for centolla has been closed since 1999; therefore, the entry of new vessels is only exceptionally possible. Also, in that year the species was declared fully exploited.

The following testimony evidences the difficulty faced by fishers due to the closing of the RPA:

The minimum that a person can aspire, a household's head, is to give dignity to their home, education to their children and with that system (referring to the RPA) they (the government) tied our arms. Hence, who in their right mind can believe that there will be no illegal fishing? If the State is not giving us guarantees, the State is closing the doors to us, it is taking us to an abyss (Fisher' leader, Punta Arenas, 2017).

Thus, the general perceptions regarding governmental regulations are not merely the result of immediate personal interests or frustrations but are founded on a negotiated collective rationality. Therefore, it would perhaps be a reasonable hypothesis that the perceived moral obligation to comply with these regulations is weak or absent.

Over time, and paradoxically, illegal fishing rather than a social dilemma seems to depict a win-win situation with no losers, where illegal practices are justified on moral and solidarity basis. As the sector is mainly dominated by fishing families and networks of close friends, fishers without RPA are "helped out" by legal fishers, as one fisher stated: *It is over 10 years that the records [RPA] are closed and that super fisher,*[11] *who does it out of solidarity—because many times he does not even earn money—is getting all the blame today. But these super fishers are sometimes relatives who are helping those who are not registered and have to eat*" (*Fisher' leader, Port Natales, 2017*).

This could explain why they do not denounce, which is strengthened by kinship, a typical trait of artisanal fishing communities. No denunciation can also be explained by the fact that illegal and legal operations do not interfere with one another since king crab is perceived as still abundant and there is really no clear distinction of legal and illegal fishers (Nahuelhual et al. 2018a). As stated by one interviewee: '*there are legal, legal-illegal, and illegal fishers*' (*Intermediary, Punta Arenas, 2017*).

Access regulation (RPA closing) is ineffective as long as prices and demand remain high, and structural factors (e.g., the complex geography of the fishing grounds which complicates surveillance) prevent fishers from compliance (Fig. 3). As vessels without RPA continue to enter the fishery, fishing effort increases and

[11] Super fishers are authorized fishers (vessel-owners and carriers with RPA) landing unreported illegal capture as legal. In other words, they are the ones undertaking the whitewashing. The local label of "super fishers" refers to the fact that their catches exceed their landing capacity (vessel size, crew size, and fishing gear) (Nahuelhual et al. 2018a).

captures increase for an unspecified time. The whitewashing[12] of unreported captures helps sustain the export chain and earnings across the value chain increase, promoting further illegal entrance. The precariousness of the fishing activity and its informality are reinforcing factors.

At present, most fishers assert that illegal fishing is not affecting the size of the stock but, at the same time, the causes of declining captures (Fig. 4) are unknown and attributed to very disparate causes such as the decrease in illegal fishing or climate change, denying in most of the cases, the possibility of declining stocks. The lack of information and the scarcity of financial resources interact to prevent effective enforcement. In turn, in the absence of information, the State's only path is to consolidate a deterrence model for combating illegal fishing, deploying more resources and stringent regulations, which are very unlikely to solve the problem.

Soon enough, an apparently win-win, yet undesirable situation from the point of view of the SES as a whole, might turn into a dilemma or a trap in view of the sustained decreases in landings. In this state of affairs, it becomes highly complex to foresee the effects of management measures. For fishers, the most recurrently proposed solution to illegal fishing is the recognition of "historical rights," a question that implies from their point of view, opening the RPA to those currently marginalized. That is, to legalize the illegals, a solution that is considered unfeasible for State regulators since it contradicts the precautionary vision of FAGA and the RPA itself, as it could lead to an increase in fishing effort. Such a measure could set a dubious example for other regions where centolla is caught, with uncertain administrative outcomes. Another alluded solution is the establishment of a quota; yet, the design of such an instrument, in view of unknown stocks and the uncertain magnitude of illegal fishing, seems highly controversial.

4 Discussion

Successful conservation of natural resources and collective well-being depend as much on people working together as it does on sound science and good governance. Collaboration, and therefore, SESs sustainable outcomes, may fail when: (1) individual and group benefits are in conflict (social dilemmas) or (2) SESs become caught in problem-causing and problem-enhancing feedbacks (SESs traps; Cumming 2018). The two cases presented in this chapter depict situations that share these features. Panguipulli clearly portrays an inequality trap reinforced by past and present poverty conditions. Persistent cycles of resource dependence, overexploitation, acquisition, and concentration of land and unsustainable forest practices can be observed (e.g., high-grading), which are aggravated by illegal logging, extractivist

[12] Whitewashing: This transaction involves vessel-owners and intermediaries and it occurs in three different ways. The first involves the transfer at sea of crabs from vessels without RPA to fishing or carrier vessels with RPA which "whitewash" the undeclared catch and land it and deliver it to processing plants as legal (Nahuelhual et al. 2018a).

projects, and non-native tree plantations expansion. Government policies surrounding land and forest tenure since the imposition of colonial rule and the modern State have interacted with other factors to concentrate economic power in large landowners, marginalize small peasants, and weaken customary management institutions, as it has been reported in other developing countries (Kasanga and Kotey 2001; Kuusaana and Gerber 2015). Contrarily to situations where weak property rights are pointed as main factors of management syndromes (Weingart and Kirk 2011), in the southern landscapes of Chile, secure private ownership has been functional to the consolidation of the inequality trap.

In the Magellan case, the situation could be erroneously classified as "no dilemma" or a "win-win" situation (Cumming 2018) since no (human) losers are really identified. However, several situations could trigger a dilemma or trap, such as the certainty that the decreases in landings are due to decreasing stocks (Fig. 4). As 3 years of interviews and participant observations revealed, the apparent absence of dilemma rests on a certain confidence that "there is still resources for all" and that illegal fishing is not pressing the size of the stock, as most fishers interviewed asserted.

Although the two chosen SESs differ in many respects, they also share some commonalities. For example, in both cases juncture points and drivers are closely associated with formal institutional frameworks. In the case of Panguipulli, there were a series of laws linked with the possession of land and forests which consolidated inequalities over time. For this reason, it is a legal initiative contrary to those which could have changed the SESs trajectory away from inequality (i.e., the Agrarian Reform). In the case of the king crab fishery, formal laws (e.g., FAGA; RPA) have created rigidities in the system, which are difficult for artisan fishers to deal with. These laws have ignored social and cultural factors associated with artisanal fishing, such as their traditional extractive logic. However, it is risky and largely unfeasible to think about removing the restrictions currently in operation and, therefore, new forms of governance must be explored.

Among common drivers in the dynamics of both systems is the figure of the market. As early as the beginning of the twentieth century, the extraction of natural resources to satisfy distant markets became an axis of the country's "progress." In a more continental view, the capitalist dispossession of nature has become one of the distinctive—and tragic—signs of our time. Latin America is one of the most biodiverse regions of the planet and, not coincidentally, is one of the main grounds for privatization and commodification of natural assets by transnationals and States. However, there are numerous social resistances that emerge to defend the territories and propose forms of production and consumption that respect the living processes and the self-determination of peoples (Composto 2012). This is especially evident in the case of Panguipulli and the customary organizations that, among other, vindicate the rights of indigenous people to those natural resources from which they were deprived.

In terms of outcomes, the most relevant observation is the emergence and increase of illegal extraction practices (timber and fish). The *illegality* arises as a dictated condition, not a feature of the systems or intended behavior of the people; it is a consequence of regulations consigned in the statutory bodies established by the State. Thus, illegal practices are framed in a relational context where the interests of the State and the social base (e.g., surreptitiously of the market) intersect and conflict. In this context, it could be admitted that illegal practices constitute, on the one hand, subsistence strategies in the face of adverse scenarios (Panguipulli), and, on the other hand, material prosperity strategies based on the increase of financial benefits associated with an opportunity (Magellan region). Although poverty could be considered as a transversal conditioning factor—visible in early migration from Chiloé to Magallanes—it is only a decisive factor in the first case study.

From a cultural point of view and if culture is understood as the set of ideas that are expressed as representations of the world, conditioning its material appropriation, the extractive logic is also likely to be rooted in culture. Perhaps it can be argued that under adverse conditions (Panguipulli) or under perverse market incentives (Magellan region) those habits or cultural drivers toward extractivism become exacerbated. And at the same time, as we could see especially in the case of Panguipulli, culture admits a less pessimistic reading: customs, habits, traditions, local worldviews, social capital, among others, can act as factors of resistance and of bio-socio-cultural resilience.

The study cases illustrate the unique governance challenges that Chile faces to effectively confront dilemmas and traps in terrestrial landscapes and seascapes, which include: (1) lack of administrative capacities to design and implement sustainable State policies, (2) ongoing tensions between customary or local and State institutions, (3) ambiguity regarding management responsibility in the case of commons (including ecosystem services), (4) a clear gap between stated resource management rules and the current socioeconomic and cultural context of people, and (5) the scarce capacity of the State to put limits on the market, an issue that seems difficult to solve due to the design of the economic policy that prevails in Chile.

A short-term and urgent action toward facing these challenges is the alignment of conservation and development policies. Whereas undertaking natural resources loss and improving well-being in developing countries are stated international goals, the alignment between the two has been deficient (Roe et al. 2013). A real focus on these two goals, which transcends the rhetoric, involves acknowledging the importance of (1) the historical context as a factor modeling inequality, poverty, and vulnerability (Rodríguez-Robayo and Merino-Perez 2017), (2) the relative disempowerment of weaker groups such as small farmers and fishers (World Bank 2016, 2017), and (3) past injustices (Jerneck et al. 2011), and also extends to the need of designing public action to promote greater "equality of agency" (Rao 2006) with respect to existing social hierarchies.

Acknowledgments This work has been funded by Grants FONDAP 15150003 and FONDECYT 1190207.

References

Albert F (1911) La necesidad urgente de crear una inspección general de bosques, pesca y caza. Ministerio de Industria, Sección de Aguas y Bosques, Santiago de Chile

Almonacid F (2009) El Problema de la Propiedad de la Tierra en el Sur de Chile (1850-1930). Revista Historia 42:45–56

Angosto-Fernández L (2018) Neoextractivism and class formation: lessons from the Orinoco mining arc project in Venezuela. Lat Am Perspect 46:190–211

Armesto J, Villagrán C, Donoso C (1994) La historia del bosque templado chileno. Amb Des 10:66–72

Armesto J, Manuschevich D, Mora A et al (2010) From the Holocene to the Anthropocene: a historical framework for land cover change in southwestern South America in the past 15,000 years. Land Use Policy 27:148–160

Azariadis C, Stachurski J (2004) Poverty traps. In: Aghion P, Durlauf S (eds) Handbook of economic growth, 1st edn. North Holland, pp 295–384

Baker D, Murray G, Agyare A (2018) Governance and the making and breaking of social-ecological traps. Ecol Soc 23:28–38

Barrena J, Hernando M, Rojas F (2016) Antecedentes históricos sobre el Complejo Forestal y Maderero Panguipulli, provincia de Valdivia, Centro-sur de Chile. Bosque 37:473–484

Barrett C, Swallow B (2006) Fractal poverty traps. World Dev 34:1–15

Barry B, Bateman T (1996) A social trap analysis of the management of diversity. Acad Manag Rev 21:757–790

Bengoa J (1990) Historia social de la agricultura chilena. Tomo II: Haciendas y campesinos. Ediciones SUR, Santiago

Bengoa J (1999) Historia de un conflicto: El Estado y los Mapuches en el siglo XX. Planeta/Ariel, Santiago

Bennett A (2008) Process tracing: a bayesian perspective. In: Box-Steffensmeier J, Brady H, Collier D (eds) The oxford handbook of political methodology. United Kingdom, Oxford, pp 702–721

Benra F, Nahuelhual L (2019) A trilogy of inequalities: land ownership, forest cover and ecosystem services distribution. Land Use Policy 82:247–257

Bitrán E (1989) El desafío pesquero chileno: la explotación racional de nuestras riquezas marinas. In: Hachette (ed), Santiago, p 412

Bize C (2017) El otoño de los raulíes. In: Tiempo Robado (ed) Poder popular en el complejo Forestal y Maderero Panguipulli. (Neltume 1967-1973). Santiago, pp 41–55

Boonstra W, Björkvik E, Haider L et al (2016) Human responses to social-ecological traps. Sustain Sci 11:877–889

Bourguignon F, Ferreira F, Walton M (2007) Equity, efficient and inequality traps: a research agenda. J Econ Inequal 5:235–256

Camus P (2006) Ambiente, bosques y gestión forestal en Chile 1541–2005. In: LOM (ed) Centro de Investigaciones Barros Arana, Santiago

Capraro V (2013) A model of human cooperation in social dilemmas. PLos One 8:e72427

Ceballos M, Ther F (2011) Transformation in Today's artisan fisheries economies: the case of Cucao and Tenaún (province of Chiloé, Los Lagos region, Chile). Cuadernos de Goegrafía Revista Colombiana de Geografía 2:61–75

Cinner J (2011) Social-ecological traps in reef fisheries. Global Environ Chang 21:835–839

Clapp R (1998) Regions of refuge and the agrarian question: peasant agriculture and plantation forestry in Chilean Araucania. World Dev 26:571–589

Comaroff J, Comaroff J (2009) Ethnicity, Inc. University of Chicago Press, Chicago

Composto C (2012) Acumulación por despojo y neoextractivismo en ámerica latina. Una reflexión crítica acerca del estado y los movimientos socio-ambientales en el nuevo siglo. Astrolabio (ed) 8:323–352

Coomes O, Takasaki Y, Rhemtulla J (2016) Forests as landscapes of social inequality: tropical forest cover and land distribution among shifting cultivators. Ecol Soc 21:1–20
Corporación de Promoción y Defensa de los Derechos del Pueblo, CODEPU (1991) Chile: Recuerdos de la Guerra. Valdivia – Neltume – Chihuio – Liquiñe, vol 2, de la Serie Verdad y Justicia, Santiago
Corporación Nacional Forestal (CONAF) (1998) Catastro de uso vegetacional para la región de los Lagos 1998
Corporación Nacional Forestal (CONAF) (2013) Catastro de uso vegetacional para la región de los Ríos 2013
Costanza R (1987) Social traps and environmental policy. Bioscience 37:407–412
Cumming G (2018) A review of social dilemmas and social-ecological traps in conservation and natural resource management. Conserv Lett 11:1–15
Enfors E (2013) Social-ecological traps and transformations in dryland agro-ecosystems: using water system innovations to change the trajectory of development. Global Environ Chang 23:51–60
Escobar A (2008) Territories of difference: place, movements, life, redes. Duke University Press, Durham
Fairhead J, Leach M, Scoones I (2012) Green grabbing: a new appropriation of nature? J Peasant Stud 39:237–261
García Canclini N (1990) Culturas híbridas. Estrategias para entrar y salir de la modernidad. Grijalbo SA, Barcelona
Garrido J, Guerrero C, Valdés M (1988) Historia de la reforma agraria en Chile. Universitaria (ed) Santiago
George A, Bennett A (2005) Case studies and theory development in the social sciences. MIT Press, London
Gómez S, Echeñique J (1991) La Agricultura Chilena, las Dos Caras de la Modernización. Flacso-Agraria, Santiago
González P (2003) Colonialismo Interno (Una redefinición). La teoría Marxista hoy. Revista Rebeldía N° 12
Haider L, Boonstra W, Peterson G et al (2018) Traps and sustainable development in rural areas: a review. World Dev 101:311–321
Harvey D (2005) A brief history of neoliberalism. Oxford University Press, Oxford
Holmes G (2015) Markets, nature, neoliberalism, and conservation through private protected areas in southern Chile. Environ Plan A 47:850–866
Huber A, Iroumé A, Mohr C et al (2010) Efecto de las plantaciones de *Pinus radiata* y *Eucalyptus globulus* sobre el recurso agua en la Cordillera de la Costa de la región del Biobío, Chile. Bosque 31:219–230
IFOP (Instituto de Fomento Pesquero) (1966) The fishing of the king crab in the province of Magallanes. Observations on the current state of the industry. Circular 2 fisheries economy. IFOP, Santiago
INE (Instituto Nacional de Estadísticas) (1955) Censo agropecuario y forestal para la región de los Ríos 1965. INE, Santiago
INE (Instituto Nacional de Estadísticas) (1976) Censo agropecuario y forestal para la región de los Ríos 1976. INE, Santiago
INE (Instituto Nacional de Estadísticas) (1997) Censo agropecuario y forestal de para la región de los Ríos 1997. INE, Santiago
INE (Instituto Nacional de Estadísticas) (2007) Censo agropecuario y forestal de para la región de los Ríos 2007. INE, Santiago
INE (Instituto Nacional de Estadísticas) (2017) Resultados Censo para la Región de Magallanes y Antártica Chilena. INE, Santiago
Ingold T, Palsson G (2013) Biosocial becomings: integrating social and biological anthropology. Cambridge University Press, Cambridge

ITTO (International Tropical Timber Organization) (2002) Directrices de la OIMT para la restauración, ordenación y rehabilitación de bosques tropicales secundarios y degradados. Serie de políticas forestales No 13

ITTO (International Tropical Timber Organization) (2005) Criterios e indicadores revisados de la OIMT para la ordenación sostenible de los bosques tropicales con inclusión de un formato de informes. Serie de políticas forestales No 15

Jarvis L, Wilen J (2014) The political economy of the Chilean Nearshore fisheries reform. In: UCLA (ed) University of California

Jerneck A, Olsson L, Ness B et al (2011) Structuring sustainability science. Sustain Sci 6:69–82

Jiao X, Smith-Hall C, Theilade I (2015) Rural household incomes and land grabbing in Cambodia. Land Use Policy 48:317–328

Kasanga K, Kotey N (2001) Land management in Ghana: building on tradition and modernity. International Institute for Environment and Development, London

Klubock T (2014) La Frontera. Forest and ecological conflict in Chile's frontier territory. Duke University Press, Durham, Durham

Kronenberg J, Hubacek K (2013) Could payments for ecosystem services create an "ecosystem service curse"? Ecol Soc 18:1–10

Kuusaana E, Gerber N (2015) Institutional synergies in customary land markets-selected case studies of large-scale land acquisitions (LSLAs) in Ghana. Land 4:842–868

Lansing D (2014) Unequal access to payments for ecosystem services: the case of Costa Rica. Dev Change 45:1310–1331

Lara A, Veblen T (1993) Forest plantations in Chile: a successful model? In: Mather A (ed) Afforestation policies, planning and progress. Belhaven Press, London, pp 118–139

Lebel L, Manuta J, Garden P (2011) Institutional traps and vulnerability to changes in climate and flood regimes in Thailand. Reg Environ Chang 11:45–58

Mahoney J (2001) Path-dependent explanations of regime change: Central America in comparative perspective. Stud Comp Int Dev 36:111–141

Martinic M (1999) Migración Chilota a Magallanes. In: Anales del Instituto de la Patagonia (ed) Instituto de la Patagonia, Punta Arenas, vol 27. pp 27–47

McCarter M, Budescu D, Scheffran J (2011) The give-or-take-some dilemma: an empirical investigation of a hybrid social dilemma. Organ Behav Hum Decis Process 116:83–95

McDermott M, Mahanty S, Schreckenberg K (2013) Examining equity: a multidimensional framework for assessing equity in payments for ecosystem services. Environ Sci Pol 33:416–427

Miller M (2006) Opportunities and challenges for migrant and migrant-background youth in developed countries. In: Report submitted to the United Nations Social and Economic Council

Montalba R, Carrasco N (2005) ¿Desarrollo sostenible o eco-etnocidio? El proceso de expansión forestal en territorio Mapuche-nalche de Chile. Revista de Estudios sobre Despoblación y Desarrollo Rural 4:101–133

Muñoz J, Zamora E (1975) De Chiloé a Magallanes. El inmigrante chilote en la Patagonia Magallánica. Memoria de título en Geografía. Universidad de Valparaíso, Valparaíso

Nahuelhual L, Saavedra G, Blanco G et al (2018a) On super fishers and black capture: images of illegal fishing in artisanal fisheries of southern Chile. Mar Policy 95:36–45

Nahuelhual L, Saavedra G, Henriquez F et al (2018b) Opportunities and limits to ecosystem services governance in developing countries and indigenous territories: the case of water supply in southern Chile. Environ Sci Pol 84:11–18

Peña-Torres J (1997) The political economy of fishing regulation: the case of Chile. Mar Resour Econ 12:239–248

Platt J (1973) Social traps. Am Psychol 28:642–651

Rao N (2006) Land rights, gender equality and household food security: exploring the conceptual links in the case of India. Food Policy 31:180–193

Reyes R, Blanco G, Lagarrigue A, Rojas F (2016) Ley de Bosque Nativo: Desafíos Sociales para su Implementación. Instituto Nacional Forestal (INFOR), Santiago

Rivas R (2006) Desarrollo forestal de Neltume; Estado y trabajadores (1924–1990). Tesis de Profesor de Historia, Geografía y Educación Cívica. Facultad de Filosofía y Humanidades. Universidad Austral de Chile, Valdivia

Rodríguez-Robayo K, Merino-Perez L (2017) Contextualizing context in the analysis of payment for ecosystem services. Ecosyst Serv 23:259–267
Roe D, Yassin E, Porras I et al (2013) Linking biodiversity conservation and poverty reduction: de-polarizing the conservation-poverty debate. Conserv Lett 6:162–171
Rosenblitt J, Correa M, Hajek E (2001) La modernización de la agricultura Chilena. Pobreza y medio ambiente después de la reestructuración. Revista de humanidades y Ciencias Sociales 50:163–191
Rudel T, Katan T, Horowitz B (2013) Amerindian livelihoods, outside interventions, and poverty traps in the Ecuadorian Amazon. Rural Sociol 78:167–185
Schkolnik S (1994) Los mapuches: una investigación multidisciplinaria en reducciones indígenas de Chile. In: CEPAL (ed) Estudios sociodemográficos de pueblos indígenas. Series históricas. pp 147–169
Schoon M, Van der Leeuw S (2015) The shift toward social-ecological systems perspectives: insights into the human-nature relationship. Nat Sci Soc 23:166–174
Servicio Nacional de Pesca y Acuicultura (SERNAPESCA) (2018) Anuario Estadistico de Pesca. SERNAPESCA, Santiago
Skewes J, Solari ME, Guerra D et al (2012) Los paisajes del agua: naturaleza e identidad en la cuenca del río Valdivia, Sur de Chile. Chungara 46:651–668
Steneck R, Hughes T, Cinner J et al (2011) Creation of a gilded trap by the high economic value of the Maine lobster fishery. Conser Biol 25:904–912
Stewart N, Dawson N (2013) Forest thinning: a Landowner's tool for healthy woods. Extension bulletin EB-407. University of Maryland Extension, Maryland
SUBPESCA (Subsecretaría de Pesca y Acuicultura) (2004) Informe Técnico (R. Pesq.) N° 102 suspensión temporal del acceso a la pesquería de Centolla de la XII región 2005–2009
Susaeta E (1989) La sustentabilidad en el desarrollo forestal chileno. Amb Des 5:13–27
Tidball K, Frantzeskaki N, Elmqvist T (2016) Traps! An introduction to expanding thinking on persistent maladaptive states in pursuit of resilience. Sustain Sci 11:861–866
Torres R, Azócar G, Rojas J et al (2015) Vulnerability and resistance to neoliberal environmental changes: an assessment of agriculture and forestry in the biobio region of Chile. Geoforum 60:107–122
Tura H (2018) Land rights and land grabbing in Oromia, Ethiopia. Land Use Policy 70:247–255
Urbina R (1998) Chiloé, foco de migraciones. In: Fliman-Grinberg X (ed) Chiloé y su influjo en la XI Región. II Jornadas territoriales. Universidad de Santiago, Santiago, pp 31–46
Van Lange P, Joireman J, Parks C et al (2013) The psychology of social dilemmas: a review. Organ Behav Hum Decis Process 120:125–114
Virapongse A, Brooks S, Covelli E et al (2016) A social-ecological systems approach for environmental management. J Environ Manag 178:83–91
Weingart A, Kirk M (2011) Escaping poverty traps? Collective action and property rights in post-war rural Cambodia. In: Mwang E, Markelova H, Suseela R (eds) University of Pennsylvania Press. Published for the International Food Policy Research Institute. pp 328–356
World Bank (2016) Poverty and shared prosperity: taking on inequality. World Bank, Washington
World Bank (2017) World development report 2017: governance and the law main messages. World Bank, Washington
Zavala J (2008) Los colonos y la escuela en la araucanía: los inmigrantes europeos y el surgimiento de la educación privada laica y protestante en la región de la Arauicanía (1887-1915). Revista Universum 23:268–286

Ecosystem Services from a Multi-Stakeholder Perspective: A Case Study of a Biosphere Reserve in Central Chile

Claudia Cerda and Iñigo Bidegain

Abstract When assessing ecosystem services, it is essential to identify which services are relevant to different stakeholders as well as their perception of those services, in order to make informed decisions about land use management. Divergent social interests may lead to conflicts over the use of a territory, stressing the tension between conservation needs and economic activities, which may be productive but nevertheless threatening the achievement of conservation goals. In this chapter, we present an assessment of social preferences for ecosystem services in a globally relevant biosphere reserve in South America; a region that requires more research on how to conserve ecosystems while incorporating human needs and values. Using a semi-structured approach, we found differences among stakeholders about the importance they attribute to different ecosystem services. On one side, local farmers and members of local organizations give higher value to provisioning services and cultural services of symbolic plants. On the other, scientists, environmentally concerned people, teachers, NGOs, and employees of the local government lend more importance to regulating and cultural services, revealing contrasting preferences for ecosystem services.

Keywords Social-ecological systems · Latin America · Complexity · Chile · Ecosystem services · Stakeholders · Biosphere reserve

C. Cerda (✉) · I. Bidegain
Department of Forest Management, Faculty of Forest Sciences and Conservation of Nature, University of Chile, Santiago, Chile
e-mail: clcerdaj@uchile.cl

L. E. Delgado, V. H. Marín (eds.), *Social-ecological Systems of Latin America: Complexities and Challenges*, https://doi.org/10.1007/978-3-030-28452-7_19

1 Introduction

In recent decades, research into ecosystem services (ES), driven by the potential of this approach to guide decision-making in favor of more sustainable territories and with greater social justice, has had important impacts in both scientific and political forums (Hevia et al. 2017; Kok et al. 2017; Seppelt et al. 2011). ES are defined as *the aspects of ecosystems that are used (actively or passively) to maintain human well-being* (Fischer and Turner 2008). From an ES point of view, humans and their environment are tightly linked, and human well-being depends on "healthy" ecosystems (Menzel and Teng 2010). This means that ecosystems should be maintained or brought to a state in which they support human well-being. Thus, under the concept of ES, *human well-being* is the most important driver of resource management and conservation (Menzel and Teng 2010). In this regard, organizations such as the Intergovernmental Platform on Biodiversity and Ecosystem Services (IPBES 2018), or the United Nations Agenda 2030 for Sustainable Development (UN 2015), stimulate assessments and valuations of ecosystem services in different regions of the world that incorporate ecological theory and also social sciences from the perspective of understanding human values (García-Llorente et al. 2015; Jacobs et al. 2016; Díaz et al. 2018). Key issues in resource management and conservation are: (1) understanding how different social actors perceive, value and use ES in the target area, and (2) the strategies they describe and analyze to ensure their access to certain services (e.g., Tallis et al. 2008; Ramírez-Gómez et al. 2015). Confronting these tasks, identifying which services are relevant to different stakeholders is essential to make informed decisions about land use management (Carrasco et al. 2016; Menzel and Teng 2010). ES matter differently to different stakeholders because they have different relationships with the same ecosystem (Martín-López et al. 2012; López-Santiago et al. 2014). Therefore, diverging interests of social actors emerge because one ecosystem may be valued differently by different stakeholders in relation to the ecosystem's capacity to provide services that fulfill their own interests (Martín-López et al. 2012: 1). Thus, the relevance that different actors attribute to different services makes it possible to identify which social trade-offs need to be addressed when making decisions regarding land use management (Seppelt et al. 2011). Diverging social interests may lead to social actors disputing or entering into conflict over the appropriation and use of ecosystem services (O'Brien and Leichenko 2003) and they must therefore be considered to achieve sustainability and avoid conflicts. Analysis of social groups' perceptions provides information on the social dynamics that originate around ecosystems (Howe et al. 2014). Establishing the different points of view and interests of social actors can identify the welfare gain and loss relationships resulting from changes in the territory and in the provision of ES (López-Santiago et al. 2014).

Diverging social interests for ecosystem services have been documented in the literature. For example, Martín-López et al. (2012) found that specific ES such as air purification, microclimate regulation, esthetic value, tourism activities, and environmental education are more valued by urban people. However, ES essential for life, such as food, are less perceived by urban people, despite their increasing dependence on these essential provisioning services. Other studies indicate that large market-oriented landowners value agro-ecosystem services differently from subsistence-oriented farmers (Daw et al. 2011; Díaz et al. 2011). López-Santiago et al. (2014) found that environmentalists are more interested in guaranteeing the flow of regulating ecosystem services. Unless these different perspectives are integrated into ES assessments, it is unlikely that the resulting management decisions will adequately address all the issues and trade-offs (Díaz et al. 2018). Understanding such trade-offs allows for a fuller characterization and representation of diverse ecosystem values in research and practice (Chan et al. 2012).

The need to incorporate multiple social value systems when using the ES approach has also been recognized for Latin America (Balvanera et al. 2012; Mastrangelo et al. 2015; Laterra et al. 2017). The spatial representation of the supply and consumption of ES by different stakeholders to inform land planning processes has been recognized as one of the essential challenges for the region (Laterra et al. 2017; Martínez-Harms and Balvanera 2012). However, understanding the conflicts and disputes over these services from the perspective of multiple stakeholders remains poor. In Chile for example, the State has promoted policies of incorporating citizens in protected areas management for the last 20 years (Estévez et al. 2019). However, the participatory processes are not always diverse, representative, and plural as required for successful participation in conservation contexts. Estévez et al. (2019) recognize that the challenges of participation in conservation management are enormous, especially considering the diversity of social values involved. Such challenges must be recognized, for example, for the implementation of the new methodologies for protected areas planning based on the Open Standards mechanism (Sepúlveda et al. 2015), which includes not only protecting biological diversity but also guaranteeing the flow of ES in such a way that protected areas may contribute to the well-being of local human communities. In this chapter, we provide an example of an assessment of social preferences for ecosystem services from a multi-stakeholder perspective, considering different social actors of the Campana-Peñuelas Biosphere Reserve in central Chile.

The literature on ES has exposed different methodological approaches to investigate social preferences for ecosystem services by taking a variety of actors into consideration. For example, a detailed analytical framework to characterize different methods for assessing social preferences can be found in Villamor et al. (2014) and in López-Santiago et al. (2019). The authors emphasize that the level of available information on social preferences for ES in the decision-making process, the typology of social actors and their knowledge systems, and the typology of values that social actors communicate are relevant aspects to analyze to ultimately decide the methodological approach to be used for the assessment of social preferences for ES. When there is not much information about them, one can start using lists of ES (e.g., Table 1) that are presented to different local actors to obtain their social perception of the ES from the list on the same scale of analysis (López-Santiago et al. 2019). This approach can be very useful, given that as ES of different natures are included (i.e., ES with or without market value, or cultural ES), it allows for the

Table 1 Ecosystem services that flow into the study area, identified from an extensive literature review

Type	Services
Provisiong	Food derived from traditional agriculture
	Food derived from organic agriculture
	Food from cattle (milk, meat)
	Forage (trees and shrubs that are useful for cattle/grazing)
	Food from hunting (hunting of wild animals for human consumption)
	Mushroom hunting for human consumption
	Beekeeping
	Wild fruits (for human and animal consumption)
	Medicinal plants (leaves, bark, roots)
	Genetic resources (e.g., wild species used in breeding programs)
	Seeds
	Plants for fibers/handcrafts
	Industrial use of animals and plants
	Drinking water
	Water for agriculture
	Water for industrial use
	Wood fuel
	Coal
	Wood for building
	Organic compost
	Soil litter extraction
Regulating	Genetic pool of the plant communities in Central Chile with global relevance
	Fresh air and climate change control
	Soil fertility for agricultural crops and pasture
	Water regulation and retention
	Erosion control
	Pest and disease control
	Pollination
Cultural	Educational value: possibilities of developing educational programs about local wildlife
	Conservation activities carried out by different actors motivated by iconic endangered animal and plant species (conservation value)
	Rural tourism
	Resort tourism
	Cultural tourism
	Nature tourism
	Sport hunting
	Possibilities to develop research (e.g., genetic patterns in plants, effects of invasive species on the dynamics of Chilean palm (*Jubaea chilensis*) relicts)
	Local ecological knowledge
	Identity and sense of place
	Spiritual and religious values
	Symbolic animals
	Symbolic plants

incorporation of a wide range of people–nature connections on the same analytical level. In this regard, Martín-López et al. (2012); López-Santiago et al. (2014), and Iniesta-Arandia et al. (2014) provide useful examples to assess social preferences for ecosystem services using multivariate techniques. Results are comprehensive images of preferences involving different ecosystem services and different stakeholders. This approach is accessible to a wide array of people who are not always comfortable with methods that involve a high degree of verbal abstraction or group participation processes (López-Santiago et al. 2019). Results from these approaches can provide a first look at the social preferences for different ecosystem services from a multi-stakeholder perspective, identifying contrasting or synergic preferences among different actors. In addition, this approach paves the way to stronger participatory techniques.

We present our example in a reserve of the biosphere, located in the Mediterranean climate zone of Chile. Mediterranean ecosystems in central Chile are considered a global priority for conservation (Mittermeier et al. 2005) as they have high concentrations of endemic species and are experiencing accelerated rates of habitat destruction (Ceballos and Ehrlich 2006; Myers et al. 2000), land use conversion processes and land cover changes such as deforestation, urbanization, and agricultural abandonment (Carmona et al. 2010). Thus, given the highly anthropized matrix in which protected areas are embedded, the incorporation of public perceptions about natural areas and the ES they provide is a critical component of Mediterranean ecosystem conservation management (Cerda et al. 2013).

An important aspect of our example is that we conducted our study in a globally relevant reserve of the biosphere. Biosphere reserves have been recognized as sustainability models, corresponding to large tracts of land where biodiversity conservation is practiced in conjunction with local people living, working, and striving for sustainable livelihoods (Stoll-Kleemann and O'Riordan 2017). Landscapes of biosphere reserves are constantly changing and their habitats need active management, which must involve safeguarding the flow of ES for the well-being of local human communities. Thus, it is essential to integrate the surrounding communities and the different local actors in the assessment of social preferences for ES with the purpose of enriching production while fulfilling the conservation function.

2 Case Study: Social Perceptions Toward Ecosystem Services in the Campana-Peñuelas Biosphere Reserve

The Campana-Peñuelas Biosphere Reserve is located in the central zone of the Mediterranean region of Chile (Fig. 1). It was created in 1984 covering 170 km^2, and in 2009 it was expanded to 2380 km^2. La Campana National Park represents the core area of the Reserve and is recognized as an international icon of biotic conservation with abundant vegetation supporting a variety of fauna (Moreira and Barsdorf 2014). Its main ecosystems are Mediterranean sclerophyllous forests and scrublands; together, these ecosystems constitute a globally relevant biodiversity hotspot

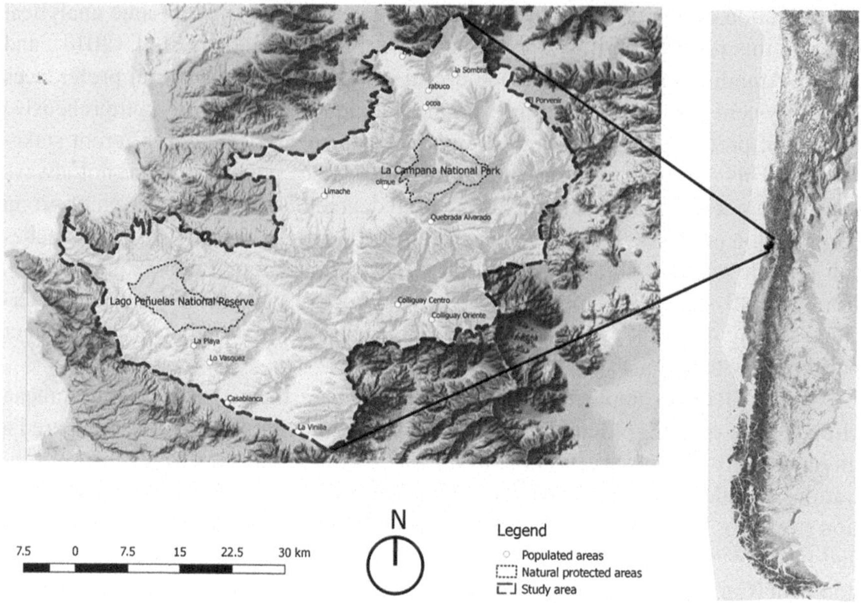

Fig. 1 Study area and public protected areas

(Mittermeier et al. 2005). Additionally, more than 80% of the population lives in Chile's central zone, which has been intensely affected by human activities, causing a significant loss and fragmentation of natural habitats (Blondel and Fernández 2012). The Reserve's biological diversity is threatened by various factors, including forest fires, domestic livestock in protected areas, the illegal extraction of native palms, the presence of non-native species such as rabbits (*Oryctolagus cuniculus*) and blackberries (*Rubus* spp.), (Smith-Ramírez et al. 2005) and urbanization (real estate) projects. The Reserve also includes a diverse mosaic of human groups that interact differently with ecosystem services and wildlife. Thus, it is a reserve with a strong component of urban pressure and conflicts that threaten the real application of the functions of a biosphere reserve in practice (Moreira and Barsdorf 2014). Complex relationships in an urban–rural interaction among diverse actors and ES support the need to assess social preferences for such services in order to illuminate potential conflicts or synergies in the territory for the use or valuation of ES.

The provision of ES in the Biosphere Reserve and the value that social actors attribute to such services has been recognized in the literature. For example, Cerda et al. (2013) found that human water consumption, the value of species such as birds and endemic herbaceous plants as well as landscape beauty are benefits clearly valued by visitors to the Biosphere Reserve. These visitors are local residents and people from other regions of Chile. Cerda and Losada (2013) report that ecosystems that harbor relicts of Chilean palm (*Jubaea chilensis*) are highly valued for their contribution to the identity of the region (cultural ecosystem service) and also for ecosystem services of direct use such as honey.

Catalán (2015) indicates that at present the exploitation of the native forest has decreased, as have traditional agricultural activities. Nevertheless, some of the inhabitants of local communities—the *arrieros*—maintain productive practices related to the native forest, particularly breeding cattle and horses on the hill, evidencing, therefore, the flow of ecosystem provisioning services.

At the planning level, the current Management Plan for La Campana National Park, which accounts for the Open Standards mechanism (Sepúlveda et al. 2015), identifies ES provided by the park that are relevant to different dimensions of human well-being. Examples of such services are water for irrigation and consumption, clean air, pest control by raptors and carnivores, recreation, the cultural relevance of biological attributes, leisure and tranquility, spiritual development, research possibilities, and sense of belonging. The fact that the current conservation management planning process inside the Biosphere Reserve incorporates human well-being makes it necessary to assess the social preferences for ES of multiple stakeholders.

Study Design

We selected key local actors with different interests in the Biosphere Reserve territory. We selected actors with a strong connection to decision-making in the area (i.e., employees of the local government with no environmental focus in their management = 7, managers/owners of businesses related agronomics, farming and real state = 8), with strong interests at stake in local nature management (i.e., small farmers = 12; representatives of small local organizations such as neighborhood boards, foundations or indigenous communities and are not focused on the conservation of biodiversity $n = 8$) and tourism workers ($n = 6$). Educators in schools and colleges as actors who can influence the long-term valuation of local nature were also included ($n = 6$), as well as scientists working in the area on conservation and environmental topics, as they might influence decision-making processes ($n = 6$). We also included NGO members ($n = 10$), a group composed of locals with environmental concerns. Employees of the Chilean National Forestry Corporation (CONAF) ($n = 7$), the park rangers and administrators of the public protected areas located in the Biosphere Reserve, were also included as their work focuses directly on conserving biodiversity and they are in charge of the administration of protected areas inside the Biosphere Reserve.

The method we used for data collection was consultative in nature (Villamor et al. 2014), using semi-structured interviews. This type of methodological approach is recommended when there is a scarcity of information on the research topic, as in this case. A semi-structured interview was chosen because it is known to provide reliable and comparable data in a way that can be understood by different audiences (Crouzat et al. 2016).

Using a previously designed list of ES, participants were asked to select the five most important ecosystem services and to explain why those services were important to them (Table 1). The list was built through an extensive literature review of the area. Each respondent was given the same list of ecosystem services.

We used the types of ecosystem services proposed by the Millennium Ecosystem Assessment for classification (MEA 2005). The Common International Classification of Ecosystem Services (CICES) was also used to facilitate the description of some ecosystem services in the list (Source: Bidegain et al. 2019)

From the five most important selected ecosystem services, we also asked respondents about the perceived vulnerability of the chosen services to changes in the future using a 5-point Likert scale (from 1, not at all vulnerable, to 5, extremely vulnerable). The last section of the questionnaire included questions related to sociodemographic (e.g., occupation) and cultural characteristics of the respondents (e.g., knowledge about protection figures, rural–urban character, environmental organization membership, a visitor to protected areas, and recycling habits). Participants were identified through key contacts and through snowball sampling. We used descriptive and multivariate statistics to analyze contrasting perceptions for ecosystem services among different respondents. Multivariate analysis allowed us to relate preferences of ecosystem services and characteristics of respondents; thus, it allowed us to identify contrasting perceptions among respondents.

Results

The ranking of ecosystem services shows that 11 services from the list (Table 1) have higher-than-mean importance-vulnerable index scores ($\bar{x}$ =108.2; Table 2). The most important service was drinking water (importance value = 124), followed by fresh air and climate change control (90), the genetic pool of the plant communities in central Chile (89), educational value (e.g., possibilities of developing educational programs and books about local wildlife) (89), conservation activities motivated by iconic endangered animal and plant species (79) and water regulation and retention (74). Results for the complete list of ecosystem services (Table 1) are shown in Annex I.

The order of ecosystem services in the first column follows the importance-vulnerability index value from highest to lowest (Modified from Bidegain et al. 2019).

The services with the highest perceived vulnerability scores were the conservation value of endangered animal and plant species ($\bar{x} = 4.421$), water regulation and retention ($\bar{x} = 4.343$), food from agriculture ($\bar{x} = 4.050$) and drinking water ($\bar{x} = 4.022$).

Using a principal component analysis (PCA), these 11 services were analyzed in order to reduce variables for further analysis. From this analysis, we selected nine services with the largest square cosine in those components with an eigenvalue greater than 1. Selected variables were analyzed using redundancy analysis (RDA) in order to explore their association with the respondents' sociodemographic and cultural characteristics and to identify contrasting perceptions of stakeholders

Table 2 Ecosystem services with the higher-than-mean importance-vulnerable index scores

Ecosystem service	Importance value	Mean vulnerability	Importance-vulnerability index
Drinking water	124.0	4.0	498.7
Fresh air and climate change control	90.0	3.9	352.1
Conservation activities motivated by iconic endangered animal and plants species	79.0	4.4	349.3
Genetic pool of the plant communities in Central Chile with global relevance	89.0	3.7	332.1
Water regulation and retention	74.0	4.3	321.4
Educational value: possibilities of developing educational programs about local wildlife	89.0	2.7	240.0
Water for agriculture	49.0	3.6	176.4
Food derived from traditional agriculture	36.0	4.1	145.8
Medicinal plants (leaves, bark, roots)	39.0	3.5	136.5
Symbolic plants	32.0	3.6	116.5
Beekeeping	33.0	3.5	113.9
Mean	**108.2**		

regarding the ES. Resulting factor scores from the RDA are shown in Annex II. Figure 2 presents the biplots obtained from the RDA.

The first five axes explained 83% of the total variance. Based on the explained variance and eigenvalues, we focused on the first two axes because they showed the most important trends in terms of explaining the differences in stakeholders' perceptions of important ES.

Figure 2 shows the association between the ES and the different variables that characterize the respondents. Axis 1 (which explained 32,31% of the variance) shows contrasting perceptions of importance between provisioning services (e.g., food derived from traditional agriculture, water for agriculture, drinking water and beekeeping) and the cultural service represented by symbolic plants on the positive side of the axis juxtaposed against water regulation and conservation activities (on the negative side of axis 1 and the positive side of axis 2) and educational value on the negative sides of axes 1 and 2. We found that occupation is a variable that affects social perceptions of the importance of ES. In this regard, members of local organizations, small farmers and business-related stakeholders believe that provisioning services and symbolic plants are more important than another ES on the list. This view is contrary to that of NGO members, employees of CONAF, and employees of the local government and schools, as they all place a higher value on regulating services such as fresh air and climate change control and on the educational value of cultural services. Scientists and tourism workers prefer water regulation and retention and the possibilities of developing conservation activities related to endangered wildlife.

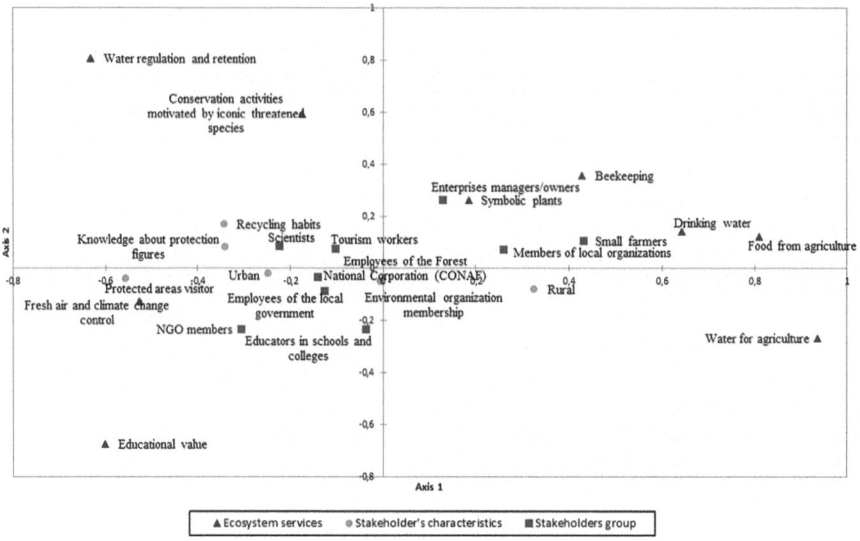

Fig. 2 Biplots obtained from the RDA. Adapted from Bidegain et al. (2019)

Furthermore, we see that the perception of importance is also guided by other variables; for example, preferences for provisioning ecosystem services are associated with stakeholders that identify with a rural lifestyle, who also value symbolic plants more than other stakeholders. On the other hand, people who recycle, visit protected areas, and are familiar with protection figures tend to favor regulating ecosystem services and cultural ecosystem services associated with endangered wildlife.

The axis 2 (which explained 17.78% of the variance) showed contrasting perceptions between the importance of water retention and regulation and the possibilities of developing conservation activities juxtaposed against the educational value and fresh air and climate change control. The characteristics of stakeholders and respondents were not strongly associated with their preferences for ecosystem services on this axis, indicating that there must be some variables, not included in this study, that explain this dichotomy.

3 Discussion

Biosphere reserves are territorial management models that aim to achieve sustainability by protecting biological attributes and contributing positively to human well-being. Commonly in biosphere reserves, different stakeholders have different views on how the territory should be managed, and the ecosystem services approach contributes to incorporating the different perspectives of local actors into the decision-making process.

This is particularly relevant for the Campana-Peñuelas Biosphere Reserve that currently faces a planning process in which conservation of biological diversity must reason with human well-being (Sepúlveda et al. 2015). In Chile, progress has been made in understanding how the ecosystem services paradigm can contribute to achieving more sustainable territories (e.g., Delgado et al. 2013; Alfonso et al. 2017; Zorondo-Rodríguez et al. 2019). However, the social dimensions of ecosystem services and the different ways that different users have of accessing the services still require further research. At a biosphere reserve level, the investigation into the social preferences for ecosystem services of people living in them is key to advancing the understanding of synergic or contrasting interests in ecosystem services. By considering stakeholders' differing views, potential social conflicts can be identified and different ways of life in the territory can be better understood. Such an understanding is crucial to effective conservation management. The relation between people and ecosystem services or the benefits people perceive from nature determine behaviors that contribute to maintaining the flow of ecosystem services in the long term (Asah et al. 2014).

The main motivation of this work was to analyze contrasting social perceptions toward ecosystem services. In this regard, we provide a first look at social preferences for ecosystem services where different ecosystem services and social actors are analyzed on the same scale of analysis. With the information generated here, we can start visualizing webs of relationships between different stakeholders and nature on the same level of analysis.

Our results show that the respondents' sociocultural variables affect the value that participants attribute to different ecosystem services. Synergic perceptions are observed between small farmers and members of local organizations who assign more importance to provisioning ecosystem services and to the cultural value related to the existence of symbolic plants than the other groups of stakeholders. The preferences of these groups of stakeholders, contrast with those of scientists, CONAF employees, NGO workers, and employees of the public administration who tend to prefer regulating ecosystem services and cultural ecosystem services associated with endangered species. Thus, contrasting preferences are observed between decision makers and local actors that depend directly on ecosystem services. The fact that environmental professionals differ in their perception of the importance of ecosystem services to local actors such as farmers requires a dialog to bring about a deeper understanding of these contrasting perceptions in decision-making processes to avoid conflicts among stakeholders and achieve the goals of sustainable regional development. There is also a rural–urban contrast where an urban lifestyle tends to favor the protection of endangered species and regulating ecosystem services, whereas more rural actors prefer services associated with the provision of water and food, and the cultural aspects emphasize native plants for their symbolic value. Similar tendencies of social preferences for ecosystem services have been found in other studies conducted in the developed world (e.g., Martín-López et al. 2012), where environmental professionals and urban actors tend to value endangered species and regulating ecosystem services such as those related to fresh air and climate change control, water regulation, and environmental education over ecosystem provisioning services.

We are aware that our results represent a first look at different perceptions for ecosystem services; however, our approach may pave the way to delve more deeply into the relationship between different users and ecosystem services. Stronger participatory techniques such as group valuation (Wilson and Howarth 2002), qualitative interviews, and deliberative participation (Asah et al. 2014) are practically mandatory to address effective conservation. For example, our results show that regulating services do not seem to be the most important services for local actors such as farmers, although their activities strongly depend on these services (e.g., those related to water and soil). Other studies have found that local actors with local ecological knowledge recognize regulating services such as erosion control or water regulation because traditional agricultural practices directly depend on them. In our case, the Biosphere Reserve is constantly changing for reasons such as real estate and climate change, among others, which may have affected the protection of local ecological knowledge. To explore this aspect more deeply, qualitative interviews can reveal richer information about the relationships of small farmers and regulating services.

Another important aspect of our study is the classification of services into different typologies. Traditionally in Chile, attention is paid to and highlights the profitability that local actors obtain from provisioning services (Razeto et al. 2019), but when one explores the testimonies of local actors more deeply, it may occur that in their narratives their proposals are not of the production for the market. For example, Razeto et al. (2019) found that local honey producers related to forestry systems in central-southern Chile attribute more relevance to the recognition of their identity forged in their relationships with bees than the relative utilities they can obtain in a determined period. In our case, analyzing the arguments that respondents give to justify the importance they attribute to the services they selected from the list, multiple value dimensions emerge for different ecosystem services. For example, local farmers recognize the food service from agriculture as an ecosystem provisioning service but also as a cultural ecosystem service because they emphasize that their identity has been forged through their relationship with agricultural products. Respondents recognize the cultural value associated with traditional farming activities and also the decline of this cultural service. These results suggest the need to conduct strong qualitative research to examine the relationship between local actors and ecosystem services in such a way that enables the full range of benefits people get from ecosystems to be explored.

Furthermore, we identify contrasting social perceptions of the importance of ecosystem services that flow into the area among different stakeholders. Such a contrast is affected by variables like occupation, urban–rural character, and other environment-related characteristics of the respondents (visitor to protected areas, knowledge about protection figures, recycling habits). These results were obtained using a semi-structured interview, which gives the research a preliminary consultative character. In summary, our methodological approach is characterized by (a) identification of ecosystem services that flow into the area through an extensive literature review, (b) identification of local stakeholders that manage, use or enjoy ecosystem services, (c) the creation of a list of ecosystem services in which such services are classified using international classification typologies, (d) the design of a semi-structured interview where the list of ecosystem services is presented, (e) the application of the questionnaire, and (f) the analysis of the information using multivariate techniques. From our point of view, our quantitative approach is very useful for grouping stakeholders not only according to their socioeconomic characteristics but also regarding their links to ecosystem services, their views on the subject, and their type of knowledge (Villamor et al. 2014). In addition, given that in our case information on social preferences from multi-stakeholders for ecosystem services in the study area is scarce; our approach contributes to building the history of links between people and ecosystem services through a comprehensive investigation of social preferences identifying a priori divergent interests. Different ecosystem services are presented on the same scale of analysis that facilitates the transmission of information to decision makers. In a second phase, crucial aspects such as synergies and divergences between actors in relation to their preferences for ecosystem services can be identified. Divergent interests could represent potential conflicts among actors and thus a deeper assessment of such a result should be performed. Another advantage of our approach is that all kinds of stakeholders might participate in the process because no formal knowledge is needed. Thus, methodological procedures as they are used here can uncover the main values of different social groups and be complemented with other qualitative and more participatory/deliberative tools.

In Chile, the conservation of biological resources has been defined as a technical issue, and although the efforts for local actors to participate in the decision-making process have increased, the notions associated with the rational management of natural resources persists (Razeto et al. 2019). We think that the involvement of different stakeholders including scientists and decision makers in decisions regarding ecosystem services is critical to achieving conservation goals together with human well-being. If current conservation planning strategies in Chile aim to reason with human well-being (e.g., protected areas management), local actors that use the ecosystem services must be included. Establishing participatory dialogues among different actors beyond the merely consultative is, in our opinion, one of the main challenges to achieving sustainable territories. Here we have provided, within the framework of ecosystem services, a way to begin unchaining links between people and these services, identifying relevant services for different actors and preliminary contrasts of perception of importance to them. This strategy allows us to visualize needs for deeper participation strategies, aspects that emerge as relevant, such as the contrast of views between farmers and scientists, the synergies between scientists and decision makers, and the invisibility of regulating services from the perspective of local actors.

4 Conclusions

- Sociocultural variables affect the way ecosystem services are valued. Local actors with multiple interests at stake but little power in decision-making processes prefer provisioning and some cultural services, whereas groups of stakeholders with some power in decision-making prefer regulating and cultural

services. The rural–urban dichotomy was also identified in ecosystem services valuation showing an urban interest in recreational benefits and a rural interest in benefits that sustain a traditional way of life.
- Divergences in ecosystem service valuation between stakeholder groups making it possible to identify divergences and synergies between stakeholders open the possibility to getting ahead of possible conflicts in decision-making processes by identifying winners and losers in specific management scenarios.
- The assessment of social preferences for ecosystem services integrates multiple value dimensions for different ecosystem services. That multiple dimensions do not always fit with ecosystem services classifications suggests that more qualitative research and discourse analysis on ecosystem contributions to human well-being are needed.
- Biosphere reserves represent ideal socioecological systems to study relations between biological conservation and human well-being and to explore sustainability scenarios with multi-stakeholder approaches.

Acknowledgements We thank Sebastián Saa for contributing to editing the manuscript. Funding was provided by Fondecyt Grant N° 1151063 (CONICYT-Chile).

Annex I

Ranking of ecosystem services according to importance score, mean vulnerability value, and importance-vulnerability index. The order of ecosystem services in the first column follows the importance-vulnerability index value from highest to lowest (Bidegain et al. 2019).

Ecosystem service	Importance value	Mean vulnerability	Importance-vulnerability index
Drinking water	124.0	4.0	498.7
Fresh air and climate change control	90.0	3.9	352.1
Conservation activities motivated by iconic endangered animal and plants species	79.0	4.4	349.3
Genetic pool of the plant communities in Central Chile with global relevance	89.0	3.7	332.1
Water regulation and retention	74.0	4.3	321.4
Educational value: possibilities of developing educational programs about local wildlife	89.0	2.7	240.0
Water for agriculture	49.0	3.6	176.4
Food derived from traditional agriculture	36.0	4.1	145.8
Medicinal plants (leaves, bark, roots)	39.0	3.5	136.5

Ecosystem service	Importance value	Mean vulnerability	Importance-vulnerability index
Symbolic plants	32.0	3.6	116.5
Beekeeping	33.0	3.5	113.9
Mean	**108.2**		
Identity and sense of place	38.0	2.7	100.7
Food derived from organic agriculture	36.0	2.8	100.2
Local ecological knowledge	24.0	3.8	90.4
Erosion control	31.0	2.7	84.2
Food from cattle (milk, meat)	24.0	3.4	80.4
Forage (trees and shrubs that are useful for cattle/browse)	18.0	3.4	61.2
Nature tourism	34.0	1.8	60.7
Possibilities to develop research	21.0	2.2	46.2
Spiritual and religious value	14.0	2.0	28.0
Rural tourism	19.0	1.3	25.3
Wild fruits (for human and animal consumption)	14.0	1.5	21.0
Soil fertility for agricultural crops and pasture	8.0	1.7	13.7
Seeds	9.0	1.4	12.6
Symbolic animals	7.0	1.8	12.6
Organic compost	8.0	1.5	12.0
Pest and disease control	9.0	1.3	11.7
Pollination	5.0	1.6	8.0
Genetic resources (e.g., wild species used in breeding programs)	3.0	1.8	5.4
Water for industrial use	5.0	1.0	5.0
Cultural tourism	5.0	0.8	4.0
Wood fuel	4.0	0.6	2.4
Industrial use of animals and plants	5.0	0.4	2.0
Plants for fibers/handcrafts	0.0	0.0	0.0
Food from hunting	0.0	0.0	0.0
Sport hunting	0.0	0.0	0.0
Coal	0.0	0.0	0.0
Wood for building	0.0	0.0	0.0
Resort tourism	0.0	0.0	0.0
Mushroom hunting for human consumption	0.0	0.0	0.0
Soil litter extraction	0.0	0.0	0.0

Annex II

Resulting Factor Scores from RDA Eigenvalues and variance explained by the analysis. Biplots were created using these data (Bidegain et al. 2019)

	F1	F2	F3	F4	F5
Eigenvalue	1.1848	0.6521	0.5248	0.3582	0.3321
Variance explained	32.3132	17.7860	14.3137	9.7684	9.0579
Cumulative %	32.3132	50.0992	64.4128	74.1813	83.2391
Ecosystem services					
Food from traditional agriculture	0.8107	0.1215	−0.1225	0.1268	0.2686
Symbolic plants	0.1845	0.2637	0.9502	0.0380	0.2586
Drinking water	0.6418	0.1419	0.3256	−0.3149	−0.5759
Water for agriculture	0.9353	−0.2668	−0.1897	−0.0946	0.1714
Conservation activities motivated by iconic endangered species	−0.1748	0.5939	0.2854	−0.2653	0.2822
Fresh air and climate change control	−0.5265	−0.1262	0.3017	0.6479	−0.2602
Water regulation and retention	−0.6316	0.8070	−0.3941	−0.1554	0.0149
Beekeeping	0.4279	0.3556	−0.1597	0.5196	0.2805
Educational value	−0.6016	−0.6751	0.1320	−0.2576	0.4231
Stakeholder sociodemographic characteristics (occupation)					
Scientists	−0.2248	0.0846	−0.1922	−0.1017	−0.0622
Employees of the Chilean National Forestry Corporation (CONAF)	−0.1428	−0.0339	−0.0815	−0.3130	0.0066
Business managers/owners	0.1284	0.2622	0.1959	0.0596	0.2710
Educators in schools and colleges	−0.0376	−0.2354	0.3878	−0.0017	−0.0309
Employees of the local government	−0.1272	−0.0867	0.0720	0.1731	−0.0838
NGO members	−0.3064	−0.2345	−0.0573	0.0999	0.0245
Members of local organizations	0.2588	0.0720	0.0063	0.0472	−0.1835
Small farmers	0.4303	0.1052	−0.2064	0.0977	0.0203
Tourism workers	−0.1031	0.0746	−0.0843	−0.1435	0.0250
Other sociodemographic and cultural characteristics independent of occupation					
Knowledge about protection figures	−0.3401	0.0822	0.0846	−0.1387	0.0842
Rural	0.3255	−0.0801	0.0536	0.0641	0.0963
Urban	−0.2469	−0.0217	0.0404	0.0747	−0.1136
Environmental organization membership	−0.0044	−0.0512	0.1283	−0.0289	−0.0467
Protected areas visitor	−0.5553	−0.0396	0.0307	0.0129	0.1122
Recycling habits	−0.3425	0.1701	0.1351	0.1060	0.0434

References

Alfonso A, Zorondo-Rodríguez F, Simonetti J (2017) Perceived changes in environmental degradation and loss of ecosystem services, and their implications in human well-being. Int J Sust Dev World Ecol 24:561–574

Asah ST, Guerry AD, Blahna DJ et al (2014) Perception, acquisition and use of ecosystem services: human behavior, and ecosystem management and policy implications. Ecosyst Serv 10:180–186

Balvanera P, Uriarte M, Almeida-Leñero L et al (2012) Ecosystem services research in Latin America: the state of the art. Ecosyst Serv 2:56–70

Bidegain I, Cerda C, Catalán E et al (2019) Social preferences for ecosystem services in a biodiversity hotspot in South America. PLoS One 14(4):e0215715

Blondel M, Fernández I (2012) Efectos de la fragmentación del paisaje en el tamaño y frecuencia de incendios forestales en la zona central de Chile. Revista de Conservación Ambiental 2(1):7–16

Carmona A, Nahuelhual L, Echeverría C et al (2010) Linking farming systems to landscape change: an empirical and spatially explicit study in southern Chile. Agric Ecosyst Environ 139:40–50

Carrasco LR, Papworth SK, Reed J et al (2016) Five challenges to reconcile agricultural land use and forest ecosystem services in Southeast Asia. Conserv Biol 30:962–971

Catalán E (2015) Relaciones humano-ambiente en el Parque Nacional La Campana. Una trayectoria de encuentros y desencuentros entre Comunidades Locales y el Área Protegida. Universidad de Chile, Santiago, Memoria

Ceballos G, Ehrlich PR (2006) Global mammal distributions, biodiversity hotspots, and conservation. Proc Natl Acad Sci 103:19374–19379

Cerda C, Ponce A, Zappi M (2013) Using choice experiments to understand public demand for the conservation of nature: a case study in Central Chile. J Nat Conserv 21:143–153

Cerda C, Losada T (2013) Assessing the value of species: a case study on the willingness to pay for species protection in Chile. Environ Monit Assess 185:10479–10493

Chan KM, Shaw MR, Cameron DR et al (2012) Conservation planning for ecosystem services. PLoS Biol 4:E379

Crouzat E, Martín-López B, Lavorel S (2016) Disentangling trade-offs and synergies around ecosystem services with the influence network framework: illustration from a consultative process over the French Alps. Ecol Soc 21(2):32

Daw T, Brown K, Rosendo S, Pomeroy R (2011) Applying the ecosystem services concept to poverty alleviation: the need to disaggregate human well-being. Environ Conserv 38(4):370–379

Delgado L, Sepúlveda MB, Marín VH (2013) Provision of ecosystem services by the Aysén watershed, Chilean Patagonia, to rural households. Ecosyst Serv 5:102–109

Díaz S, Quétier F, Cáceres D et al (2011) Linking functional diversity and social actor strategies in a framework for interdisciplinary analysis of nature's benefits to society. Proc Natl Acad Sci USA 108:895–902

Díaz S, Pascual U, Stenseke M et al (2018) Assessing nature's contributions to people. Science 359:270–272

Estévez R, Martínez P, Sepúlveda M et al (2019) Gobernanza y participación en la gestión de las áreas silvestres protegidas del Estado de Chile. In: Cerda C, Silva-Rodríguez E, Briceño C (eds) Naturaleza en Sociedad: Una mirada a la dimensión humana de la conservación de la biodiversidad. Ocho Libros, Santiago, pp 381–403

Fisher B, Turner RK (2008) Ecosystem services: classification for valuation. Biol Conserv 141(5):1167–1169

García-Llorente M, Iniesta-Arandia I, Willaarts BA et al (2015) Biophysical and sociocultural factors underlying spatial trade-offs of ecosystem services in semiarid watersheds. Ecol Soc 20:39

Hevia V, Martín-López B, Palomo S et al (2017) Trait-based approaches to analyze links between the drivers of change and ecosystem services: synthesizing existing evidence and future challenges. Ecol Evol 7:831–844

Howe C, Vira B, Switch H et al (2014) Creating win-wins from trade-offs? Ecosystem services for human well being: a meta-analysis of ecosystem service trade-offs and synergies in the real world. Global Environ Chang 28:263–275

Iniesta-Arandia I, García-Llorente M, Aguilera PA et al (2014) Socio-cultural valuation of ecosystem services: uncovering the links between values, drivers of change, and human well-being. Ecol Econ 108:36–48

IPBES (2018) Summary for policymakers of the assessment report on the intergovernmental science-policy platform on biodiversity and ecosystem services on regional assessment for the Americas. Secretariat of the Intergovernmental Science-Platform on Biodiversity and Ecosystem Services, Bonn

Jacobs S, Dendoncker N, Martín-López B et al (2016) A new valuation school: integrating diverse values of nature in resource and land use decisions. Ecosyst Serv 22:213–220

Kok MTJ, Kok K, Peterson GD et al (2017) Biodiversity and ecosystem services require IPBES to take novel approach to scenarios. Sustain Sci 426:177–181

Laterra P, Martin-López B, Mastrangelo M, Garibaldi L (2017) Servicios Ecosistémicos en Latinoamérica. De la investigación a la acción. Ecol Austral 27:94–98

López-Santiago C, Oteros-Rozas E, Martín-López B et al (2014) Using visual stimuli to explore the social perceptions of ecosystem services in cultural landscapes: the case of transhumance in mediterranean Spain. Ecol Soc 19

López-Santiago C, Aguado M, González-Novoa JA et al (2019) Evaluación sociocultural del paisaje: Una necesidad para la planificación y gestión sostenible de los sistemas socioecológicos. Aportaciones y utilidad de los métodos visuales. In: Cerda C, Silva-Rodríguez E, Briceño C (eds) Naturaleza en Sociedad: Una mirada a la dimensión humana de la conservación de la biodiversidad. Ocho Libros, Santiago, pp 107–141

Martínez-Harms MJ, Balvanera P (2012) Methods for mapping ecosystem service supply: a review. Int J Biodivers Sci Ecosyst Serv Manag 8:17–25

Martín-López B, Iniesta-Arandia I, Garcia-Llorente M et al (2012) Uncovering ecosystem service bundles through social preferences. PLoS One 7:E38970

Mastrangelo ME, Weyland F, Herrera LP et al (2015) Ecosystem services research in contrasting socio-ecological contexts of Argentina: critical assessment and future directions. Ecosyst Serv 16:63–73

Menzel S, Teng J (2010) Ecosystem services as a stakeholder-driven concept for conservation science. Conserv Biol 24:907–909

Mittermeier R, Gil P, Hoffman M et al (2005) Hotspots revisited: Earth's biologically richest and most endangered terrestrial ecoregions. Conservation International and Agrupation Sierra Madre, Monterrey

Moreira A, Barsdorf A (2014) Reservas de la Biósfera de Chile: Laboratorios para la Sustentabilidad. Academia de Ciencias Austríaca, Pontificia Universidad Católica de Chile, Instituto de Geografía, Santiago de Chile. Serie Geolibros N° 17

Myers N, Mittermeier RA, Mittermeier CG et al (2000) Biodiversity hotspots for conservation priorities. Nature 403:853–858

O'Brien K, Leichenko R (2003) Winners and losers in the context of global change. Ann Am Assoc Geogr 93:99–113

Razeto J, Skewes JC, Catalán E (2019) Prácticas de conservación, sistemas naturales y procesos culturales: apuntes para una reflexión crítica desde la etnografía. In: Cerda C, Silva-Rodríguez E, Briceño C (eds) Naturaleza en Sociedad: Una mirada a la dimensión humana de la conservación de la biodiversidad. Ocho Libros, Santiago, Chile, pp 75–106

Ramírez-Gómez S, Torres-Vitolas C, Schreckenberg K et al (2015) Analysis of ecosystem services provision in the Colombian Amazon using participatory research and mapping techniques. Ecosyst Serv 13:93–107

Seppelt R, Dormann CF, Eppink FV et al (2011) A quantitative review of ecosystem service studies: approaches, shortcomings and the road ahead. J Appl Ecol 48:630–636

Sepúlveda M, Estévez RA, Silva-Rodríguez E (2015) Manual para la planificación del manejo de las áreas silvestres protegidas del SNASPE. Programa de Naciones Unidas para el Desarrollo (PNUD), Santiago

Smith-Ramírez C, Armesto JJ, Valdovinos C (2005) Historia, Biodiversidad y Ecología de los Bosques Costeros de Chile. Editorial Universitaria, Santiago

Stoll-Kleemann S, O'Riordan T (2017) The challenges of the anthropocene for biosphere reserves. Parks 23(1):89–100

Tallis H, Kareiva P, Marvier M et al (2008) An ecosystem services framework to support both practical conservation and economic development. Proc Natl Acad Sci U S A 105(28):9457–9464

United Nations General Assembly (UN) (2015) Resolution 70/1. https://www.un.org/ga/search/view_doc.asp?symbol=A/RES/70/1&Lang=E. Accessed 26 Feb 2019

Villamor GB, Palomo I, López-Santiago C et al (2014) Assessing stakeholders' perceptions and values towards social-ecological systems using participatory methods. Ecol Process 3:22

Wilson MA, Howarth RB (2002) Discourse-based valuation of ecosystem services: establishing fair outcomes through group deliberation. Ecol Econ 41:431–443

Zorondo-Rodríguez F, Carrasco-Oliva G, Alfonso A et al (2019) Vinculando bienestar humano y servicios ecosistémicos: Ganancias y pérdidas de bienestar de comunidades rurales por cambios ecosistémicos. In: Cerda C, Silva-Rodríguez E, Briceño C (eds) Naturaleza en Sociedad: Una mirada a la dimensión humana de la conservación de la biodiversidad. Ocho Libros, Santiago, pp 207–239

Restoration of Riverine Forests: Contributions for Fisheries Management in the Pichis River Watershed of the Selva Central Region of Peru

Magaly Aldave, Edgardo Castro, Percy Summers, and Pedro Tipula

Abstract The Pichis river basin is located in the Selva Central region of Peru; it has large biodiversity of flora and fish that are of importance in local food security. However, deforestation of riparian forests directly affected their presence and the quality of the ecosystem. In this chapter, we discuss different techniques for the restoration of riparian forests, which would allow decision-makers at the community and government level to reach agreements for an integrated management of fisheries. The methodology integrated local indigenous knowledge and scientific research. The main results were the "Conservation Agreements for the restoration of riverine forests," with the restoration of up to 55% of the ecosystem by 2015. The most common techniques used in the recovery of the riparian forest were the management of natural regeneration, reforestation, and plant succession, using about 40 species of native flora. The other relevant result was the creation of the first "Fishing Surveillance Committee," confirming the reappearance of fish species. The local and regional government implemented and strengthened environmental regulations in favor of managing fisheries and riparian forests, including an economic fund to strengthen environmental monitoring and education in the Pichis river basin.

Keywords Social-ecological systems · Latin America · Complexity · Peru · Restoration · Amazonia · Fisheries

---

M. Aldave (✉)
Jardín Botánico de Missouri-Oxapampa, Pasco, Peru

Instituto del Bien Común, Jesús María, Lima, Peru
e-mail: mauraldave@gmail.com

E. Castro · P. Tipula
Instituto del Bien Común, Jesús María, Lima, Peru
e-mail: ecastro@ibcperu.org; ptipula@ibcperu.org

P. Summers
Instituto del Bien Común, Jesús María, Lima, Peru

Conservation International - Peru, Magdalena, Lima, Peru

Center for Biodiversity Outcomes - Arizona State University, Tempe, AZ, USA
e-mail: psummers@conservation.org

L. E. Delgado, V. H. Marín (eds.), *Social-ecological Systems of Latin America: Complexities and Challenges*, https://doi.org/10.1007/978-3-030-28452-7_20

1 Introduction

The Pichis river watershed, located in the Selva Central region of Peru is composed by a mosaic of native primary forests, secondary growth (degraded areas), natural protected areas, indigenous communities, and inland waterways. Furthermore, it is located in a transitional zone from lower primary forests to Andean cloud forests, which provides a suite of microclimates and biodiversity. By the year 2010, it was evident that more than half of the riverine forests were being deforested and inadequate fishing practices were threatening the presence, size and weight, articulating the use of local indigenous knowledge and environmental regulations. By 2015, different restoration techniques such as the management of natural regeneration, reforestation and plant succession, and conservation agreements allowed the restoration of riverine forests by up to 55% of previously degraded areas. Of 888 species of flora registered in the riparian forests, 112 species have ethnobotanical use and 40 had high priority in forest restoration, evidencing a high environmental quality. Forests stabilize climate, harbor biodiversity, and sustain local communities, providing goods and services that contribute to sustainable development (IUCN 2018); among the different tropical forest ecosystem types, one of the most overlooked but critical in providing these services are the riverine or gallery forests. These occupy the areas of rivers, streams, and other bodies of water, in between the minimum level of the waters and the maximum they reach during the rainy season and/or when rivers flood (ANA 2010; Aldave and Summers 2014). They differ structurally and in composition from other forest ecosystems for its strong interrelation with the aquatic ecosystem. Rich in vegetation species that serve as sediment filters and fixating soil, they are responsible for maintaining river water health and quality for aquatic fauna and fisheries. Riverine vegetation offers multiple other benefits for animals and people that live in the waterfront; terrestrial and aquatic wildlife find refuge and food, and use them as corridors to move from one place to the another (Aldave et al. 2013). People use the provisioning (primary materials, genetic resources, medicinal plants, ornamental species), and the regulation services (hydrology, microclimate, edaphic retention and formation, biological buffer) (Carrasco et al. 2014). Consequently, riverine forests do not only generate well-being for local communities but also form cultural corridors that conserve information for the quality of life of people, animals, and hydrobiological management (Aldave and Summers 2014).

Despite its ecological and socioeconomic importance, by the year 2010, more than half (50–60%) of the riverine forests of the Pichis river basin have been cut and replaced by agricultural crops, pastures, and secondary growth. As a consequence, erosion, the widening of the current channel, and the fishing and aquatic fauna that use this vegetation as corridors have been altered (Aldave et al. 2013).

Previous studies have estimated that fisheries contribute with around 62% of the protein for the average household in the Amazon (McClain and Llerena 1998). Most pressing threats that affect Amazonian fisheries are the modifications in the river's hydrology and the loss of riverine forests (IBC 2012; McClain and Llerena 1998).

On the other hand, the importance of forests with respect to the water cycle is clear; it slows down water flow, which infiltrates gradually through the soil, guaranteeing a stable supply of water throughout the year, even in during the driest months. Furthermore, forests filter water that flows into rivers, lakes, streams, and groundwater rivers, increasing its quantity and quality (ANA 2018). There is a clear need to restore these spaces, considering the restoration as the process that provides support to recover an area, ecosystem, or landscape that has been degraded, damaged, or destroyed (SERFOR 2018a).

With the objective to contribute to fisheries management from a social-ecological approach, we have integrated different research techniques based on traditional knowledge of the local population, scientific research and action research. In this chapter, we describe and analyze restoration experiences of riverine vegetation (between the years 2008 and 2012), hydrobiological resource research and management processes of fisheries at different levels (communal, municipal, and regional, between the years 2002 and 2018). This permitted the characterization of the social-ecological relations and the design of interventions for an integrated restoration of the Pichis river watershed. This study was developed by the Commons Institute (Instituto del Bien Común or IBC for its acronym in Spanish), riverine indigenous communities, through the Ashaninkas Nationalities of the Pichis River Association (ANAP for its acronym in Spanish) and the support of the Missouri Botanical Garden. The final objective is to restore and conserve riverine forests and contribute to the resilience of the aquatic ecosystem to secure a healthy ecosystem that fulfills the key functions that provide adequate conditions for fisheries, and therefore, for the well-being of the native communities through improved food security and empowering them in environmental governance for their development.

2 Methods

Study Area

The Pichis basin is located in the central zone of Peru, Pasco region, Oxapampa province, Puerto Bermúdez district (Fig. 1), on the eastern slope of the Andes and is part of the Andean-Amazon basin of the Pachitea River, a tributary of the Ucayali River. It includes a drainage area of 10,250 km^2, its altitude ranges between 216 m (confluence with the Palcazu River) and 1770 m (Cordillera de la Reserva Comunal El Sira). Geographically, it limits on the west with San Matías mountain range, on the south with San Carlos mountain range (both mountain ranges belong to the category of Protected Forest of the National Service of Natural Protected Areas by the State—SERNANP for its acronym in Spanish), on the east with the mountain range El Sira (also categorized as a Communal Reserve by the SERNANP), and the north with the lower basin of the Palcazu River. Based on the national climatic map, climates vary between humid tropical forest and very humid tropical (SENAMHI 2018).

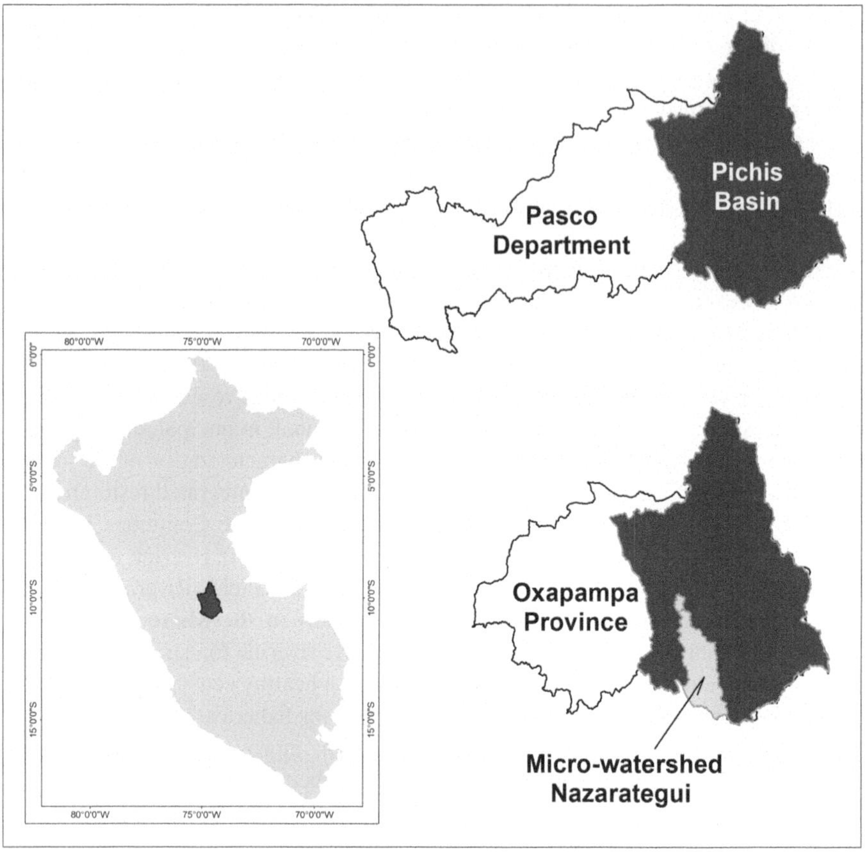

Fig. 1 The study area is located in the Pichis river basin, district of Puerto Bermúdez and Ciudad Constitución, province of Oxapampa, Department of Pasco in Central Peru

Rainfall ranges between 1812 and 4274 mm per year and the average temperature is 25.9°°C.

The basin houses a population made up of Asháninka and Ashéninkas indigenous people, who are grouped into approximately 90 native communities, migrant colonists from the coast and the Andean mountains and descendants of settlers of Austro-German origin. The indigenous people constitute approximately 80% of the population on the basin, and they are distributed throughout the basin dedicating themselves to activities such as agriculture, hunting, and fishing. Colonist from the coast and mountain range reside mostly in the capital of Puerto Bermúdez and are mainly engaged in trade, agriculture, and livestock. Colonists of Austro-German descent reside in the capital of the district and its surroundings; their main economic activity is livestock (Pinedo 2008).

The Nazarategui river micro basin, one of the 5 micro basins of the Pichis river basin, where 73% is protected by SERNANP through the Bosque de Protección San Matías San Carlos was selected for establishing the pilot fisheries management model. We consider information on the fishery at the level of the Pichis river basin, because the seasonality of the fish is very dynamic, both for feeding and spawning. The Nazarategui River has an average length of 130 linear km, which includes 14 native communities and three migrant colonist villages. Fourty-four percent of the riverside forests of the Nazarategui River have favorable navigability conditions, so the riverbanks are cleared to install houses, where the families that make subsistence agriculture and pastures are constituted (Fig. 2a). In addition, the Nazarategui River is a continuous source of food (native fish), water for consumption, and serves as a means of communication between the communities and the district capital. Also, it is important as a space for recreation and integration by the native communities (Fig. 2b).

Fishery in the Pichis River Basin

The Pichis basin is an important reference point in the area of ichthyological research in the Peruvian Amazonian Andes. The research component of the fishery resource had two stages; in the first stage (2002–2011) the main focus was to expand

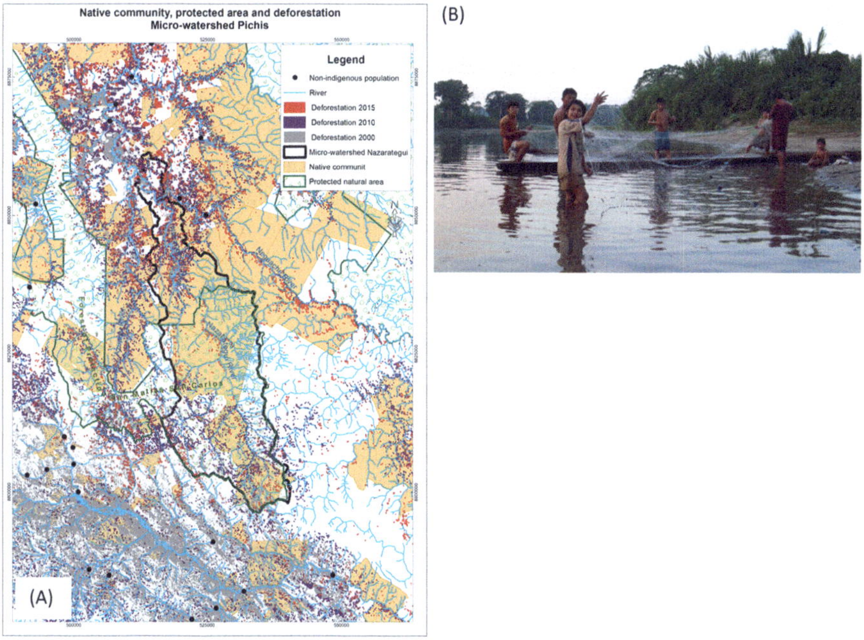

Fig. 2 (**a**) Deforestation in the Pichis basin from 2000 to 2015 and location of the Nazarategui river micro basin (Source: IBC). (**b**) Family fishing with nets in the Nazarategui River. Back, on the right margin, a secondary forest with a young secondary vegetable structure from 1 to 3 years old and on the left margin, an adult secondary forest

our knowledge of the fishing resource as well as create the baseline of the resource and fishing activity in the area. In the second stage (2014–2018), the methodology was modified to seek greater participation of the communities in the actions of monitoring the fishery resource, its management and use.

Inventory of Fish Between 2002 and 2006, we did exhaustive inventories of fish species covering the diverse habitats of representative aquatic ecosystems of the basin. Data collections took place at both low water and the intermediate rising water season (transition toward flooding season), using small trawl nets of various sizes (3–7 m) and with mesh openings of 0.7, 0.5, and 0.4 cm., ½″ and 1 ½″ mesh nets and hand nets (cal-cal).

Fishing Structured Surveys Structured surveys on fishing and fish consumption were applied in 2005 and 2008, applying random-stratified sampling. As a first step, the number of representative samples was established from the following formula: $\boldsymbol{n} = (F^2 \times N)/(F^2 + (2 \times p)^2 \times (N - 1))$, where $\boldsymbol{F}$ is the factor for a level of reliability at 95% (standard value = 1.96), $\boldsymbol{N}$ is the total population that is subject to the study, and $\boldsymbol{p}$ represents the margin of error at 5% (standard value = 0.05). The minimum sample of the number of households was distributed proportionally considering the hydrological cycle (low water, transition to rising water, and transition to low water), and variables such as the population type (indigenous or colonist), the type of access to the village (river or road) and the type of settlement (urban or rural). Finally, the entire population was surveyed randomly in each group.

Fish Monitoring Between the years 2003 and 2011, fish catches were recorded, in biannual intervals and coinciding with the most important times within the biology of migratory fish: "mijano" start time (term of rising or falling waters), maximum "mijano" (lowest water levels), start of spawning (beginning of crescent rising waters). The fishing was done taking into consideration the most representative habitats, such as backwaters, beaches, streams, and pools of rivers. Large nets (40–80 m) were used, applying methodologies for trawling and waiting.

Participatory Monitoring of Fishing Activity During the years in which fishery management was promoted in the Pichis, many indigenous people and settlers have participated in activities and responsibilities within the framework of the management of aquatic resources, both as environmental coordinators, in biological research work, and as members of the fisheries surveillance committees. Also known as "monitors" and distributed throughout the Pichis basin, they have been recording daily the fishing activity of their respective families since 2014. For this purpose, they use a unique registration form in which they record the date, the areas of fishing, the type of gear, and/or method and the effort applied. The collected fish are separated by species and identified with the help of a photographic catalog. The individuals of each species are weighed by a mechanical balance and measured separately, using 3 types of measures such as standard length (LS), fork length (LH), and total length (LT). In addition, in a complementary manner and to correlate it with the catches, the fish consumption of the family is recorded, noting the type of fish (species), the number of specimens of each type and the sizes, discerning whether it corresponds to breakfast, lunch, or dinner.

The Restoration of Riparian Forests in the Pichis River Basin

For the restoration activities, 44.31 linear km of deforested areas were prioritized, where eight native communities and two migrant colonists live. The following criteria were met:

Definition of the Size of Riparian Forests The Law on Water Resources of the National Water Authority, Law No. 29338, mentions that the riverbank is the area of land located between the minimum level of its waters and the highest floods, the marginal vegetation, are the immediate areas above the riverbanks (ANA 2010). The National Institute of Natural Resources determines, through the Forestry and Wildlife Law, Law No. 27308, the riverside forests as forests in protected lands by the biotic and abiotic characteristics (INRENA 2000). These two concepts were used as references for the delimitation of riparian forests in each native and migrant colonist communities through awareness workshops, according to a methodology adapted from the Farmers Field School (FFSs). The objective of these workshops was that the community, as the owner of the territory, including the spaces of the riverbanks and its adjacent vegetation, reversed current uses for agriculture and pastures, back to native vegetation.

Inventory of the Riparian Flora Two types of inventories were made, botanical and ethnobotanical, and applied at different times. The botanical inventory (2008) was carried out in three of the five micro basins that exist in the basin, prioritized by being in the headwaters. The vegetation of seven natural pools of the rivers was evaluated. Each edge of the pools of rivers had a temporal plot of 500 m^2 installed by a transect perpendicular to the edge of the river 100 × 5 m. The characterization consisted in diagramming in the 100 linear meters (1) *the geoform,* which is influenced by the topography, soil type, and floristic composition, and (2) *the plant structure*, which considers the different forms of life or habits present, and the natural and/or anthropogenic disturbance. Also, botanical collections of all the species present were made within the plot. The abundance of species as well as their life forms were analyzed through relative abundance (RA) and the mixing ratio. The Shannon Weaver index (H′) was used to determine the species diversity of the flora and the Morisite Horn index for the similarity of species present among river pools.

The second inventory (2009) considered botanical and ethnobotanical collections. The botanical inventory consisted of identifying the ecological interrelations of the riparian flora and the feeding of fish. The ethnobotanical inventory was made as a rescue of ancestral information and revaluation of riparian forests. Temporary plots of 1000 m^2 were installed on each bank of the river, 100 m upstream of each natural pond. In total, eight natural pools of rivers located between the head and mouth of the Pichis river basin were evaluated. In each 100 m linear transect, sub transects of 50 × 2 m were done in an angle of 90°, separated every 25 m, and distributed alternately.

For the ethnobotanical information of the flora present in the riparian forests, interviews and structured surveys were used with local indigenous people and experts on the most important and used plant species including categories such as food (for people and animals), traditional medicine and technology (use of wood, crafts, fodder, toxic, ornamental). With the list of prioritized species, botanical material was collected for the determination of the botanical nomenclature. The information was processed and in a second field trip, the information was validated (names in language, scriptures, and relationship with the photographs) through community workshops.

Identification of the Ecosystems in Riparian Forests Fisheries research and cartographic information determined the type of ecosystems present in the banks of the Pichis river basin. Considering the dynamic distribution of fish in the basin, it was necessary to carry out inventories by ecosystem types and not only on the waters above natural pools of rivers. The third inventory (2010) consisted of recognizing the flora present in each river system, and its connectivity within the broader context of a riparian landscape corridor. For this, we prioritized the area with the greatest negative impact on the riparian forests, that is, the micro basin of the Nazarategui River where 44.3 linear km coincide with the area of greatest navigable accessibility and deforestation.

The inventory consisted of differentiating each ecosystem of the riverbanks by (1) *the structural characteristics of the vegetation*, and (2) *its age*. A technical sheet was used to identify and georeference the start and end points of each ecosystem, recognizing the species present according to the common names of the plants. In the case of the riparian vegetation, both at the beginning (head of the river) and at the end (mouth of the river) of the Nazarategui River, a temporary plot of 1000 m^2 was installed in each ecosystem representative of the micro basin. Botanical collections were made in each ecosystem type.

During the years 2011, 2012, 2014, and 2015, the riparian systems continued being monitored, in a participative way with the native communities, in each margin of the Nazarategui River.

Resource Use and Threat Mapping Methodology Using maps as a tool for gathering information and managing the territory, participatory mapping was carried out in each of the native and colonist communities of the Nazarategui river micro basin. Based on a proposal of the range of different types of use of riparian ecosystems and the threats present in the area, a legend and a set of symbols were developed for the registration of the information related to the use of the territory and the threats to the aquatic ecosystems. Threats included any activity that weakens soils such as logging, grazing, and burning.

Through workshops, that included mapping exercises with the local population, information was gathered in each native and colonist community. Baseline maps with information such as contour lines, rivers and streams with their respective toponymy, and limits of the communal territory were used. Information collected in the field through the mapping exercises was processed using the ArcGIS software. With the first draft of the mapping of the use of resources and threats to the aquatic ecosystems processed, each native and colonist community was visited again to validate them. Finally, the maps, corrected and validated by the community, were processed and a final map of riverine ecosystems, threats and uses were drawn for each native and colonist community of the Nazarategui River.

Techniques for the Restoration of River Forests Depending on the type of ecosystem, level of soil/vegetation degradation, connectivity and current use, different techniques for the restoration of riparian forests were prioritized: (1) *recovery by plant succession,* which used progressive plant succession allowing the ecosystem to recover naturally, (2) *vegetative enrichment* to strengthen the spaces that already have secondary vegetation requiring the incorporation of additional plants that help accelerate the recovery of the ecosystem; this technique considers enrichment by natural regeneration, seeds, vegetative propagation, and seedlings produced in nurseries, and (3) *recovery by reforestation*, especially in highly degraded areas, for which the installation of a temporary nursery and seeds for the propagation of native vegetation seedlings was required. For the three techniques, different distribution models of plant species were tested, with ancestral knowledge of the population being key for the recovery of the forests and the results obtained from botanical and ethnobotanical inventories.

For the selection of the species to be used in restoration activities, we considered the following aspects: formation of the root system of the plants, soil stabilization, form of plant growth, provision of shade, food production, mainly for fish, and socioeconomic interest for the local population, in addition to the ecological association of the species. In all the techniques, we designed and promoted a participatory monitoring system of the growth and development of plants by community members.

Governance for Fisheries Management

Several levels of governance and governability were worked on in parallel, to promote regulations and initiatives that provided institutional support to the strategies promoted for the management of the fishery. We use an action-research approach, where the activities proposed in the different levels of governance are studied and adapted as part of a continuous learning process that is nourished, at the same time, by the results of the studies of the fishery resource and the restoration of the riparian vegetation. Additionally, due to the existence of multiple decision-making levels, some formal and others informal, we adopted a polycentric governance approach

(Ostrom 2009), where instead of recognizing a top-down hierarchical governance system, we promoted one where the different decision-making spaces compete and cooperate, interact and learn from each other, and responsibilities are adjusted to the scale where the services are provided and the rules are exercised more efficiently/effectively.

Community Agreements for the Restoration and Conservation of Riparian Forests, and Fishing Surveillance Native communities use the Community Assemblies, where the entire population comes to make decisions and agreements of common interest, which are subscribed in *Actas comunales* for its subsequent implementation, as the decision-making space for natural resource management issues. Community Assemblies were the management space to discuss and determine the sizes of the riparian forests to be conserved, the restoration techniques, the monitoring of the riparian forests, the surveillance and control of the forms of fishing, and the strategies for the management of the fishing.

In order to reach the agreements, each family was first informed and sensitized through adapted Farmer Field Schools (FFSs) that provided alignment for reaching a consensus at the community level. Priorities and needs of each community were systematized, discussed, and approved in assemblies at the micro-watershed level. Mechanisms were adapted to ensure adequate communication between the community and the intermediary institution, in this case, the non-governmental organization (NGO) *Instituto del Bien Común* (IBC). Each community had a *community environmental coordinator* to give support on issues of restoration and conservation of riparian forests and a *community vigilant* to strengthen fisheries management. Both coordinated with their community leaders and later informed and discussed with the whole community. Once agreements were approved at the community level, they were then proposed to their local and regional governments.

Decision-Making Based on Governance The Municipal Environmental Commission (Comisión Ambiental Municipal or CAM for its acronym in Spanish) is a tool that seeks to organize environmental management in Peru, setting priorities and establishing collaboration procedures between the public and private sectors and civil society, considering three levels of action: national, regional, and local. In this last level, the Local Environmental Management Systems (Sistema Local de Gestión Ambiental or SLGA for its acronym in Spanish) is a fundamental tool for local governments to organize themselves, plan concertedly, and address environmental needs in rural areas (Soria and Rodríguez 2008; Rodriguez and Collado 2013). The CAM is one of the actionable spaces of the SLGA, and it constitutes an interinstitutional participatory space for making relevant decisions for environmental management, being led by the Mayor of the Municipality.

At the district level, the strategies were aimed at institutionalizing the SLGA in the district municipality of Puerto Bermúdez, promoting by Municipal Ordinance the creation of the CAM and its environmental management instruments to ensure regulations that promote fisheries management and the conservation of riverine forests.

At the regional level, through advocacy and creating awareness among the Regional Government authorities of the Pasco Regional Government (GOREPA for its acronym in Spanish), Ordinances and financial sustainability instruments were sought as strategies for implementing the management of fisheries, including the conservation and restoration of riparian forests ecosystems. As part of this integrated management system, native and colonist communities are recognized as key actors in this process.

3 Results

Fisheries in the Pichis River Basin

From Fish Inventories Research results show that the ichthyofauna in the Pichis River is represented by 241 species, which are included in 10 orders, 30 families, and 144 genera. At the level of orders, the Characiformes are the best represented with 121 species (50%), followed by the Siluriformes with 88 species (37%), Perciformes with 13 species (5%), Gymnotiformes (electric fish) with eight species (3%), and other less diverse orders such as Myliobatiformes, Beloniformes, Clupeiformes, Cyprinodontiformes, Synbranchiformes, and Pleuronectiformes, which together comprise nine species. At the family level, the most representative groups correspond to Characidae with 92 species, Loricariidae with 39 species, Pimelodidae with 24 species, and Cichlidae with 12 species.

The main channel of the Pichis River and its small tributaries, such as streams, registered the presence of 209 species, of which 78 are frequently consumed by the local population. According to the major tributaries, the largest number of species corresponded to those located to the south of the basin, such as Azupizú and Nazarategui with 113 and 73 species, whose drainage comes from the San Matías—San Carlos mountain range. The Characiformes are represented mainly by species of small size whose level of economic advantage is low; nevertheless, they are very frequent and abundant, emphasizing *Bryconamericus beta*, *Astyanax bimaculatus*, and *Creagrutus changae*, among others. Those of medium size (approximately 40 species), make up a group of high importance for local consumption, 10 of them are commercially significant, and most are characterized by moderate migrations with medium distance displacements (Barthem et al. 1997; Barthem and Goulding 2007; Usma et al. 2010), highlighting *Prochilodus nigricans*, *Brycon hilarii*, *Salminus iquitensis*, *Mylossoma duriventre*, *Leporinus friderici*, and *Schizodon fasciatus*. These species have a wide distribution in the Pichis basin, reaching some of them at 500 m above sea level. They are characterized by high fecundity, spawning in the bed of the Pichis River and surrounding areas at the beginning of the rainy season, although some of them may extend until the end of the flood. Another group of Characiformes corresponds to those species that make short migrations with local displacements, such as *Serrasalmus rhombeus*, *Hoplias malabaricus*, and

Cynopotamus amazonus, among others. These species are characterized by low fecundities, parceled spawning and parental care, can present broad distributions such as *H. malabaricus*, reaching 450 m of altitude, or moderate distributions such as *S. rhombeus* and *C. amazonus* (Castro 2013).

The Siluriformes are represented by species with a wide range of sizes, more than 40 species are consumed quite frequently, and among them, 15 are of commercial importance in the Pichis. Among the family Pimelodidae, species such as *Brachyplatystoma rousseauxii* and *Brachyplatystoma vaillanti* are registered, which are categorized within the group of large migrators (Barthem and Goulding 2007). Both species are not abundant in the basin, have a range of reduced distribution, not exceeding 240 masl. Some field observations suggest that they spawn below the confluence of the Pichis and Palcazú rivers (Pachitea River) during the growing season. Other Pimelodidae are categorized within the group that performs moderate migrations; standing out among them are *Pseudoplatystoma punctifer*, *Pseudoplatystoma tigrinum*, and *Zungaro zungaro*. These three species are characterized by having high fecundity, spawning in the bed of the river Pichis during the growing season, and having moderate to large (*P. tigrinum*) distributions reaching altitudes below 380 m. Another important group corresponds to the family Loricariidae, whose species make short or local migrations. This group, as in other Andean-Amazon zones, constitutes a diverse and widely distributed group in the Pichis, adapted to environmental variations and extreme conditions; They are also an important part of the diet of indigenous populations. Genera such as *Chaetostoma*, *Hypostomus*, and *Panaque* are very well represented (Castro 2013).

Regarding some scientific developments, at least five species were reported that have not been described to date, constituting possible new species for science, including a variety of the genus *Hemibrycon* (Characidae) recorded in the Quirishiari stream, another of the *Moenkhausia* genus (Characidae) in Quebrada Kuashironi and Quirishiari, *Imparfinis* (Heptapteridae) in Quebrada Carhuaz. Two possible new genera of the families Characidae and Heptapteridae, registered in the Apurucayali basin and the Karanganao stream respectively, were also reported.

Monitoring and Natural State of Fisheries Resources The information collected since 2004 and 2011 was analyzed in order to evaluate the behavior of the communities and fish populations, resulting in some changes regarding the composition, abundance, and size structure of some species in the Pichis basin.

Medium-sized species such as *Brycon hilarii* "sábalo cola roja," *Salminus affinis* "sábalo macho" and *Pimelodus ornatus* "manitoa," showed a decrease in catches over the years, to the point that they were not recorded at the end of the monitoring period (2011). Other large species such as *Pseudoplatystoma punctifer* "dondella" and *Pseudoplatystoma tigrinum* "puma zúngaro" also showed a decrease in catches and their registration is occasional. However, the presence of small species such as *Triportheus angulatus* "sardina," *Pimelodus blochii* "cunshi" and *Centromochlus heckelii* "aceitero" increased in the last years of monitoring (Castro 2013).

The abundances recorded by the catch index per unit of effort (CPUE) in two habitat types showed a tendency to decrease in recent years. In Presvi beach

(Pichis River), the CPUE dropped from 7.6 kg/trawl in 2004 to 0.27 kg/trawl in 2011. Likewise, in Pozo Charapa (Azupizú River) the CPUE decreased from 17.6 kg/trawl in 2004 to 271 g/carryover in 2011. For both cases, there was a fairly accelerated decrease in abundances between the 2004 and 2006 period (Castro 2013).

Regarding the sizes, *Prochilodus nigricans* "chupadora," a species of high importance for local consumption, recorded a decrease in the average size of 27.3 cm (total length) in 2002 to 19 cm in 2010; likewise for *Pimelodus blochii* "cunshi," whose average size decreased from 22.3 cm in 2004 to 18.2 cm in 2010. Other species such as *Curimata aspera* "yahuarachi" and *Mylossoma duriventre* "palometa" showed the same tendency, although not as marked (Castro and Orihuela 2011).

Restoring Riparian Forests

Types of Ecosystems Uses and Threats, and Definition of the Size of Riparian Forests The forest inventories of the years 2008 and 2009 identified the types of riparian ecosystems, being: (1) *by its geoform*, such as flooded forests, beach vegetation, vegetation of barriada or barrizal, tahuampa forests and restinga forests; and non-flood forests, which include gallery forests; and (2) *by the plant structure* such as secondary forests, secondary forests associated with agricultural crops, and pastures. The studies from 2010 to 2012 on the ecosystems, on both banks of the Nazarategui River, identified 11 ecosystems, being those that have the most negative impact, or that interrupt the connectivity of the riparian forests: the secondary forests, from 2 to 4 years of age, with 27% and 25% presence, temporary and permanent agriculture with 13% and 20% and pastures with 4% and 5% presence. The composition of habit types present in these ecosystems is mainly represented by herbs (35.2%), shrubs (33.9%), trees (29.9%), and reeds and bamboo (26.4%).

The methodology of mapping the use of resources and threats allowed identifying and reflecting on each community and colonist population the form of occupation and use of the river banks as well as the threats that put their food security at risk, the quality of the waters, and the fish. It recognized as threats and main agents in the fragmentation of the forests, the activities of grazing, felling and burning of native vegetation to start agriculture and pasture, among others. As a threat to fisheries, it identified fishing with explosives, fishing nets with small sizes, and the use of toxic plants. In addition, weakening of soils in riverside forests was caused by the increase in the establishment and use of ports or boat stations for each family, a cultural change pattern, since before there was only one in each community and now they are expanding their use to one per household.

Through the FFSs, the regulations of the Water Authority Agency (ANA for its acronym in Spanish) were socialized and analyzed, and the needs of the communities were identified, in order of priority (1) food security, through native fish, (2) water quality of the rivers for human consumption, and (3) navigability. By communal agreements, each native community decided to physically identify and paint, the marginal strip or riverbank forest, to prohibit the felling and burning of riverine vegetation, to minimize the number of ports and to limit agriculture activities behind the marginal strip. The delimitation of the riverine forests was by painting the largest trees or reforesting, using it as a boundary to designate spaces for protection and conservation. The definition of the size varied in each native community, being between 18 and 25 m from the edge of the river to the community. The monitoring of these conservation agreements for the restoration of riparian ecosystems was carried out by the community members themselves during the years 2011 and 2012.

Diversity of Species and Restoration Techniques Botanical inventories made in 2008 shows that four of the seven river pools located at the head of the Pichis river basin have greater H' diversity on both banks of the river, with an average between 2.48 and 3.50. The greatest similarity was found among the species present in the right margin of the pools (four pools) with respect to the species present in the left margin, where only two pools have the greatest similarity (Aldave et al. 2010).

Botanical and ethnobotanical inventories carried out between 2008 and 2010, in a total of 0.04 km^2 of temporary plots; 120 botanical families were registered, distributed in 434 genera and 888 botanical species, of which 112 species are ethnobotanical (Aldave and Summers 2014). The 10 families with the highest number of species in descending order are: Fabaceae (53 species), which is what was expected and registered for the lower jungle by Vásquez and Phillips (2000) and Vásquez et al. (1997) followed by the Piperaceae family (35 species), Araceae (29 species), Rubiaceae (26 species), Poaceae and Malvaceae (22 species each), Moraceae and Acanthaceae (21 species each); Melastomataceae (20 species) and Asteraceae (19 species).

Of the 112 ethnobotanical species recognized by the native communities, for having some use or local benefit, it is observed that the ribereña population gives greater importance to the use of medicinal plants (77.9%), followed by species for technology purposes (10.4%), and food (9.6%) mainly for animals such as monkeys, parrots, and fish. It was also observed that 4.5% of the species have multiple uses, while 95.5% are of exclusive use. The five families that have greater use are Acanthaceae (13%), Rubiaceae and Fabaceae (8% each) and, Araceae and Bignoniaceae (5% respectively). The assignment of the same name for different species was observed, for example "piri piri" which refers to the family Poaceae in different genera and species, or the "leaves to vaporize" that includes the families Rubiaceae and Acanthaceae.

Techniques for the Restoration and Conservation of Riparian Forests Recovery by plant succession and enrichment was the technique most used by the communities for the young secondary ecosystem or locally known as *purmas*, which are classified by the age of the vegetation, being: purma from 1 to 3 years, as the first phase of succession and purma from 4 to 6 years as the second phase of plant succession. In the first phase of the succession the most common species were herbs such as: *Pfaffia paniculata, Echinodorus* sp., *Dichorisandra hexandra, Carludovica palmata, Lantana camara,* and *Urera baccifera*, among others. To accelerate the recovery of this phase, we enriched through reforestation with species such as *Calliandra angustifolia, Cecropia membranácea, Guarea macrophylla, Zygia longifolia, Inga acranea* that provide shade, has extensive root system, and the seeds are used as food by the fish and other animals. On average, between 400 and 625 plants per km were introduced, where only 1 m of each seedling was cleaned to avoid competition with other species. In the second phase of the plant succession, the pioneer bushes and trees take more presence than the herbs, such as: *Virola calophylla, Trichilia pallida, Acalypha sp, Inga acreana, Ficus trigona, Neea divaricata,* and *Triplaris americana*, among other species. Species such as *Hevea brasiliensis, Geonoma stricta, Matisia cordata,* and *Cecropia membranacea* were planted on these species. In this second phase, the soils begin to show the recovery of organic matter. Plants are spaced between 400 and 320 plants per km.

The adult secondary ecosystem or late secondary forest, comprised of plant species 7 years of age and older, is considered as the third phase of plant succession. In this phase, the diversity of species increases in the same proportion as the productivity of the soils. The most common species are: *Acalypha macrostachya, Erythrina edulis, Chrysochlamys weberbaueri, Dichorisandra hexandra, Trichilia maynasiana, Carludovica palmata, Psychotria* sp., *Ficus* sp., *Trichilia maynasiana, Piper marowynense, Brosimum utile, Cecropia membranácea,* and *Zygia longifolia*, among other species. At this stage, species are no longer introduced to enrich forests since diversity and competition are greater.

In the case of permanent and temporary agriculture, and pasture ecosystems, the technique used was reforestation. The distribution of the species was designed by rows parallel to the edge of the river considering as well the geoforms on each bank. For example, the first row comprised by the first 10 m from the edge of the river to the community, if the geoform was gallery forest with slight slopes and relative heights between 10 and 15 m above the river level, we used species such as *Gynerium sagittatum, Calliandra angustifolia, Zygia longifolia,* and *Paspalum* sp. This is due to the structure of the roots that form a network to avoid erosion during the flood of the rivers and because they are small plants, like herbs. If, the geoform of the riverbank was a bajial forest, which are floodplain forest vegetation, we used species such as *Gynerium sagittatum, Calliandra angustifolia, Zygia longifolia, Swartzia simplex, Cecropia membranaceae, Hevea guianensis,* or *Paspalum spp*, that protect the soil and stabilize it due to their proportions and ecological habits.

In the second row, the next 10 m of the bank adjacent to the edge of the river, we placed shrubs and trees, which in addition to stabilizing the soil provide shade generating an ideal microclimate for fish; species included *Inga edulis, Sanchezia ovatam Guarea pterorhachis,* and *Swartzia simplex,* among others. The third row and onwards medium-sized arboreal species were selected that contributed other uses for local populations such as for timber: *Jacaranda glabra, Andean Eschweilera, Helicostylis tomentosa, Dendrocalamus asper, Erythrina edulis, Matisia odorata,* or *Ormosia*. These species, having a larger size, also become a biological corridor for other animal species such as birds and mammals.

Permanent crops on the banks included species such as *Carica papaya, Bixa orellana, Musa paradisiaca,* and *Theobroma cacao.* These crops were enriched as agroforestry systems, promoting the reforestation of forest species such as *Trichilia maynasiana, Guazuma crinita, Inga edulis,* or *Cedrela odorata.* We promoted the enrichment of tree species to a total of 278 to 321 plants per hectare. Herbs such as *Eirmocephala brachiata, Solanum appressum,* or *Paspalum* sp., were kept and promoted within this agroforestry ecosystem to avoid erosion during the increase of the water flow in rainy seasons.

Regarding temporary crops, these occur mainly on beaches, which are spaces of accumulation of sand and nutrients carried by the river current and left in the beaches when the water recedes. It coincides with the dry season, which means between the months of April to September. These spaces are used to grow *Zea maize, Manihot esculenta, Arachis hypogaea, Citrullus lanatus, Saccharum officinarum, Calathea allouia,* or *Ipomoea batatas.* As they are temporary crops, their harvesting cycle is short, so their affectation on beaches is null, due to their rapid natural recovery.

It is worth mentioning that for the reforestation, the production of seedlings in temporary nurseries was carried out, which includes a series of cultural techniques that are not typical of indigenous cultures, requiring greater technical support once the technique has been introduced. The supply of seeds has to be planned to follow the seasonal phenology of riparian vegetation species and depending on whether these are obtained through fruits, natural regeneration, and/or stakes (for vegetative propagation).

Governance and Environmental Governability

The Communal Agreements In this space, it was the women who led the initiative and the process, conducting dialogues in each native and colonist community, to finally hold a meeting in the micro basin with the local and regional government authorities, such as PRODUCE, ANA, Direccion Regional de Agricultura (DRA), District Municipality, National Police, and ANAP. Subsequently, the communal agreements were socialized and the first Local Committee of Fisheries Vigilance—COLOVIPE—was formed in the Pichis river basin, integrated by a member of each native community, among the Environmental Coordinators and Fisheries Vigilantes and representatives of the settlers of the Nazarategui River. This committee was recognized by PRODUCE, had financing from the local government for its operation, and was trained by both PRODUCE and the Ecological Police. Its responsibility was to monitor the forms of fishing and activities that can impact riparian forests, identify and train offenders and inform the competent authorities. The Ecological Police sanctioned if reiterative.

Within the communal agreements, the delimitation of the riverside forests was considered for their restoration, being prohibited the rubbing, felling, and burning. In the ecosystems of permanent agricultural crops, no more crops were developed, but instead, they waited until they reached their stage of production (varying between 3 and 15 years), installing Agroforestry Systems for the protection of the soil. At the same time, the native communities started the production of crops behind the delimited riverine forests. The installation of ports was reduced, the recovery techniques were agreed, the plant succession being the one most in accordance with its culture, and fishing forms were regulated, mainly the use of explosives and the size of the opening of the nets.

Decision-Making Based on Governance Agreements for fisheries management between the native communities and the local and regional authorities took place using the spaces of the CAM, where environmental promoters and community vigilantes informed and demonstrated the efficiency/efficacy of their communal agreements to the local authorities. The SLGA of the district municipality of Puerto Bermúdez included in its Local Plan of Environmental Action 2011–2021, in Goal 4 of Forest and Climate Change, the reduction of the degradation of the soils of the marginal belt of the Pichis river basin, having as responsible parties ANA, MDPB, ANAP, INRENA, SERNANP, IBC, and the Regional Agrarian Directorate (DRA). Likewise, in Goal 5 of Biological Diversity, the promotion of responsible subsistence fishing, contributing to the conservation of water ecosystems in the basin, with DIREPRO, ANA, DRA, INRENA, IBC, ANAP, and MDPB as the responsible parties (MDPB 2011).

With the creation of the first COLOVIPE, the MDPB issues an ordinance prohibiting the use of dynamite and agrochemicals for fishing declares the intangibility of the riparian areas and generates a fund that allows for monitoring, as well as serving as an advisory board for community members of the COLOVIPE.

4 Discussion

The Pichis river watershed, with 241 aquatic fish species, represents approximately 28% of all registered species for the Peruvian Amazon. It can be considered an area with high biodiversity when compared with other larger watershed like the Madre de Dios River, which reports 287 species, Urubamba watershed with 232 species, Pastaza River (Peru-Ecuador) with 312 species, and the Napo River (Peru) with 242 species. The taxonomic composition of the ichthyofauna reported for the Pichis River is related to the fish composition registered for other continental waters of Peru (Ortega et al. 2011).

The Pichis river watershed is one of the most important areas of Andean foothills in the Peruvian Amazon for the bioecological processes (reproduction, feeding, growth) of numerous migratory commercial fish species. It's very likely that some species maintain genetic flow with populations in proximate hydrographic systems (Palcazu and Pachitea rivers), as well as with the larger Ucayali hydrological system, and probably with some large Pimelodidae, traveling as far as the Amazon estuary as reported by Barthem and Goulding (1997). About their reproduction aspects, both, the group that does moderate migrations (Characidae and Pimelodidade) as well as the large migratory species (Pimelodidae), spawn between the start of the rains and the end of the flooding season. Similar behavior has been registered for many researchers for the Peruvian and Brazilian Amazon, as well as for aquatic systems in Colombia (Montreuil et al. 2001; García 2016; Barthem et al. 1995; Lasso et al. 2011).

The patterns of fishing activities in the Pichis River, and in general in the Amazon, are usually different throughout the year (seasonality), presenting differences in the fishing localities, gear and methods used, species captured and sociocultural aspects of fishermen (Rodríguez et al. 2018; Barthem et al. 1995; Castro et al. 2008). In general, the gear used in Amazonia is diverse, and its use is related to a specific environmental condition and type of species; and especially the fishing nets (cast net, mesh nets, traps, etc.), as well as creels, bow and arrows, are all important in artisanry fisheries (Barthem et al. 1997). The use of prohibited techniques, as the use of explosives and natural toxic substances are not only reported for the Pichis River, but also are common in other areas of the Pachitea River (Castro et al. 2008), as well as in other areas of the Amazon where artisanry fisheries are developed (Barthem et al. 1997, Van Damme et al. 2008). In relation to the consumption of fishes per capita reported for the Pichis River (0.15 kg d^{-1}), this is in line with other values obtained for other regions of the Peruvian Amazon, including values obtained for the Pachitea River (Castro et al. 2008).

The demarcation of riverine forests for the Amazon, was one of the most important but poorly addressed issues at the policy level, due to its poor applicability to the reality of land tenure in indigenous communities, which can lead to territorial conflicts. This specific case had the support of Peru's water agency (ANA for its acronym in Spanish) to provide guidance on how to treat the vegetation strip and associated goods, recommending 50 m. However, since most of these areas are used for subsistence agriculture, communities agreed in an initial phase, to cede between 20 and 30 m of riverine vegetation to initiate the restoration processes.

The use of combined techniques for the restoration of forests, like the enrichment of native species and natural ecological succession is the most efficient way to restore forests (Kometter and Gálmez Márquez 2018). The presence of mammals, birds, and insects that have an important role in the dispersal of seeds, as well as fishes that contribute to the dispersion, demonstrate the importance of the maintenance, and management of riverine social-ecological systems (Cortijo 2012). Additionally, this technique is more aligned with the traditional way of forest management of indigenous communities. However, they are the least used in Peru (4–6%), considering forest plantations as the most commonly reported with 72% (SERFOR 2018b). Other tools used included in the planning of restoration activities were the participatory mapping processes that allowed visualizing and community ownership for managing their territories (IBC 2012).

Native communities identified and prioritized close to 40 species of vegetation from the 888 species registered in these ecosystems. These were further categorized into reproduction by seeds, natural regeneration, or cuttings for vegetative propagation. This number is higher than those mentioned by SERFOR (2018b) that reports

15 species for forest restoration, and for the specific case of riverine forests only mentions the use of *Guadua angustifolia*. Uses are limited to erosion protection and economic use, highlighting the need for further scientific research, including botanical collections, of the different types of riverine forests and associated goods and services it provides. The interrelations between vegetation types and human uses are further areas of research needed to understand the real dimensions of riverine forests as social-ecological systems.

The restoration of riverine forests situated in between the lower Amazon foothills and the higher cloud forests are needed to protect erosion of river banks, provide shade and vegetation along rivers, etc. In this respect, Cortijo (2012) mentions that only examining feeding habits of fishes like *Brycom hilarii*, in 67 stomachs collected, found up to 27 botanical families that include 24 genera, and in the case of the *Leporinus friderici*, from 53 stomachs collected, we found 26 botanical families and 18 genera. We found that the feeding index of these fishes was more important in the plantae kingdom (58.4%), reporting fruits and seeds of the Fabaceae, Arecacear, Celastraceae, and Connaraceae families as the most consumed (82.5%). Hence, we can consider the diversity of vegetation in the restoration of riverine forests as a key element of restoring for fisheries management. Furthermore, the importance of these for ethnobotanical uses by indigenous people provides riverine forests with additional value for conserving and restoring them.

While the experience of IB with the restoration of riverine forests was developed between 2008 and 2012, young secondary forests ecosystems recovered by 10% between the years of 2010 and 2012, following a restoration monitoring system done in 2010 and 2012. Follow-up monitoring of restoration activities indicates that by the year 2015, 55.5% of riverine forests were restored in the micro basin of the Nazarategui River, especially in the areas of indigenous communities (Fig. 2a). Furthermore, fisheries species with a high demand for environmental quality were reported in higher numbers in the monitoring of fisheries in 2014. These numbers confirm that the higher levels of environmental governance and exhaustive scientific research of fish species, riverine vegetation ecosystems, and fisheries combined, can lead to a more effective fisheries management system.

The present case study demonstrates the advantages of the social-ecological approach to promote the restoration of riverine ecosystems. The integrated analysis of riverine vegetation and fisheries, coupled with social data about the use of these resources, the use of the broader space in which riverine forests exist and the current and future threats to riverine ecosystems, permits a more holistic view for the design of restoration activities and its governance. Furthermore, the polycentric governance approach, in addition to the active participation of local populations in the restoration and monitoring activities of riverine forests, permitted to further mitigate threats and secure the restoration of the riverine ecosystem. We suggest that a following step should be to design a conceptual framework from which this and similar experiences of riverine vegetation restoration can draw generalizable lessons to reduce its costs and efforts. While, at the same time, amplify the impact of these activities that can be adapted in similar restoration programs for rivers and streams across the Andean-Amazon region.

References

Aldave M, Summers P (2014) Plantas útiles de las riberas del Pichis Selva Central del Perú. Instituto del Bien Común, Lima, pp 11–15

Aldave M, Vásquez R, Summers P (2010) Caracterización Ecológica del hábitat ribereño en la cuenca del río Pichis, Pasco. Resumen presentado en XIX Reunión Científica ICBAR, Universidad Nacional Mayor de San Marcos, Lima 26-27 de agosto de 2010

Aldave M, Chuco ME, Perez R (2013) Recuperando nuestros bosques ribereños. Instituto del Bien Común, Lima

ANA (Autoridad Nacional del Agua) (2010) Reglamento de la Ley N° 29338 DS N° 001-2010-AG. Título V, Capítulo III, Artículos. p 111–114

ANA (Autoridad Nacional del Agua) (2018) Una mirada a la naturaleza: Agua y Bosque. http://repositorio.ana.gob.pe/handle/ANA/2056. Accessed 1 Dec 2018

Barthem R, Goulding M (1997) Os Bagres Balizadores: Ecologia, Migração e Conservação de Peixes Amazónicos. Sociedade Civil Mamirauá/CNPq, Tefé

Barthem R, Goulding M (2007) Un ecosistema inesperado, La Amazonía revelada por la pesca. Asociación para La Conservación de La Cuenca Amazónica/Amazon Conservation Association, Lima

Barthem R, Guerra H, Valderrama M (1995) Diagnóstico de los Recursos Hidrobiológicos de la Amazonía. Tratado de Cooperación Amazónica, Lima

Barthem R, Petrere M, Isaac V et al (1997) A pesca na Amazônia: problemas e perspectivas para o seu manejo. In: Valladares-Padua C, Bodmer RE (eds) Manejo e conservação da vida silvestre no Brasil. Sociedade Civil Mamirauá, Belém, pp 173–184

Carrasco S, Hauenstein E, Peña-Cortés F et al (2014) Evaluación de la calidad de vegetación ribereña en dos cuencas costeras del sur de Chile mediante la aplicación del índice QBR, como base para su planificación y gestión territorial. Gayana Bot 71:1–9

Castro E (2013) Situación de la pesca y los recursos pesqueros en la cuenca del río Pichis, Oxapampa, Perú. In: Collado L, Castro E, Hidalgo M (eds) Hacia el manejo de las pesquerías en la Cuenca Amazónica, perspectivas transfronterizas. Instituto del Bien Común, Lima, pp 39–49

Castro E, Orihuela D (2011) La pesca y el conocimiento bioecológico de los peces como herramienta para la gestión de los recursos acuáticos en la cuenca del río Pichis. Instituto del Bien Común, Oxapampa

Castro E, Borios S, Summers P (2008) La pesca en la cuenca andino- amazónica del río Pachitea, Perú. In: Pinedo D, Soria C (eds) El manejo de las pesquerías en ríos tropicales de Sudamérica. Centro Internacional de Investigaciones para el Desarrollo (IDRC), Mayol Ediciones S.A, Ottawa, pp 39–74

Cortijo AM (2012) Alimentación natural de peces Characiformes: Brycon hilarii y Leporinus friderici en la parte baja del río Palcazú (Oxapampa-Pasco). Disertación, Universidad Nacional Mayor de San Marcos

García A (2016) Evaluación de los parámetros reproductivos de palometa *Mylossoma duriventre* como base para el manejo sostenible de su pesquería en la región Loreto-Perú. (Loreto-Perú). Disertación, Universidad Nacional de la Amazonía Peruana

IBC (Instituto del Bien Común) (2012) Frente a las amenazas a los ecosistemas acuáticos. In: Magaly A (ed) Cartografía socio ambiental en la gestión de la Amazonía peruana. IBC, Lima, pp 25–28

INRENA (Instituto Nacional de Recursos Naturales) (2000) Ley Forestal y de Fauna Silvestre 27308, Título II, Artículo 8, Inc 3

IUCN (2018) Informe anual 2017. https://www.iucn.org/es/acerca-de-la-uicn/programa-de-trabajo-e-informes/informes-anuales. Accessed 1 Dec 2018

Kometter R, Gálmez Márquez V (2018) La restauración de bosques andinos y sus vínculos con el agua. Orientaciones para una comunidad campesina de Apurimac. https://www.researchgate.net/publication/319316938. Accessed 1 Aug 2018

Lasso CA, Agudelo E, Jiménez-Segura LF et al (eds) (2011) Catálogo de los Recursos Pesqueros Continentales de Colombia. Serie Editorial Recursos Hidrobiologicos y Pesqueros Continentales de Colombia. Instituto de Investigacion de Recursos Biologicos Alexander von Humboldt (IAvH), Bogotá

McClain ME, Llerena C (1998) El manejo de cuencas en la selva: De los Andes a la Amazonía. Agua y Riego 11:11–13

MDPB (Municipalidad Distrital de Puerto Bermudez) (2011) Sistema Local de Gestión Ambiental Distrito de Puerto Bermúdez. Plan Local de Acción Ambiental

Montreuil V, Garcia A, Rodriguez R (2001) Biología reproductiva del boquichico *Prochilodus nigricans*, en la Amazonia Peruana. Folia Amazónica 12(1–2):IIAP

Ortega H, Hidalgo M, Correa E et al (2011) Lista anotada de los peces de aguas continentales del Perú: estado actual del conocimineto, distribucion, usos y aspectos de conservacion. Ministerio del Ambiente. Direccion General de Diversidad Biologica-Museo de Historia Natural. UNMSM

Ostrom EA (2009) Polycentric approach for coping with climate change. Policy research working paper 5095

Pinedo D (2008) La orfandad de los peces: Uso consuetudinario de los recursos pesqueros en la cuenca del río Pichis, Perú. In: Pinedo D, Soria C (eds) El manejo de las pesquerías en ríos tropicales de Sudamérica. Centro Internacional de Investigaciones para el Desarrollo (IDRC), Lima, pp 14–75

Rodriguez V, Collado L (2013) Hacia la gobernanza colaborativa pesquera en la Amazonia Peruana. In: Collado L, Castro E, Hidalgo M (eds) Hacia el manejo de las pesquerías en la Cuenca Amazónica, perspectivas transfronterizas. Instituto del Bien Común, Iquitos-Perú, pp 116–134

Rodríguez V, Castro E, Collado L (2018) Gobernanza colaborativa en la pesca: Modelo participativo para la gestión pesquera en la Amazonía Peruana. Instituto del Bien Común, Lima, p 320

SENAMHI (2018). https://www.senamhi.gob.pe/?p=mapa-climatico-del-peru. Accessed 22 Oct 2018

SERFOR (Servicio Nacional Forestal y de Fauna Silvestre) (2018a) Lineamientos para la Restauración de Ecosistemas Forestales y otros Ecosistemas de Vegetación Silvestre. Aprobado mediante R.D.E N° 083–2018-MINAGRI-SERFOR-DE. Lima

SERFOR (Servicio Nacional Forestal y de Fauna Silvestre) (2018b) Bioversity—Perú, ICRAF (2018). Experiencias de restauración en el Perú. Lecciones aprendidas. Lima, Perú. https://www.researchgate.net/publication/329464353. Accessed 2 Jan 2019

Soria C, Rodríguez V (2008) El marco legal formal y consuetudinario de la pesca de subsistencia en el río Pichis, Perú. In: Pinedo D, Soria C (eds) El manejo de las pesquerías en ríos tropicales de Sudamérica. Centro Internacional de Investigaciones para el Desarrollo, Lima, pp 143–204

Usma JS, Valderrama M, Escobar M et al (2010) Peces dulceacuícolas migratorios en Colombia. In: Plan Nacional de las especies migratorias. WWF. Dirección de Ecosistemas, Bogotá, pp 103–131

Van Damme PA, Carvajal Vallejos FM, Pouilly M et al (2008) Amenazas para los peces y pesquerías de la Amazonía boliviana. In: Van Damme PA, Carvajal-Vallejos FM, Molina Carpio J (eds) Los peces y delfines de la Amazonía Boliviana. Imprenta Unicornio, Cochabamba, BO, pp 327–366

Vásquez M, Phillips O (2000) Allpahuayo: floristics, structure, and dynamics a high diversity forest in Amazonian Peru. Ann Mo Bot Gard 87:499–527

Vásquez R, Taylor CM, Rudas Lleras A (1997) Missouri botanical garden. In: John D. and Catherine T. MacArthur Foundation (ed) Flórula de las reservas biológicas de Iquitos, Perú: Allpahuayo-Mishana, Explornapo Camp, Explorama Lodge. Missouri Botanical Garden, St. Louis

Environmental Governance for the Coastal Marine Ecosystem Services of Chiloé Island (Southern Chile)

Luisa E. Delgado, Víctor H. Marín, Rodrigo Asún, Claudia Zúñiga, Claudia Natenzon, Ricardo Castro-Díaz, Lorenna D. Paredes, and Franco Capriroli

Abstract The Chiloé Island, located in Southern Chile, is worldwide known for its culture, originated in the syncretism between ancestral groups and colonizers, and molded through the interactions with the isolated ecosystems and extreme weather conditions. In this chapter, we analyze the state of the island's social-ecological systems, the drivers (direct and indirect), and the impacts. We also analyze the

L. E. Delgado (✉)
Escuela de Ciencias, Facultad de Ciencias, Universidad de Chile, Santiago, Chile

Fundación Centro Transdisciplinario de Estudios FES-Sistémicos, Santiago, Chile
e-mail: ldelgado@uchile.cl

V. H. Marín · L. D. Paredes · F. Capriroli
Laboratorio de Modelación Ecológica, Departamento de Ciencias Ecológicas, Facultad de Ciencias, Universidad de Chile, Santiago, Chile
e-mail: vmarin@uchile.cl; franco.capriroli@ug.uchile.cl

R. Asún
Departmento de Sociología, Facultad de Ciencias Sociales, Universidad de Chile, Santiago, Chile
e-mail: rasun@uchile.cl

C. Zúñiga
Facultad de Psicología, Universidad de Chile, Santiago, Chile
e-mail: cczuniga@u.uchile.cl

C. Natenzon
Programa de Investigaciones en Recursos Naturales y Ambiente (Natural Resources and Environmental Research Program), Instituto de Geografía, Facultad de Filosofía y Letras, Universidad de Buenos Aires, Buenos Aires, Argentina
e-mail: natenzon@filo.uba.ar

R. Castro-Díaz
Autonomous University of Entre Rios, Concordia, Argentina

National Scientific and Technical Research Council, Buenos Aires, Argentina
e-mail: ircastrod@unal.edu.co

L. E. Delgado, V. H. Marín (eds.), *Social-ecological Systems of Latin America: Complexities and Challenges*, https://doi.org/10.1007/978-3-030-28452-7_21

current modes of governance related to coastal marine ecosystems and the social effects of an event that occurred during the year 2016, known as Chilote's May.

Keywords Social-ecological systems · Latin America · Complexity · Chile · Environmental governance · Chiloé

1 Introduction

The reflection over contemporaneous environmental governance should start by understanding that societal changes, at both world and local scales, have produced impacts redefining public actions. Thus, traditional government systems based upon hierarchies and unilaterality (e.g., centralized governments) are not enough to face the new challenges from a society with social-ecological risks, emerging from diverse areas and impacting unevenly on the social and ecological subsystems (Calame 2009). Consequently, the environmental governance can be understood, contemporaneously, as a process that implies the participation of governmental and non-governmental actors, who through negotiation processes face a problem and make mutually beneficial decisions (see also chapter "A New Environmental Governance").

The environmental governance in Latin America is understood as an opportunity to participate in a public space, where actions generated by a group of human beings can manage a vital resource (e.g., drinking water or fishing resources) equitably (Córdova Montúfar 2018). On the other hand, the general conditions to generate a local self-governance correspond to locations where people are socioeconomically vulnerable, depending strongly on a natural resource or ecosystem service for subsistence and good quality of life and where, in geographically isolated places, there are not government representatives to decrease risks or threats (Natenzon and Ríos 2015).

In this chapter, we analyze the environmental governance of the ecosystem services of the coastal marine areas of Chiloé Island in the south of Chile (Fig. 1), centering in the social-ecological phenomenon known as "Chilote's May" (occurred during the year 2016). During this event, different social groups went out to the streets, protesting against their political vulnerability (derived from centralized and hierarchical decisions) on the one hand, and their socioeconomic vulnerability on the other, as related to appearance of harmful algal blooms (HAB) in the coastal marine area of Chiloé (Delamaza et al. 2017).

This social event generated a wide social discussion and opposing opinions, making clear, at a local level, the inequality and inefficiency of the governmental management of the territories, from the standpoint of public policies in Chile. Furthermore, it was clearer than ever the threats of the local populations, especially those belonging to the category of poor, depending strongly on the direct use of provisioning coastal ecosystem services.

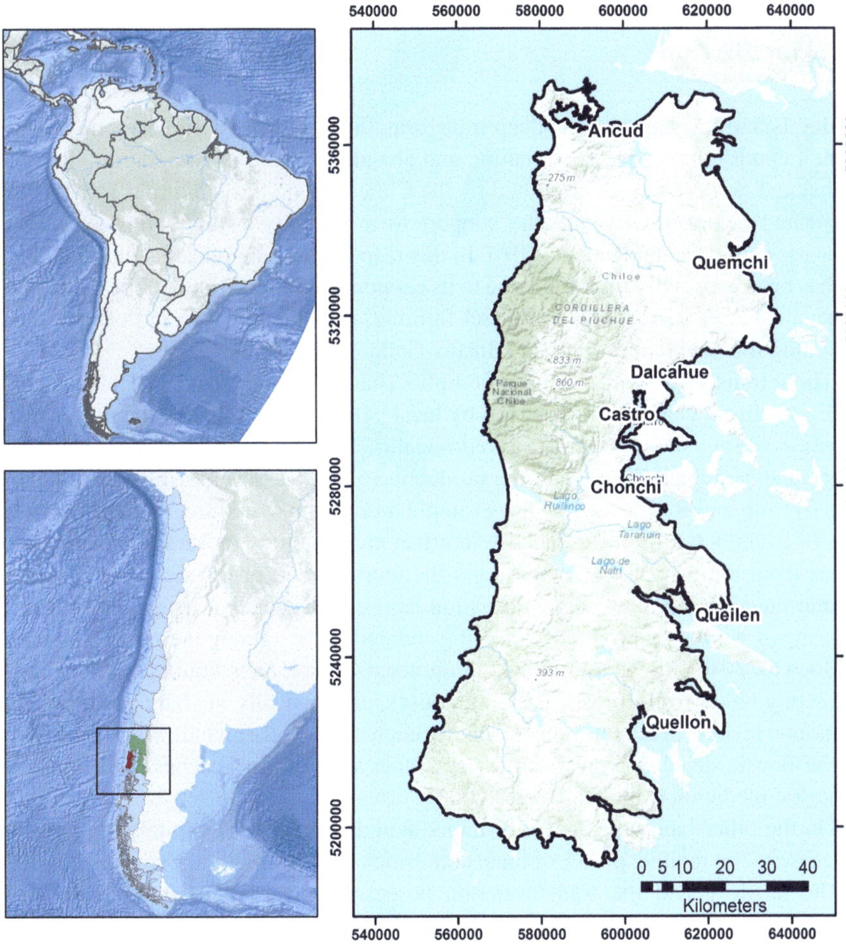

Fig. 1 Geographic location of Chiloé Island (red polygon in the lower left map and right map). Geographic Projection: UTM 18S/WGS84

The high pressure on ecosystem services and coastal marine ecosystems at Chiloé Island has resulted in non-sustainable complex social-ecological systems, generating diverse conflicts related to the control, protection, and use of ecosystem services, leaving uncovered several social, territorial, and political problems. These conflicts sometimes manifest themselves violently, but in other cases, they are invisible to the outside world. They are mediated by the established power relationships, but they also represent an opportunity, from the perspective of the environmental governance of ecosystem services, to renegotiate them.

2 The Social-ecological System of Chiloé Island Coastal Marine Zone

Chiloé Island has experienced deep transformations in the last 40 years, transiting from a condition of poverty, isolation, and abandonment to its integration into our global and modern world, promoted by the Chilean State through improvements in the island's connectivity and the support to a diversity of extracting-exporting industries (Barton and Román 2016). In this respect, the salmon aquaculture industry has been especially important due to its economic relevance and public visibility. Other industries correspond to mussel farming, mining, forestry, and energy also including the irruption of tourism (Bustos-Gallardo and Prieto 2019).

These transformations produced complex changes in the island's social system, such as cultural changes experienced by local people (Miller 2018), the power distribution among social groups (Saavedra Gallo et al. 2016), demographic changes (Aros and Marchant 2018), and the productive activities developed by its inhabitants (Barton and Román 2016). For example, in relation to culture, there has been a loss of island's traditions, especially in urban areas, getting closer to consumption styles from the rest of the country, and the modification of gender's roles toward greater equality. The latter transformation has been especially significant after the entering of women in productive activities related to the salmon industry (Saavedra Gallo et al. 2016). In the traditional culture from Chiloé, men would go out fishing, extracting wood from forests, building houses, and basically in charge of bringing economic resources to their households. Women, on the other hand, would take care of the house, develop wool handicraft, collect seafood on beaches, and practice domestic medicine (Mansilla Torres 2006).

On the other hand, the arrival of national and international capitals triggered a decrease in the relative power of local authorities and communities, shifting control of the territory and the transformation processes toward the central State and national and multinational corporations (Fløysand et al. 2010). Furthermore, from a demography point of view, the island became a center for immigrants from Chile and Argentina, looking for better life opportunities. Immigrants, both qualified and non-qualified workers, hoped to find jobs in the new industries installed in Chiloé, but without the cultural traditions from the local people (Aros and Marchant 2018). Finally, new industries generated an increase in the proletariat in Chiloé, also decreasing the attractiveness of traditional economic activities such as artisan fisheries, agriculture, and cattle farming (Bustos-Gallardo and Prieto 2019; Aros and Marchant 2018).

This contradictory set of transformations, that occurred in the lapse of a generation, not only have had a strong impact on the ecological systems and their capacity to absorb impacts but has also generated latent social tensions that several times have been translated into explicit conflicts (Fouilloux 2018). It has also made difficult the generation of governances promoting the sustainable development of the territory and its ecosystem services since local social actors have heterogeneous positions regarding the changes (Barton and Román 2016). Also, actors with high decision power (national and international capitals and the central State) not always have knowledge about the territory and its ecosystems (Mascareño et al. 2018). However, research shows that there have lately been important changes in the most aggressive productive-industrial activities opening spaces for a more sustainable environmental governance. This seems to have occurred due to the ecological crisis of the extractive industry (Irarrázaval and Bustos-Gallardo 2019) and the conflicts between local actors and the community (sometimes allied with foreign actors) fighting in defense of their territories (Delamaza et al. 2017).

For example, the most important changes of the salmon industry in Chiloé occurred when the Chilean State, the businessmen, and the financial system faced the consequences of the infectious salmon anemia (ISA) virus outbursts during 2007. These events endangered the industry's sustainability facilitating the acceptance of a better control and regulation system by the involved actors (Bustos-Gallardo and Irarrázaval 2016). Of course, there is still disagreement upon whether the implemented changes will allow a truly sustainable productive system, non-aggressive toward the ecosystem. However, there is a consensus about the impact of the crisis in the regulatory and control systems that the industry has to go through (Bräuning Wistuba and Rivera Mercado 2017).

3 A Governance for Ecosystem Services

There is a renewed interest, in Chile and in other Latin American countries, for new forms of environmental governance, including multiple levels and actors (see also chapter "A New Environmental Governance"). There is a diversity of studies within this field of knowledge, centered in the variety of approaches, including hybrid schemes, with the participation of the community and market actors, and hierarchical schemes (Sattler et al. 2018). The challenges for Latin America are multiple, given the variety of social-ecological systems and the heterogeneity of actors with divergent interests, given the cultural wealth of native peoples.

The fundamental idea for an environmental governance strategy is the participation of all social actors, that decisions should be taken acknowledging their consequences and their environmental, social, and economic effects regarding the use of natural resources, the biodiversity, and the ecosystems (Yu Iwama and Delgado 2018). The environmental governance can be defined as the set of regulations, processes, mechanisms, and organizations through which sociopolitical actors influence on environmental results and actions (Lemos and Agrawal 2006). The governance of the ecosystem services has recently been identified as one of the key actions related to the sustainability of social-ecological systems (Costanza et al. 2017).

The urgency and need for social participation on environmental issues has been more evident in the last two decades, both at an international level (e.g., The

Convention on Biodiversity[1]) and at national levels (e.g., Chilean Law for the environment[2]). Indeed, participation in Chile has been transformed into a frequently used term, but with several connotations (Estévez et al. 2019). For example, all environmental projects must include participation elements, with links to the community. However, since there is not a social-ecological approach for environmental problems, participation can either be just informing about a given process, or active deliberations, not necessarily reflecting the perspectives and opinions of local social actors.

In this chapter, we have used the proposals from Primmer et al. (2015) who characterize the different environmental governance approaches based on the processes and/or arrangements between actors and state's structures, the government's educational institutions, and civil society and markets. The authors propose the following types of environmental governance:

(a) *Hierarchical governance,* based on the transference of ideas from higher to lower political levels
(b) *Scientific-technical governance,* that emphasizes the transference of knowledge, and uncertainty, from scientists to local social actors
(c) *Collaborative-adaptive governance,* pointing to the inclusion of social actors (bottom-up governance), collective learning, and adaptation of social-ecological systems
(d) *Strategic governance,* self-organized networks with common good objectives

4 A Social-ecological Approach to Ecosystem Services

Interdisciplinary studies involving society and ecosystems have had a fructiferous development in the last years. Several sciences (e.g., ecology, anthropology, psychology, sociology, and political sciences) have contributed to this progress based on the theoretical concept of social-ecological systems (Delgado et al. 2019) and the proposals for a governance for ecosystem services. Costanza and Daly (1992) were the first to introduce the idea of natural capital, related to the natural ecosystems and their contributions to human well-being. The authors propose a model that considers nature and its resources as the main supporting structure for humans through fluxes of goods and services from ecosystems. These services provided by nature are today known as "ecosystem services" (ES), (Daily 1997), that is, "the benefits that society receives from ecosystems" (MEA 2005). The concept incorporates the idea of human-nature dependency social-ecological interactions, contributing to the well-being of people creating, in turn, an opportunity for the inter- and transdisciplinary study and valuation of the biophysical components of complex social-ecological systems.

[1] https://www.cbd.int/

[2] *https://www.sea.gob.cl/documentacion/...ambientales/normativa-ambiental-aplicable*

Coastal marine areas provide important ES contributing to human well-being and the world's economy (Costanza et al. 1997; MEA 2005; De la Barrera et al. 2015). Despite its importance, there are few studies related to the services and environmental governance of these areas and its incorporation in legal bodies has been slow (Ruiqian et al. 2017; De la Barrera et al. 2015; Delgado and Marín 2015). Furthermore, González Loguercio (2016) states that there is not enough data in Chile to value the contribution of coastal areas to human well-being. In this chapter, we analyze, through a literature review, the ecosystem services of Chiloé Island (Fig. 1) and their governance modes.

5 The Study Area

Our study area, the Chiloé Island (Fig. 1), is known worldwide by the traditions of its people and its cultural and natural richness. If we use the ES classification system from CICES (2019), the island provides provisioning (PES), regulation (RES), and cultural (CES) services. The coastal zones of Chiloé, especially its interior sea, provide numerous direct and indirect services. These include tourism of special interests (e.g., whales, dolphins, and birds watching) and rural and historical tourism (CES), subsistence fisheries and macroalgae extraction (PES), salmon and mussel's aquaculture for export (RES) and a center for scientific analysis and cultural inspiration (CES). So, the island's coastal zone is an area where multiple ES converge, those used by local people (i.e., subsistence economy and exportations) and international industries (e.g., salmon farming). Consequently, the lack of coastal zone integrated management generates tensions between the local people and those that do not live on the island but develop productive activities on it (Paredes 2019). If we add to that an absence of participative regional/local management strategies, we have all the ingredients for environmental conflicts given the diversity of nature's valuations (Skewes et al. 2012; Paredes 2019).

6 Methods

The ecosystem services provided by the coastal of Chiloé Island were identified using bibliographic references (Pérez-Orellana 2019; Paredes 2019; Delgado et al. Submitted) and expert's meetings. The generated list of services was then validated in situ during the development of a social-ecological survey conducted on the island during January–February 2019. Yet, we do not include results from the survey in this chapter. The bases for ES were De Groot et al. (2012), MEA (2003), and CICES (2019).

The relationships between the social-ecological and political characteristics of the island and the use of ecosystem services were studied by means of the DPSIR framework (see chapter "Social Valuation of Ecosystem Services at Local Scale:

Challenges for the Management of a Multiple-Use Coastal and Marine Protected Area (MU-CMPA): Isla Grande de Atacama: Chile" for details about the framework). Using the bibliographic references cited above, we synthesized the five main elements of the framework (Drivers, Pressures, State, Impacts, and Responses) applied to the island.

Finally, we studied the modes of governance that could be applied to the ES management using the instrument matrix proposed by Sarkki (2017). The matrix was filled using literature references, related to Chiloé Island, to characterize the four main elements proposed by Sarkki: policy, markets, civil society, and science and knowledge production.

7 Results

Chiloe's ES

The ecosystem services provided by Chiloé Island and used by the local communities covered four of the six sections defined by CICES V5.1 and 15 classes of a total of 95 (Table 1). Services spread equally in marine and terrestrial ecosystems agreeing with the land-sea association that characterizes Chiloé's culture (Skewes et al. 2012).

The Components of a DPSIR Model for Chiloé Island

The analysis of Chiloé's social-ecological system by means of the DPSIR framework considered indirect and direct drivers, the pressures generated by those drivers, the impacts of those pressures and their effect on the state of the ecosystems and governmental responses. We describe them in the following paragraphs.

Indirect Drivers

They correspond to the socioeconomic and political processes that have evolved outside the study area (i.e., global and national). In this context, the main indirect driver for Chiloé has been the Chilean economic development model, which has been based on the export of natural resources (Saavedra 2014). An additional driver was the insertion of Chile in the global economy as related to salmon farming (Estay and Chávez 2015) which is one of the dominant economic drivers in Chiloé (Ramírez et al. 2009). Furthermore, the lack of local laws, norms, and programs has resulted in the application of top-down, national, legal frameworks that do not consider the local context (Delgado et al. Submitted).

Table 1 Ecosystem services (terrestrial and marine) provided by Chiloé Island

Section	Ecosystem type	Service class
Provisioning (biotic and abiotic)	Marine	Wild animals for nutrition (fishes, mollusks, crustacea), both industrial and artisan
		Wild plants for nutrition and materials (macroalgae)
	Terrestrial	Fibers and other materials from wild plants (e.g., *Sphagnum* spp.)
		Mineral substances used for nutrition (forestry, prairies)
		Animals reared for nutritional purposes (cattle, birds)
		Wild plants as a source of energy (e.g., wood)
Regulation and maintenance (biotic)	Marine	Pest and disease control (harmful algal blooms)
		Gene pool protection (mussels)
		Maintenance of habitats (salmon and mussel farming)
	Terrestrial	Pest and disease control (applied to agriculture and cattle raising)
Cultural (biotic)	Terrestrial and marine	Interactions with nature (tourism and recreation)
		Spiritual interactions with nature (e.g., Chiloé's myths)
		Traditional ecological knowledge (e.g., mingas, medanes)
		Experiential interactions with nature (e.g., fishing corrals)
		Intellectual interactions with nature (e.g., scientific work)

Services classification was based on the Common International Classification of Ecosystem Services (CICES) V5.1 (Haines-Young and Potschin 2018)

Direct Drivers

Changes in the last decades have been characterized by rural-urban migrations within Chiloé due to the lack of opportunities in rural areas (Amtmann and Blanco 2001). Also, the investment's dynamic of the salmon farming industry and cultural tourism has generated incentives to leave, invest, and visit Chiloé. As a result of this dynamics, there has been a revaluation of land prices, generating incentives for real state businesses (Andrade et al. 2000) further increasing tourism of special interests (Gajardo et al. 2017). One of the consequences of the operation of these drivers has been social-ecological impacts derived from water shortages during the summer period. In this regard, it is important to note that the island does not have high mountains; consequently, water is not accumulated as snow during winter.

Pressures

The main pressures in Chiloé relate to the increase in salmon production since Chile became the second largest salmon producer worldwide (Estay and Chávez 2015). There also has been an increase in monoculture of exotic species, both in terrestrial and in marine ecosystems, and the frequent use of antibiotics to fight against the infectious salmon anemia virus. Other pressures include the expansion of urban areas in the oriental coast of the island with a concomitant increase in the human population and an increase in the exploitation of native forests (Barton et al. 2013).

Impacts

The main impacts of the above-described pressures have been an increase in visual pollution of coastal marine ecosystems, derived mostly from the use of plastic as parts of the aquaculture systems (Arenas et al. 2001) and heavy metal pollution (Sandoval et al. 2015). Likewise, in terrestrial ecosystems, there have been impacts due to the use of native forests by a constantly increasing human population.

Rebolledo (2012) states that the boom and the crisis of the salmon industry in Chiloé have changed the traditions and customs of the community making them closer to modernity. On the other hand, the positive impact once predicted as a potential result of the installation of salmon farms, it was only temporary because the generation of new jobs in the island was characterized by offering precarious, transitory, conditions affecting directly the local economy (Fløysand et al. 2010).

An additional, locally important, social impact was the event called "Mayo Chilote" (Chilote's May). During the month of May of 2016, a series of protests developed derived from a long-lasting and widespread red tide phenomenon, that forced the Chilean State to declare the area as a "zone of catastrophe" (MINSAL 2018). Although they were not the first ecologically related protests on the island, they were indeed the most massive and long-lasting (Fouilloux 2018).

State

The main changes in the island's coastal marine ecosystems have been related to their ecological structure with a decrease in native species (Claude et al. 2000). Currently, the main threats for the state of these ecosystems are (a) the salmon anemia virus, (b) their eutrophication due to nutrient increases (FAO 2011), and (c) harmful algal blooms, especially during summer months (Carrasco 2015). In the case of terrestrial ecosystems, the main changes relate to an increase in fragmentation mainly in the northern and eastern sectors and a decrease in the surface areas of wetlands, prairies, and native forest (Capriroli Aguirre 2019).

Responses

Most responses, despite Chilote's May, have been governmental. They have included (a) increase in governmental subsidies, (b) micro-entrepreneurship for families affected by the development of red tides (CONADI 2016), (c) and the structuring of coastal marine areas for original peoples (SUBPESCA 2018). The salmon industry, on the other hand, has responded improving their installations using copper nets and water mass displacers (DMA) with the purpose of recovering the bottom environment below salmon aquaculture centers (Aqua 2018).

The Governance of Chiloé Coastal Marine Ecosystem Services

Our analysis of the governance modes currently developed around Chiloé coastal ecosystems showed a dominance of hierarchical and technocratic modes (Table 2). Thus, island's political authorities should increase their efforts to embrace adaptive collaborative governance, so there may be more participation of local actors (e.g., cooperatives, tourism operators, fishermen) to co-produce knowledge and adaptive learning. But, if government people in charge of decision-making and local scientists (the most influential given our results) do not see the need for the participation of local actors, it's going to be difficult and/or time-consuming to embrace such a governance strategy. Indeed, we found a few elements to support collaborative and strategic governance modes within Chiloé (Table 2). Those found relate to the civil society (self-organized networks that appeared during Chilote's May) and non-environmental organizations (CECPAN 2018).

8 Discussion

Outeiro and Villasante (2013) propose that the general perception of the local community in Chiloé Island is that the ecosystems have been degraded due to the expansion of salmon farming. According to Burkhard et al. (2011), the intensity of the human use of natural resources correlates negatively with the provision of ecosystems services (ES). The reason for this relationship is that intensive use may generate abrupt and/or long-lasting changes in the dynamic of ecosystems, finally affecting entire societies. Indeed, several studies have shown that the ecosystems and the services they provide may be transformed by human actions into less productive systems, generating impacts in human societies (MEA 2003).

On the other hand, there currently is confusion in several Latin American countries, including Chile, between the concept of environmental governance and governability. We understand governability as the degree in which a political system is institutionalized, that is, the process through which governmental organizations acquire value and stability. Conversely, the environmental governance studies the

Table 2 Ecosystem service governance instrument matrix for the coastal ocean of Chiloé Island (modified from Sarkki 2017)

	Hierarchical	Technocratic	Collaborative	Strategic
Policy	S: Top-down policies	S: Experts from governmental institutions	*Absent in the study area*	Absent in the study area
	P: Decision processes dominated by those in power	P: Technocratic science-policy interactions		
	Ex: Centralized public policies (e.g., Sernapesca 2018)	Ex: National forestry corporation (Marín and Delgado 2013)		
Markets	S: Environmental certificates for exportation of products by Chilean government and other countries	*Absent in the study area*	*Absent in the study area*	*Absent in the study area*
	P: Economic interests			
	Ex: National and international environmental certificates (SalmonChile 2018, Decree 320, 2001)			
Civil society	S: Hierarchically led NGOs	S: Nonprofit organizations from civil society	*Absent in the study area*	S: Self-organized networks
	P: Conservation projects without attention to local agendas	P: Voluntary associations seek to shape the governance		P: Use of mass media
	Ex: Whale conservation projects (CCC 2018)	Ex: National federation of Artisan fishermen (CONAPACH 2018)		Ex: Protests during "Mayo Chilote" (Vargas 2018)
Science and knowledge production	S: Hierarchical science-policy interfaces	S: Science advisory committees	*Absent in the study area*	S: Interest groups
	P: Knowledge production based on normal science	P: Policy advice dominated by few experts		P: Mission driven issue advocacy
	Ex: The ethos of PNS (Kønig et al. 2017)	Ex: Chilean red tide commission (CMR 2016); this study		Ex: Local environmental organizations (CECPAN 2018)

S structures; *P* processes; *Ex* examples

modes social actors are co-responsible for the common good based on their capacities, interactions, and external conditions. The crisis we see at Chiloé Island indeed relates to environmental governance because the different structures that characterized the governance modes (Table 2) do not communicate between them (see also Calame 2009).

In fact, the low participation levels and options in environmental issues in Chile are inadequate, given the evolution of social actors and society, with processes that are both local and global (e.g., salmon market and its local influence of local ecosystems). This mixture of global-local processes generates low possibilities for the adaptation of local communities heavily depending on the services of coastal social-ecological systems. According to Pérez-Orellana (2019), Chiloé started a social-ecological collapse phase during the year 2007 (the ISA virus outburst), characterized by high vulnerability within a context of low resilience. Under these conditions, connectivity can be disintegrated either by a social or ecological crisis with the result that social-ecological organization is lost. According to Calame (2009), the insertion of a local economy into the global market destroys social bonds, generating impoverishment of the social capital. Local results in Chiloé seem to have been a low or null social resilience (Pérez-Orellana 2019) and a land-sea culture dissociation affecting Chiloé's culture (Paredes 2019). Anderies et al. (2004) state that social-ecological resilience, or adaptive capacity, occurs when humans change their practices adjusting them to the new ecosystem dynamics. Our results show that the current governance of Chiloé's ecosystem services attempts against this adaptation due to several reasons:

1. The dominance of formal governmental institutions.
2. A lack of collective learning.
3. Low participation.
4. A lack of consideration about the scales where social-ecological processes operate (global, national, regional, local).
5. Slow political responses.
6. The persistence of centralized and hierarchical actions.
7. A sectorial (disciplinary) institutional management.
8. Strong market influences (lack of real sustainability incentives).
9. A centralized decision-making scheme.

Delgado et al. (2013) have proposed that the problem that faces ES in Chile could be even worse if we consider the lack of conscience about their role as the basis of the rural population's livelihood. If we add a generalized lack of ES studies about coastal zones, we then have all the ingredients for un-sustainability (Delgado and Marín 2015; De la Barrera et al. 2015). Therefore, we propose that the ES historically used by human societies could be a good sustainability indicator for social-ecological systems (see also Delgado and Marín 2017). However, in order to achieve reliable indicators, inter- and transdisciplinary studies integrating biophysical, social, economic, and political aspects are necessary (Delgado et al. 2019). The next step should then be a compromise of all involved actors with integrated management of ecosystem services. However, this may not be an easy task because (1) actor's interests differ whether they are suppliers of services or users and (2) several actors have multiple and sometimes opposite perceptions. That is why opening debates regarding the governance of ecosystem services are important, including social valuations, their distribution within the community, and the local use in order to inform policy-making (Lienhoop and Schröter-Schlaack 2018; Calame 2009).

Acknowledgements The work contained in this chapter was financed by CONICYT-Chile FONDECYT Grant N° 1170532 awarded to L. E. Delgado.

References

Amtmann CA, Blanco G (2001) Efectos de la salmonicultura en las economías campesinas de la Región de Los Lagos, Chile. Revista Austral de Ciencias Sociales 5:93–106

Anderies JM, Janssen MA, Ostrom E (2004) A framework to analyse the robustness of social-ecological systems from an institutional perspective. Ecol Soc 9(1):18–35

Andrade C, Arenas F, Quense J (2000) Caracterización ambiental aplicada y ordenamiento del territorio: la costa oriental de la Isla Grande de Chiloé. Revista de Geografía Norte Grande 27:123–132

Aqua (2018) DMA: El desarrollo chileno que permite la recuperación de los fondos marinos y lacustres. Aqua Acuicultura + Pesca. http://www.aqua.cl/2018/01/17/dma-el-desarrollo-chileno-que-permite-la-recuperacion-de-los-fondos-marinos-y-lacustres/. Accessed 7 Jun 2019

Arenas F, Andrade B, Qüense J (2001) La valorización de un espacio periférico: El caso de la costa oriental de la Isla Grande de Chiloé. Revista de Geografía Norte Grande 28:79–90

Aros F, Marchant C (2018) Transformaciones en la identidad territorial en Chiloé desde la llegada de la industria salmonera: el caso de la localidad de Quemchi. Espacios 7(13):47–71

Barton JR, Román Á (2016) Sustainable development? Salmon aquaculture and late modernity in the archipelago of Chiloé, Chile. Isl Stud J 11(2):651–672

Barton J, Pozo R, Román Á et al (2013) Reestructuración urbana de un territorio globalizado: una caracterización del crecimiento orgánico en las ciudades de Chiloé, 1979-2008. Revista de geografía Norte Grande 56:121–142

Bräuning Wistuba M, Rivera Mercado C (2017) Reconversión de industrias ligadas a recursos naturales: La promesa incumplida de la industria salmonera en Chile. Revista Venezolana de Gerencia 22(77):121–136

Burkhard B, Fath BD, Müller F (2011) Adapting the adaptive cycle: hypotheses on the development of ecosystem properties and services. Ecol Model 222:2878–2890

Bustos-Gallardo B, Irarrázaval F (2016) Throwing money into the sea: capitalism as a world-ecological system. Evidence from the Chilean Salmon industry crisis, 2008. Capital Nat Social 27(3):83–102

Bustos-Gallardo B, Prieto M (2019) Nuevas aproximaciones teóricas a las regiones-commodity desde la ecología política. Revista EURE-Revista de Estudios Urbano Regionales 45(135):153–176

Calame P (2009) Hacia una revolución de la gobernanza: reinventar la democracia. LOM Ediciones, Santiago

Capriroli Aguirre FI (2019) Desarrollo de un modelo conceptual socio-ecológico de los cambios de uso de suelo de la isla grande de Chiloé en las últimas dos décadas. Facultad de Ciencias Agronómicas, Universidad de Chile, Santiago

Carrasco D (2015) Floraciones algales nocivas (FAN), un Fenómeno Natural Presente en Nuestras Costas. Laboratorio de Toxinas Marinas Instituto de Ciencias Biomédicas, Facultad de Medicina, Universidad de Chile, Santiago. http://labtox.cl/?p=984. Accessed 7 June 2019

CCC (2018) Centro de Conservación Cetacea. http://www.ccc-chile.org/. Accessed 7 Jun 2019
CECPAN (2018) Centro de Estudios y Conservación del Patrimonio Natural. http://wwwcecpan.org. Accessed 7 Jun 2019
CICES (2019) Common international classification of ecosystem services. European Environment Agency, Copenhagen. https://CICES.EU. Accessed 7 Jun 2019
Claude M, Oporto J, Ibáñez C et al (2000) La ineficiencia de la salmonicultura en Chile: aspectos sociales, económicos y ambientales. Terram Publicaciones, Santiago
CMR (2016) Informe final. Comisión Marea Roja. http://www.economia.gob.cl/wp-content/uploads/2016/11/InfoFinal_ComisionMareaRoja_24Nov2016-1.compressed.pdf. Accessed 7 Jun 2019
CONADI (2016) Concursos o Licitaciones CONADI. https://sistemas.conadi.cl/repositorio/gobiernotransparente/Concursos_o_Licitaciones.html. Accessed 22 Aug 2019
CONAPACH (2018) Confederación Nacional de Pescadores Artesanales de Chile. http://www.conapachcl/. Accessed 7 Jun 2019
Córdova-Montúfar M (2018) Gobernanza y políticas públicas. La seguridad ciudadana de Bogotá y Quito. Editorial Universidad del Rosario/Flacso, Quito
Costanza R, Daly H (1992) Natural capital and sustainable development. Conserv Biol 6(1):37–46
Costanza R, de Groot R, Braat L et al (2017) Twenty year of ecosystem services: how far have we come and far do we still to go? Ecosyst Serv 28:1–16
Costanza R, d'Arge R, de Groot R et al (1997) The value of the world's ecosystem services and natural capital. Nature 387:253–260
Daily G (1997) What are ecosystem services? In: Daily G (ed) Nature's services: societal dependence on natural ecosystems. Island Press, Washington, DC, pp 1–10
De Groot R, Brander L, VanderPloeg S et al (2012) Global estimates of the value of ecosystems and their services in monetary units. Ecosyst Serv 1:50–61
De la Barrera F, Bachmann-Vargas P, Tironi A (2015) La investigación de servicios ecosistémicos en Chile: una revisión sistemática. Investigaciones Geográficas 50:3–18
Delamaza G, Maillet A, Neira CM (2017) Socio-territorial conflicts in Chile: configuration and politicization (2005-2014). Eur Rev Lat Am Car Stud 104:23–46
Delgado LE, Marín VH (2015) Ecosystem services: where on earth? Ecosyst Serv 14:24–26
Delgado LE, Marín VH (2017) Human well-being and historical ecosystems: the environmentalist's paradox revisited. Bioscience 67:5–6
Delgado LE, Sepúlveda MB, Marín VH (2013) Provision of ecosystem services by the Aysén watershed, Chilean Patagonia, to rural households. Ecosyst Serv 5:102–109
Delgado LE, Tironi-Silva A, Marín VH (2019) Sistemas socioecológicos y servicios ecosistémicos: modelos conceptuales para el humedal del Río Cruces (Valdivia, Chile). In: Cerda CL, Silva-Rodríguez E, Briceño C (eds) Naturaleza en sociedad: Una mirada a la dimensión humana de la conservación de la biodiversidad. Editorial Ocho Libros, Santiago, pp 177–205
Delgado LE, Zúñiga CC, Asún RA et al (Submitted) Ecosystem services of Chiloé Island marine coastal zones and integrated management: a social-ecological analysis. Lat Am J Aquat Res
Estay M, Chávez C (2015) Decisiones de localización y cambios regulatorios: el caso de la acuicultura en Chile. Lat Am J Aquat Res 43:700–717
Estévez RA, Martínez P, Sepúlveda M et al (2019) Gobernanza y participación en la gestíon de las áreas silvestres protegidas del Estado de Chile. In: Cerda CL, Silva-Rodríguez E, Briceño C (eds) Naturaleza en sociedad: Una mirada a la dimensión humana de la conservación de la biodiversidad. Editorial Ocho Libros, Santiago, pp 381–403
FAO (2011) Desarrollo de la acuicultura. 4. Enfoque ecosistémico a la acuicultura. http://www.fao.org/docrep/014/i1750s/i1750s.pdf. Accessed 8 Jun 2019
Fløysand A, Barton JR, Román Á (2010) La doble jerarquía del desarrollo económico y gobierno local en Chile: El caso de la salmonicultura y los municipios chilotes. EURE (Santiago) 36(108):123–148

Fouilloux I (2018) Movimientos sociales y comoditización del territorio: el caso de los movimientos sociales ante el desarrollo de la industria salmonera en el archipiélago de Chiloé (1980–2016). Tesis para optar al título de Sociólogo. Universidad de Chile, Santiago

Gajardo P, Mondaca E, Santibáñez P (2017) La minería industrial como una nueva amenaza al espacio marino costero de Chiloé: Bahía de Cucao como caso de estudio. RIVAR 3(10):110–138

González Loguercio N (2016) Creación de un Sistema Nacional Integral de Áreas Protegidas para Chile: Estructura financiera y operacional—Informe final 2008–2015. Programa de las Naciones Unidas para el Desarrollo (PNUD). Fondo para el Medio Ambiente Mundial (GEF). Ministerio del Medio Ambiente, Gobierno de Chile, Santiago

Haines-Young R, Potschin M (2018) Common international classification of ecosystem services (CICES) V5.1. Guidance on the application of the revised structure. https://cices.eu/content/uploads/sites/8/2018/01/Guidance-V51-01012018.pdf. Accessed 8 Jun 2019

Irarrázaval F, Bustos-Gallardo B (2019) Global Salmon networks: unpacking ecological contradictions at the production stage. Econ Geogr 95:159–178

Kønig N, Børsen T, Emmeche C (2017) The ethos of post-normal science. Futures 91:12–24

Lemos MC, Agrawal A (2006) Enviromental governance. Annu Rev Environ Resour 31:297–325

Lienhoop N, Schröter-Schlaack C (2018) Involving multiple actors in ecosystem service governance: exploring the role of stated preference valuation. Ecosyst Serv 34:181–188

Mansilla Torres S (2006) Chiloé y los dilemas de su identidad cultural ante el modelo neoliberal chileno: la visión de los artistas e intelectuales. Revista Alpha (Osorno) 23:9–36

Marín VH, Delgado LE (2013) From ecology to society and back: the (in)convenient hypothesis syndrome. Int J Sust Dev 16:46–65

Mascareño A, Cordero R, Azócar G et al (2018) Controversies in social-ecological systems: lessons from a major red tide crisis on Chiloe Island, Chile. Ecol Soc 23(4)

MEA (Millenium Ecosystem Assessment) (2005) Ecosystems and human well-being: synthesis report. Millennium Ecosystem Assessment. Island Press, Washington, DC

MEA (Millennium Ecosystem Assessment) (2003) Ecosystems and human Well-being: a framework for assessment. A report of the conceptual framework working group. Island Press, Washington, DC

Miller JC (2018) No fish, no mall. Industrial fish produce new subjectivities in southern Chile. Geoforum 92:125–133

MINSAL (Ministerio de Salud) (2018) Marea Roja en Chiloé: Se amplió zona de riesgo para extraer mariscos hasta comuna de Queilen. https://www.minsalcl/marea-roja-en-chiloe-se-amplio-zona-de-riesgo-para-extraer-mariscos-hasta-comuna-de-queilen/. Accessed 8 June 2019

Natenzon C, Ríos D (2015) Una revisión sobre catástrofes, riesgo y ciencias sociales. In: Natenzon C, Ríos D (eds) Riesgos, catástrofes y vulnerabilidades: aportes desde la geografía y otras ciencias sociales para casos argentinos. Ediciones Imago Mundi, Buenos Aires, pp 1–27

Outeiro L, Villasante S (2013) Linking salmon aquaculture synergies and trade-offs on ecosystem services to human wellbeing constituents. Ambio 42(8):1022–1036

Paredes LD (2019) desarrollo de un modelo conceptual para el manejo de servicios ecosistémicos costeros: Isla Grande de Chiloé. Proyecto de grado presentado como parte de los requisitos para optar al grado de Magíster en Gestión y Planificación Ambiental. Facultad de Ciencias Forestales y de la Conservación de la Naturaleza, Universidad de Chile, Santiago

Pérez Orellana DC (2019) Uso histórico de los servicios ecosistémicos en la isla grande de Chiloé, Región de Los Lagos, Chile: una aproximación conceptual. Seminario de Título entregado a la Universidad de Chile en cumplimiento parcial de los requisitos para optar al Título de Biólogo ambiental. Facultad de Ciencias, Universidad de Chile, Santiago

Primmer E, Jokinen P, Blicharska M et al (2015) Governance of ecosystem services: a framework for empirical analysis. Ecosyst Serv 16:158–166

Ramírez E, Modrego F, Macé JC et al (2009) Dinámicas territoriales en Chiloé central: La fuerza de las coaliciones extraterritoriales. Documento de trabajo/Programa Dinámicas Territoriales Rurales N° 54. RIMISP-Centro Latinoamericano para el Desarrollo Rural, Santiago

Rebolledo L (2012) Resistencia y cambios identitarios en trabajadores/as del salmón en Quellón. Polis (Santiago) 11(31):223–239
Ruiqian L, van den Brink M, Woltjer J (2017) Market-based instruments for the governance of coastal and marine ecosystem services: an analysis based on the Chinese case. Ecosyst Serv 23:71–81
Saavedra E (2014) El modelo económico-político de Chile: Desarrollo Institucional en la encrucijada. Economía y Política (Universidad Adolfo Ibáñez) 1:117–148
Saavedra Gallo G, Mardones Leiva K, Torres Zamora MP (2016) La esquizofrenia del desarrollo: Un análisis semántico-discursivo de las relaciones entre salmonicultura y pesca artesanal en el sur-austral de Chile. Cultura-hombre-sociedad 26(2):71–105
SalmonChile (2018) Certificaciones. http://www.salmonchilecl/es/certificaciones-productophp. Accessed 8 Jun 2019
Sandoval C, Paredes E, Ulloa M (2015) Intoxicación por metales pesados. http://www.vehice.cl/paper/Intoxicaciones-metales-peces.pdf. Accessed 8 Jun 2019
Sarkki S (2017) Governance services: co-producing human well-being with ecosystem services. Ecosyst Serv 27:82–91
Sattler C, Loft L, Mann C et al (2018) Methods in ecosystem services governance analysis: an introduction. Ecosyst Serv 34:155–168
SERNAPESCA (2018) Servicio Nacional de Pesca y Acuicultura. Gobierno de Chile. http://www.sernapescacl. Accessed 8 Jun 2019
Skewes C, Álvarez R, Navarro M (2012) Usos consuetudinarios, conflictos actuales y conservación en el borde costero de Chiloé insular. Magallania (Punta Arenas) 40(1):109–125
SUBPESCA (Subsecretaría de Pesca y Acuicultura) (2018) Espacios Costeros Marinos para Pueblos Originarios (ECMPO). http://www.idecl/descargas/capas/subpesca/ECMPOzip. Accessed 8 Jun 2019
Vargas D (2018) Reflexiones en torno al mayo chilote. https://www.lemondediplomatique.cl/Reflexiones-en-torno-al-mayo.html. Accessed 8 Jun 2019
Yu Iwama A, Delgado LE (2018) Acción: Participación Comunitaria en Procesos de decisión en la Conservación del territorio. Cuadernos del Pensamiento Crítico Latinoamericano 56:1–3

Social Valuation of Ecosystem Services at Local Scale: Challenges for the Management of a Multiple-Use Coastal and Marine Protected Area (MU-CMPA): Isla Grande de Atacama: Chile

Marcela Torres-Gómez, Enrique Calfucura, and Eugenio Figueroa B.

Abstract This chapter analyzes the contribution of Participative Management Plans (PMP) for the identification of ecosystem services and the protection of conservation objects from the multiple-use protected coastal marine areas (**MU-CMPA**). The objective of these areas is to conserve the natural capital and cultural patrimony without restricting traditional productive activities such as fishing, mollusks and algae extraction, and energy resources. There are ten MU-CMPAs areas in Chile, but their implementation has been slow and 14 years after the first areas were legally declared, some of them still do not have management plans. Here we analyze the experiences of Isla Grande de Atacama **MU-CMPA** (**MU-CMPA** IGA) in the north of Chile, including the complexities of implementing PMPs and the challenges and opportunities of generating an ecosystem perspective in the management plans for protected areas. Administrative problems and conflicts of interest have worn social relationships generating little community participation regarding the design of a

M. Torres-Gómez (✉)
Fundación Centro Transdisciplinario de Estudios FES-Sistémicos, Santiago, Chile

Department of Economics and Center of Environmental and Natural Resource Economics (CENRE), School of Economics and Business, University of Chile, Santiago, Chile
e-mail: mtorresg@fen.uchile.cl

E. Calfucura
Facultad de Economía y Empresa, Universidad Diego Portales, Santiago, Chile

Center for Intercultural and Indigenous Research (CIIR), Pontifical Catholic University of Chile, Santiago, Chile
e-mail: enrique.calfucura@udp.cl

E. Figueroa B.
Department of Economics and Center of Environmental and Natural Resource Economics (CENRE), School of Economics and Business, University of Chile, Santiago, Chile
e-mail: efiguero@fen.uchile.cl

L. E. Delgado, V. H. Marín (eds.), *Social-ecological Systems of Latin America: Complexities and Challenges*, https://doi.org/10.1007/978-3-030-28452-7_22

management plan. Nevertheless, there is a consensus among local social actors about the benefits of the ecosystems of the **MU-CMPA** IGA due to the high economic and social values given by the community to the services provided by the area.

Keywords Social-ecological systems · Latin America · Complexity · Chile · Ecosystem services · Social valuation · Atacama

1 Introduction

Coastal marine ecosystems provide several services, contributing to human well-being. Some of these services are flood control, contaminant detoxification, and vital food resources for millions of people both living in coastal areas and far away from them (Peterson and Lubchenco 1997; Holmlund and Hammer 1999; Worm et al. 2006). Moreover, those benefits in Chile are important actives for the national economy, given the large latitudinal extension of the coastal zone (8000 km from north to south; Figueroa and Calfucura 2006).

Despite the valuable contribution to humankind and its well-being, marine, coastal, and estuarine ecosystems are among the most exploited and threaten the systems of the planet (Barbier et al. 2011; Halpern et al. 2008; Lotze et al. 2006). The transformation of coastal marine ecosystems, product of economic activities and global climate change, is affecting the provision of ecosystem services through several mechanisms such as marine pollution due to the industrial activity or the change in sea temperature due to global warming among others (Doney et al. 2011). Furthermore, their ecosystem services have been less studied than their terrestrial counterparts (Delgado and Marín 2015).

One of the consequences of the international concern over coastal marine ecosystems is the worldwide encouragement for the generation of coastal marine protected areas (CMPA), a conservation tool located within the VI biodiversity management category according to the International Union for the Conservation of Nature (IUCN). CMPAs are legally chosen, delimited, and designated sites generated with the purpose of maintaining, conserving, and restoring coastal species and habitats in relatively small areas or ecosystem fragments. These areas may not be self-sustainable requiring active management interventions in order to ensure the survival of the protected objects (Dudley 2008; Lambert 2003).

Chile started creating multiple-use protected coastal marine areas (**MU-CMPA**) during the year 2004. These areas, according to the IUCN, correspond to "spaces that include portions of water and marine bottom, rocks beaches and fiscal beaches' terrains, flora and fauna, historical and cultural resources, that the law or other efficient methods grant the 'on reserve' category to protect all or part of the delimited area." They have a double objective; on the one side they contribute to biodiversity conservation, the reduction of conflicts over the use of resources and to generate research and education. On the other, they serve to guard the historical patrimony of the communities of inhabitants and the development of low-impact activities such as tourism, artisan fisheries, and recreation (Delgado 2014; Praus et al. 2011; MMA 2011).

The first **MU-CMPA**s in Chile were decreed in the year 2005, in the framework of the GEF/World Bank project "Conservation of Biodiversity of International Importance along the Chilean Coast" (2005–2010). However, this category has weak judicial support. Although its quality as a judicial and national instrument is recognized under the Law 20.417/2010, there are not, until today, guidelines to follow in order to define and subsequently administrate these areas, only sectorial, insufficient, instruments (CENRE 2017).

So, the process that started with the GEF project did not accomplish the proposed objectives in terms of infrastructure and planning. Since the year 2010 the Chilean Ministry of the Environment has retaken the challenge to generate and execute the management plan for the MU-CMPA of Isla Grande de Atacama (MU-CMPA IGA). However, a consultation process conducted during 2018 showed that the task has not yet been finished. The result has been a mistrust from social actors, putting in check the credibility of whether the plan can be executed under its current form (CENRE 2017). Under these circumstances, keeping the participation of the local populations in the generation of the administration plans, and their subsequent commitment toward the measures to be executed, is a rather complex objective to fulfill.

This chapter is a case study of the generation process of the general administration plan for the **MU-CMPA IGA** in the northern Chilean coast. Its main objective was the management of the services provided by the ecosystems protected in the area; they are key services to support some crucial traditional economic activities carried out by local communities. The used methodology was based on the theoretical and methodological proposals for ecosystem management (Christensen et al. 1996); they specifically incorporate the human society as an ecosystem component recognizing the strong interactions between the natural and social systems. An ecosystem, under this perspective, is a dynamic system that changes not only due to natural processes but also as the result of social trends (e.g., ideologies and cultures) and political decisions, as elements influencing the configuration of a territory (Torres-Gómez 2019; Figueroa and Aronson 2006). Nevertheless, the complexity of the interactions and the dynamics of social-ecological systems may prohibit having access to the whole ecological and social information necessary to evaluate the system's evolution, especially in the presence of environmental conflicts and urgent decisions. Under these conditions, it is necessary to consider alternative forms of knowledge, including scientific-technical, local and traditional, with the purpose of generating a contextualized and inclusive decision-making process with shared responsibilities and compromises (Berkes et al. 2000).

This strategy requires, in the first place, an action plan oriented to identify local social actors that relate with the territory on a daily basis since they are its main transformers. This is especially relevant for **MU-CMPA**s in developing countries where there are strong economic-dependence relationships for human well-being, especially in high natural and patrimonial valued ecosystems. Despite the many problems, described above, we propose that it is possible to generate conceptual strategies through flexible methods allowing to capture the local needs, concerns, and social perspectives related to the future management of the AMCP-MU IGA.

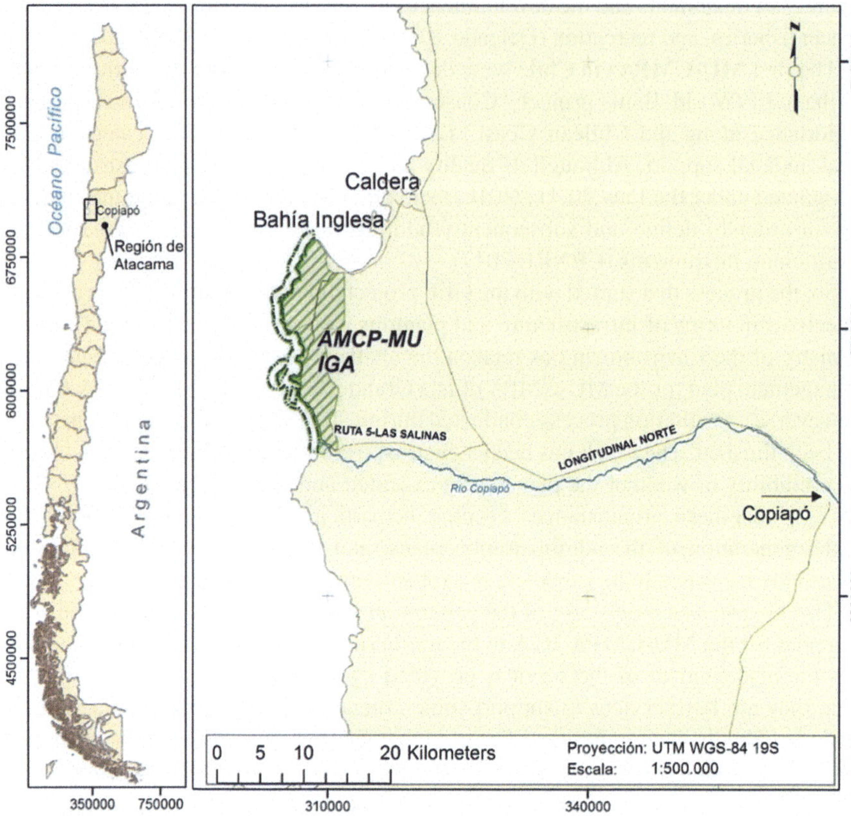

Fig. 1 Geographic location of the MU-CMPA IGA

2 Methods

Characterization of the Study Area

The **MU-CMPA** IGA was created in the year 2004, fulfilling the commitments subscribed by Chile with the Convention for Biological Diversity (CBD), with the purpose of "establishing an integrated environmental management and an in-situ conservation modality of the ecosystems and natural habitats"[1] in the coastal zone between Punta Morro and the mouth of the Copiapó River in Caldera (Atacama region, Chile; Fig. 1). The area occupies 39.93 km^2 of marine and beach sectors and 81.02 km^2 of the adjacent continental territory.

This area is known not only due to its value as habitat for marine and terrestrial (Gaymer et al. 2008) but also due to its geological peculiarities, paleontological

[1] Decree N° 360/2004; https://goo.gl/sLQ4TG

richness (Cabello et al. 2010), and its scenic beauty. The marine coastal zone is an important place for the reproduction of species with high ecological and economic value, a characteristic determined by physical factors: (1) the rocky structure of the bottom allows for the establishment of kelp forests, which act as a refuge and food for a large variety of species and (2) an upwelling system that increases water nutrients. The terrestrial sector of the **MU-CMPA** IGA also has a big richness of species of flora and fauna, with typical flowering desert species. Furthermore, there is a costal wetland in the southern boundary which is unique within the desert matrix with a fundamental role in the maintenance of biological communities including migratory bird species (UCN 2014; Vásquez 2007). Due these characteristics, and the large scientific and educational value for the earth sciences, the area is being proposed to become a UNESCO global geopark (Castro and Zuñiga 2007).

Regarding the social system, local communities have historically used this area; although currently it does not have permanent residents, it is used in the development of economic and recreational activities, specifically fishing and tourism. The fishery is exclusively artisan, a traditional activity developed by native peoples since ancient times (Vásquez 2007).

Incorporation of Social Actors in Participative Planning

The consideration of ecosystem services in the generation of the administration plan for a protected area necessarily requires a participative process, where the beneficiaries identify and value the services with which they relate most. This is particularly important in Latin America, where local communities frequently show a high dependency on natural resources.

Identification and Classification of Local Social Actors

We consider a social actor, for the purposes of this chapter, as "any individual, community, group or organization with an interest in the outcome of a program, either as a result of being affected by it positively or negatively, or by being able to influence the activity in a positive or negative way" (Dearden et al. 2003: p. 2.1). The actor's identification was accomplished by means of secondary sources, including social organization's databases, publications, and reports associated with the generation of the management plan since 2005. The collected information was then validated and completed by means of interviews to already identified actors using a snow-ball methodology.

Most of the social actors contacted already knew the project, having participated in the workshops when the area was created. They immediately expressed their discomfort due to the lack of results in the generation of a general management plan. So, they showed little interest in participating. Consequently, we developed a methodology for the classification and involvement of social actors, which allowed us focusing the participative work with key social actors as a function of the interests of each group, a strategy used in social-environmental conflict situations (Delgado et al. 2009).

Social actors were classified using an interest/power or Meadow matrix (Dearden et al. 2003). This methodology allows classifying actors in three categories: key, primary, and secondary (see chapter "Social-ecological Systems and Human Well-Being"). The interest/power matrix is built assigning a qualitative value (from 1 to 10) to the degree of interest and power that actors have on the analyzed system. Power is defined as the degree on which actors are capable of influencing or coercing other actors in decision-making or to follow courses of action through the resources they control, such as politics, wealth, organization, and negotiation capabilities.

Interest, on the other hand, is related to the consequences that an actor could face as a result of a given project or policy that may modify the perceived benefits they receive (Mayers 2005). We considered, for the purposes of our work, that the actors with more interest will be those who have a larger dependence from the benefits of the **MU-CMPA** IGA and that the loss of ecosystem services may significantly affect their quality of their lives and the continuation of their activities. Finally, actors with the highest power are those capable of influencing the decisions of government actors.

Identification and Participative Valuation of Ecosystem Services

We conducted participative workshops, using methods for conflict situations (Delgado et al. 2009), with the identified/classified social actors with the objective of prioritizing the ecosystem services and the threats derived from their daily use. The methodology is based on the idea of working with small groups, moderated by a neutral person with knowledge on intermediation rules with the objective of allowing all opinions, in an environment where a person does not question the value of other person's opinions.

We developed two specific activities (1) an individual valuation of ecosystem services, using a matrix with a Likert scale (Likert 1932), which is a methodology widely used for the social-cultural valuation of ecosystem services (Calvet-Mir et al. 2012; Liu et al. 2013; López-Santiago et al. 2014: Martín-López et al. 2012), and (2) small group workshops where individual matrices were checked further, and afterwards, the most relevant ecosystem services were recognized and their supply and demand sectors were identified on a map, following the methodologies of Burkhard et al. (2012). The same map was used to identify and pinpoint the main threats to ecosystem services.

Economic Valuation of Ecosystem Services

Prior to making the valuation of the ecosystem services associated with the **MU-CMPA** IGA, we evaluated the available methods for economic valuation in order to choose the appropriate procedures considering the reduced availability of primary and secondary information of Chilean desert ecosystems. This condition generated a series of limitations such as lack of data on the biomass of the different species, number of visitors, and extraction rates, among others. We then used four methodologies: market price (for provisioning and habitat services), hybrid (for recreational services), opportunity cost (for cultural services such as scientific research), and cost of replacement (for abiotic provisioning and cultural services such as paleontological materials). The details of the methods used are provided in Annex I.

The results of the economic valuation shown in this study are conservative. That is, we adopted the criterion to present the lowest values of all the obtained results. Our objective was to avoid overestimating the ecosystem services given the variety of methodological difficulties we encountered. Furthermore, information availability was low preventing more comprehensive estimations in all ecosystems.

Problems in the Implementation of the Management Plan

The problems in the implementation of the management plan for the **MU-CMPA** IGA were analyzed using the DPSIR framework (driving forces—pressures—state—impacts—responses). This analytical framework, proposed by the Organization for Economic Cooperation and Development and the United Nations in 1994, has been widely used to synthesize and structure information on social-ecological relationships (Bradley and Yee 2015; Sarkki et al. 2017). It allows explaining society-nature relationships in an easy way to understand the context under which a response occurs to a shift in the state generated by the detected impacts.

Although the DPSIR framework is considered useful for the description of social-ecological systems, its layout has a linearity bias that simplifies complexities of the systems. So, in order to analyze the limitations of the development and execution of the **MU-CMPA** IGA management plan, we complemented the DPSIR framework with three additional components derived from a model designed to study complex social systems (CSS) proposed by López and Sánchez (2000). These components make explicit the nonlinear condition of these systems, defining that the changes are processes corresponding to change and resistance cycles (Fig. 2).

Resistance forces refer to oppositions to system changes derived from external and internal pressures. The mechanism that would trigger internal resistance is the need to maintain the identity; that is, the image that human communities have about

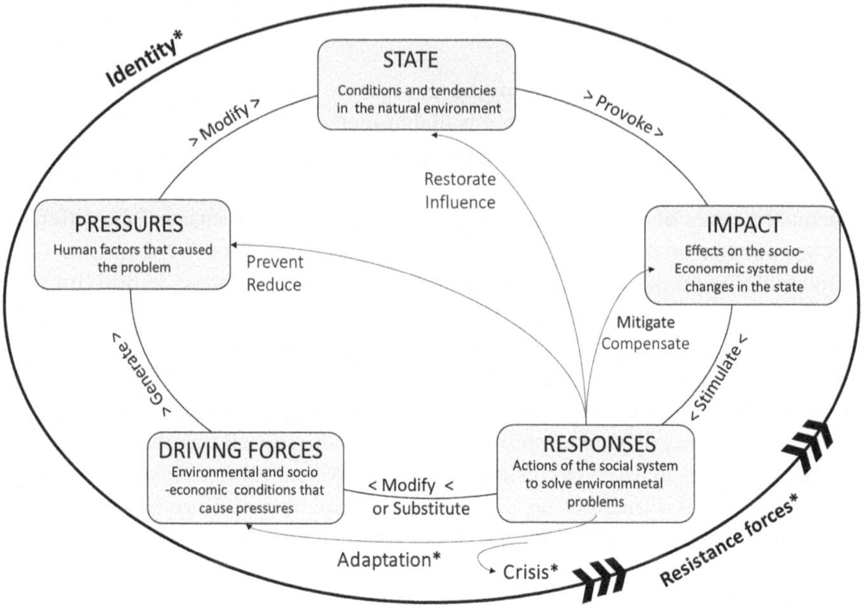

Fig. 2 Adapted DPSIR framework complemented with the concepts derived from the Complex Social System's model. (Source: modified from López and Sánchez (2000) and Sarkki et al. (2017))

their selves derived from their social structure, customs, and other factors. This is the mechanism that generates resilience to the social system (López and Sánchez 2000). Thus, resistance and identity act in synergy when the system faces changes, with undesired consequences.

Finally, responding to these changes, under the proposed framework, actions can go through diverse ways (dotted lines, Fig. 2):

1. Responses to pressures, seeking to take into account the factors unleashing the impacts; thus, oriented to prevent or reduce changes.
2. Responses to the state of the system, oriented to restore the system structure in order to maintain the desired state.
3. Responses to impacts, a scenario accepting that the state has changed, consequently requiring responses to compensate or mitigate impacts.
4. The lack of responses can also be considered an action, with consequences derived from a lack of planning, potentially generating a crisis when the system comes out of the identity framework that produces resilience and adaptation capacities.

3 Results

Identification and Classification of Social Actors

We identified four groups of social actors: (1) representatives of public services related to the management/inspection of protected areas and their biodiversity and also those related with territorial management, (2) scientists and specialists with technical or traditional knowledge, recognized by other actors, or having conducted studies in the area, (3) organized beneficiaries, including syndicates, trade associations, and other professional and civil society's groups, such as neighbor associations or groups of local people with common interests (e.g., environmental protection and recreational fisheries, among others), and (4) independent beneficiaries, specially referred to people who visit the area with low frequency either as tourists or to perform an economic activity. Although we did not have statistical information to quantify group 4, it was evident, due to its characteristics and diversity, that it was the most numerous. Non-organized fishermen, illegal users, and tourists, both frequent (summer camp visitors) and casual, belong to this group. **MU-CMPA** IGA actors are shown in Fig. 5 (Annex I).

Key social actors are those organized users with high interest in the maintenance of ecosystem services due to their direct use, having an economic dependency on them. They include trade associations, user's syndicates, fisheries management tables, and chamber for tourism. These actors have high power, due to the fact that they correspond to formalized organizations with valid interlocutors in the area. They also include all regional and local government organizations with a responsibility in the management and administration of the area.

Primary actors with low interest and high power corresponded to two groups: (1) social organizations or individuals without economic dependency on the ecosystem services of the area, which include NGOs (e.g., conservation and neighbors' meetings) and scientists developing research in the area and (2) users who depend economically on the ecosystem services but without organizations or doing it illegally. This group of actors should not be ignored in participative processes since due to their high interest, they may be allies of the key actors in the achievements of the management objectives and if their opinion is not considered, they may hinder some of the steps of the plans. The actors from this group, according to Dearden et al. (2003), must be included in decision-making, and besides, special involvement strategies should be designed since their decision power is low, they could affect the execution of the management plan. Thus, unless specific regulatory strategies, oriented to discourage this type of practices, are generated, conservation activities may fail also increasing the conflicts between legal and illegal users.

Table 1 Percentages of workshop attendees with respect to the number of invited people

Actors	Invitations	Confirmations	Attendees
Public services	16	10	(10) 62%[a]
Fisheries	18	14	(5) 28%
Tourism	24	8	(3) 12%
Organizations	11	6	(5) 45%[a]
Academics	14	9	(7) 50%
Total	83	47	34%

[a]Includes people who sent the interviews via web platform

Identification and Valuation of Ecosystem Services

The actor's participation was lower than expected, attending only 34% of the people invited to the workshops (Table 1). However, all groups invited were represented.

The prioritization of ecosystem services was based on the results of the Likert scale matrices responded by the participants. We summed the results in order to obtain the services that actors considered the ones providing the greatest benefits, transforming the values to a percentual scale of relative importance (Fig. 3). The most valued services, as a group, were cultural, especially contemplative landscape use which is related to the scenic beauty. The most valued group of services related to regulation corresponded to habitat for wild species and the maintenance of archaeological and paleontological remains. Finally, the most valued provisioning service was fisheries.

Economic Valuation of Ecosystem Services

An annual synthesis of the economic valuation of ecosystem services showed that their total value was close to USD 35.6×10^6 for the year 2017 (Table 2). The value with the largest participation in the summed value corresponded to the maintenance of archaeological and paleontological remains (57.0%) and habitat maintenance and reproduction (39.6%), both corresponding to regulation services. Two of the prioritized services are of economic importance, but with a relevant and indirect relationship with the users of other services. They correspond to (1) habitat maintenance and reproduction and (2) maintenance of archaeological and paleontological remains. These regulation services are key to sustain everybody's use of the AMCP-MU IGA. That is, these are essential for the provision and cultural services of the area. However, their social valuation was low (Fig. 3). This characteristic in intrinsic to regulation services because they are intangible.

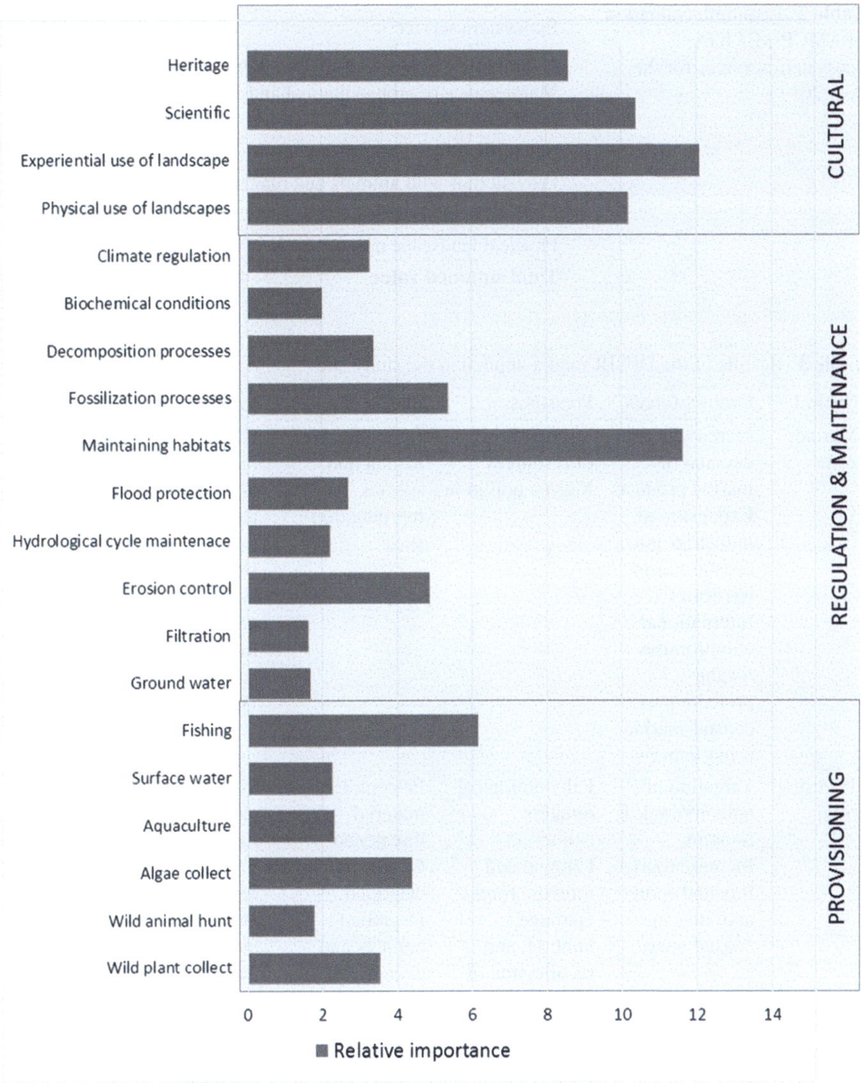

Fig. 3 Relative importance of ecosystem services given by social actors

Problems in the Implementation of the Management Plan

The drivers, pressures, state, and impacts for the **MU-CMPA** IGA could be treated with multiple instruments, as shown in the last column of Table 3. However, one of the biggest problems in the implementation of **MU-CMPAs** in Chile is the lack of robust legislation and experience in the creation of instruments that incorporate the complexity of social-ecological systems. Although the general administration plan

Table 2 Economic valuation of AMCP-MU IGA ecosystem services for the year 2017

Ecosystem service	Annual value (USD)
Habitat maintenance and reproduction	14,091,892
Maintenance of archaeological and paleontological remains	20,271,297
Scientific information	286,820
Provision of wild animals and plants	248,441
Landscape experiential use	427,740
Physical landscape use	247,552
Total summed value	35,573,742

Table 3 Results of the DPSIR model applied to the study site

Cycle 1	Driving forces	Pressures	State	Impacts	Responses
Marine zone	Increase in the demand of marine products Expansion of industries into coastal-marine territories International compromises for the protection of coastal-marine ecosystems	Overexploitation of resources Marine pollution	Degradation of habitat and species reproduction sites	Decrease in biodiversity Decrease in natural resources Decrease of landscape quality Increase in communities' vulnerability	Creation of protected coastal marine zone areas Protection of traditional activities
Terrestrial zone	Valuation of paleontological remains Increase tourist flux and sport activities in coastal areas	Paleontological remains extraction Unregulated touristic routes Garbage, hunting, and recollection of native species	Paleontological material threatened. Geoforms degraded. Degraded habitats and decrease in the stock of resources	Loss of cultural patrimony Landscape degradation Decrease in biodiversity and hydrological resources Decrease of scientific value	Coupling of the terrestrial zone to the AMCP-MU IGA

is not ready yet, there is a series of institutional/legal responses developed nationally to protect the biodiversity that should be included within the plan:

- A decree for areas of exploitation and management of benthic marine resources
- Closed seasons for extraction of marine species
- Individual fishing quota
- Sanction values
- Protected species

These instruments are oriented mainly to protect species of commercial interest, in order to maintain the reproductive cycles. These procedures are a response to the existing pressures on the species, oriented to prevent or reduce the damage to their populations (Fig. 4). Yet, the limitation of these measures lies in the lack of control capacity and the weak enforcement of the state regarding the use of coastal marine resources in Chile. Indeed, Fernández and Castilla (2005) point out that a chronic problem in the regulation of Chilean fisheries is the lack of efficient small-scale control.

On the other hand, the guidelines of the management plans point to generating strategies oriented to maintain or restore the state of the ecosystems and consequently the fluxes of ecosystem services (Fig. 4). The strategies are oriented to control the access into the area as a function of the carrying capacity of the ecosystems. However, under the current conditions, controlling the entrance into the territory is difficult, unless money is invested in actions oriented to the adaptation of the local communities, including education and the generation of economic alternatives.

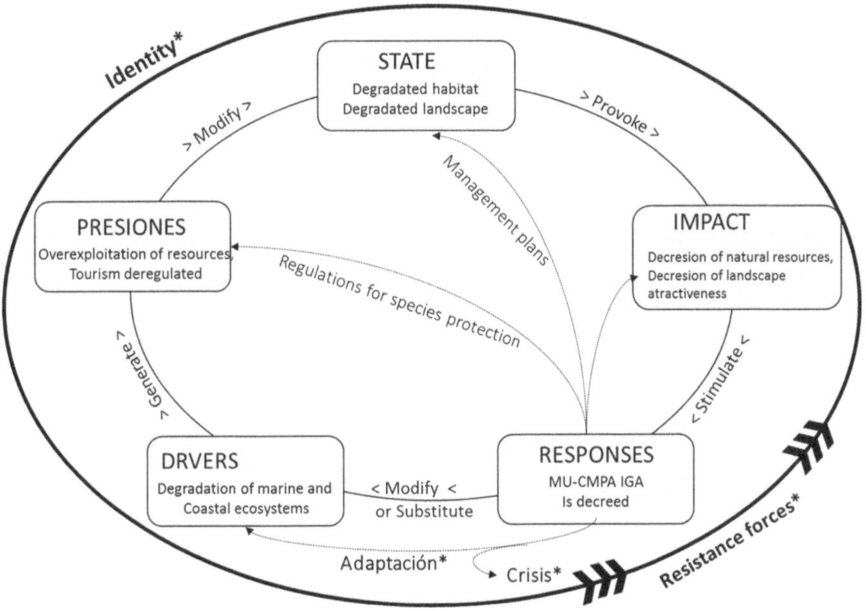

identity	Resistance forces
Associated to artisanal fisheries	Related to the possibility of future restrictions in resources extraction.
Asociated to camp summer visitors	It is considered a traditional activity of families living nearby.

Fig. 4 AMCP-MU IGA DPSIR model

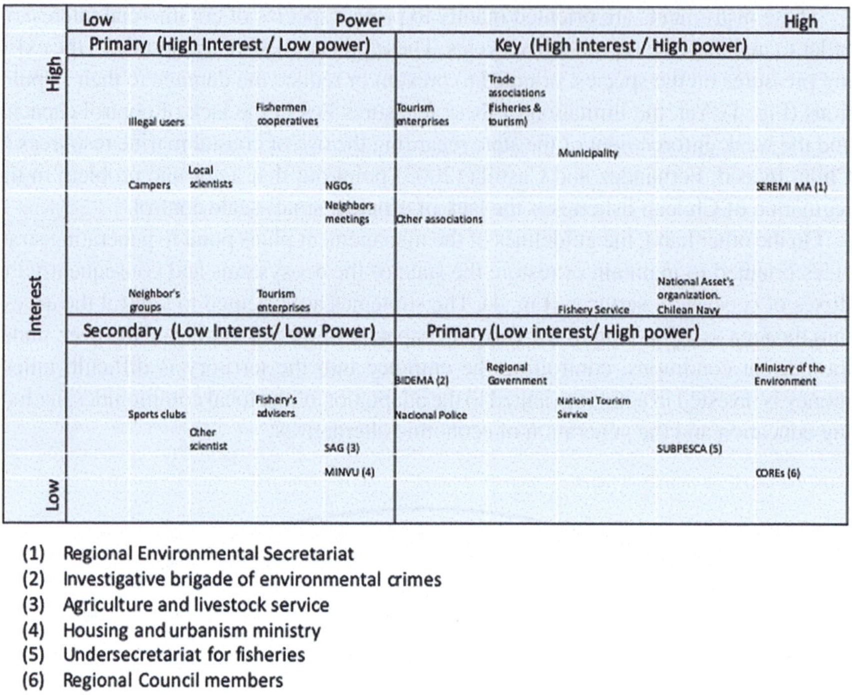

Fig. 5 AMCP-MU IGA actor's matrix resulting from the interests/power analysis in relation to ecosystem services

In summary, the DPSIR model allowed identifying a fundamental condition, not incorporated before, to advance into the concretion of the management plan. That is, the definition of adaptation strategies for the local population allows them to decrease their resistance to changes, resulting from their high dependency on the natural resources (Fig. 4).

Challenges for Ecosystem Management

What is the utility of our results in terms of public policies? Although natural resources are important, protection measures are not always implemented. The reasons seem to be that there are other needs with higher priority (e.g., health and education) and that the value of these resources is not always clear. However, the present and future well-being of the fishermen and tourism operators (and their families) from the studied area depend upon the services provided the local ecosystems. Thus, there are important challenges in the management of the **MU-CMPA** IGA and other protected areas in Chile:

(a) Attracting capitals to the area, in order to develop the infrastructure necessary to administer tourism as an alternative.
(b) Closing the terrestrial access to the AMCP-MU and to encourage charging access fees would help in the sustainability of the ecosystem.
(c) Implementation of satellite surveillance technologies in order to control illegal fishing. This should include financing to generate legislations and regulations to guarantee the environmental enforcement of the AMCP.
(d) Encouraging the creation of the first geo-paleontological park for tourism.
(e) Developing a social-ecological approach with higher participation of local actors.
(f) A more transparent management of public policies, incorporating ecological and traditional knowledge (see also chapter "A New Environmental Governance").

4 Final Considerations

The generation and implementation of the **MU-CMPA** IGA management plan has been a long and complex process. Because, in Chile, MU-CMPAs were originally conceived to be implemented in areas caracterized by quite diverse actors, who additionally had few previous relationships among them. This was a problem because, as it is explained by Berkes et al. (2000). The sustainable management of natural resources and complex social-ecological systems requires the knowledge derived from shared local experiences and traditions. However, in practice participation processes are many times non-inclusive, without empowerment of local communities leaving aside the necessary locally oriented participation processes necessary for the community's adaptation to new sociopolitical and environmental conditions. This is especially relevant for Latin American countries where there is still a strong centralized tradition regarding environmental and natural resource management. In the Chilean case, the GEF project that originated the AMCP-MU network was a top-down instrument, rather than a local community biodiversity conservation project (Calfucura 2018).

It is not easy, in non-inclusive societies, as many in Latin America, to modify the top-down approach when designing policies and strategies for the conservation of ecosystems and their services. This trend seems to be related to our historical development (Robinson and Acemoglu 2012) and the empowerment mechanisms (López 2019). Thus, it is important, within the region, the regulatory activity of the State in order to face changes into more resilient social-ecological systems (López and Figueroa 2016). The local communities within the **MU-CMPA** IGA critically depend upon the ecosystem-derived resources, requiring empowering participation mechanisms so they may generate their own destiny.

Annex I: Methodologies Used in the Economic Valuation of Prioritized Ecosystem Services

Ecosystem service	Methodology	Data and assumptions
Plants and animal provisioning, algae and their products	Market price	Marine resources were valued using the 2016 and 2017 "beach prices," obtained from the National Fisheries Services (SERNAPESCA). The estimation of extraction/ collection rates was based on information from UCN (2014) since there was not recent data
Habitat maintenance and reproduction	Market price	Coastal marine species were valued using the penalty value established by the Undersecretariat for Fisheries (SUBPESCA), under the assumption that it reflects the value that the society assigns to the conservation of those species. Biomass values necessary to estimate service fluxes were obtained from secondary sources for the 16 species in the area. There is no biomass data for terrestrial species. So, we used a willingness to pay value for conservation expenses from Figueroa and Calfucura (2017). This corresponds to a base value of 120 USD ha^{-1}
Experiential use of plants, animals, and landscapes in different environmental conditions	Hybrid	Tourist fluxes and the prices of tours were obtained from surveys to tourism operators. The valuation method estimates the expenses of a visitor in activities such as snorkeling, trekking, geological/paleontological observations, and flora and fauna watching. We estimated that an average typical visitor would spend USD 4.5 in travel and USD 13.9 in stay (total = USD 18.4). Mediation expenses would amount to 15.8 USD per person while those associated sport activities are close to USD 27.9 per person
Scientific research	Cost of opportunity	The value of scientific research was estimated on the basis of academic publications. We conducted a search, by means of Google Scholar, of publications developed within AMCP-MU IGA during 2007 and 2016, finding a total of 46 publications. The value of each publication was obtained from the public fund that financed the research[a] divided by the number of publications. We obtained a value of USD 41,477 per article. In the case of master thesis, we used the amount of money that the National Commission for Science and Technology (CONICYT) provides to each student (USD 8924 per student)

Ecosystem service	Methodology	Data and assumptions
Maintenance of archaeological and paleontological remains	Cost of replacement	The soil of "Bahía Inglesa formation" is part of the area allowing for the conservation of paleontological remains due to the high concentration of phosphorite. So, we valuated this service estimating the cost that it would correspond to replicate the natural conservation conditions in the area. Bahía Inglesa formation has a 30 cm thickness within the AMCP-UM IGA,[b] and a density close to 1.2 tons m^{-3}. Consequently, a 1 m^2 surface would require 0.36 tons of phosphorite. Since the international price of phosphorite is USD 90 per ton, the cost would be USD 32.4 m^{-2}

[a]Through the National Fund for Scientific and Technological Development (FONDECYT) and the Fund for Financing Research Center of Priority Areas (FONDAP)
[b]Information gathered in direct conversations with technicians from the Paleontological Museum of Caldera

References

Barbier EB, Hacker SD, Kennedy C et al (2011) The value of estuarine and coastal ecosystem services. Ecol Monogr 81(2):169–193

Berkes F, Folke C, Colding J (2000) Linking social and ecological systems: management practices and social mechanisms for building resilience. Cambridge University Press, Cambridge

Bradley P, Yee S (2015) Using the DPSIR framework to develop a conceptual model: technical support document. US Environmental Protection Agency, Office of Research and Development, Atlantic Ecology Division, Narragansett, RI. EPA/600/r-15/154

Burkhard B, Kroll F, Nedkov S, Müller F (2012) Mapping ecosystem service supply, demand and budgets. Ecol Indic 2:17–29

Cabello G, González C, Garrido F (2010) Revisitando la secuencia cronológica cultural de la Región de Atacama. In: Lorca M (ed) Identidades en Diálogo: Articulando actores y construyendo realidades. Estudio Fortalecimiento de la Identidad de Atacama. CEAZA y MyA Consultores. p 31–49

Calfucura E (2018) Land, governance and distribution: a discussion on the political economy of community-based conservation. Ecol Econ 145:18–26

Calvet-Mir L, Gómez-Baggethun E, Reyes-García V (2012) Beyond food production: ecosystem services provided by home gardens. A case study in Vall Fosca, Catalan Pyrenees, northeastern Spain. Ecol Econ 74:153–160

Castro C, Zuñiga A (2007) Informe Geomorfológico AMCP-MU Isla Grande de Atacama. Región de Atacama, Comuna de Caldera

CENRE (2017) Identificación, valoración y propuesta de conservación de los servicios ecosistémicos del Área Marina Costera Protegida de Múltiples Usos Isla Grande de Atacama Santiago, Chile Final Report http://www.islagrandeatacama.cl/descarga-nuestras-publicaciones/. Accessed 15 Apr 2019

Christensen NL, Bartuska AM, Brown JH et al (1996) The report of the ecological Society of America Committee on the scientific basis for ecosystem management. Ecol Appl 6(3):665–691

Dearden P, Jones S, Sartorius R (2003) Tools for development. A handbook for those engaged in development activity. Development. Tools for development. A handbook for those engaged in development activity. https://webarchive.nationalarchives.gov.uk/+/http:/www.dfid.gov.uk/Documents/publications/toolsfordevelopment.pdf. Accessed 5 May 2019

Delgado E (2014) Plan de manejo de visitantes del area marina y costera protegida de multiples usos, Isla Grande de atacama. Proyecto de grado para optar al grado de Magíster en Gestión y Planificación Ambiental, Universidad de Chile

Delgado LE, Marín VH (2015) Ecosystem services: where on earth? Ecosyst Serv 14:24–26

Delgado LE, Marín VH, Bachmann-Vargas P et al (2009) Conceptual models for ecosystem management through the participation of local social actors: the Río cruces wetland conflict. Ecol Soc 14(1):50

Doney SC, Ruckelhaus M, Emmett J et al (2011) Climate change impacts on marine ecosystems. Ann Rev Mar Sci 4:11–37. https://www.annualreviews.org/doi/pdf/10.1146/annurev-marine-041911-111611. Accessed 5 May 2019

Dudley N (2008) Guidelines for applying protected area management categories. IUCN, Gland, Switzerland

Fernández M, Castilla J (2005) Marine conservation in Chile: historical perspective, lessons, and challenges. Conserv Biol 19(6):1752–1762

Figueroa BE, Aronson J (2006) New linkages for protected areas: making them worth conserving and restoring. J Nat Conserv 14(3–4):225–232

Figueroa E, Calfucura E (2006) Sectores Productivos y Biodiversidad. In: Rovira J, Ugalde J, Stutzin (eds) Biodiversidad de Chile: Patrimonio y Desafíos. Comisión Nacional del Medio Ambiente, Santiago, pp 434–461

Figueroa E, Calfucura E (2017) Chile: análisis económico de la Estrategia Nacional de Cambio Climático y Recursos Vegetacionales 2017-2025. Report prepared for The World Bank

Gaymer CF, Rojas U, Squeo FA et al (2008) AMCP-MU Isla Grande de Atacama: Flora y Fauna marina y terrestre. In: Squeo FA, Arancio G, Gutiérrez RJ (eds) Libro Rojo de la Flora Nativa y de los Sitios Prioritarios para su Conservación: Región de Atacama. Ediciones Universidad de La Serena, La Serena, pp 223–249

Halpern BS, Walbridge S, Selkoe KA et al (2008) A global map of human impact on marine ecosystems. Science 319:948–952

Holmlund CM, Hammer M (1999) Ecosystem services generated by fish populations. Ecol Econ 29:253–268

Lambert A (2003) Valoración económica de los humedales: un componente importante de las estrategias de gestión de los humedales a nivel de las cuencas fluviales. https://www.portalces.org/sites/default/files/migrated/docs/1118.pdf. Accessed 10 May 2019

Likert R (1932) A technique for the measurement of attitudes. Arch Psychol 140:1–55

Liu S, Crossman ND, Nolan M et al (2013) Bringing ecosystem services into integrated water resources management. J Environ Manag 129:92–102

López RE (2019) Power in economics: growth, inequality and politics. Department of Economics, University of Chile, Santiago

López RE, Figueroa E (2016) On the nexus between fiscal policy and sustainable development. Sustain Dev 24:201–219

López J, Sánchez M (2000) Acerca del cambio en los sistemas complejos. In: Estebaranz A (ed) Construyendo el cambio: Perspectivas y Propuestas de Innovación Educativa. Universidad de Sevilla, Sevilla, pp 1–27

López-Santiago CA, Oteros-Rozas E, Martín-López B et al (2014) Using visual stimuli to explore the social perceptions of ecosystem services in cultural landscapes: the case of transhumance in Mediterranean Spain. Ecol Soc 19(2):27

Lotze HK, Lenihan HS, Bourque BJ et al (2006) Depletion, degradation, and recovery potential of estuaries and coastal seas. Science 312:1806–1809

Martín-López B, Iniesta-Arandia I, García-Llorente M et al (2012) Uncovering ecosystem service bundles through social preferences. PLoS One 7(6):e38970

Mayers J (2005) Stakeholder power analysis. Power tools series. International Institute for Environment and Development, London

MMA (2011) Las áreas protegidas de Chile, antecedentes, institucionalidad, estadísticas y desafíos. http://bibliotecadigital.ciren.cl/handle/123456789/6990. Accessed 10 June 2019

Peterson CH, Lubchenco J (1997) Marine ecosystem services. In: Daily GC (ed) Nature's services: societal dependence on natural ecosystems. Island Press, Washington, DC, pp 177–194

Praus S, Palma M, Domínguez R (2011) La situación jurídica de las actuales áreas protegidas de Chile Proyecto. GEF-PNUD-MMA. Andros Impresores, Santiago de Chile

Robinson JA, Acemoglu R (2012) Why nations fail. Crown Publishers Group, Random House, New York

Sarkki S, Ficko A, Wielgolaski F et al (2017) Assessing the resilient provision of ecosystem services by social-ecological systems: introduction and theory. Clim Res 73:7–15

Torres-Gómez M (2019) Multifuncionalidad de los espacios rurales: impacto territorial de los instrumentos públicos y propuesta de indicadores de seguimiento para una planificación rural sostenible en la Región de los Ríos (Chile). PhD thesis, Geography, Universitat de Barcelona

UCN (2014) Diseño y ejecución de una plan de monitoreo de los objetos de conservación del Área Marina Costera Protegida de Múltiples Usos Isla Grande de Atacama, Coquimbo. Coquimbo

Vásquez J (2007) Pesquería de algas pardas en la Región de Atacama y manejo de macroalgas pardas al interior del AMCP-MU Isla Grande de Atacama, Región de Atacama. Coquimbo

Worm B, Barbier EB, Beaumont N et al (2006) Impacts of biodiversity loss on ocean ecosystem services. Science 314:787–790

Part IV
Future Challenges

Social-ecological Challenges for a Complex Latin-American Future

Víctor H. Marín and Luisa E. Delgado

Abstract This final chapter corresponds to a summary, from the perspective of the editors of this book, the main concepts, and applications shown by the diverse authors of the previous chapters. The main goal is to identify future challenges in the study of social-ecological systems in Latin America and their potential use for the management of complex issues in a risk society.

Keywords Social-ecological systems · Latin America · Complexity · Scientific challenges · Social challenges · Public policies

1 Introduction

If this chapter was a letter, it would probably start like this: Dear Reader, we hope that you have arrived at this chapter after traveling through several Latin-American countries (our 12 case studies from Part III), equipped with a backpack full of concepts (Part I) derived from epistemological (chapter "Postnormal Science and Social-ecological Systems"), systemic (chapter "Simplifying the Complexity of Social-ecological Systems with Conceptual Models"), and social-ecological (chapters "Social Actors and Participation in Environmental Issues in Latin America" and "Social-ecological Systems and Human Well-Being") foundations. We know that the backpack was also modified by us, when you entered into Latin America (Part

V. H. Marín (✉)
Laboratorio de Modelación Ecológica, Departamento de Ciencias Ecológicas, Facultad de Ciencias, Universidad de Chile, Santiago, Chile
e-mail: vmarin@uchile.cl

L. E. Delgado
Escuela de Ciencias, Facultad de Ciencias, Universidad de Chile, Santiago, Chile

Fundación Centro Transdisciplinario de Estudios FES-Sistémicos, Santiago, Chile
e-mail: ldelgado@uchile.cl

L. E. Delgado, V. H. Marín (eds.), *Social-ecological Systems of Latin America: Complexities and Challenges*, https://doi.org/10.1007/978-3-030-28452-7_23

II) since we wanted you to understand our region from within its many contexts. That is why we developed contextual views on theoretical concepts such as social-ecological systems viewed from Latin-American social scientists (chapter "Studying Social-ecological Systems from the Perspective of Social Sciences in Latin America"), how do we understand poor people and their environment (chapter "Environmentalism of the Poor: Environmental Conflicts and Environmental Justice"), how do we view participation and governance (chapter "A New Environmental Governance"), how should we value our ecosystems (chapter "A Hierarchical Approach for the Evaluation of Multiple Ecosystem Services"), should we embrace novelty? (chapter "Social-ecological Complexities and Novel Ecosystems"), and the complexities of our economies in relation to nature (chapter "Social-ecological Systems and the Economics of Nature: A Latin American Perspective"). If you already went through all that, please prepare yourself for a little more complexity as we discuss potential futures.

Throughout the book, we have shown the scientific development of Latin America, how nations have matured and how citizens have evolved into social actors. However, we continue to assist to the development of conflicts associated with our economic development models, unregulated markets, and centralized, hierarchical, and reductionist's public policies. Unfortunately, these processes persist in several Latin American countries, generating risks, threats, and vulnerabilities to our social-ecological systems.

Human well-being and traditional cultures in Latin America are strongly related to the state of ecosystems and their provision of goods and services. Indeed, today rural people in many countries obtain directly services both from historical and novel ecosystems whether or not in a poverty state. However, although living in historical ecosystems generates satisfaction and peace, geographic isolation generates problems when searching for basic social services such as health (Delgado and Marín 2016). Furthermore, as discussed by Marín and Finlayson (chapter "Social-ecological Complexities and Novel Ecosystems" of this book), only 14.26% of Latin American ecosystems can be considered historical, the rest of them are novel. Thus, we may have to learn how to live with this reality.

Scientists, experts, resource managers, and international organizations have complexed local social-ecological systems, further justifying the late responses to social problems to the large-scale, difficult to handle, phenomena such as climate change. However, local people require urgent answers to their social-ecological problems and our acceptance of their diversity of perspectives and values, in other words, a postnormal approach to science (Marín et al., chapter "Postnormal Science and Social-ecological Systems" of this book). Traditional Latin American cultures, formed by the syncretism of original peoples, those arriving from other continents and the ecological subsystems, have built a culture deeply marked by the social-ecological attributes of their territories which we are only beginning to understand. One example is the social-ecological traps that may move such systems to undesirable states (Nahuelhual et al., chapter "Exploring Traps in Forest and Marine Socio-Ecological Systems of Southern and Austral Chile" of this book).

We are confident that a holistic view and inter- and transdisciplinary cooperation will help us to remove the veil from our eyes when focusing on social-ecological systems. Yet, the clearer we view them, the more complex they become. In other words, a conundrum! However, there is no way out; complexity should be embraced using, for example, systemic tools such as conceptual models (Delgado et al., chapter "Simplifying the Complexity of Social-ecological Systems with Conceptual Models" of this book). Furthermore, the successful conservation of our natural resources and our collective well-being will depend upon working together, improving the cooperation and exchange of knowledge between scientists, governments, and local communities. We are confident that if 65 human beings could get together and join forces to generate a book about the social-ecological systems of Latin America, other people in other regions can do the same and indeed some have already done it (e.g., Forbes et al. 2006; Lovecraft and Eicken 2011). We have still a long way to go, but as stated by Carlos Dittborn, Chilean football administrator: "Because we have nothing, we will do everything."[1] In the following paragraphs, we identify some of the future challenges in the study of social-ecological systems in Latin America, from a postnormal perspective or science guided to action.

2 Scientific Challenges

First and foremost, we need to make more efforts to develop studies using inter- and transdisciplinary science. Many science funding institutions still accept proposals mostly through disciplinary channels,[2] and although there are initiatives to implement networks, funds are scare (<USD25,000 for a group of three or more scientists for 2 years[3]). We have to accept that extreme climatic events are a reality, affecting the region in several ways and requiring the combined efforts of climate experts, sociologists, ecologists, and political scientists (Freire and Natenzon, chapter "Analyzing Social Vulnerability to Natural Disasters in Northeast Brazil: Catastrophic Flooding Cycles at Alagoas Littoral Zone" of this book). They do and will affect the agricultural production and the availability of ecosystem services being used for subsistence within the region.

The social-ecological system concept has represented an enormous jump ahead in terms of transdisciplinary studies. However, as shown by Marín et al. (chapter "Postnormal Science and Social-ecological Systems" of this book), only 1% of the published scientific articles on social-ecological systems correspond to Latin America. Therefore, we need to increase our efforts including the four main conditions that affect the territorial societal responses to ecological changes: (a) biophysi-

[1] https://en.wikipedia.org/wiki/Carlos_Dittborn

[2] https://www.conicyt.cl/fondecyt/grupos-de-estudios/

[3] https://www.conicyt.cl/pci/files/2019/05/REX-5492-2019_-aprueba-bases-Concurso-REDES-2019.pdf

cal components and processes, (b) social structures, (c) economic forces, and (d) political configurations.

Finally, we need to explore social-ecological systems using spatial-dynamic models to analyze the future of ecosystem services including the participation of local social actors. Indeed, Flórez et al. (chapter "Water Supply Valuation of High Andes Wetlands, Chinchiná River High Watershed, Colombia" of this book) show that the participation of social actors was vital when evaluating the water supply in a Colombian watershed. In fact, four chapters of this book deal with the water problem (chapters "Spatial Modelling of Social-ecological Systems of Hydrological Environmental Services in Las Conchas Creek Basin, Argentina", "Analyzing Social Vulnerability to Natural Disasters in Northeast Brazil: Catastrophic Flooding Cycles at Alagoas Littoral Zone", "Water Supply Valuation of High Andes Wetlands, Chinchiná River High Watershed, Colombia", and "Integrated Evaluation of the Effects of the Payment for Hydrological Environmental Services Program in Ajusco, Mexico City"). This issue not only requires a transdisciplinary group but also the generation of models to study "what-if" scenarios, since classical experimental science is simply not possible.

3 Institutional and Political Challenges

More Participative Public Policies

We urgently need more participative risk management in order to act over the social vulnerability during normal times and to decrease the uncertainty when facing preventive decision-making in cases of anomalous events (Freire and Natenzon, chapter "Analyzing Social Vulnerability to Natural Disasters in Northeast Brazil: Catastrophic Flooding Cycles at Alagoas Littoral Zone" of this book). Participative public policies should allow the consideration of a plurality of interests and perspectives of social actors and groups (Funtowicz and Ravetz 1993).

The strengthening of the governability should be based in a learning process that motivates participation, and it should include an agreement for practicing it under local governments. The premise is that good governance shall strengthen the governability, improves social trust, increases the effectivity of environmental management, and makes sustainability possible within the spheres of territorial development (Gutiérrez and Morales 2017).

If we now turn to environmental governance, it is necessary to combine market strategies and participation, favoring the consolidation of the social capital in local areas through incentives for reciprocal relationships among social actors. These processes seem to be currently absent in several Latin American areas (e.g., Delgado et al., chapter "Environmental Governance for the Coastal Marine Ecosystem Services of Chiloé Island (Southern Chile)" of this book).

It is not easy, in non-inclusive societies such as many from Latin America, to change the top-down political processes and the strategies for the conservation of the ecosystems and their services (Pérez-Orellana et al., chapter "Social Actors and Participation in Environmental Issues in Latin America" of this book). This trend seems to be related to our historical development (Acemoglu and Robinson 2012) and the empowering mechanisms (López 2019). So, it is important both the regulating activity of the states in order to help to adapt social-ecological systems to changes (López and Figueroa 2016) and the participation of the communities that depend on the services provided by the ecosystems as shown by Torres-Gómez et al. (chapter "Social Valuation of Ecosystem Services at Local Scale: Challenges for the Management of a Multiple-Use Coastal and Marine Protected Area (MU-CMPA): Isla Grande de Atacama: Chile" of this book).

A short-term and urgent action to face these challenges is the necessary alignment of conservation and development policies. Our use of natural resources and the improvement in the well-being of developing nations are internationally established objectives, yet their alignment has been deficient (Roe et al. 2013). If we focus on both issues at the same time, beyond pure rhetoric, we should acknowledge the importance of (1) the historical context, as a factor modeling inequality, poverty, and vulnerability (Rodríguez and Merino 2017), (2) the lack of power of local people such as farmers and fishermen (World Bank 2016, 2017), and (3) past injustices (Jerneck et al. 2011). Thus, the main political challenge regarding social-ecological systems is to generate structures and actions with a participatory perspective having the ecosystem's resilience as a goal and the sustainability of human societies as a concurrent goal.

Institutional Responses

The governmental responses to extreme events, which science predicts will occur more often,[4] can no longer be "The river is guilty" or "…fortuitous event of nature and the divine" (Freire and Natenzon, chapter "Analyzing Social Vulnerability to Natural Disasters in Northeast Brazil: Catastrophic Flooding Cycles at Alagoas Littoral Zone" of this book). If disasters are determined by the structure of the social-ecological system, then we need to anticipate the best way to react to them. This action again requires the exchange of ideas and knowledge between scientists, governmental personnel, and local social actors.

The problem and challenge we face in this case are that knowledge advances slowly and its incorporation into society even slower (Marín and Finlayson, chapter "Social-ecological Complexities and Novel Ecosystems" of this book). So, as we have discussed several times in this book, we have to ask ourselves: are we already running out of time? We hope that the answer is no. But the only way to contribute is to open Latin American academy to societal and governance problems.

[4] https://www.sciencedaily.com/releases/2018/03/180321130859.htm

4 Dichotomies

Sociocultural variables affect the way in which ecosystem services are valued. For example, Delgado et al. (Submitted) have shown for the case of Chiloé Island in the South of Chile that social actors with multiple interests and low decision power concentrate on provisioning and cultural services while those with decision power use regulating services also acknowledging cultural conditions. This dichotomy impinges on what to do when managing social-ecological systems. Do we pay attention to rural environments, where traditions are alive? Or do we integrate both types of environments, rural and urban? This is certainly a challenge since we have less than 15% of Latin America structured as historical ecosystems; yet, they are a deep source of environmental concern. But, 85% of Latin Americans live in urban areas that also deserve our attention.

The rural–urban dichotomy generates divergencies when valuing ecosystem services with the likelihood of generating conflicts between different types of social actors (Torres-Gómez et al., chapter "Social Valuation of Ecosystem Services at Local Scale: Challenges for the Management of a Multiple-Use Coastal and Marine Protected Area (MU-CMPA): Isla Grande de Atacama: Chile" of this book; Amarilla et al., chapter "San Rafael Reserve, Paraguay: Key Social Stakeholder and Sustainability Scenarios Through Environmental Governance Approaches" of this book). So, improving our knowledge about the valuations that different social actors give to nature opens the possibility of anticipating to conflicts, improving management scenarios. This road requires a mixture of qualitative and quantitative analyses, which is one of the strengths of the social-ecological analysis (Berkes et al. 2003).

5 Latin American Ecosystems: Conservation, Valuation, and Concerns

The report of the Millennium Ecosystem Assessment (MEA 2005) clearly shows the degradation of natural resources and ecosystems at alarming rates. In Latin America, the degradation of forests (tropical and dry), the Amazonian ecosystem, pampas, wetlands, and coastal marine ecosystems generate deep concern, some of which are analyzed in our case studies (Part III of this book). Latin America hosts nearly 40% of all living species, it has the largest reserves of tropical forests on earth, and it is the second region in terms of water availability (nearly 25%).

The disorganized expansion of agriculture and the associated inequities (e.g., Freire and Natenzon, chapter "Analyzing Social Vulnerability to Natural Disasters in Northeast Brazil: Catastrophic Flooding Cycles at Alagoas Littoral Zone" of this book) is a phenomenon threatening the natural and cultural patrimony of the region. Indeed, as discussed by Gligo (2006), food security and the protection of the natural capital weight much less than short-term commercial interests when making political decisions about agriculture. If we turn to coastal marine ecosystems, we are witnessing overexploitation of fish resources, the introduction of invasive species that are affecting their structure and functions including the goods and services used by local populations (Toledo and Catillo 1999; Sepúlveda and Ibrahim 2009). This is clearly a big social-ecological challenge for the future of our region. The good news is that there are examples that show we can indeed do better if we see the earth from an integrated perspective of human/nature.

We would like to finish this chapter with a reflection. We are convinced that human societies should focus on embracing a transdisciplinary perspective at all possible learning levels. The goal should be to raise awareness and to prepare new generations of scientists, resource managers, and social actors to face global and local problems and risk threats in order to improve the sustainability of Latin American social-ecological systems.

References

Acemoglu D, Robinson JA (2012) Why nations fail. Crown Publishers, Random House, New York

Berkes F, Colding J, Folke C (2003) Introduction. In: Berkes F, Colding J, Folke C (eds) Navigating social-ecological systems. Building resilience for complexity and change. Cambridge University Press, Cambridge, pp 1–29

Delgado LE, Marín VH (2016) Well-being and the use of ecosystem services by rural households of the Río cruces watershed, southern Chile. Ecosyst Serv 21:81–91

Delgado LE, Zúñiga CC, Asún RA et al. (Submitted) Ecosystem services of Chiloé Island marine coastal zones and integrated management: a social-ecological analysis. Lat Am J Aquat Res

Forbes BC, Bölter M, Müller L et al (eds) (2006) Reindeer management in northernmost Europe: linking practical and scientific knowledge in social-ecological systems. Ecological studies book, vol 184. Springer, Berlin

Funtowicz S, Ravetz J (1993) La ciencia posnormal. Ciencia con la gente. Grupo Editor de América Latina, Buenos Aires

Gligo N (2006) Estilos de desarrollo y Medio Ambiente en América Latina. Un cuarto de siglo después. Serie Medio ambiente y desarrollo. CEPAL, Santiago de Chile

Gutiérrez A, Morales T (2017) Evaluación de la gobrnanza ambiental local en Risaralda. Revista Luna Azul. No. 45. https://doi.org/10.17151/luaz.2017.45.1

Jerneck A, Olsson L, Ness B et al (2011) Structuring sustainability science. Sustain Sci 6:69–82

López RE (2019) Power in economics: growth, inequality and politics. Department of Economics, Universidad de Chile, Santiago

López RE, Figueroa E (2016) On the nexus between fiscal policy and sustainable development. Sustain Dev 24:201–219

Lovecraft AL, Eicken H (eds) (2011) North by 2020: perspectives on Alaska's changing social-ecological systems. University of Alaska Press, Fairbanks

MEA (Millennium Ecosystem Assessment) (2005) Ecosystems and human well-being: opportunities and challenges for business and industry. World Resources Institute, Washington, DC

Rodríguez K, Merino L (2017) Contextualizing context in the analysis of payment for ecosystem services. Ecosyst Serv 23:259–267

Roe D, Yassin E, Porras I et al (2013) Linking biodiversity conservation and poverty reduction: de-polarizing the conservation-poverty debate. Conserv Lett 6:162–171

Sepúlveda CL, Ibrahim (eds) (2009) Políticas y sistemas de incentivos para el fomento y adopción de buenas practicas agrícolas como una medida de adaptación al cambio climático en América

Central. Serie Técnica Informe técnico 37. Centro Agronómico Tropical de Investigación y Enseñanza, Turrialba
Toledo V, Catillo A (1999) La ecología en Latinoamerica: Siete Tesis para una ciencia pertinente en una región en Crisis. Interciencia 24(3):157–168
World Bank (2016) Poverty and shared prosperity: taking on inequality. World Bank, Washington, DC
World Bank (2017) World development report 2017: governance and the law main messages. World Bank, Washington, DC

Correction to: Social-ecological Systems of Latin America: Complexities and Challenges

Luisa E. Delgado and Víctor H. Marín

Correction to: L. E. Delgado, V. H. Marín (eds.), *Social-ecological Systems of Latin America: Complexities and Challenges*, https://doi.org/10.1007/978-3-030-28452-7

The following corrections are made to this Book:

The Acknowledgement section is added to Front Matter.

Chapter 13:

The name of the first author of the chapter entitled "San Rafael Reserve, Paraguay: Key Social Stakeholders and Sustainability Scenarios Through Environmental Governance Approaches" should be: Stella Mary Amarilla R.

The updated online version of this chapter can be found at
https://doi.org/10.1007/978-3-030-28452-7_13
https://doi.org/10.1007/978-3-030-28452-7

L. E. Delgado, V. H. Marín (eds.), *Social-ecological Systems of Latin America: Complexities and Challenges*, https://doi.org/10.1007/978-3-030-28452-7_24

Index

L. E. Delgado, V. H. Marín (eds.), *Social-ecological Systems of Latin America: Complexities and Challenges*, https://doi.org/10.1007/978-3-030-28452-7

D

Q

R

The manufacturer's authorised representative in the EU is Springer Nature Customer Service Centre GmbH, Europaplatz 3, 69115 Heidelberg, Germany. If you have any concerns regarding our products, please contact ProductSafety@springernature.com

Printed and bound by CPI Group (UK) Ltd, Croydon, CR0 4YY
30/09/2026
02225657-0015